CARBOCATION CHEMISTRY

CARBOCATION CHEMISTRY

EDITED BY

George A. Olah
G. K. Surya Prakash
Loker Hydrocarbon Research Institute
University of Southern California

A JOHN WILEY & SONS, INC., PUBLICATION

Published by John Wiley & Sons, Inc., Hoboken, New Jersey.
Published simultaneously in Canada.

Library of Congress Cataloging-in-Publication Data:

Carbocation chemistry / edited by George A. Olah and G. K. Surya Prakash.
p. cm.
Includes bibliographical references and index.
ISBN 0-471-28490-4 (cloth : alk. paper)
1. Carbocations. I. Olah, George A. (George Andrew), 1927– II. Prakash, G. K. Surya.
QD305.C3C37 2004
547′.1372–dc22 2004002265

Printed in the United States of America

10 9 8 7 6 5 4 3 2 1

CONTRIBUTORS

Annette D. Allen, Department of Chemistry, University of Toronto, 80 St. George Street, Toronto, Ontario, Canada M5S 3H6

Gennady I. Borodkin, N. N. Vorozhtsov Novosibirsk Institute of Organic Chemistry, 9 Lavrentiev Avenue, Novosibirsk, 630090, Russia

Zhongfang Chen, Center for Computational Chemistry and Department of Chemistry, University of Georgia, Athens, GA 30602-2525

Alain Goeppert, Laboratoire de Physico-Chimie des Hydrocarbures, Université Louis Pasteur, UMR 7513, 4, rue Blaise Pascal, 67070 Strasbourg, France

Bruce N. Hietbrink, Department of Chemistry, California State University, Northridge, Northridge, CA 91330-8262

Kendall N. Houk, Department of Chemistry and Biochemistry, University of California, Los Angeles, 405 Hilgard Avenue, Los Angeles, CA 90095-1569

Jean Claude Jacquesy, UMR 6514, Laboratorie De Chimie XII, Faculty of Sciences, Universite de Poitiers, 40 avenue du Recteur Pineau-86022, Potiers Cedex, France

Olga Kronja, Faculty of Pharmacy and Biochemistry, University of Zagreb, Kovacica 1, POB 156, 10000 Zagreb, Croatia

Kenneth K. Laali, Department of Chemistry, Kent Sate University, Kent, OH 44242

Herbert Mayr, Department of Chemistry, Ludwig Maxmilian University, Münich, Butenandstr. 5-13, Haus F, D-81377 Münich, Germany

Craig A. Merlic, Department of Chemistry and Biochemistry, University of California, Los Angeles, 405 Hilgard Avenue, Los Angeles, CA 90095-1569

Armin R. Ofial, Department of Chemistry, Ludwig Maxmilian University, Münich, Butenandstr. 5-13, Haus F, D-81377 Münich, Germany

George A. Olah, Loker Hydrocarbon Research Institute, Department of Chemistry, University of Southern California, 837 Bloom Walk, Los Angeles, CA 90089-1661

Keith A. Porter, Department of Chemistry, Technical University, Münich, D-85747 Garching, Lichtenbergstrasse 4, Münich, Germany

G. K. Surya Prakash, Loker Hydrocarbon Research Institute and Department of Chemistry, University of Southern California, 837 Bloom Walk, Los Angeles, CA 90089-1661

V. Prakash Reddy, Department of Chemistry, University of Missouri–Rolla, Rolla, MO 65401

Martin Saunders, Department of Chemistry, Yale University, P. O. Box 208107, 225 Prospect Street, New Haven, CT 06520-8107

Paul von Ragué Schleyer, Center for Computational Chemistry and Department of Chemistry, University of Georgia, Athens, GA 30602-2525

Hubert Schmidbaur, Department of Chemistry, Technical University, Münich, D-85747 Garching, Lichtenbergstrasse 4, Münich, Germany

Vitalij D. Shteingarts, N. N. Vorozhtsov Novosibirsk Institute of Organic Chemistry, Siberian Division of Russian Academy of Sciences, 630090 Novosibirsk, Russia

Vyacheslav G. Shubin, Head of Laboratory, N. N. Vorozhtsov Novosibirsk Institute of Organic Chemistry, 9 Lavrentiev Avenue, Novosibirsk 630090, Russia

Jean Sommer, Laboratoire de Physico-Chimie des Hydrocarbures, Université Louis Pasteur, UMR 7513, 4, rue Blaise Pascal, 67070 Strasbourg, France

Peter J. Stang, Department of Chemistry, University of Utah, 315 S. 1400 E., Salt Lake City, UT 84112-0850

Dean J. Tantillo, Department of Chemistry, University of California, Davis, Davis, CA 95616

Thomas T. Tidwell, Department of Chemistry, University of Toronto, 80 St. George Street, Toronto, Ontario, Canada M5S 3H6

Chaitanya S. Wannere, Center for Computational Chemistry and Department of Chemistry, University of Georgia, Athens, GA 30602-2525

CONTENTS

PREFACE

The chemistry of carbocations reached its centennial, a significant landmark indeed! This book developed from a Loker Hydrocarbon Research Institute–sponsored Kimbrough Research Symposium commemorating the "100 Years of Carbocations" held in January 4–5, 2001. Carbocations are the key intermediates in a great variety of acid-induced reactions and processes involving petroleum cracking and refining, coal processing, polymerization chemistry, synthetically important reactions, isomerization and rearrangements, addition reactions, and aromatic substitutions, and even in a number biosynthetic transformations. Consequently, this area of study is of great general importance, ranging from basic understanding of the nature and behavior of positively charged organic intermediates (carbocations) to the many reactions and processes in which they participate.

Since the original discovery of triphenylmethyl cation in 1901, the field of carbocations continues to grow into maturity. Stereochemical and kinetic studies of acid-catalyzed processes led to their indirect identification in the 1940s and 1950s. Development of superacidic stable ion conditions by one of us in the early 1960s enabled us to study them directly. Since that time, a great variety of diverse carbocations (of trivalent carbenium and higher coordianate carbonium ion types) have been investigated by varied spectroscopic means [nuclear magnetic resonance (NMR), infrared (IR), ultraviolet–visible spectrum (UV-vis), electron spectroscopy for chemical analysis (ESCA), X-ray)], theoretical calculations, and chemical studies.

We thank all the authors who participated in this project. Although not all the topics of the 2001 symposium could be covered in the book, the 14 chapters give a representative and broad view of the field, including its history, preparative aspects, characterization techniques, empirical and theoretical treatments, and applications in synthesis. We regret the inevitable delay in the publication and thank Ms. Amay Romano, our Wiley editor, for her efforts in bringing the project to fruition.

It is a real pleasure for us to join with our colleagues to pay tribute to 100 Years of Carbocation Chemistry with this volume. We believe that the field will continue to be of major significance for years to come.

Los Angeles, California

George A. Olah
G. K. Surya Prakash

The Olah grandchildren, Peter (10), Kaitlyn (8), and Justin (3), enlighten this otherwise unused page.

1

HISTORICAL PERSPECTIVE

Peter J. Stang
Department of Chemistry
University of Utah
Salt Lake City, Utah

Carbocations are ubiquitous and arguably the most important reactive intermediates in all of organic chemistry. They are certainly the most important and occupy center stage at this special 2-day symposium on "100 years on carbocations." An indication of just how ubiquitous they are is the fact, as we learned from Professor Olah in the symposium with reference to Pierre Vogel's book (*Carbocation Chemistry*, Elsevier, New York, 1985), that carbocations are even present in all red wines. So, from now on, every time you enjoy a glass of red wine you may think of and remember carbocations.

We can divide a century of carbocation chemistry into approximately three equal periods: the beginning, from 1900 to the late 1930s, the golden age from the 1940s to the mid-1970s and the modern period since then. I will attempt to provide a brief overview of each period. In the process I have relied on several sources, among them Jack Roberts' wonderful scientific autobiography, *The Right Place at the Right Time* and others in the ACS Series on Profiles, Pathways and Dreams, edited by Jeff Seeman; Jerry Berson's recent little tome on "chemical creativity" and most of all the late Costin Nevitzescu's informative chapter "Historical outlook," which constitutes Chapter 1 of the five-volume series "Carbonium Ions" edited by our host George Olah and my former postdoctoral mentor Paul Schleyer and published

Carbocation Chemistry, Edited by George A. Olah and G. K. Surya Prakash
ISBN 0-471-28490-4

by Wiley between 1968 (Vol. I) and 1976 (Vol. V), which are still considered the definitive monographs on carbocations. I only regret that, at the time of the symposium, I did not have available George's new *Autobiographical Reflections*.

Briefly reviewing years of exciting chemistry, inevitably omissions of important contributions and/or individuals occur. My reflections are highly personal and certainly not comprehensive.

The independent observations and reports of J. F. Norris in the *American Chemical Journal* and F. Kehrmann in *Chemische Berichte*, both in 1901, on the heels of Moses Gomberg's discovery of the triphenylmethyl radical in 1900, are generally credited with the discovery of carbocations and hence our 100-year anniversary this year. They independently observed deep yellow solutions on the dissolution of the colorless triphenylcarbinol and triphenylmethyl chloride, respectively, in concentrated sulfuric acid. It was, however, Adolf Baeyer in Munich who in 1902 recognized and reported the saltlike character of compounds formed in solutions of triphenylcarbinol in sulfuric acid and the correlation between the formation of salt and the appearance of color.

There ensued a period of great activity in the field involving such luminaries as Baeyer, Gomberg, Walden, and many others, including a fairly acrimonious debate over nomenclature with Baeyer favoring the term *carbonium salts* while Gomberg suggested *carbyl salts*. Just as an example, Walden, of the Walden inversion fame, demonstrated as early as 1902, as reported in *Chemische Berichte* Vol. 35, p. 2018, the electrolyte character of triphenylmethyl chloride and bromide in sulfur dioxide solution by conductivity measurements. This was all the more remarkable for the early use of liquid SO_2, a nonaqueous solvent with a very high ionizing power. A. Hantzch, as early as 1907, established the ionic nature of the species formed by measuring the molal freezing point depression of a solution of triphenylcarbinol in concentrated sulfuric acid. These measurements were confirmed and refined by Louis Hammett in the early 1930s as reported in the *Journal of the American Chemical Society* (JACS) in 1933. Ziegler and coworkers measured the conductivity of a series of substituted triarylcarbonium perchlorates in SO_2 in the 1920s as reported in *Annalen der Chemie* during 1925–1930. These measurements were refined and extended by Lichtin and Paul Bartlett and their coworkers in the early 1950s and allowed them to measure the equilibrium constants between undissociated triarylchloride, their ion pairs, and the dissociated ions.

Meanwhile, two seemingly unrelated studies emerged to further develop and enhance the field. The pinacol rearrangement was first reported by R. Fitting in *Annalen* in 1860. Likewise, during the period 1896–1900, Georgh Wagner discovered the ring rearrangement reaction of certain terpenes, in particular, when borneol and isoborneol are dehydrated to camphene. It was, however, Hans Meerwein who recognized and demonstrated during 1910–1925 that cationic carbon intermediates were involved in these reactions and in particular in the camphene hydrochloride isobornyl chloride change, as first reported by Meerwein in *Berichte* in 1920. These rearrangements, proceeding through carbocations, have of course become known to every organic chemist as the *Wagner–Meerwein rearrangement*, a topic I shall return to later.

These early and remarkable results set the stage for the work of Hughes, Ingold, Hammett, and others and what might be called the "birth of physical organic chemistry" during the 1930s and 1940s that in turn led to the golden age of mechanistic organic chemistry in the 1950s through the mid-1970s. The players and contributors to this golden age of mechanistic, physical organic chemistry, centered around understanding reaction pathways and the identification and characterization of reactive intermediates by all available chemical and physical means and methods, read like a *Who's Who* in organic chemistry and they are among the titans of twentieth-century chemistry. Significant contributors to carbocation chemistry were Saul Winstein, Paul Bartlett, Bill Doering, Jack Roberts, H. C. Brown, N. C. Deno, Don Cram, Gardner Swain, George Olah, Jerry Berson, Ernie Grunwald, Andy Streitwieser, Paul Schleyer, Clair Collins, Chuck De Puy, Martin Saunders, William Saunders, Ned Arnett, Ron Breslow, Paul Gassman, Don Noyce, Jay Kochi, Ted Sorensen, Ken Wiberg, Fred Bordwell, Stan Cristoll, Ted Lewis, W. G. Young, Bob Taft, Frank Anet, Frank Westhheimer, Michael Dewar, Jack Shiner, Mike Karabatsos, Happy Hogoween, and Victor Gold. I can only hope that I did not miss any of the important contributors to the field.

Such topics as nonclassical ions, electrophilic aromatic substitutions, all manner of rearrangements, aromatic and homoaromatic cations, heteroatom-stabilized carbocations, addition–elimination reactions, solvolytic reactions, allylic and homoallylic ions, cyclic and bicyclic species, and so on ad infinitum were all extensively and rigorously investigated, and a treasure trove of knowledge, that greatly enriched organic chemistry, was acquired.

In fact, mechanistic studies and investigations of carbocations were carried out even by E. J. Corey, the paragon of synthetic organic chemistry, whom one would seldom associate with physical organic chemistry. In a series of studies with Joe Casanova. Jr., Corey reported on "the formation of carbocations by oxidative decarboxylation of carboxylic acids, with lead tetraacetate," in JACS in 1963. He even has a paper titled "On the norbornyl cation problem" in JACS, Vol. 85, pp. 169–173 (1963).

Obviously it would take a whole lot of time to discuss in only modest detail the elegant, seminal, and significant contributions of just the persons participating in this symposium: Jack Roberts, Ned Arnett, Paul Schleyer, Martin Saunders, Ted Sorensen, and of course Professor Olah, let alone all the people mentioned previously.

Hence, I will very briefly mention only the highlights: Jack Roberts' seminal studies on the cyclopropylcarbinyl-cyclobutyl systems. There is something special about history when narrated by an active participant. Thank you, Jack! Ned Arnett's very important calorimetric investigations; Paul Schleyer's careful solvolytic studies and investigations of bridgehead cations, not to mention his numerous, insightful computational contributions; Martin Saunders' ingenious isotope perturbation method that I'll return to again; Jerry Berson's incredibly erudite investigations of rearrangements; Ted Sorensen's novel NMR studies of transannually bridged systems; and, of course, the seminal contributions of our host Professor Olah, are all milestones in the rich history of carbocation chemistry.

Focusing more on George Olah's contributions, we now know of course that the Friedel-Crafts reaction, reported by the French chemist Charles Friedel, and the American chemist James Craft, in 1877 in *Comptes Rendus* and the *Bulletin de Société Chimie de France*, and related electrophilic aromatic substitutions, proceed via carbocations, involving π and σ complexes. Their formation was not recognized until well into the mid-1900s. Moreover, although various ionic arene complexes were extensively investigated over a period of more than half a century by numerous chemists including H. C. Brown, it was not until the late 1950s that Olah and coworkers were able to isolate several crystalline salts of various methylbenzenonium tetrafluoroborates, hexafluorophosphates, and hexafluoroantimonates and thereby unequivocally establish their nature. They also extensively and vigorously investigated Friedel–Crafts and other electrophilic aromatic substitution reactions. I can refer to a marvelous account of these reactions in the four-volume definitive monographs on this topic, *Friedel–Crafts and Related Reactions*, edited by George Olah and published by Wiley-Interscience in 1963. I particularly call attention to the wonderful historical chapter by George Olah and Robert Dear on Friedel and Craft that includes actual reproductions from the notebook entries of both Friedel and Craft.

George's knowledge and admiration of Hans Meerwein and his work, with whom he also carried out an extensive correspondence; his success in isolating and studying arenium salts; and his early realization that carbocations are very powerful Lewis acids that cannot long survive in the presence of any nucleophile or even very weakly basic solvents, all played a role in his subsequent seminal discoveries regarding carbocations. He hit upon the brilliant and ingenious idea of using powerful Lewis acids, not only to generate but also as the medium in which to observe carbocations. While still at Dow Chemical, he systematically screened nearly all high-valence Lewis acid fluorides and in the process came upon antimony pentafluoride, SbF_5. In the summer of 1962, at the Brookhaven Organic Reaction Mechanisms Conference, Olah literally shocked the organic chemistry community by his report that he was able to observe by NMR the *tert*-butyl cation as the SbF_6 salt that was stable enough for both spectroscopic and chemical study and kept around at low temperature for many hours. This and Olah's and his coworkers' extensive subsequent NMR studies of carbocations were reported in a series of papers in JACS and elsewhere and served as a magnificent capstone on the numerous and extensive studies of carbocations that began at the dawn of the twentieth century. As the Nobel Foundation was to observe some 30 years later, "George Olah gave the cations of carbon longer life," longer by about a trillion-fold. These studies were definitely one of the major highlights of the golden age of mechanistic physical organic chemistry.

That brings us to the "modern" period of carbocation investigations, approximately since the mid-1970s. Here, I would like to but briefly mention half a dozen recent (as of 2004) studies that illustrate the continued interest and vitality of carbocations. A major forward step in this field and in particular the NMR observation of carbocations was, as I already mentioned, Martin Saunders' development and application of the method of isotopic perturbation of equilibrium, observable in the carbon-13 (^{13}C) NMR spectrum. The application of this ingenious technique in the 1980s provided compelling evidence that the parent 2-norbornyl cation, under strongly acidic conditions, indeed has a bridged nonclassical structure and is not a rapidly equilibrating mixture of two classical ions. This was elegantly confirmed by Yanoni's solid-state NMR studies at 5 Kelvin, reported in JACS in 1985. Likewise, one must mention the very nice X-ray structure determinations of several carbocations, including some substituted norbornyl ions and the *tert*-butyl cation as reported by Thomas Laube in a series of papers in *Angewandte* and JACS from the late 1980s through the mid-1990s.

Similarly, Jay Kochi's marvelous comprehensive study, by a judicious combination of X-ray crystallography, NMR, time-resolved UV-vis spectroscopy, and kinetics, elucidates in exquisite details the precise structure of the aforementioned π and σ complexes in electrophilic aromatic substitutions. As we now know, these intermediates can have lifetimes from femtoseconds to hours, and hence one needs all available techniques and methods to thoroughly investigate them in order to understand them even better. A "perspective" on Jay Kochi's elegant studies since 1990 or so was featured in the October 2002 issue of the *Journal of Organic Chemistry* (JOC). Most important, of course, is the cover of issue 9 of JOC in 2001 honoring 100 years of carbocations.

Among the important recent developments in the carbocation field of course are George Olah's studies on C—H and C—C bond activation; in other words, protonation under strong, superacid conditions of saturated hydrocarbons, the parent ion being, of course, CH_5^+.

Likewise, there are, of course, Professors Olah and Prakash's elegant studies on superelectrophiles that include the protonitronium dication, protohydronium dication, protosulfonium dication, trialkyloxonium dications, protoacetylium dication, and others. These are fascinating, new species whose structure, reactivity, and exact role we are just beginning to understand; likewise, of course, Helmut Schwartz' elegant gas-phase studies and Herbert Mayer's nucleophilicity scale.

Finally, I would like to conclude by briefly discussing the importance of carbocations in bioorganic and natural product chemistry. As mentioned earlier, Wagner discovered, and Meerwein recognized, the role of carbocations in the acid-catalyzed rearrangements of the components of turpentine, molecules we now know as terpenes or isoprenoids. At that early point of development, monumental work and insight was required to merely deduce the structures of the products of these complex reactions. Concepts such as biosynthetic pathways, enzyme catalysts, and storage of information in DNA were still years away. The early work by the German group formed the basis for Ruzika's idea about how isoprenoid molecules were assembled in nature by carbocation alkylations and rearrangements. However, it was several decades before chemists and biochemists began to test these ideas with enzymes in the isoprenoid biosynthetic pathway. The biggest hurdle, getting sufficient quantities of the enzymes for mechanistic studies, was only solved during the last decade as a "spinoff" of the rapid developments in genetics and protein engineering.

What has emerged, largely through the efforts of my colleague Dale Poulter and his coworkers at Utah, is incontrovertible evidence that carbocations reign king in the isoprenoid pathway. The fundamental chain elongation reactions that assemble long isoprenoid chains from the simple five-carbon building blocks are alkylations of carbon–carbon double bonds by electrophilic allylic carbocations. The joining of two isoprenoid chains in a non-head-to-tail fashion, as, for example, in the biosynthesis of cholesterol, is an electrophilic alkylation followed by a complex set of carbocation rearrangements. And the attachment of isoprenoid chains to other molecules to make switches that control cell division or redox couples needed to burn glucose are all examples of carbocation alkylations. Of course, we still don't understand how proteins catalyze these reactions without themselves being alkylated, nor has anyone taken an NMR spectrum of an enzyme-bound carbocation.

As this very brief, highly subjective history indicates, it has been an exciting 100 years. In a century we have come from the simple observation of color changes on the dissolution of triarylcarbinols and halides in concentrated sulfuric acid to the involvement of carbocations in complex biological processes. Along the way an entire new subfield of chemistry—mechanistic, physical organic chemistry—was developed, to fully understand these and other reactive intermediates.

As we stand at the threshold of the twenty-first century, who knows what exciting new developments will occur involving carbocations. Just as even the most erudite, brilliant, and imaginative chemists like Baeyer, Wagner, or even Meerwein in 1900 could not foresee, let alone predict, the particulars of the many marvelous advances in carbocations, such as their direct observation by NMR and X ray, none of us at this symposium can foretell the exact details of further advances, to be made in the twenty-first century. Thus, it is difficult to even begin to guess where the future of carbocations lies. What is certain is that there will be many unanticipated discoveries and surprises that will challenge and stimulate many of the best chemists in the first century of this new millennium just as they did during the last 100 years. May the next 100 years of carbocation chemistry be as adventurous, exciting, and fruitful as the last 100 years have been.

2

100 YEARS OF CARBOCATIONS AND THEIR SIGNIFICANCE IN CHEMISTRY*

George A. Olah

Loker Hydrocarbon Research Institute and Department of Chemistry
University of Southern California
Los Angeles, California

*Reprinted with permission from *J. Org. Chem.*, **66** (18), 5943–5957, September 7, 2001. Copyright 2001, American Chemical Society.

Carbocation Chemistry, Edited by George A. Olah and G. K. Surya Prakash
ISBN 0-471-28490-4 Copyright © 2004 John Wiley & Sons, Inc.

The centennial of the first report of a stable (persistent) carbocation, the triphenylmethyl cation, serves as the occasion to give a perspective based on recollections of my half a century of their search.[1] This included development of general methods to prepare persistent long-lived carbocations in superacidic media, realization of the general concept of carbocations including trivalent and five or higher coordinated ions, as well as their role and significance in chemistry.

2.1 EARLY STUDIES OF TRIPHENYLMETHYL CATIONS

Many elements readily form ionic compounds, for example, sodium chloride, in which the cationic sodium and anionic chlorine are held together by electrostatic, ionic bonding. Carbon, however, was long considered to lack the ability to form similar ionic compounds, except in very specific, highly stabilized systems, such as triphenylmethyl dyes.

In 1899 the University of Chicago chemist Stieglitz, while studying salts of imido ethers, raised the possibility of ionic hydrocarbon compounds.[2] It was in 1901 that Norris,[3] as well as Kehrman and Wentzel,[4] independently discovered that colorless triphenylmethyl alcohol gave deep yellow solutions in concentrated sulfuric acid. Triphenylmethyl chloride similarly formed orange complexes with aluminum and tin chlorides.

Baeyer should be credited for having recognized in 1902 the saltlike character of the compounds formed.[5] He then suggested a correlation between the appearance of color and salt formation, the so-called halochromy. Gomberg (who had just shortly before discovered the related stable triphenylmethyl radical), as well as Walden, contributed to the evolving understanding of the structure of related cationic dyes such as malachite green:[6]

$$(R_2NC_6H_4)(C_6H_5)C{=}C_6H_4{=}\overset{+}{N}R_2 \longleftrightarrow (R_2NC_6H_4)(C_6H_5)\overset{+}{C}{-}C_6H_4{-}NR_2$$

Whereas the existence of ionic triarylmethyl and related dyes was thus established around the turn of the twentieth century, the more general significance of carbocations in chemistry long went unrecognized. Triarylmethyl cations were considered an isolated curiosity of chemistry, not unlike Gomberg's triarylmethyl radicals. Not only were simple hydrocarbon cations believed to be unstable; even their fleeting existence was doubted.

2.2 KINETIC AND STEREOCHEMICAL STUDIES INDICATING CARBOCATIONS

One of the most original and significant ideas in organic chemistry was the suggestion by Meerwein that carbocations (as we now call all the positive ions of carbon compounds) might be intermediates to the course of reactions that start from nonionic reactants and lead to nonionic covalent products.

In 1922, Meerwein, while studying the Wagner rearrangement of camphene hydrochloride to isobornyl chloride with van Emster, found that the rate of the reaction increased with the dielectric constant of the solvent.[7] Furthermore, he found that certain Lewis acid chlorides such as $SbCl_5$, $SnCl_4$, $FeCl_3$, $AlCl_3$, and $SbCl_3$ (but not BCl_3 or $SiCl_4$) as well as dry HCl, which promote the ionization of triphenylmethyl chloride by formation of carbocationic complexes, also considerably accelerated the rearrangement of camphene hydrochloride to isobornyl chloride. Meerwein concluded that the isomerization actually does not proceed by way of migration of the chlorine atom but by a rearrangement of a cationic intermediate:

Camphene hydrochloride

Isobornyl chloride

Hence, the modern concept of carbocationic intermediates was born. Meerwein's views were, however, greeted with much skepticism by his peers in Germany, discouraging him from following up on these studies.

Ingold, Hughes, and their collaborators in England, starting in the late 1920s, carried out detailed kinetic and stereochemical investigations on what became known as "nucleophilic substitution at saturated carbon and polar elimination reactions."[8] Their work relating to unimolecular nucleophilic substitution and elimination, called S_N1 and El reactions, in which formation of carbocations is the slow rate-determining step, laid the foundation for the role of electron-deficient carbocationic intermediates in organic reactions.

Frank Whitmore in the United States in the 1930s in a series of papers generalized these concepts to include many other organic reactions.[9] Carbocations, however, were generally considered to be unstable and transient (short-lived) because they could not be directly observed. The concept of carbocations slowly grew to maturity through kinetic, stereochemical, and product studies of a wide variety of reactions. Leading investigators such as P. D. Bartlett, C. D. Nenitzescu, S. Winstein, D. J. Cram, M. J. S. Dewar, J. D. Roberts, P. v. R. Schleyer, and others contributed fundamentally to the development of modern carbocation chemistry. The role of carbocations as one of the basic concepts of modern chemistry has been well reviewed.[10–12] With the advancement of mass spectrometry, the existence of gaseous carbocations was proved, but this could not give an indication of their structure or allow extrapolation to solution chemistry. Direct observation and study of stable, long-lived carbocations, such as alkyl cations in the condensed state, remained an elusive goal.

2.3 MY SEARCH FOR PERSISTENT, LONG-LIVED ALKYL CATIONS

My search for cationic carbon intermediates started in Hungary while I was studying Friedel–Crafts-type reactions with acyl[13] and subsequently alkyl fluorides catalyzed by boron trifluoride.[14] In the course of these studies I observed (and, in some cases, isolated) intermediate complexes of either donor–acceptor or ionic nature:

$$RCOF + BF_3 \rightleftharpoons \underset{F}{\overset{\delta+}{RCO}} \longrightarrow \overset{\delta-}{BF_3} \rightleftharpoons RCO^+ BF_4^-$$

$$RF + BF_3 \rightleftharpoons \overset{\delta+}{RF} \longrightarrow \overset{\delta-}{BF_3} \rightleftharpoons R^+ BF_4^-$$

The idea that ionization of alkyl fluorides to stable alkyl cations could be possible with an excess of strong Lewis acid fluoride that also serves as solvent first came to me in the early 1950s while studying the boron trifluoride–catalyzed alkylation of aromatics with alkyl fluorides. In the course of these studies I attempted to isolate $RF{:}BF_3$ complexes. Realizing the difficulty of finding suitable solvents that would allow ionization but at the same time would not react with developing, potentially highly reactive alkyl cations, I condensed alkyl fluorides with neat boron trifluoride at low temperatures. I had, however, no access at the time to any modern spectrometers.

All I could do at the time with the $RF–BF_3$ complexes was to measure their conductivity. The results showed that methyl fluoride and ethyl fluoride complexes gave low conductivity, whereas the isopropyl fluoride and *tert*-butyl fluoride complexes were highly conducting.[14] The latter systems, however, also showed some polymerization (from deprotonation to the corresponding olefins). The conductivity data thus must have been to some degree affected by acid formation:

$$R{-}F + BF_3 \rightleftharpoons R{-}\overset{\delta+}{F} \longrightarrow \overset{\delta-}{BF_3} \rightleftharpoons R^+ BF_4^-$$

My work on long-lived (persistent) carbocations started in late 1950s while working for Dow Chemical in Sarnia, Ontario, Canada and resulted in the first direct observation of alkyl cations.

In Friedel–Crafts chemistry it was known that when pivaloyl chloride is reacted with aromatics in the presence of aluminum chloride, *tert*-butylated products are obtained in addition to the expected ketones. These were assumed to be formed by decarbonylation of the intermediate pivaloyl complex or cation. In the late 1950s I returned to my earlier investigations of Friedel–Crafts complexes and extended them by using IR and NMR spectroscopy. I studied isolable complexes of acyl fluoride with Lewis acid fluorides, including higher-valence Lewis acid fluorides such as SbF_5, AsF_5, and PF_5.[15–17] In the course of these studies, it was not entirely unexpected that the generated $(CH_3)_3CCOFSbF_5$ complex showed a substantial tendency toward decarbonylation. What was exciting, however, was that it was possible to follow this process by NMR spectroscopy and to observe what turned out to be the first stable, long-lived alkyl cation salt, namely, *tert*-butyl hexafluoroantimonate:

$$(CH_3)_3CCOF + SbF_5 \longrightarrow (CH_3)_3CCO^+SbF_6^- \xrightarrow{-CO} (CH_3)_3C^+SbF_6^-$$

This breakthrough was first reported in 1962 and was followed by further studies that led to methods for preparing various long-lived alkyl cations in solution.

During a prolonged, comprehensive study of attempted ionization of alkyl halides, including fluoride with numerous Lewis acid halides, I finally hit on antimony pentafluoride. It turned out to be an extremely strong Lewis acid and, for the first time, enabled the ionization of alkyl fluorides to stable, long-lived alkyl cations. Neat SbF_5 solutions of alkyl fluorides are viscous, but diluted with liquid sulfur dioxide, the solutions could be cooled and studied at $-78°C$. Subsequently, I also introduced even lower-nucleophilicity solvents such as SO_2ClF or SO_2F_2, which allowed studies at even lower temperatures. Following up the observation of the decarbonylation of the pivaloyl cation that gave the first spectral evidence for the tertiary butyl cation, *tert*-butyl fluoride was ionized in excess antimony pentafluoride. The solution of the *tert*-butyl cation turned out to be remarkably stable, allowing chemical and spectroscopic studies alike.[16,18]

In the late 1950s the research director of our laboratory was not yet convinced of the usefulness of NMR spectroscopy. Consequently, we had no such instrumentation of our own. Fortunately, the Dow laboratories in Midland, Michigan just 100 miles away, had excellent facilities run by E. B. Baker, a pioneer of NMR spectroscopy, who offered his help. To probe whether our SbF_5 solution of alkyl fluorides indeed contained alkyl cations, we routinely drove in the early morning to Midland with our samples and watched Ned Baker obtain their NMR spectra. *tert*-Butyl fluoride itself showed a characteristic doublet in its ^1H NMR spectrum due to the fluorine–hydrogen coupling ($J_{H\text{-}F} = 20$ Hz). In SbF_5 solution, the doublet disappeared and the methyl protons became significantly deshielded from about δ 1.5 to 4.3. This was very encouraging but not conclusive proof of the presence of the *tert*-butyl cation. If one assumes that with SbF_5, *tert*-butyl fluoride forms only a polarized donor–acceptor complex, which undergoes fast fluorine exchange (on the NMR timescale), the fluorine–hydrogen coupling would be "washed out," while a significant deshielding of the methyl protons would still be expected. The differentiation of a rapidly exchanging polarized donor–acceptor complex from the long-sought-after ionic $t\text{-}C_4H_9^+SbF_6^-$ thus became a major challenge:

$$(CH_3)_3C{-}F + SbF_5 \rightleftharpoons (CH_3)_3C\cdots F\cdots SbF_5 \text{ (bridged complex)} \quad \text{or} \quad (CH_3)_3C^+\ SbF_6^-$$

Ned Baker, himself a physicist, showed great interest in our chemical problem. To solve it, he devised a means to obtain the ^{13}C NMR spectra of our dilute solutions, an extremely difficult task at the time before the advent of Fourier transform NMR techniques. Labeling with ^{13}C was possible at the time only to about a 50% level (from $Ba^{13}CO_3$). When we prepared 50% ^{13}C-labeled *tert*-butyl fluoride for ionization in SbF_5, we could, however, obtain at best only a 5% carbocationic solution in SbF_5. Thus the ^{13}C content in the ionic solution was tenfold diluted. However, Baker, undaunted, devised what became known as the *internuclear double resonance (INDOR) method.* Using the high sensitivity of the proton signal, he was able with the double-resonance technique to observe the ^{13}C shifts of our dilute solutions—a remarkable achievement around 1960. The ^{13}C shift of the tertiary carbon atom in $(CH_3)_3CFSbF_5$ of δ 335.2 turned out to be more than 300 ppm deshielded from that of the covalent starting material. Such very large chemical deshielding (the most deshielded ^{13}C signal at the time) could not be reconciled with a donor–acceptor complex. It indicated rehybridization from sp^3 to sp^2 and at the same time showed the effect of significant positive charge on the carbocationic carbon center.[19]

Besides the *tert*-butyl cation, we also succeeded in preparing and studying the related isopropyl and the *tert*-amyl cations; the isopropyl cation was of particular relevance:

$$(CH_3)_2CHF + SbF_5 \longrightarrow (CH_3)_2CH^+SbF_6^-$$
$$(CH_3)_2CFCH_2CH_3 + SbF_5 \longrightarrow (CH_3)_2C^+CH_2CH_3SbF_6^-$$

Whereas in the *tert*-butyl cation the methyl protons are attached to carbons that are only adjacent to the carbocationic center, in the isopropyl cation a proton is directly attached to the center. When we obtained the proton NMR spectrum of the i-C_3H_7F–SbF_5 system, the CH proton showed up as an extremely deshielded septet at 13.0 ppm, ruling out a polarized donor–acceptor complex and indicating the formation of the $(CH_3)_2CH^+$ ion. The ^{13}C NMR spectrum was also conclusive, showing a very highly deshielded (by $\Delta\delta > 300$) C^+ atom ($\delta^{13}C$ 320.6). The spectrum of the *tert*-amyl cation showed an additional interesting feature due to the strong long-range H–H coupling of the methyl protons adjacent to the carbocationic center with the methylene protons. If only the donor–acceptor complex were involved, such long range coupling through an sp^3 carbon would be small (1–2 Hz). Instead, the observed significant coupling ($J_{H\cdot H} = 10$ Hz) indicated that the species studied indeed had an sp^2 center through which the long-range H–H coupling became effective. Figure 2.1 reproduces the ^1H NMR spectra of the *tert*-butyl, *tert*-amyl, and isopropyl cations. These original spectra are framed and hang in my office as a memento, as are the ESCA spectra of the *tert*-butyl and of the norbornyl cation (see text below).

IR spectroscopic studies on *tert*-butyl cation was also carried out in neat SbF_5 at room temperature long before the advent of the Fourier transform methods. Subsequently, in 1968–1970 with DeMember and Commeyras in Cleveland, we were able to carry out more detailed IR and laser Raman spectroscopic studies of alkyl

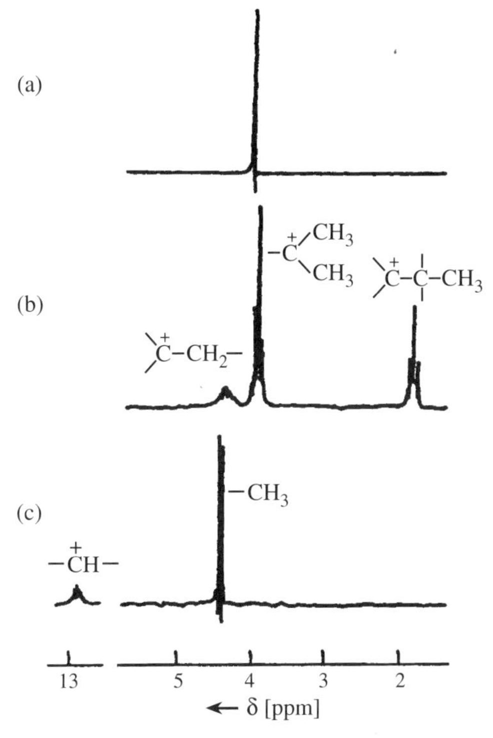

Figure 2.1 ^{1}H NMR spectra of (a) the *tert*-butyl cation [trimethylcarbenium ion, $(CH_3)_3C^+$], (b) the *tert*-amyl cation [dimethylethylcarbenium ion, $(CH_3)_3C^+$-C_2H_5], (c) the isopropyl cation [dimethylcarbenium ion, $(CH_3)_2C^+H$] (60 MHz, in SbF_5/SO_2ClF solution, −60°C).

cations.[20] Comparison of the data of unlabeled and deuterated *tert*-butyl cations with those of isoelectronic trimethylboron proved the planar structure of the carbocation:

$$(CH_3(D_3))_3C^+ \qquad (CH_3(D_3))_3B$$

Our studies also included IR spectroscopic investigation of the observed ions (Fig. 2.2). John Evans, who was at the time a spectroscopist at the Midland Dow laboratories, offered his cooperation and was able to obtain and analyze the vibrational spectra of our alkyl cations. It is rewarding that, some 30 years later, Fourier transform infrared (FTIR) spectra obtained by Sunko and his colleagues in Zagreb[21] with low-temperature matrix deposition techniques and Schleyer's calculations of the spectra showed good agreement with our early work. This was also an early example of the realization that for nearly all carbocations there exists a neutral isoelectronic isostructural boron analog, which later proved itself so useful in the hands of my colleagues Williams, Prakash, and Field.

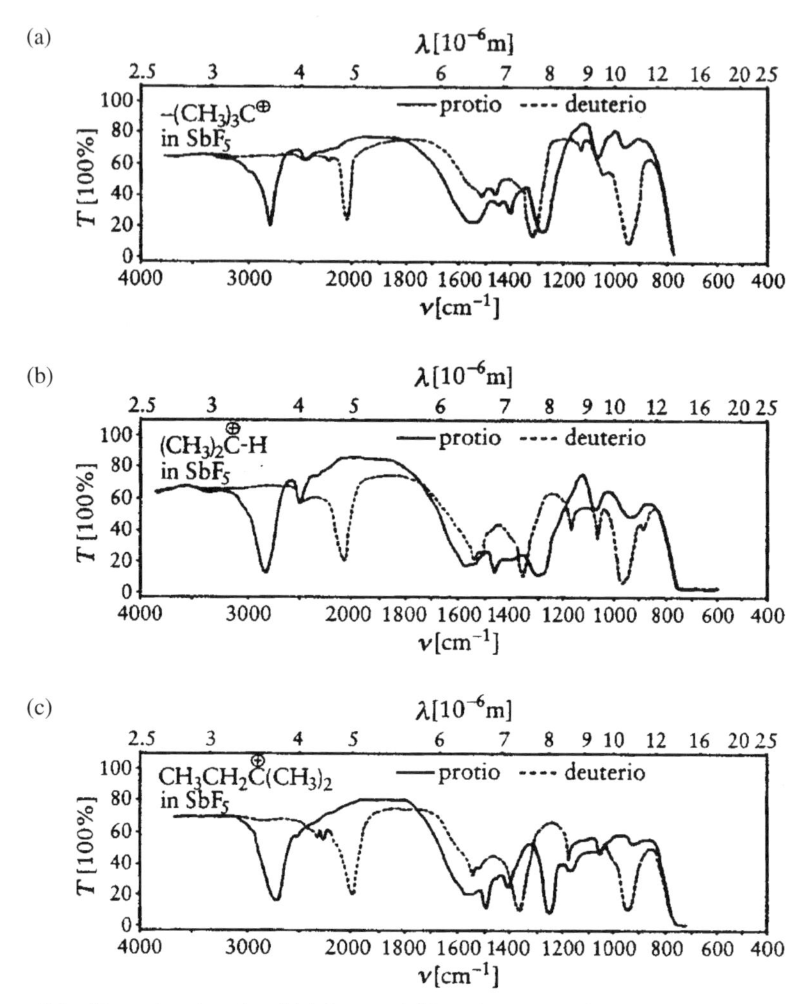

Figure 2.2 IR spectra of *tert*-butyl (a), isopropyl (b), and *tert*-amyl (c) cations (T = transmission).

The chemistry of stable, long-lived (or persistent) carbocations, as they became known, thus began. Many have contributed since to the study of long-lived carbocations. The field rapidly expanded and allowed successful study of practically any carbocationic system. My talented and hard-working former associates and students deserve the lion's share of credit for our work, as do the many researchers around the world who joined in and contributed so much to the development of the field (their work can be found in the references listed at the end of the chapter). I would like, however, to mention particularly the pioneering work of Brouwer and Mackor,[22] as well as their colleagues at the Shell Laboratories in Amsterdam in the 1960s and 1970s. They contributed fundamentally to the study of long-lived carbocations and related superacidic hydrocarbon chemistry. The first publication from the Shell laboratories on alkyl cations appeared in 1964, closely following my initial reports of 1962/63.

Subsequent to my study of alkyl cations a wide spectrum of carbocations as long-lived species was studied using antimony pentafluoride as an extremely strong Lewis acid and later using other highly acidic (superacidic) systems.

Until this time alkyl cations were considered only transient species. Their existence had been indirectly inferred from kinetic and stereochemical studies, but no reliable spectroscopic or other physical measurements of simple alkyl cations in solution or in the solid state had been obtained.

2.4 THE ROLE OF SUPERACIDS

It was not fully realized until my breakthrough using extremely strong superacids that, to suppress the deprotonation of alkyl cations to olefins and the subsequent formation of complex mixtures by reactions of olefins with alkyl cations, such as alkylation, oligomerization, polymerization, and cyclization, acids much stronger than those known and used in the past were needed:

$$(CH_3)_3C^+ \rightleftharpoons H^+ + (CH_3)_2C{=}CH_2$$

Finding such acids (called "superacids")[23] turned out to be the key to obtaining stable, long-lived alkyl cations and, in general, carbocations. If any deprotonation were still to take place, the formed alkyl cation (a strong Lewis acid) would immediately react with the formed olefin (a good π-base), leading to the abovementioned complex reactions.

In the course of my studies, it became increasingly clear that a variety of highly acidic systems besides the originally used antimony pentafluoride systems are capable of generating long-lived, stable carbocations. The work was thus extended to a variety of other superacids. Protic superacids such as FSO_3H (fluorosulfuric acid) and CF_3SO_3H (triflic acid) as well as conjugate acids such as $HF–SbF_5$, $FSO_3H–SbF_5$ (magic acid), $CF_3SO_3H–SbF_5$, and $CF_3SO_3H–B(O_3SCF_3)_3$ were extensively used. Superacids based on Lewis acid fluorides such as AsF_5, TaF_5, and NbF_5 and other strong Lewis acids such as $B(O_3SCF_3)_3$ were also successfully introduced. The name "magic acid" for the $FSO_3H–SbF_5$ system was given by Joe Lukas, a German postdoctoral fellow working with me in Cleveland in the 1960s, who after a laboratory Christmas party put remainders of a candle into the acid. The candle dissolved, and the resulting solution gave a clear NMR spectrum of the *tert*-butyl cation. This observation understandably evoked much interest, and the acid used was named "magic." The name stuck in our laboratory. I think it was Ned Arnett who learned about it during one of his visits and subsequently introduced the name into the literature, where it came into quite general use.

I would like to credit especially the fundamental contributions of Ron Gillespie to the inorganic chemistry of strong acids (superacids)[23a,b] which greatly contributed to the development of superacid chemistry. His longstanding interest in fluorosulfuric acid and our studies of SbF_5-containing systems thus found common ground in studies of $FSO_3H–SbF_5$ (magic acid) systems.[24]

Until the late 1950s chemists generally considered mineral acids, such as sulfuric, nitric, perchloric, and hydrofluoric acids, to be the strongest acid systems in existence. This has changed considerably as extremely strong acid systems—many billions or even trillions of times stronger than sulfuric acid—have been discovered.

The acidity of aqueous acids is generally expressed by their pH, which is a logarithmic scale of the hydrogen ion concentration (or, more precisely, of the hydrogen ion activity). pH can be measured by the potential of a hydrogen electrode in equilibrium with a dilute acid solution or by a series of colored indicators. In highly concentrated acid solutions or with strong nonaqueous acids the pH concept is no longer applicable, and acidity, for example, can be related to the degree of transformation of a base to its conjugate acid (keeping in mind that this will depend on the base itself). The widely used so-called Hammett acidity function H_0 relates to the half-protonation equilibrium of suitable weak bases. The Hammett acidity function is also a logarithmic scale on which 100% sulfuric acid has a value of $H_0 - 11.9$. The acidity of sulfuric acid can be increased by the addition of SO_3 (oleum). The H_0 of HF is -11.0 (however, when HF is completely anhydrous, its H_0 is -15, but even a slight amount of water drops the acidity to -11, as shown by Gillespie).

Perchloric acid ($HClO_4$; $H_0 - 13.0$), fluorosulfuric acid (HSO_3F; $H_0 - 15.1$), and trifluoromethanesulfonic acid (CF_3SO_3H; $H_0 - 14.1$) are considered to be superacids, as is truly anhydrous hydrogen fluoride. Complexing with Lewis acidic metal fluorides of higher valence, such as antimony, tantalum, or niobium pentafluoride, greatly enhances the acidity of all these acids.

In the 1960s Gillespie suggested calling protic acids stronger than 100% sulfuric acid "superacids." This arbitrary but most useful definition is now generally accepted. It should be mentioned, however, that the name "superacid" goes back to J. B. Conant of Harvard, who used it in 1927 in a paper in the *Journal of the American Chemical Society* to denote acids such as perchloric acid, which he found stronger than conventional mineral acids and capable of protonating such weak bases as carbonyl compounds.[25] My book on superacids, published in 1985 with Surya Prakash and Jean Sommer,[23c] was appropriately dedicated to the memory of Conant. Few of today's chemists are aware of his contributions to this field. Conant subsequently became the president of Harvard University and gave up chemistry, which may explain why he never followed up on his initial work on superacids.

In a generalized sense, acids are electron pair acceptors. They include both protic (Brønsted) acids and Lewis acids such as $AlCl_3$ and BF_3 that have an electron-deficient central metal atom. In extending the concept of superacidity to Lewis acid halides, those stronger than anhydrous aluminum chloride (the most commonly used Friedel–Crafts acid) are considered super Lewis acids. These superacidic Lewis acids include such higher-valence fluorides as antimony, arsenic, tantalum, niobium, and bismuth pentafluorides. Superacidity encompasses both very strong Bronsted and Lewis acids and their conjugate acid systems.

Friedel–Crafts Lewis acid halides form with proton donors such as H_2O, HCl, and HF conjugate acids such as $H_2O–BF_3$, $HCl–AlCl_3$, and $HF–BF_3$, which ionize to $H_3O^+BF_3OH^-$, $H_2Cl^+AlCl_4^-$, and $H_2F^+BF_4^-$, and similar compounds. These conjugate Friedel–Crafts acids have H_0 values from about -14 to -16. Thus they are much stronger than the usual mineral acids. Even stronger superacid systems are $HSO_3F–SbF_5$ (magic acid), $HF–SbF_5$ (fluoroantimonic acid), and $CF_3SO_3H–B(O_3SCF_3)_3$ (triflatoboric acid). The acidity of anhydrous HF, HSO_3F, and CF_3SO_3H increases drastically on addition of Lewis acid fluorides such as SbF_5, which form large complex fluoroanions facilitating dispersal of the negative charge:

$$2\,HF + 2\,SbF_5 \rightleftharpoons H_2F^+Sb_2F_{11}^- \qquad \text{fluoroantimonic acid}$$

$$2\,HSO_3F + 2\,SbF_5 \rightleftharpoons H_2SO_3F^+Sb_2F_{10}(SO_3F)^- \qquad \text{magic acid}$$

$$2\,CF_3SO_3H + B(O_3SCF_3)_3 \rightleftharpoons CF_3SO_3H_2^+B(O_3SCF_3)_4^- \qquad \text{triflatoboric acid}$$

The acidity function of HSO_3F increases on addition of SbF_5 from -15.1 to -23.0, the acidity of 1 : 1 $FSO_3H–SbF_5$ (magic acid). Fluoroantimonic acid is even stronger; with 4 mol% SbF_5, the H_0 value for $HF–SbF_5$ is already -21.0, a thousand times stronger than the value for fluorosulfuric acid with the same SbF_5 concentration. The acidity of the 1 : 1 $HF–SbF_5$ system or those with even higher SbF_5 concentrations reaches H_0 -28. Thus these superacidic systems can be 10^{16} times stronger than 100% sulfuric acid! (A trillion is 10^{12}.)

Related superacid systems in which SbF_5 is replaced by AsF_5, TaF_5, NbF_5, and related compounds are of somewhat lower acidity but are still extremely strong acids. So is $HF–BF_3$, a very useful superacid that will not cause oxidative side reactions. Ternary superacid systems including, for example, $FSO_3H–HF$ or $CF_3SO_3H–HF$ with Lewis acid fluorides are also known and used.

Acids are not limited to liquid (or gaseous) systems. Solid acids also play a significant role. Acidic oxides such as silica and silica–alumina are used extensively as solid acid catalysts. New solid acid systems that are stronger than those used conventionally are sometimes called *solid superacids*. However, it must be kept in mind that when carbocations are developing on solid surfaces or zeolitic systems, they will coordinate (bind) to nucleophilic sites (oxygen, etc.) and thus only highly stable carbocations will be observable as persistent species on such systems. This, of course, does not mean that at the involved higher temperatures some equilibrium with carbocationic species could not exist, provoking carbocationic reactions.

Solid perfluorinated resin sulfonic acid catalysts, such as those based on the acid form of DuPont's Nafion ionomer membrane resin, and some higher perfluoroalkanesulfonic acids, such as perfluorodecanesulfonic acid, more closely resemble liquid fluorinated superacids.

2.5 GENERALIZATION OF THE STUDY OF CARBOCATIONS

After I returned to academia in Cleveland in 1965, my research continued and extended the study of carbocations in varied superacidic systems as well as exploration of the broader chemistry of superacids, involving varied ionic systems and reagents. In Cleveland, a main aspect of my research was directed at exploration of the chemistry of these persistent cations of carbon compounds (carbocations) and the fascinating new area of chemistry opened up by superacids. Particular interest was generated as a great variety of carbocations were found to be readily generated and studied in these enormously strong acid systems.

Over a decade of research, we were able to show that practically all conceivable carbocations could be prepared under what became known as "stable ion conditions" using various very strong acid systems (see discussion of superacids) and low-nucleophilicity solvents (SO_2, SO_2ClF, SO_2F_2, etc.).[26] A variety of precursors could be used under appropriate conditions, as shown, for example, in the preparation of the methylcyclopentyl cation:

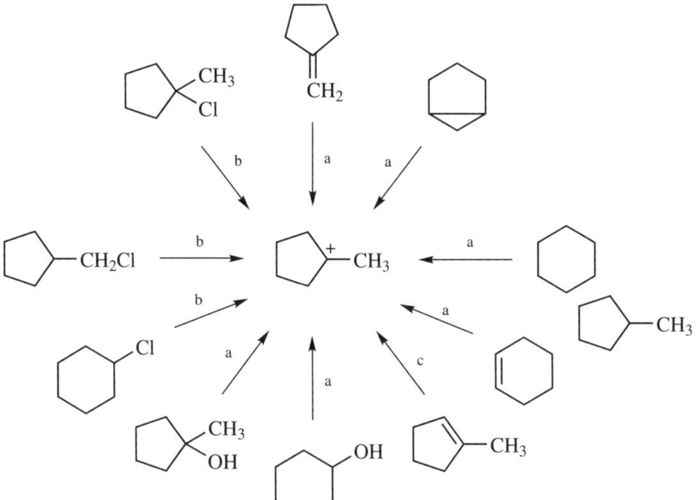

^{a}FSO$_3$H-SbF$_5$ ^{b}SbF$_5$/SO$_2$ ^{c}HF/SbF$_5$/SO$_2$

A wide variety of carbocations and carbodications, including those that are aromatically stabilized as well those as stabilized by heteroatoms, were reported in the nearly 200 publications on the topic during my Cleveland years:

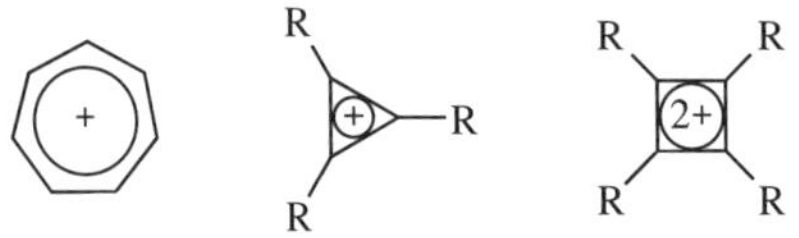

$R-\overset{+}{X}-R'$

R, R′ = alkyl or aryl
X = Br, Cl, I

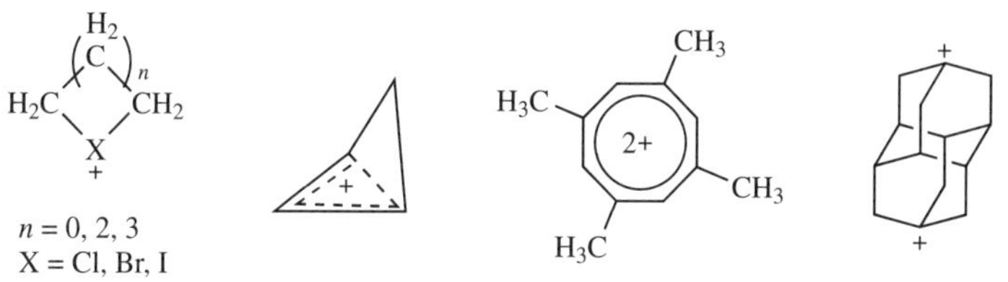

$n = 0, 2, 3$
X = Cl, Br, I

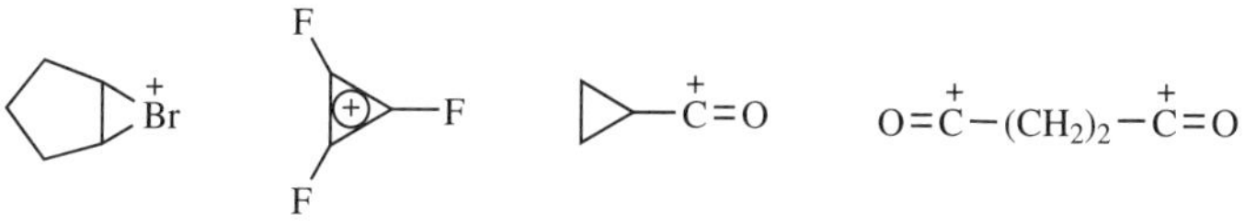

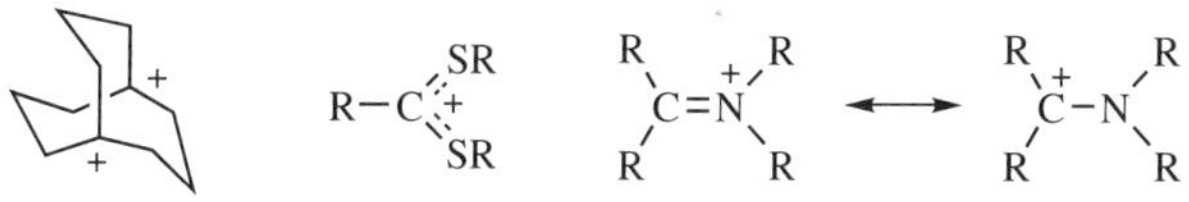

During my Cleveland years in 1969 I organized a symposium on carbocation chemistry. It was attended by many of the major investigators in the field (Nenitzescu, Brown, Winstein, Dewar, Schleyer, Gillespie, Saunders, and others). We had a lively discussion, as was the case at the symposium in fall 1977 after my move to Los Angeles at USC (University of Southern California) and at a number of symposia I attended in different locations such as at La Grand Motte (France), Bangor (UK), and eventually Seattle, Washington.

2.6 THE NONCLASSICAL ION CONTROVERSY AND ITS SIGNIFICANCE

During the study of carbocations much effort was put into studying whether certain carbocations represent rapidly equilibrating trivalent carbenium ions or static (bridged, delocalized) five-coordinate carbonium ion systems. This question became the focus of the rather heated controversy that centered around the question of the norbornyl cation and became known as the nonclassical ion controversy.[27]

The controversy in which I was inadvertently involved started over the structure of a deceptively simple seven-carbon-containing bicyclic carbocation, the 2-norbornyl (bicyclo[2.2.1]heptyl) cation. The involvement of the ion in the long-recognized rearrangement of norbornyl systems prevalent in natural terpenes was first suggested in 1922 by Meerwein. The remarkable facility of skeletal rearrangements in norbornyl systems attracted the early interest of chemists. Wagner realized first in 1899 the general nature of these rearrangements and related them to the rearrangement that takes place during the dehydration of pinacolyl alcohol to tetramethylethylene. Sommer later found some tricyclanes in the products of the

Wagner rearrangements of terpenes. In 1918 Ruzicka suggested a tricyclane-type mechanism without realizing the ionic nature of the process. Meerwein reconsidered the mechanism in 1922 and made the farsighted suggestion that the reaction proceeds through an ionic intermediate, namely, the norbornyl cation. Hence this type of transformation is now known as the Wagner–Meerwein rearrangement.

The structure of the norbornyl cation became controversial in the "nonclassical ion" controversy following Wilson's original suggestion in 1939 of a mesomeric, α-delocalized, carbocationic intermediate in the camphene hydrochloride–isobornyl chloride rearrangement. From 1949 to 1952, Winstein and Trifan reported a solvolytic study of the *exo*- and *endo*-2-norbornyl brosylates (*p*-bromobenzenesulfonates) and postulated a σ-delocalized, symmetrically bridged norbornyl ion intermediate. The *endo* reactant was found to solvolyze in various solvents such as acetic acid, aqueous acetone, and aqueous dioxane to give substitution products of exclusively *exo* configuration. The *exo*-brosylate also gave exclusively *exo* product and was markedly more reactive in acetolysis than the *endo*, by a factor of 350.

Winstein, one of the most brilliant chemists of his time, concluded that "it is attractive to account for these results by way of the bridged (nonclassical) formulation for the *norbornyl cation involving accelerated rate of formation from the exo precursor* [by anchimeric assistance; emphasis added]." His formulation of the norbornyl cation as a σ-bridged species stimulated other workers in the solvolysis field to interpret results in a variety of systems in similar terms of σ-delocalized, bridged carbonium ions.

H. C. Brown (the outstanding pioneer of hydroboration chemistry) in contrast, concluded that in solvolysis both 2-*exo* and 2-*endo* norbornyl esters (brosylates, etc.) undergo anchimerically unassisted ionization and that the singular rate and product characteristics of the system are attributable to steric effects, in particular, hindrance to ionization of the *endo* isomers. Explaining the results of the extensive solvolytic studies, he suggested that high *exo/endo* rate and product ratios do not necessitate σ participation as an explanation. In other words, *exo* is not fast; *endo* is slow. His suggestion for the structure of the norbornyl cation was that of a rapidly equilibrating pair of regular trivalent ions (classical ions), and he compared the process to that of a windshield wiper. However, at the same time none of his studies ever showed that σ participation cannot be involved.

In 1962, H. C. Brown lodged his dissent against the σ-bridged 2-norbornyl cation and, for that matter, other nonclassical carbocations. He has maintained his position virtually unchanged over the years and has continued to present his views forcefully. In arguing against carbon σ bridging, he took the position, despite his pioneering work in structurally closely related boranes, that if carbon were to participate in bridging, novel bonding characteristics must be attributed to it.

In 1965 he stated, "On the other hand, the norbornyl cation does not possess sufficient electrons to provide a pair for all of the bonds required by the proposed bridged structures. *One must propose a new bonding concept, not yet established in carbon structures*" (with my emphasis added).

In 1967 he again wrote, "The second subclass consists of ions such as the bicyclobutonium and the norbornyl cation in its σ-bridging form, which do not possess sufficient electrons to provide a pair for all the bonds required by the proposed structures. *A new bonding concept not yet established in carbon structures is required*" (emphasis added).

The Brown–Winstein nonclassical ion controversy can be summed up as differing explanations of the same experimental facts (which were obtained repeatedly and have not been questioned) of the observed significantly higher rate of the hydrolysis of the 2-*exo*- over the 2-*endo*-norbornyl esters. As suggested by Winstein, the reason for this is participation of the C1–C6 single bond leading to delocalization via the bridged "nonclassical" ion. In contrast, Brown maintained that the cause was only steric hindrance to the sterically hindered *endo* side involving rapidly equilibrating "classical" trivalent ions.

Nonclassical ions, a term first used by John Roberts,[28] were defined by Paul Bartlett as containing too few electrons to allow a pair for each "bond"; thus, they must contain delocalized σ electrons.[27a] This is where the question stood in the early 1960s. The structure of the intermediate 2-norbornyl ion could only be suggested indirectly from rate (kinetic) data and observation of stereochemistry; no direct observation or structural study was possible at the time.

My own involvement with the norbornyl ion controversy goes back to 1960–1962, when as discussed I succeeded in developing a general method of preparing and studying persistent (long-lived) alkyl cations. My interest subsequently extended to the study of various carbocations, including the controversial 2-norbornyl cation. Whereas previous investigators were able to study carbocations only indirectly (by kinetic and stereochemical studies), my newly discovered methods allowed their preparation and direct study as persistent (long-lived) species.

At the 1962 Brookhaven Mechanism Conference, where I first reported on long-lived carbocations in public I was called aside separately by Winstein and Brown, both towering and dominating personalities of the time, who cautioned me that a young man should be exceedingly careful in making such claims. Each pointed out that most probably I was wrong and could not have obtained long-lived carbocations. Just in case, however, my method turned out to be real, I was advised to obtain further evidence for the "nonclassical" or "classical" nature (depending on who was giving the advice) of the much-disputed 2-norbornyl cation.

Because my method indeed allowed me to prepare practically any carbocation as long-lived species, clearly the opportunity was there to experimentally decide the norbornyl ion question through direct observation of the ion. At the time of the Brookhaven conference I had already obtained the proton NMR spectrum of 2-norbornyl fluoride in SbF_5, but only at room temperature, which displayed a single broad peak indicating complete equilibration through hydride shifts and Wagner–Meerwein rearrangement (well known in solvolysis studies and related transformations of 2-norbornyl systems). However, my curiosity was aroused, and when I moved to Dow's Research Laboratory in the Boston area in 1964, the work was further pursued in cooperation with Paul Schleyer from Princeton (who became a lifetime friend) and Martin Saunders from Yale.[29] Using SO_2 as solvent, we were able to lower the temperature of the solution to $-78°C$, and we also prepared the ion by ionization of β-cyclopentenylethyl fluoride or by protonation of nortricylene in

$FSO_3H;SbF_5/SO_2ClF$.[30] The three separate routes (representing σ, π, or bent σ participation) gave the identical ion:

As mentioned, we were able to obtain NMR spectra of the norbornyl ion at −78°C, where the 3,2-hydride shift was frozen out. However, it took until 1969, after my move to Cleveland, to develop more efficient low-temperature techniques using solvents such as SO_2ClF and SO_2F_2.[30a] We were eventually in 1982 able to obtain high-resolution ^{1}H and ^{13}C NMR spectra (using ^{13}C-enriched precursor) of the 2-norbornyl cation down to −159°C (Fig. 2.3).[30b] Both the 1,2,6-hydride shifts and the Wagner–Meerwein rearrangement were frozen out at such a low temperature,

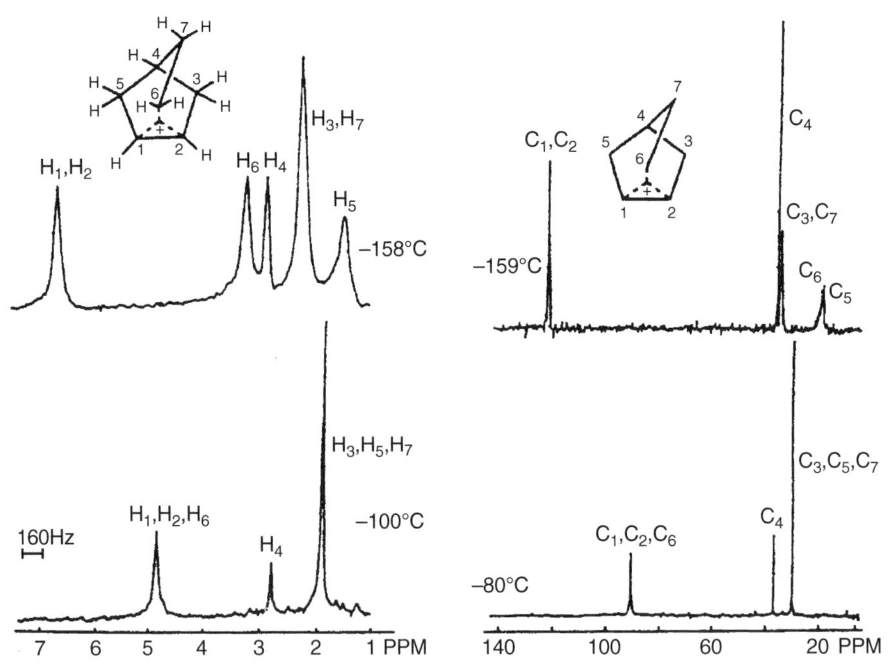

Figure 2.3 Left: 395-MHz ^{1}H NMR spectra of the 2-norbornyl cation in $SbF_5/SO_2ClF/SO_2F_2$ solution. Right: 50-MHz proton decoupled ^{13}C NMR spectra of the 2-norbornyl cation (^{13}C-enriched) in $SbF_5/SO_2ClF/SO_2F_2$ solution.

allowing us to observe the static, bridged ion, which I first reported at the Salt Lake City Organic Symposium in 1969.

The differentiation of bridged nonclassical from rapidly equilibrating classical carbocations based on NMR spectroscopy was difficult because NMR is a relatively slow physical method. We addressed this question in our work using estimated NMR shifts of the two structurally differing ions in comparison with model systems. Later, this task became greatly simplified and more precise by highly efficient calculational methods such as individual gauge for localized orbitals (IGLO) and gauge-invariant atomic orbital (GIAO), allowing the calculation of NMR shifts of differing ions and their comparison with experimental data. It is rewarding to see, however, that our earlier results and conclusions stood up well against all the more recent advanced studies.

Siegbahn's core electron spectroscopy for chemical analysis (ESCA) was another fast physical method that we applied to further resolve the question of bridged versus rapidly equilibrating ions. We were able to study carbocations in the late 1960s by this method, adapting it to superacidic matrixes. George Mateescu and Louise Riemenschneider in my Cleveland laboratory set up the necessary instrumentation and methodology for obtaining the ESCA spectra of a number of carbocations, including the *tert*-butyl and the 2-norbornyl cation in SbF_5-based superacidic matrixes.[31] These studies again convincingly showed the nonclassical nature of the 2-norbornyl cation. No trivalent carbenium center characteristic of a "classical" ion, such as is the case for the *tert*-butyl cation, was observed in the ESCA spectrum on a timescale where no chemical equilibration process could have any effect. Subsequent ESCA studies in cooperation with Grunthaner's laboratory at Caltech's Jet Propulsion Laboratory and by Dave Clark in England fully justified our initial results and conclusions (Fig. 2.4). So did the results of ever more advanced theoretical calculations.

Additional significant experimental studies were also carried out by other laboratories. Arnett reported fundamental calorimetric studies.[32] Saunders showed the absence of deuterium isotopic perturbation of equilibrium expected for a classical equilibrating system.[33] Myhre and Yannoni, at extremely low (5 K) temperature, were able to obtain solid-state ^{13}C NMR spectra that still showed no indication of freezing out any equilibrating classical ions. The barrier at this temperature should be as low as 0.2 kcal/mol (i.e., about the energy of a vibrational transition).[34] Laube was able to carry out single-crystal X-ray structural studies on substituted 2-norbornyl cations.[35] Schleyer's advanced theoretical studies including IGLO and related calculation of NMR shifts and their comparison with experimental data contributed further to the understanding of the σ-bridged carbonium ion nature of the 2-norbornyl cation.[36] The classical 2-norbornyl cation was not even found to be a high-lying intermediate!

As the norbornyl ion controversy evolved, it became a highly public and increasingly personal and sometimes even bitter public debate. Saul Winstein unexpectedly died in the fall of 1969, shortly after the Salt Lake City symposium. To my regret, I seemed to have inherited his role in representing the bridged nonclassical ion concept in subsequent discussions.

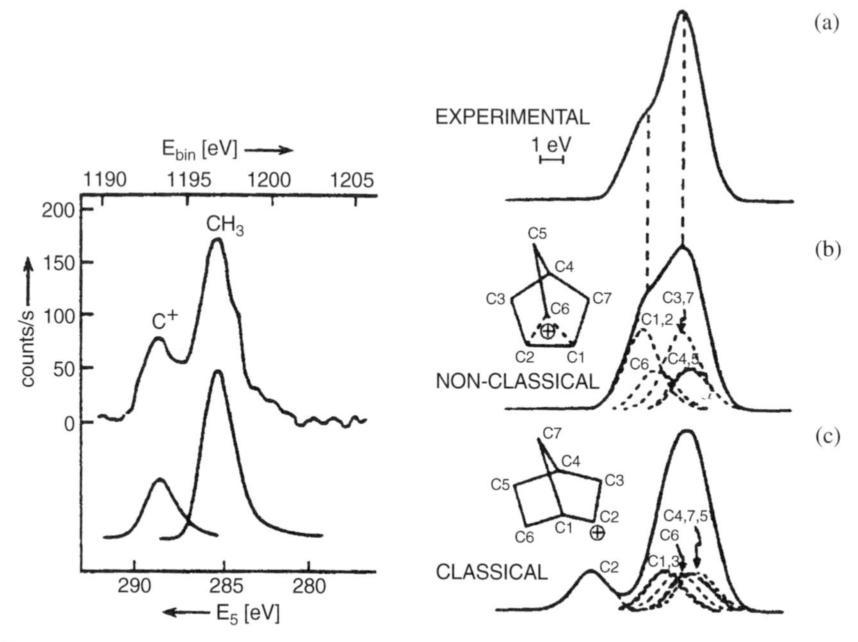

Figure 2.4 Left: carbon 1*s* photoelectron spectrum of the *tert*-butyl cation (top curve from experiment; bottom curve calculated). Right: 1*s* core–hole–state spectra for the 2-norbornyl cation (a), Clark's simulated spectra for the classical (b) and nonclassical ions (c).

The 1983 Seattle American Chemical Society symposium was the de facto "closing" of the long-running debate. Although in the heat of the debates some personal remarks were made on both sides (to which I subsequently expressed my regret) the experimental evidence at the time was already so overwhelming that I concluded my presentation saying, "I don't intend to do any more research on the matter. There is nothing further to be discussed. . . ." My lecture was subsequently published (with Prakash and Saunders) under the title "Conclusion of the classical–nonclassical ion controversy based on the structural study of the 2-norbornyl cation" in an article in the *Accounts of Chemical Research*.[37] In the same issue, Brown wrote, "The nonclassical theory is not necessarily wrong, but it has been too readily accepted" (probably an understatement concerning the enormous amount of work carried out on the topic). In any case, I kept my promise and have not done further work in the field. The chemical community seems to have accepted the closure of the debate. This is the way the much-publicized so-called nonclassical ion controversy ended. To summarize, it basically centered on the question of whether the structure of carbocations, including rapidly equilibrating "classical" ions, can be depicted adequately by using only Lewis-type two-electron two-center (2e–2c) covalent bonding or whether there are also bridged or σ-localized ions, whose structural depiction necessitates two-electron three-center (2e–3c) bonding of the involved higher-coordinate carbon. The outcome was a new understanding of the general bonding nature of carbon compounds and the electron donor ability and reactivity of single bonds of saturated hydrocarbons and σ bond in general in electrophilic reaction.

Intensive, critical studies of controversial topics also help eliminate the possibility of errors. One of my favorite quotations is by George Bekesy, a fellow Hungarian-born physicist who studied questions of the inner ear and hearing (Nobel Prize in Medicine, 1961):[38]

> [One] way of dealing with errors is to have friends who are willing to spend the time necessary to carry out a critical examination of the experimental design beforehand and the results after the experiments have been completed. An even better way is to have an enemy. An enemy is willing to devote a vast amount of time and brain power to ferreting out errors both large and small, and this without any compensation. Another trouble with enemies is that they sometimes develop into friends and lose a good deal of their zeal. It was in this way the writer lost his three best enemies. Everyone, not just scientists, need a few good enemies!

Clearly there was no lack of devoted adversaries (perhaps a more proper term than "enemies") on both sides of the norbornyl ion controversy. It is to their credit that we today probably know more about the structure of carbocations, such as the norbornyl cation, than about most other chemical species. Their efforts also resulted not only in rigorous studies but also in the development or improvement of many techniques. Although many believe that too much effort was expended on the "futile" norbornyl ion controversy, I am convinced that it eventually resulted in significant new insights and consequences to chemistry. It affected in a fundamental way our understanding of the chemical bonding of electron-deficient carbon compounds, extending Kekule's concept of the limiting ability of carbon to associate with no more than four other atoms of groups. An equally significant consequence of the norbornyl cation studies was that I was able to realize the ability of saturated C—H and C—C single bonds to act in general as σ-electron donors toward strong electrophiles such as carbocations or other highly reactive reagents in superacidic systems, not only in intramolecular but also in intermolecular reactions. The key for this reactivity lies in the ability to form two-electron three-center (2e–3c) bonds (familiar in boron and organometallic chemistry). The electrophilic chemistry of saturated hydrocarbons (including that of the parent methane) rapidly evolved following recognition of the concept of hypercoordinate, in short, hypercarbon chemistry.[39]

2.7 THE GENERAL CONCEPT OF CARBOCATIONS

Once the direct observation of stable, long-lived carbocations generated in highly acidic (superacid) systems became possible, it led me to the recognition of the general concept of hydrocarbon cations, including the realization that five (and higher)-coordinate carbocations are the key to electrophilic reactions at single bonds in saturated hydrocarbons (alkanes, cycloalkanes). In 1972, I offered a general definition of carbocations based on the realization that two distinct classes of carbocations exist.[40] I also suggested naming the cations of carbon compounds "carbocations" (because the corresponding anions were named "carbanions").[40] To my surprise, the name stuck and was later officially adopted by the International Union of Pure and Applied Chemistry (IUPAC) for general use.[41]

Trivalent ("classical") carbenium ions contain an sp^2-hybridized electron-deficient carbon atom, which tends to be planar in the absence of constraining skeletal rigidity or steric interference. The carbenium carbon contains six valence electrons; thus it is highly electron-deficient. The structure of trivalent carbocations can always be adequately described by using only two-electron two-center bonds (Lewis valence bond structures). CH_3^+ is the parent for trivalent ions.

Penta- (or higher) coordinate ("nonclassical") carbonium ions contain five or (higher)-coordinate carbon atoms. They cannot be described by two-electron two-center single bonds alone but also necessitate the use of two-electron three (or multi)center bonding. The carbocation center always has eight valence electrons, but overall carbonium ions are electron-deficient because of the sharing of two electrons among three (or more) atoms. CH_5^+ can be considered the parent for carbonium ions.

Lewis' concept that a covalent chemical bond consists of a pair of electrons shared between two atoms is a cornerstone of structural chemistry. Chemists tend to brand compounds as anomalous whose structures cannot be depicted in terms of such valence bonds alone.[42] Carbocations with too few electrons to allow a pair for each "bond" came to be referred to as "nonclassical," a name as mentioned first used by Roberts for the cyclopropylcarbinyl cation and adapted by Winstein for the norbornyl cation.[43] The name is still used, even though it is now recognized that, like other compounds, they adopt the structures appropriate for the number of electrons they contain with two-electron three (even multi)-center bonding, not unlike the bonding principle established by Lipscomb for boron compounds.[42] The prefixes "classical" and "nonclassical," I believe, will gradually fade away as the general principles of bonding are recognized more widely:

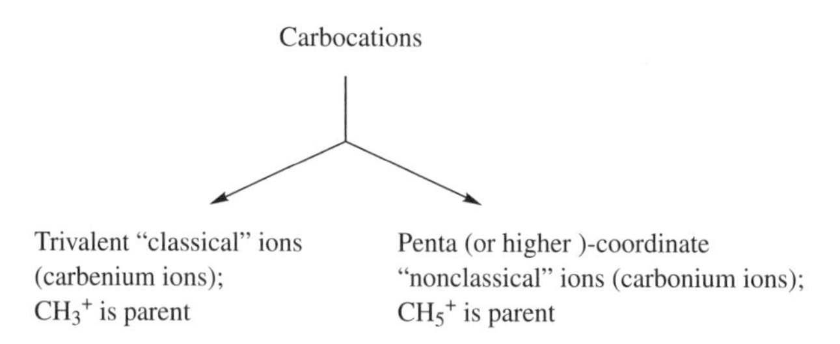

Concerning carbocations, in the previous usage nomenclature the trivalent, planar ions of the CH_3^+ type were called *carbonium ions*. If the term is considered analogous to other *onium ions* (ammonium, sulfonium, phosphonium ions), then it should relate to the higher-valency or coordination-state carbocations. These, however, clearly are not the trivalent, but the penta- or higher-coordinated, cations of the CH_5^+ type. The earlier German and French literature, indeed, frequently used the "carbenium ion" naming for trivalent cations.

Trivalent carbenium ions are the key intermediates in electrophilic reactions of π-donor unsaturated hydrocarbons. At the same time, pentacoordinated carbonium ions are the key to electrophilic reactions of σ-donor saturated hydrocarbons through the ability of C–H or C–C single bonds to participate in carbonium ion formation.

Some characteristic bonding natures in typical nonclassical ions are the following:

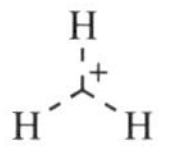

H_3^+

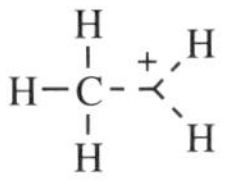

Methonium ion

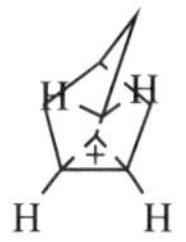

Norbornyl ion

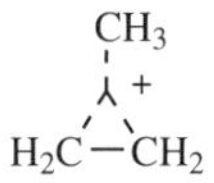

Corner-protonated cyclopropane

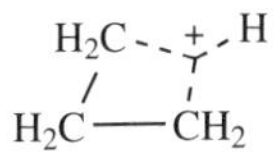

Edge-protonated cyclopropane

Expansion of the carbon octet via 3*d* orbital participation is not possible; there can be only eight valence electrons in the outer shell of carbon, a small first-row element. The valency of carbon therefore cannot exceed four. Kekulé's concept of the tetravalence of carbon in bonding terms represents attachment of four atoms (or groups) involving 2e–2c Lewis-type bonding. However, nothing prevents carbon from also participating in multicenter bonding involving 2e–3c (or multicenter) bonds.

In the 1930s, Pauling still believed that diborane had an ethanelike structure and suggested this to Kharasch during a visit to Chicago (as recalled by H. C. Brown). It was Lipscomb, Pauling's student, who in the 1950s introduced the two-electron three-center (2e–3c) bonding concept into boron chemistry, while also explaining the bridged structure of diborane.

It is remarkable that chemists long resisted making the connection between boron and electron-deficient carbon, which, after all, are analogs. I was lucky to be given the opportunity to be able to establish the general concept of five and higher coordination of electron-deficient carbon and thus be able to open up the field of hypercarbon chemistry.

Organic chemists who are dealing with carbon compounds (or perhaps more correctly with hydrocarbons and their derivatives) have long considered that 2e–3c bonding was limited to some "inorganic" or at best "organometallic" systems and to have no relevance to their field. The long-drawn-out and sometimes highly personal nonclassical ion controversy was accordingly limited to the structural aspects of some, to most chemists rather obscure, carbocations. Herbert Brown, one of the major participants in the debate and one of the great boron chemists of our time, was steadfast in his crusade against bridged nonclassical ions. He repeatedly used the argument that if such ions existed, a new yet unknown bonding concept would need to be discovered to explain them. This, however, is certainly not the case. The close relationship of electron-deficient carbocations with their neutral boron analogs has been frequently pointed out and discussed starting with my 1971 paper with DeMember and Commeyras.[20] In it we pointed out the observed close spectral (IR and Raman) similarities between isoelectronic $^{+}C(CH_3)_3$ and $B(CH_3)_3$ and emphasized the point repeatedly thereafter. My colleagues Robert Williams, Surya Prakash, and Leslie Field did a fine job in carrying the carbocation, borane, and polyborane analogy much further, and we also reviewed the topic in depth in our book, *Hypercarbon Chemistry*.[39]

One of the cornerstones of the chemistry of carbon compounds (organic chemistry) is Kekulé's concept, proposed in 1858, of the tetravalence of carbon. Although it was independently also proposed in the same year by Couper, he got little recognition for it. It is generally attributed to Kekulé that carbon can bind at the same time to not more than four other atoms or groups.

Kekulé's tetravalent carbon was explained later on the basis of the atomic concept and the "rule of eight" valence electrons. From this, G. N. Lewis introduced the electron pair concept and that of covalent shared electron pair bonding (Lewis bond), which Langmuir further developed. It was Linus Pauling and others following him, who subsequently applied the principles of the developing quantum theory to the questions of chemical bonding. I prefer to use "chemical bonding" instead of "chemical bond," because, after all, in a strict sense the chemists' beloved electron pairs do not exist. Electrons move individually, and it is only the probability that they are found paired in close proximity that justifies the practical term of covalent electron pair bonding. Pauling showed that electron pairs occupying properly oriented orbitals (which are their preferred locations, but do not exist otherwise) result in the tetrahedral structure of methane (involving sp^3 hybridization). However, neither Lewis–Langmuir nor Pauling considered that an already shared electron pair could further bind an additional atom, not just two. It was Lipscomb who revolutionalized the bonding nature of boron chemistry by introducing the concept of two-electron three-center or multicenter bonding. It was my good fortune that I was able to generalize a related concept to higher coordinate carbon compounds and their chemistry.

Whereas the differentiation of limiting trivalent and penta- or higher-coordinate carbocations serves a useful purpose in establishing the significant differences between these ions, it must be emphasized that these represent only the extremes of a continuum and that there exists a continuum of charge delocalization comprising both intra- and intermolecular interactions. This can involve participation of neighboring *n*-donor atoms, π-donor groups, or σ-donor C—H or C—C bonds.

Neighboring group participation (a term introduced by Winstein) with the vacant *p* orbital of a carbenium ion center contributes to its stabilization via delocalization, which can involve atoms with unshared electron pairs (*n* donors), π-electron systems (direct conjugate or allylic stabilization), bent σ bonds (as in cyclopropylcarbinyl cations), and C–H and C–C σ bonds (hyperconjugation).

Hyperconjugation is the overlap interaction of an appropriately oriented σ bond with a carbocationic *p* orbital to provide electron delocalization with minimal accompanying nuclear reorganization. Nuclear reorganization accompanying σ-bond delocalization can range from little or no rearrangement (hyperconjugation) to partial bridging involving some reorganization of nuclei (σ participation) and to more extensive or complete bridging. Trivalent carbenium ions, with the exception of the parent CH_3^+, consequently always show varying degrees of delocalization. Eventually in the limiting case carbocations become pentacoordinated carbonium ions. The limiting cases define the extremes of the spectrum of carbocations:

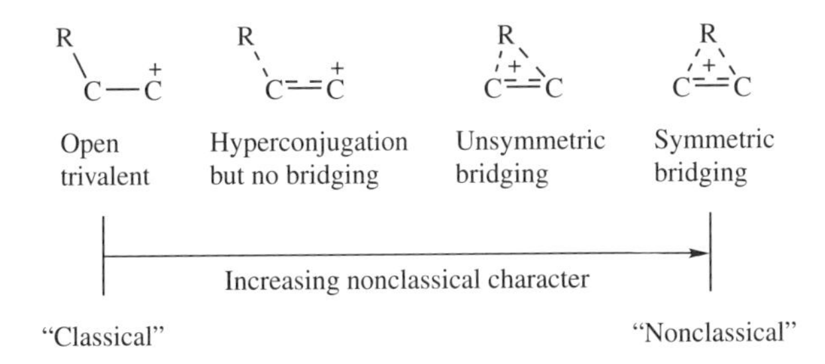

Under superacidic, low-nucleophilicity so-called stable ion conditions, developing electron-deficient carbocations do not find reactive external nucleophiles to react with; thus they stay persistent in solution stabilized by internal neighboring group interactions.

2.8 FIVE- AND HIGHER-COORDINATE CARBOCATIONS AND THEIR CHEMISTRY

On the basis of my extensive study of stable, persistent carbocations, reported in more than 300 publications, as mentioned, I was able in 1972 to develop a general concept of carbocations.[40] In higher-coordinate (hypercoordinate) carbonium ions, of which protonated methane CH_5^+ is the parent, besides two-electron two-center Lewis bonding, two-electron three-center bonding is involved.

Extensive ab initio calculations[44] reconfirmed the preferred C_s symmetric structure for the CH_5^+ cation, as we originally suggested with Klopman in 1969.[45] The structure can be viewed as a proton inserted into one of the σ C–H bonds of methane to form a 2e–3c bond between carbon and two hydrogen atoms (or CH_3^+ binding H_2 through a long, weaker bonding interaction). At the same time, we already pointed out with Klopman that ready bond-to-bond (isotopal) proton migration can take place through low barriers to equivalent or related structures that are energetically only slightly less favorable (which led more recently to Schleyer's suggestion of a fluxional, completely delocalized nature).[46]

In 1982, we studied the parent six-coordinate diprotonated methane (CH_6^{2+}),[47] which has two 2e–3c bonding interactions in its minimum-energy structure (C_{2v}).[48] On the basis of ab initio calculations, with Rasul we more recently found that the seven-coordinate triprotonated methane (CH_7^{3+}) is also an energy minimum and has three 2e–3c bonding interactions in its minimum-energy structure (C_{3v}). These results further indicate the general importance of 2e–3c bonding in protonated alkanes:

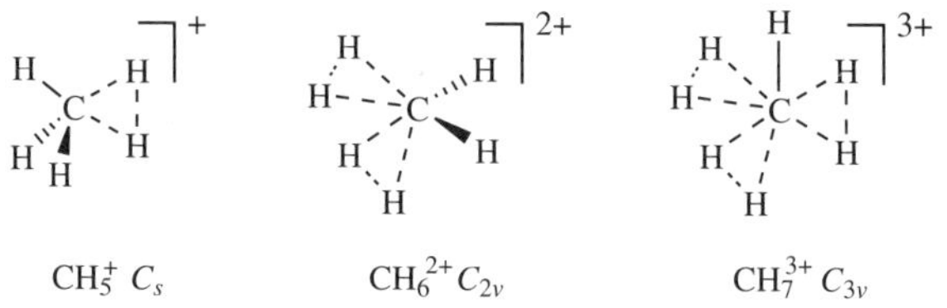

CH_5^+ C_s CH_6^{2+} C_{2v} CH_7^{3+} C_{3v}

Protonated methanes and their homologs and derivatives are experimentally indicated in superacidic chemistry by hydrogen–deuterium exchange experiments, as well as by core electron (ESCA) spectroscopy of their frozen matrixes.[49] Some of their derivatives could even be isolated as crystalline compounds. Schmidbaur has prepared gold complex analogs of CH_5^+ and CH_6^{2+} and determined their X-ray structures.[50,51] The monopositively charged trigonal bipyramidal[50] $\{[C_6H_5)_3PAu]_5C\}^+$ and the dipositively charged octahedral[51] gold complex $\{[C_6H_5)_3PAu]_6C\}^{2+}$ contain five- and six-coordinate carbons, respectively. Considering the isolobal relationship (i.e., similarity in bonding) between LAu^+ and H^+, the gold complexes represent the isolobal analogs of CH_5^+ and CH_6^{2+}:

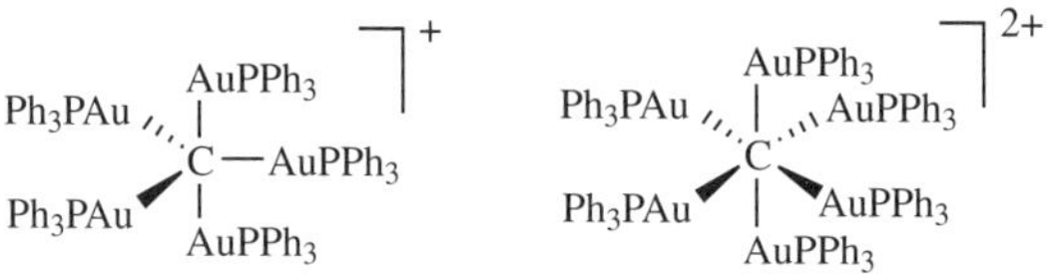

The remarkable stability of the gold complexes is due to significant metal–metal bonding. However, their isolation and structural study are remarkable and greatly contributed to our knowledge of higher-coordinate carbocations.

Boron and carbon are consecutive first-row elements. Trivalent carbocations are isoelectronic with the corresponding neutral trivalent boron compounds. Similarly, pentavalent monopositively charged carbonium ions are isoelectronic with the corresponding neutral pentavalent boron compounds. BH_5, which is isoelectronic with CH_5^+, also has a C_s symmetric structure based on high-level ab initio calculations.[52,53] Experimentally, H–D exchange was observed in our work when BH_4^- was treated with deuterated strong acids, indicating the intermediacy of isotopomeric BH_5.[54] The first direct experimental observation (by infrared spectroscopy) of BH_5 was reported in 1994.[55] The X-ray structure of the five-coordinate gold complex $[(Cy_3P^+)B(AuPPh_3)_4]$ was also reported by Schmidbaur.[56] This square pyramidal compound represents the isolobal analog of BH_5, and further strengthens the relationship of the bonding nature of higher-coordinate boron and carbon compounds.

As five- and six-coordinate CH_5^+ and CH_6^{2+} are isoelectronic with BH_5 and BH_6^+,[57] respectively, seven-coordinate tripositively charged CH_7^{3+} [48] is isoelectronic with the corresponding dipositively charged heptavalent boronium dication BH_7^{2+}.[57] We have also searched for a minimum-energy structure of tetraprotonated methane, CH_8^{4+}. However, CH_8^{4+} remains even computationally elusive because charge–charge repulsion appears to have reached its prohibitive limit. The isoelectronic boron analog BH_8^{3+}, however, was calculated to be an energy minimum.[57]

The discovery of a significant number of hypercoordinate carbocations ("nonclassical" ions), initially based on solvolytic studies and subsequently as observable, stable ions in superacidic media as well as on theoretical calculations, showed that carbon hypercoordination is a general phenomenon in electron-deficient hydrocarbon systems. The following are some characteristic nonclassical carbocations:

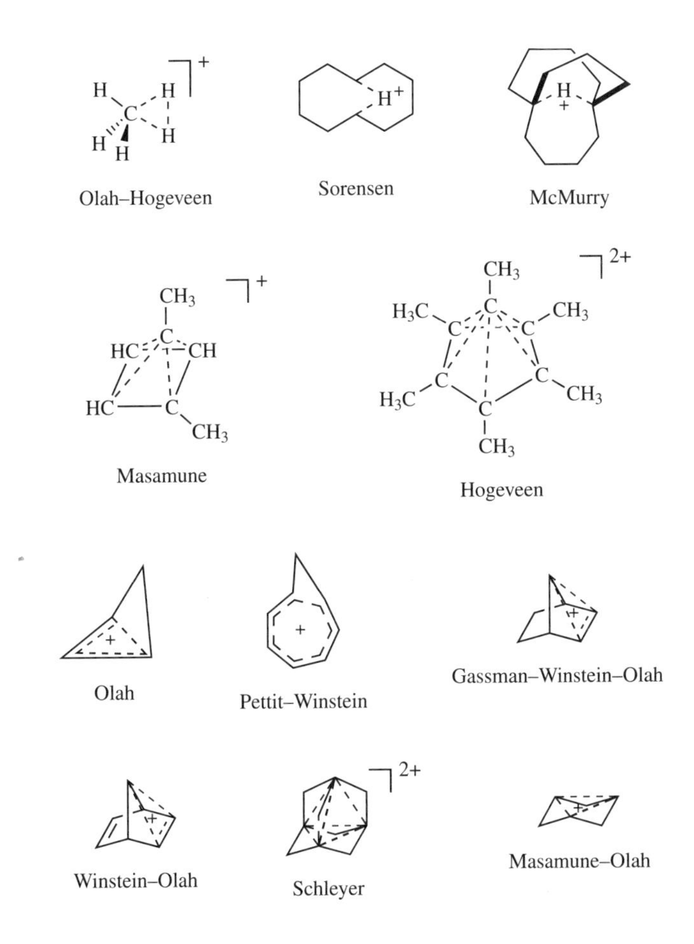

According to early theoretical calculations that Klopman and I carried[58] out in 1971, the parent molecular ions of alkanes, such as $CH_4^{+\bullet}$, observed in mass spectrometry, also can have a planar hypercarbon structure, although distortion is probable:

$CH_4^{+\bullet}$

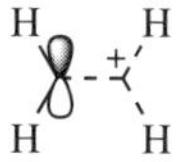

The CH_4^{2+} ion, as calculated later by Radom, however, has a completely planar C_{2v} structure.[59]

Carbon can be involved not only in single two-electron three-center bond formation but also in some carbodications simultaneously participating in two 2e–3c bonds. Diprotonated methane (CH_6^{2+})[47] and ethane ($C_2H_8^{2+}$),[47] as well as the dimer of the methyl cation ($C_2H_6^{2+}$),[47] are illustrative:

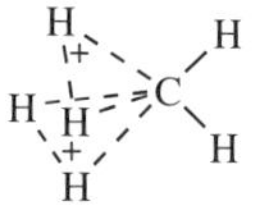

CH_6^{2+} 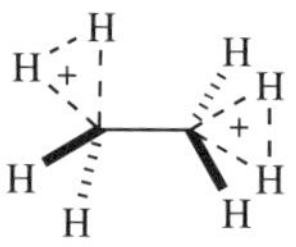 $C_2H_8^{2+}$ 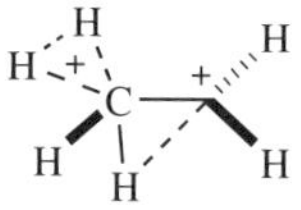 $C_2H_6^{2+}$ 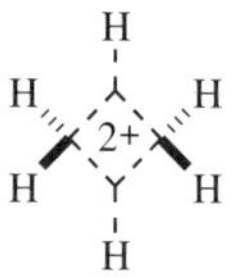 $C_2H_6^{2+}$

It was the study of hypercarbon-containing nonclassical carbocations that allowed us to firmly establish carbon's ability in a hydrocarbon system to bind simultaneously with five (or six or even seven) atoms or groups. It should be emphasized that carbocations represent only one class of hypercarbon compounds. A wide variety of neutral hypercarbon compounds, including alkyl (acyl)-bridged organometallics as well as carboranes, carbonyl, and carbide clusters, are now recognized and have been studied. They are reviewed in our book *Hypercarbon Chemistry* (written with Prakash, Williams, Field, and Wade).[39]

During my studies related to the nonclassical ion problem I realized that the formation of the σ-delocalized 2-norbornyl cation from 2-norbornyl precursors represented the equivalent of an intramolecular σ alkylation, where a covalent C1–C6 bond provided the electrons for the 2e–3c-bonded bridged ion (by σ participation):[37]

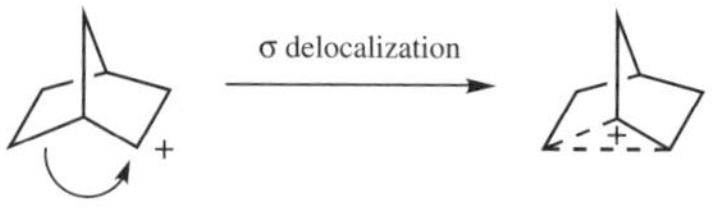

σ delocalization

The formation of the σ-delocalized norbornyl cation via ionization of 2-norbornyl precursors in low-nucleophilicity, superacidic media can be considered an analog of an intramolecular Friedel–Crafts alkylation in a saturated system. Indeed, deprotonation gives nortricyclane:[37]

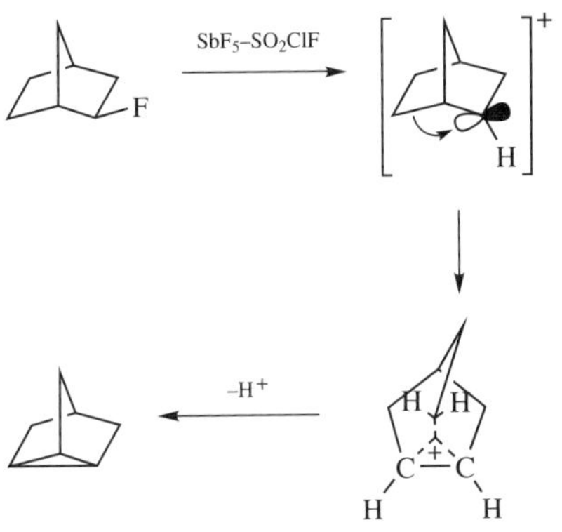

SbF5–SO2ClF

−H+

The intramolecular σ delocalization in the norbornyl system aroused my interest in studying whether similar electrophilic interactions and reactions of C—H or C—C bonds are possible in intermolecular systems. This led to the discovery of the general electrophilic reactivity of saturated hydrocarbons and single bonds in general.

The long, drawn-out nonclassical ion controversy thus led to an unexpected new chapter of chemistry. As frequently happens in science, the drive for understanding (for whatever reason) of what appear at the time to be rather isolated and even relatively unimportant problems can eventually lead to significant new concepts, new chemistry, and even practical applications. This justifies the need for exploration and study in the context of fundamental (basic) research even if initially no practical reasons or uses are indicated. The beauty of science lies in finding the unexpected, but as Niels Bohr was frequently quoted to have said, "you must be prepared for a surprise" the same time you must also understand what your findings mean and what they can be used for. To me, this is the lesson of the norbornyl ion controversy. I believe that it was not a waste of effort to pursue it, and eventually it greatly helped to advance chemistry to new areas of significance that are still emerging.

To illustrate the significance of higher coordinate carbocations in hydrocarbon chemistry,[60] some examples are illustrative. Protonation (and protolysis) of alkanes is readily achieved with superacids. The protonation of methane itself to CH_5^+, as discussed earlier, takes place readily:

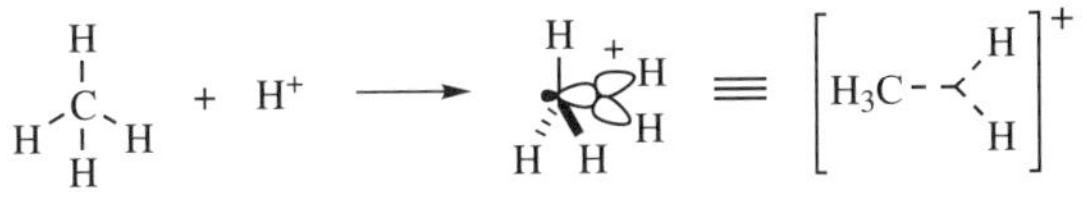

$$CH_4 + H^+ \longrightarrow \equiv \left[H_3C\text{--}\langle{}^{H}_{H} \right]^+$$

Acid-catalyzed isomerization reactions of alkanes as well as their alkylation and condensation reactions are initiated by protolytic ionization. Available evidence indicates nonlinear but somewhat bent, although not triangular transition states.[61]

$$R_3C{-}H + H^+ \longrightarrow [R_3C{\cdots}H{\cdots}H]^+ \longrightarrow R_3C^+ + H_2$$

Linear

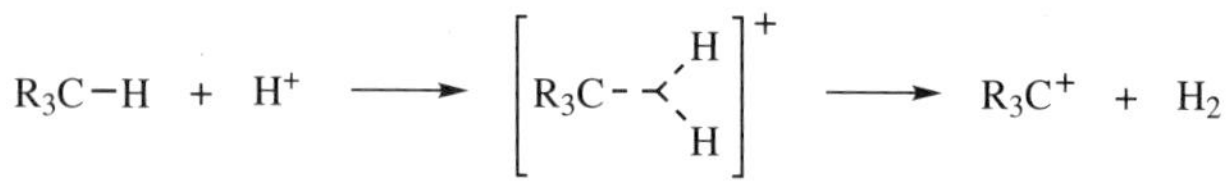

$$R_3C{-}H + H^+ \longrightarrow \left[R_3C{\cdots}\langle^{H}_{H}\right]^+ \longrightarrow R_3C^+ + H_2$$

Nonlinear

The reverse reaction of the protolytic ionization of hydrocarbons to carbocations, that is, the reaction of trivalent carbocations with molecular hydrogen giving their parent hydrocarbons, involves the same five-coordinate carbonium ions:

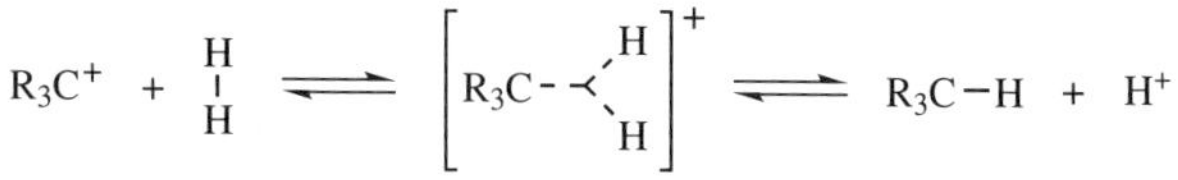

$$R_3C^+ + H{-}H \rightleftharpoons \left[R_3C{\cdots}\langle^{H}_{H}\right]^+ \rightleftharpoons R_3C{-}H + H^+$$

The isomerization of butane to isobutane in superacids is illustrative of a protolytic isomerization, where no intermediate olefins are present in equilibrium with carbocations.

The superacid-catalyzed cracking of hydrocarbons (a significant practical application) involves not only formation of trivalent carbocationic sites leading to subsequent β-cleavage but also direct C—C bond protolysis:

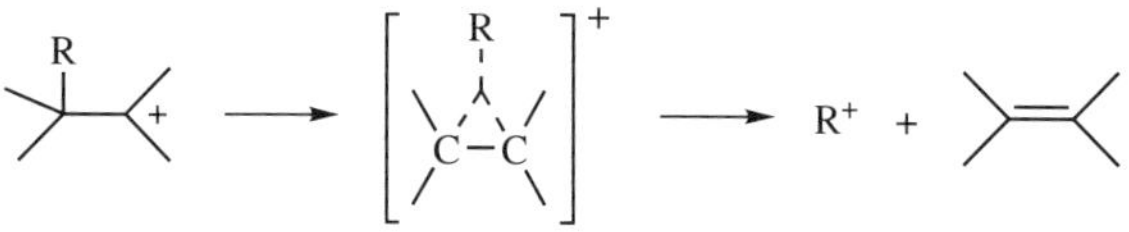

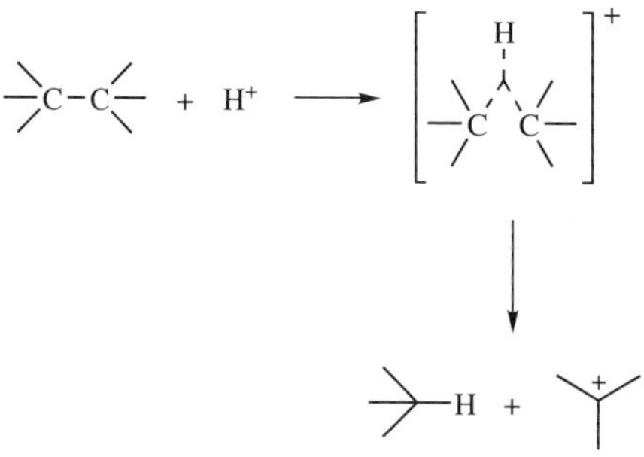

$$R^+ \quad + \quad \text{H} \quad + \quad \text{H}^+$$

Whereas superacid (HF/BF_3, HF/SbF_5, HF/TaF_5, FSO_3H/SbF_5, etc.)-catalyzed hydrocarbon transformations were first explored in the liquid phase, solid acid catalyst systems, such as those based on Nafion H, longer-chain perfluorinated alkanesulfonic acids, and fluorinated graphite intercalates, were subsequently also developed and utilized for heterogeneous reactions. The strong acidic nature of zeolite catalysts was also successfully explored in cases such as H-ZSM-5 at high temperatures.[62]

Not only protolytic reactions but also a broad range of reactions with varied electrophiles (alkylation, formylation, nitration, halogenation, oxygenation, etc.) were found to be feasible when superacidic, low-nucleophilicity reaction conditions were used.[60,63]

$$R_3C{-}H + E^+ \rightleftharpoons [R_3C\cdots(H)(E)]^+ \rightleftharpoons R_3C^+ + EH \quad \text{or} \quad R_3CE + H^+$$

$$R_3C{-}CR_3 + E^+ \rightleftharpoons [R_3C\cdots(CR_3)(E)]^+ \rightleftharpoons R_3CE + R_3C^+$$

$E = D^+, H^+, R^+, NO_2^+, Hlg^+, HCO^+$, etc.

Alkylation of isoalkanes with alkenes is of particular significance. The industrially used alkylation of isobutane with isobutylene to isooctane, is, however, de facto alkylation of the reactive isobutylene and not of the saturated hydrocarbon. Isobutane only acts as a hydride transfer agent and a source of the *tert*-butyl cation, formed via intermolecular hydride transfer. In contrast, when the *tert*-butyl cation is reacted with isobutane under superacidic conditions and thus in the absence of isobutylene, the major fast reaction is still hydride transfer, but a detectable amount of 2,2,3,3-tetramethylbutane, the σ-alkylation product, is also obtained. With sterically less crowded systems, σ alkylation becomes more predominant:[60]

$$\text{isobutylene} + H^+ \rightleftharpoons \textit{tert}\text{-}C_4H_9^+$$

$$\textit{tert}\text{-}C_4H_9^+ + \text{isobutylene} \rightleftharpoons C_8H_{17}^+$$

$$C_8H_{17}^+ + H{-}C(CH_3)_3 \rightleftharpoons C_8H_{18} + \textit{tert}\text{-}C_4H_9^+$$

$$(CH_3)_3C{-}H + \textit{tert}\text{-}C_4H_9^+ \rightleftharpoons [(CH_3)_3C\cdots(H)(C(CH_3)_3)]^+ \xrightarrow{-H^+} (CH_3)_3C{-}C(CH_3)_3$$

A fundamental difference exists between conventional-acid- and superacid-catalyzed hydrocarbon chemistry. In the former, trivalent carbenium ions are always in equilibrium with olefins, which play the key role, whereas in the latter, hydrocarbon transformation can take place without the involvement of olefins through the intermediacy of five-coordinate carbocations.[60]

The reaction of trivalent carbocations with carbon monoxide giving acyl cations is the key step in the well-known and industrially used Koch–Haaf reaction of preparing branched carboxylic acids from alkenes or alcohols. For example, in this way, isobutylene or *tert*-butyl alcohol is converted into pivalic acid. In contrast, based on the superacidic activation of electrophiles leading to superelectrophiles we found it possible to formylate isoalkanes to aldehydes, which subsequently rearrange to their corresponding branched ketones.[63] These are effective high-octane gasoline additive oxygenates. The conversion of isobutane into isopropyl methyl ketone, or isopentane into isobutyl methyl ketone is illustrative. In this reaction, no branched carboxylic acids (Koch products) are formed:

$$(CH_3)_3CH \xrightarrow[HF/BF_3]{CO} \left[(CH_3)_3C\overset{H}{\cdots}CHO\right]^+ \longrightarrow$$

$$(CH_3)_3CCHO \xrightarrow{HF\text{-}BF_3} (CH_3)_2CHCOCH_3$$

The superacid-catalyzed electrophilic oxygenation of saturated hydrocarbons, including methane with hydrogen peroxide (via $H_3O_2^+$) or ozone (via HO_3^+), allows the efficient preparation of oxygenated derivatives:[64]

$$H_2O_2 \xrightarrow{H^+} H_2\overset{+}{O}{-}OH \xrightarrow[-H_2O]{CH_4} \left[H_3C{-}\langle^{H}_{OH}\right]^+ \longrightarrow CH_3OH_2^+ \xrightarrow{-H^+} CH_3OH$$

$$O_3 \xrightarrow{H^+} O{-}O{-}\overset{+}{O}H \xrightarrow{CH_4} \begin{cases} \xrightarrow{-O_2} \left[H_3C{-}\langle^{H}_{OH}\right]^+ \\ \left[H_3C{-}\langle^{H}_{O-OOH}\right]^+ \xrightarrow{-H_2O_2} [CH_3O^+] \longrightarrow CH_2\overset{+}{O}H \xrightarrow{-H^+} CH_2O \end{cases}$$

Because the protonation of ozone removes its dipolar nature, the electrophilic chemistry of HO_3^+, a very efficient oxygenating electrophile, has no relevance to conventional ozone chemistry. The superacid-catalyzed reaction of isobutane with ozone giving acetone and methyl alcohol, the aliphatic equivalent of the industrially significant Hock reaction of cumene, is illustrative:

$$H_3C-\underset{CH_3}{\overset{CH_3}{C}}-H + \overset{+}{O}-O-OH \longrightarrow \left[H_3C-\underset{CH_3}{\overset{CH_3}{C}}-\!\!\left\langle\begin{matrix} O-O-OH \\ H \end{matrix}\right. \right]^+ \longrightarrow \left[H_3C-\underset{CH_3}{\overset{CH_3}{C}}-O^+ \right]^+$$

$$\longrightarrow H_2O_2 + (H_3C)_2C{=}\overset{+}{O}-CH_3 \xrightarrow{H_2O} CH_3OH + (H_3C)_2C{=}O$$

Electrophilic insertion reactions into C—H (and C—C) bonds under low-nucleophilicity superacidic conditions are not unique to alkane activation processes. The C—H (and C—C) bond activation by organometallic complexes, such as Bergman's iridium complexes and other transition metal systems (rhodium, osmium, rhenium, etc.), is based on somewhat similar electrophilic insertions. These reactions, however, cannot as yet be made catalytic, although future work may change this. A wide variety of further reactions of hydrocarbons with coordinatively unsaturated metal compounds and reagents involving hypercarbon intermediates (transition states) is also recognized, ranging from hydrometallations to Ziegler–Natta polymerization.[60]

In the conclusion of my 1972 paper[40] on the general concept of carbocations I wrote

> The realization of the electron donor ability of shared (bonding) electron pairs of single bonds, should rank one day equal in importance with that of unshared (nonbonding) electron pairs recognized by Lewis. We can now not only explain the reactivity of saturated hydrocarbons and in general of single bonds in electrophilic reactions, but indeed use this understanding to explore new areas of carbocation chemistry.

This was one of the few times I ever made a prediction of the possible future significance of my chemistry. More than a quarter of a century later I take some satisfaction that I was correct and that, indeed, hypercarbon chemistry has a significant place on the wide palette of chemistry.

Carbon can extend its bonding from Kekulé's[65] tetravalent limit to five-, and even higher-bonded (coordinate) hypercarbon systems. Higher-coordinate carbocations are now well recognized. They are the key to our understanding of the electrophilic reactivity of C—H or C—C single bonds and of hypercarbon chemistry in general. Some of their derivatives, such as related gold complexes, can even be isolated as stable crystalline compounds. The chemistry of higher-coordinate carbon (i.e., hypercarbon chemistry) is rapidly expanding, with many new vistas to be explored. The road from Kekulé's tetravalent carbon to hypercarbon chemistry took more than a century to travel. Carbon also unveiled other unexpected new aspects, for example, its recently discovered fullerene-type allotropes and their chemistry. There is no reason to believe that the new century and millennium just beginning will not bring much further progress in the fascinating field of hypercarbon chemistry, to which knowledge of carbocations made essential contribution.

ACKNOWLEDGMENTS

My work on carbocations would have not been possible without the fundamental contributions and hard work of the dedicated students, collaborators, and colleagues whom I was privileged to work with over the years. They became an integral part of the Olah group and my broader scientific family, many my personal friends as well. Their names are to be found in the literature citations and my recently published autobiography.[1] I would like, however, to specifically thank Professor G. K. Surya Prakash, who started with me more than a quarter century ago as a graduate student and continues as my colleague and close friend in our still continuing cooperation in the search for the wonderful challenges of carbocation chemistry.

REFERENCES

1. Based in part on my Nobel Lecture (*Angew. Chemie Int. Ed.* in English, 1995) and my autobiographical reflections in *A Life of Magic Chemistry*, Wiley-Interscience, New York, 2001.
2. J. Stieglitz, *Am. Chem. J.* **21**, 110 (1899).
3. J. F. Norris, *Am. Chem. J.* **25**, 117 (1901).
4. F. Kehrmann and F. Wentzel, *Chem. Ber.* **34**, 3815 (1901).
5. A. Baeyer and V. Villiger, *Chem. Ber.* **35**, 1189,3013 (1901).
6. (a) M. Gomberg, *Chem. Ber.* **35**, 2397 (1902); (b) P. Walden, *Chem. Ber.* **35**, 2018 (1902).
7. H. Meerwein and K. van Emster, *Chem. Ber.* **55**, 2500 (1922).
8. C. K. Ingold, *Structure and Mechanism in Organic Chemistry*, Cornell Univ. Press, Ithaca, NY, 1953, and references cited therein; 2nd ed., 1969.
9. F. C. Whitmore, *J. Am. Chem. Soc.* **54**, 3274,3276 (1932); *Ann. Rep. Progr. Chem.* (*Chem. Soc. Lond.*) 177 (1933); *Chem. Eng. News* **26**, 668 (1948).
10. G. A. Olah and P. v. R. Schleyer, eds., *Carbonium Ions*, Vols. I–V, Wiley-Interscience, New York, 1968–1976 and reviews cited therein.
11. D. Bethell and V. Gold, *Carbonium Ions*, Academic Press, London–New York, 1967.
12. P. Vogel, *Carbocation Chemistry*, Elsevier, Amsterdam, 1985.
13. G. A. Olah and S. J. Kuhn, *Chem. Ber.* **89**, 866 (1956).
14. G. A. Olah, S. J. Kuhn, and J. Opal, *J. Chem. Soc.* 2174 (1957).

15. G. A. Olah, S. J. Kuhn, W. S. Tolgyesi, and E. B. Baker, *J. Am. Chem. Soc.* **84**, 2733 (1962).
16. G. A. Olah, *Rev. Roum. Chim.* (Bucharest) **7**, 1129 (1962) (Nenitzescu issue).
17. G. A. Olah, W. S. Tolgyesi, S. J. Kuhn, M. E. Moffatt, I. J. Bastien, and E. B. Baker, *J. Am. Chem. Soc.* **85**, 1328 (1963).
18. Preliminary communications and lectures: (a) G. A. Olah, Conference Lecture at 9th Reaction Mechanism Conf., Brookhaven, NY, Aug. 1962; (b) G. A. Olah, *Abstracts 142nd National Meeting of the American Chemical Society*, Atlantic City, NJ, Sept. 1962, p. 45; (c) G. A. Olah, W. S. Tolgyesi, J. S. MacIntyre, I. J. Bastien, M. W. Meyer, and E. B. Baker, *Abstracts A, XIX International Congress of Pure and Applied Chemistry*, London, June 1963, p. 121; (d) G. A. Olah, *Angew. Chem.* **75**, 800 (1963); (e) G. A. Olah, American Chemical Society 1964 Petroleum Award Lecture, *Reprints, Division of Petroleum Chemistry, American Chemical Society* **9**(7), C31 (1964); (f) G. A. Olah, Intermediate complexes and their role in electrophilic aromatic substitutions, Conference Lecture at Organic Reaction Mechanism Conf., Cork, Ireland, June 1964 in *Special Publication* No. 19, the Chemical Society, London, 1965; (g) G. A. Olah and C. U. Pittman, Jr., in *Advances in Physical Organic Chemistry*, V. Gold, ed., Academic Press, London–New York, 1966, Vol. 4, p. 305.
19. (a) G. A. Olah, E. B. Baker, J. C. Evans, W. S. Tolgyesi, J. S. McIntyre, and I. J. Bastien, *J. Am. Chem. Soc.* **86**, 1360 (1964); (b) G. A. Olah, *Chem. Eng. News* **45**, 76 (1967); *Science* **168**, 1798 (1970).
20. G. A. Olah, J. R. DeMember, A. Commeyras, and J. L. Bribes, *J. Am. Chem. Soc.* **93**, 459 (1971) and references cited therein.
21. H. Vancik and D. E. Sunko, *J. Am. Chem. Soc.* **111**, 3742 (1989).
22. D. M. Brouwer and E. L. Mackor, *Proc. Chem. Soc.* 147 (1964).
23. (a) R. J. Gillespie, *Acc. Chem. Res.* **1**, 202 (1968); (b) R. J. Gillespie and T. E. Peel, *Adv. Phys. Org. Chem.* **9**, 1 (1972); *J. Am. Chem. Soc.* **95**, 5173 (1973); (c) G. A. Olah, G. K. S. Prakash, and J. Sommer, *Superacids*, Wiley, New York, 1985; (d) R. J. Gillespie, *Can. Chem. News* 20 (May 1991).
24. (a) J. Bacon, P. A. W. Dean, and R. J. Gillespie, *Can. J. Chem.* **47**, 1655 (1969); (b) G. A. Olah and M. Calin, *J. Am. Chem. Soc.* **90**, 938 (1968).
25. N. F. Hall and J. B. Conant, *J. Am. Chem. Soc.* **49**, 3047 (1927).
26. G. A. Olah, *Angew. Chem. Int. Ed. Engl.* **12**, 173 (1973); *Angew. Chem.* **85**, 183 (1973); G. A. Olah, *Carbocations and Electrophilic Reactions*, Verlag Chemie (Weinheim)–Wiley (New York), 1974.
27. (a) P. D. Bartlett, *Nonclassical Ions*, W. A. Benjamin, New York, 1965; (b) S. Winstein, *Quart. Rev.* (*Lond.*) **23**, 1411 (1969); (c) H. C. Brown (with commentary by P. v. R. Schleyer), *The Nonclassical Ion Problem*, Plenum Press, New York, 1977 and references cited therein.
28. J. D. Roberts and R. H. Mazur, *J. Am. Chem. Soc.* **73**, 3542 (1951).
29. M. Saunders, P. v. R. Schleyer, and G. A. Olah, *J. Am. Chem. Soc.* **86**, 5680 (1964).
30. (a) G. A. Olah, A. M. White, J. R. DeMember, A. Commeyras, and C. Y. Lui, *J. Am. Chem. Soc.* **92**, 4627 (1970); (b) G. A. Olah, G. K. S. Prakash, M. Arvanaghi, and F. A. L. Anet, *J. Am. Chem. Soc.* **104**, 7105 (1982).
31. (a) G. A. Olah, G. D. Mateescu, L. A. Wilson, and M. H. Gross, *J. Am. Chem. Soc.* **92**, 7231 (1970); (b) G. A. Olah, G. D. Mateescu, and J. L. Riemenschneider, *J. Am. Chem. Soc.* **94**, 2529 (1972); (c) S. A. Johnson and D. T. Clark, *J. Am. Chem. Soc.* **110**, 4112 (1988).

32. E. M. Arnett, N. Pienta, and C. Petro, *J. Am. Chem. Soc.* **102**, 398 (1980).
33. M. Saunders and M. R. Kates, *J. Am. Chem. Soc.* **102**, 6867 (1980).
34. C. S. Yannoni, V. Macho, and P. C. Myhre, *J. Am. Chem. Soc.* **104**, 7380 (1982).
35. T. Laube, *Angew Chem.* **99**, 580 (1987); *Angew Chem. Int. Ed. Engl.* **26**, 560 (1987).
36. P. v. R. Schleyer and S. Sieber, *Angew Chem.* **105**, 1676 (1993); *Angew Chem. Int. Ed. Engl.* **32**, 1606 (1993) and references cited therein.
37. G. A. Olah, G. K. S. Prakash, and M. Saunders, *Acc. Chem. Res.* **16**, 440 (1983).
38. G. V. Bekesy, *Experiments in Hearing*, McGraw-Hill, New York, 1960, p. 8.
39. G. A. Olah, G. K. S. Prakash, R. E. Williams, L. D. Field, and K. Wade, *Hypercarbon Chemistry*, Wiley, New York, 1987.
40. G. A. Olah, *J. Am. Chem. Soc.* **94**, 808 (1972).
41. Compendium of Chemical Terminology, *IUPAC Recommendations*, Blackwell Scientific Publication, Oxford, 1987.
42. For reviews, comparison, and additional references, see *Electron Deficient Boron and Carbon Chemistry*, G. A. Olah, K. Wade, and R. E. Williams, eds., Wiley-Interscience, New York, 1991.
43. S. Winstein and D. Trifan, *J. Am. Chem. Soc.* **74**, 1154 (1952).
44. (a) K. Raghavachari, R. A. Whitesides, J. A. Pople, and P. v. R. Schleyer, *J. Am. Chem. Soc.* **104**, 3258 (1981); (b) V. Dyczmons, V. Staemmler, and W. Kutzelnig, *Chem. Phys. Lett.* **5**, 361 (1970); (c) W. A. Lathan, W. J. Hehre, and J. A. Pople, *Tetrahedron Lett.* 2699 (1970); (d) W. A. Lathan, W. J. Hehre, and J. A. Pople, *J. Am. Chem. Soc.* **93**, 808 (1971).
45. G. A. Olah, G. Klopman, and R. H. Schlosberg, *J. Am. Chem. Soc.* **91**, 3261 (1969).
46. (a) P. R. Schreiner, S.-J. Jim, H. F. Schaefer, and P. v. R. Schleyer, *J. Chem. Phys.* **99**, 3716 (1993); (b) G. E. Scuseria, *Nature* **366**, 512 (1993).
47. (a) K. Lammertsma, G. A. Olah, M. Barzaghi, and M. Simonetta, *J. Am. Chem. Soc.,* **104**, 6851 (1982); (b) K. Lammertsma, M. Barzaghi, G. A. Olah, J. A. Pople, P. v. R. Schleyer, and M. Simonetta, *J. Am. Chem. Soc.* **105**, 5268 (1983).
48. G. A. Olah and G. Rasul, *J. Am. Chem. Soc.* **118**, 8503 (1996).
49. G. A. Olah, G. K. S. Prakash, R. E. Williams, L. D. Field, and K. Wade, *Hypercarbon Chemistry*, Wiley, New York, 1987, p. 149.
50. F. Scherbaum, G. Grohmann, G. Müller, and H. Schmidbaur, *Angew. Chem. Int. Ed. Engl.* **28**, 463 (1989).
51. F. Scherbaum, G. Grohmann, B. Huber, C. Krüger, and H. Schmidbaur, *Angew. Chem. Int. Ed. Engl.* **27**, 1544 (1988).
52. P. R. Schreiner, H. F. Schaefer, and P. v. R. Schleyer, *J. Chem. Phys.* **101**, 7625 (1994).
53. J. D. Watts and R. J. Bartlett, *J. Am. Chem. Soc.* **117**, 825 (1995).
54. G. A. Olah, P. W. Westerman, Y. K. Mo, and G. Klopman, *J. Am. Chem. Soc.* **94**, 7859 (1972).
55. T. J. Tague and L. Andrews, *J. Am. Chem. Soc.* **116**, 4970 (1994).
56. A. Blumenthal, H. Beruda, and H. Schmidbaur, *J. Chem. Soc. Chem. Commun.* 1005 (1993).
57. G. Rasul and G. A. Olah, *Inorg. Chem.* **36**, 1278 (1997).
58. G. A. Olah and G. Klopman, *Chem. Phys. Lett.* **11**, 604 (1971).
59. M. W. Wong and L. Radom, *J. Am. Chem. Soc.* **111**, 1155 (1989).

60. G. A. Olah and A. Molnar, *Hydrocarbon Chemistry*, Wiley, New York, 1995 and references cited therein.
61. G. A. Olah, G. Klopman, and R. H. Schlosberg, *J. Am. Chem. Soc.* **91**, 3261 (1969).
62. W O. Haag and R. H. Dessau, *Proc. 8th Int. Catalysis Congress*, 1984, Vol. II, p. 105.
63. G. A. Olah, G. K. S. Prakash, T. Mathew, and E. R. Marinez, *Angew Chem. Int. Ed. Engl.* 2647 (2000).
64. G. A. Olah, N. Yoneda, and D. G. Parker, *J. Am. Chem. Soc.* **98**, 5261 (1976).
65. F. A. Kekulé, *Ann.* **106**, 129 (1858); *Z. Chem.* **3**, 217 (1867).

3

ZWITTERIONIC "NEUTRAL" AND "ANIONIC" CARBOCATION ANALOGS

Chaitanya S. Wannere, Zhongfang Chen, and Paul von Ragué Schleyer

Center for Computational Chemistry and Department of Chemistry
The University of Georgia
Athens, Georgia

Carbocation Chemistry, Edited by George A. Olah and G. K. Surya Prakash
ISBN 0-471-28490-4

3.1 INTRODUCTION

That σ CC bonds can be partial and have bond lengths considerably longer than normal ($> \sim 1.7$ Å) is not well accepted by organic chemists. The conspicuous examples in carbocation structures (e.g. **1**–**5**), where such partial σ CC bonds are a consequence of the electron deficiency and delocalization associated with the positive charges,[1] are considered to be special cases and are relegated to the "nonclassical cation"[2] category. Can such structures involving long CC bonds be found in neutral species, which might be "put into a bottle" (isolable)? This chapter concerns "neutral" (charge compensated zwitterionic) carbocations, which have structural entities just like nonclassical carbocations but also incorporate negatively charged moieties elsewhere in the molecule in order to compensate the positive charge. The semantic contradictions of the title emphasize the unusual nature of these species. "Anionic" carbocations also are possible; these have two negatively charged moieties and an overall negative charge. The several known examples inspired our group to design both "neutral" and "anionic" cases computationally. We point out here the general conceptual principles involved.

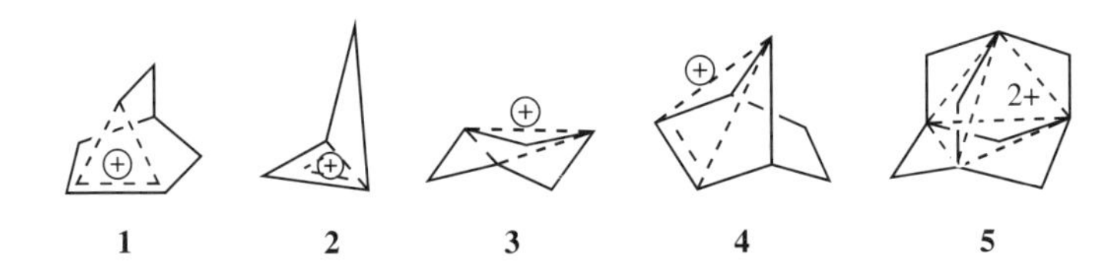

It is outside the scope of this chapter to review three other types of related neutral species, which mimic carbocationic structures and behavior to a lesser extent. We merely mention them here. Regarded as being polarized $C^{(\delta+)}$–$O^{(\delta-)}$, ketones comprise the first class. The analogies between the aromatic tropylium ion ($C_7H_7^+$) on one hand, and tropone as well as tropolone on the other hand, serve as a reminder. The second category involves the isoelectronic replacement of C^+, a carbenium carbon or one directly involved in multicenter bonding, by a boron atom. Along with borepine, $(CH)_6BH$, another tropylium ion surrogate, literature examples include the boron analogs of **3**, **4**, and **5**.[3] If the carbocation center has a $CH^{(+)}$ group, formal deprotonation results in the third category, the corresponding neutral carbenes, which, in their singlet states, also offer a vacant *p* orbital to participate in the electronic structure. But ketones, boron analogs, and carbenes can hardly be termed "carbocationic." In contrast, the neutral and anionic species described in this chapter have geometries and electronic structures quite close to their positively charged analogs.

3.2 SOME EXAMPLES OF KNOWN "NEUTRAL" CARBOCATIONIC ANALOGS

The first "neutral" nonclassical carbocationic analog involving a long CC bond was discovered serendipitously. The trimerization of 2-butyne in the presence of a Lewis acid like $AlCl_3$, to yield hexamethyl-Dewar-benzene, was formulated initially (1967) as proceeding via an intermediate complex $(CH_3\text{–}C)_4AlCl_3$, **6a** (Scheme 1).[4] Other structures (**6b**, **6c**) were postulated subsequently.[5]

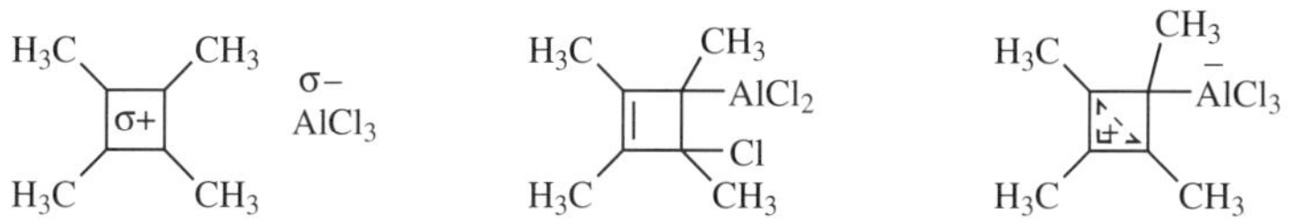

6a **6b** **6c**

Scheme 1

Detailed NMR data provided evidence for **6c**.[6] X-ray analysis by Krüger et al. in 1974 confirmed that $AlCl_3$ is attached to a carbon of the four-membered ring by a σ bond.[7] The nonclassical C2–C2* distance, 1.775 Å, is much longer than a normal C–C bond but is considerably shorter than the cross-ring separation of cyclobutane (2.170 Å). Remarkably, **6c** has a homocyclopropenyl cation moiety, **2**, and is a derivative of the simplest homoaromatic cation first proposed by Roberts;[8] both involve three center cyclic electron delocalization of the 2π systems through an intervening sp^3-hybridized carbon.

In **6c**, the pendant–$AlCl_3$ (−) group serves to neutralize the positive charge. Conceptually similar charge-compensated complexes of the $L_2B_{12}H_{10}$ (L = CO, R_3N, R_3P, R_2S) type, where the two ligands (which are formally positively charged) occupy the *ortho, meta*, or *para* positions of a 12-vertex boron cage (as depicted below), have been known in inorganic chemistry since the early 1960s. Advantage is taken of the stability of the icosahedral $(B_{12}H_{10}^{2-})$ counterion. While compounds of this type have not been described, candidates are easily designed. For example, its B3LYP/6-311 + G**–computed vibrational frequencies characterize **7** to be a minimum. Furthermore, the natural population analysis (NPA) (given by the NBO program) charge distribution indicates that significant positive charge, +0.78, is delocalized over the C_5H_4 pyramidal cation moiety while the 6-vertex *closo* carborane cage stabilizes the negative charge. Note that the pentacoordinate carbon in zwitterion **7** has five C–C bonds of nearly normal length. However, the octet rule is not violated since the four 1.588-Å bonds participate in five-center six-electron (5c–6e) multicenter bonding. We have computed many charge-compensated species with boron cage appendages of this type; some even involve "classical" carbocations.

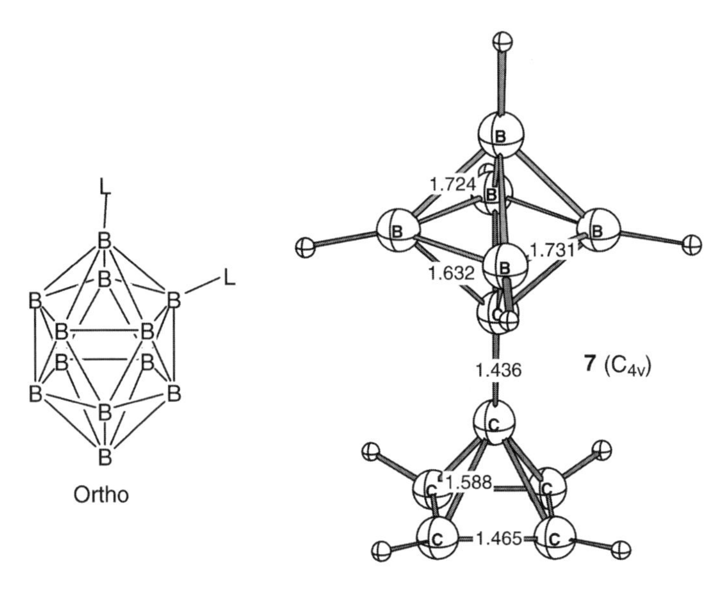

Zwitterionic "neutral" classical carbocationic analogs are also known experimentally. Appropriately enough, triarylcarbenium salts, the first known carbocations postulated a century ago (1899–1902),[10] serve as the basis for one of these. A triphenylmethyl carbocation moiety was shown by Santillan and coworkers more recently to be present in thymolsulfonephthalein, an acid–base indicator, based on X-ray crystallography and ^{1}H as well as ^{13}C NMR measurements.[11] While thymolsulfonephthalein might exist in zwitterionic, quinoidal, or lactonoid forms (Scheme 2), the X-ray geometry revealed that the zwitterion is favored because its triarylcarbenium ion fragment is stabilized by positive charge delocalization over the phenol rings.

The B3LYP/6-31G*-optimized geometries of both the parent trityl cation **8a** and the neutral trityl sulfonate model (C_1), **8b**, mirror the details of Santillan's X-ray structure. We computed the *meta*-sulfonate trityl structure, **8b**, instead of the *ortho* isomer (Scheme 2), in order to prevent lactone formation (see below). The C–C

HO OH HO O HO OH

SO_3 SO_3H

Zwitterionic Quinoid Lactonoid

Scheme 2

bond distances in the phenyl rings of **8a** and **8b** are remarkably similar. Both **8a** and **8b** have the same NPA charge on the central carbon (0.20) and similar charge distributions on phenyl rings.

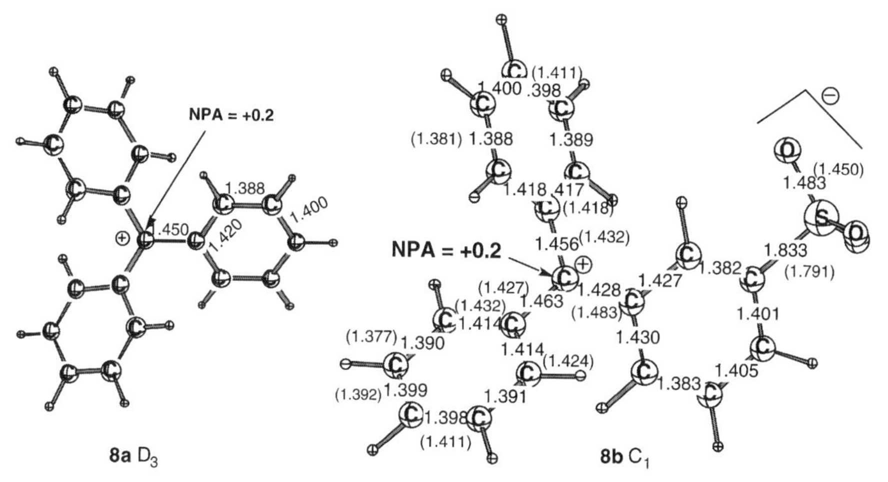

The figures in parentheses are the X-ray data for thymolsulfonephthalein

In contrast to lactonoid form postulated by Santillan, we find that the *ortho*-triaryl sulfonate (without the phenolic OH groups) optimizes to a structure with a covalent C–O bond to the phenyl ring, **9**, rather than to the triaryl tertiary carbon (as in the lactonoid form, Scheme 2). The CC bond lengths of nonaromatic ring **A** clearly indicate that this ring has only two double bonds, while phenyl rings **B** and **C** have almost similar CC bond lengths.

9

"Anionic" nonclassical carbocation analogs can be exemplified by the discovery by Evans et al. of the neutral 7-norbornadienyl cation complex **10**,[12] which can be regarded as an ion pair stabilized internally by the zwitterionic interaction of the metal-free dioxanorbornyl anion fragment (**10-anion**) with a cationic transition metal moiety. This achievement exposes a new class of structural possibilities with frameworks similar to those in nonclassical carbocation systems,[13] **4**. The X-ray structure of **10** revealed a C2–C7 distance, 1.876(4) Å, much longer than normal but typically found[14] in three-center two-electron (3c–2e) bonding arrangements. The C2–C3 distance, 1.426(4) Å, is somewhat larger than the 1.38Å separations found for the C=C moieties in alkyl-bridged 3c–2e carbocation systems.[1c]

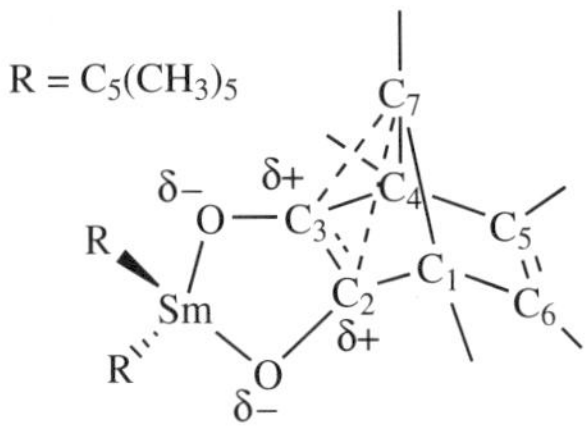

R = $C_5(CH_3)_5$

10 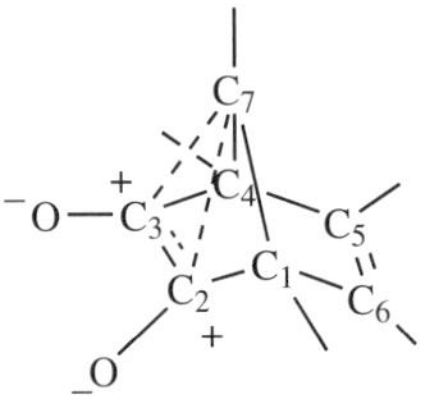 **10-anion**

In earlier days, the existence of carbocations was inferred from kinetic, stereochemical, and product analyses of varied reactions. Later, George Olah and his associates observed stable carbocations directly by spectroscopic methods (e.g., NMR and IR) with great success. However, the detailed geometries of most of these species are not available experimentally. Stable carbocation salts are difficult to crystallize and are often disordered or otherwise unsuitable for X-ray crystallographic analysis, and relatively few structures have been reported.[14] The neutrality and isolability of the charge compensated carbocations may offer further opportunities for experimental structure investigations.

3.2.1 Comments on Unusually Long C–C Bonds

Generally, the 3c–2e bridging in many "nonclassical" carbocations and in their "neutral" counterparts is characterized by long CC bond distances, in the range of 1.8–2.1 Å. When such species have threefold symmetry (as in **3**), the formal bond order of the long CC bonds (1.864 Å) is $\frac{1}{3}$. The bridgehead–bridgehead separation in the dehydroadamantdiyl dication **5** is 2.101 Å. Since two electrons are delocalized and bind four atoms tetrahedrally, the formal bond order is $\frac{1}{6}$. However, both **3** and **5** benefit from the "support" of the molecular skeleton and the CC separations are shortened as a consequence. Unsupported CC bonds longer than ~ 1.8 Å are very rare,[15] but the $^3\sum_d$ state of the ethane radical cation is a conspicuous exception. Like ethane itself, $C_2H_6(1+,2)$ favors staggered D_{3d} symmetry, but it has only a one-electron C–C bond (rCC = 1.963 Å at B3LYP/6-311+G**) and a

49.8 kcal/mol bond dissociation energy (BDE, into a methyl cation and a methyl radical). This energy is somewhat more than half the BDE of ethane (88.2 kcal/mol at the same level).

Likewise, in a paper dealing with the related question of homoaromaticity in neutral hydrocarbons, Stahl et al. examined the consequences of stretching and finally dissociating the C—C bond in ethane (at UBS-B3LYP/6-311+G**).[16] The geometry with the C—C bond elongated to 2.2 Å was only 47.5 kcal/mol higher in energy than the equilibrium structure (rCC = 1.531 Å); hence, about half of the total energy (88.2 kcal/mol) persists at this long distance.[16] The bonding interaction was still significant even at 2.5 Å! Consequently, the long C—C distances in nonclassical carbocations (typically ranging from 1.8 to 2.1 Å) correspond to partial bonds, which can be expected to have substantial bond energies (assuming that the interacting orbitals overlap well).

3.2.2 Strategies for Forming Zwitterionic "Neutral" Carbocation Analogs

"Neutral" carbocation analogs can be envisioned as consisting of two moieties: a carbocation and a counterion, typically a highly delocalized anion. The analogy between an ionic salt containing counterion pairs can be extended to these systems except that the carbocation and an anion are parts of the same molecule. The counter anion can either be (1) attached as a substituent or, more elegantly, (2) integrated into the basic structure of the system itself:

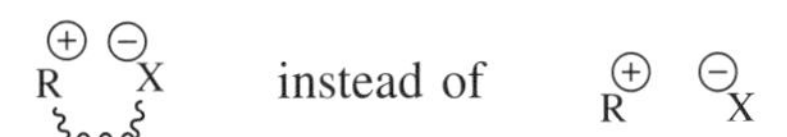

In the examples above (**6c**, **7**, **8b**, and **10**), the counteranion is attached externally as a substituent replacing hydrogen in order to counterbalance the positive charge. Wang and Schleyer[7] employed a different charge compensation strategy to help design neutral molecules with perfectly planar $C(C)_4$-type tetracoordinate carbon (ptC) arrangements computationally. (As discussed later, this strategy had been illustrated in an adamantane example earlier by Olah et al.[28] but had not been generalized). The design principle was based on the planarity of the methane dication; hence, charge-compensated dicationic ptC candidates with polycyclic "alkaplane" cages were proposed. The key idea was the replacement of two remote skeletal carbons by borons, thus converting CH_2 into $BH_2(-)$ groups, as in **11**. The two formally anionic boron units incorporated in this manner compensate the formal double-positive charge on the central ptC. Substitution of two remote hydrogens by $BH_3(-)$ groups, as in **12**, also was successful. Although species like **11** and **12** are zwitterionic conceptually natural bond orbital (NBO) analysis indicates that none of the atoms have large positive or negative charges. The charge delocalization over the whole molecule also is documented by the nature of the highest occupied molecular orbital, **11**-HOMO.

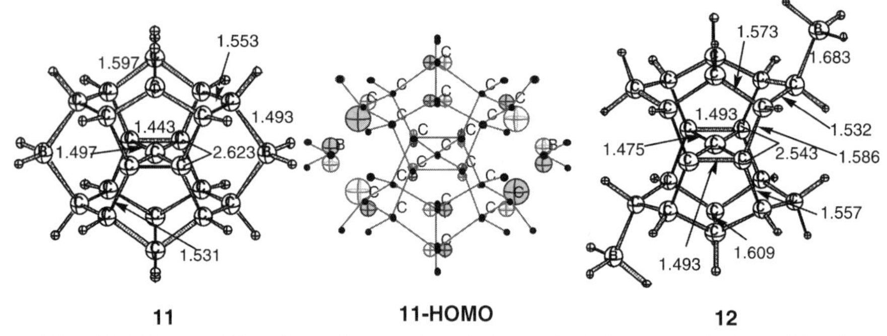

The third "neutral" carbocation analog design principle is a variation of the first, but is based conceptually on the Lewis acid complexation of "nonclassical carbenes." As discussed in more detail below, the latter species are related to the corresponding carbocations by removal of a proton (see Scheme 3). The carbene can be complexed by a Lewis acid, such as BF_3, to yield a neutral nonclassical carbocation analog. These ylid-like species also can be designed by direct hydrogen–Lewis acid replacement of the parent carbocation. The difference from the first design principle is that the substituted hydrogen is attached directly to a directly-involved, rather than a remote carbon.

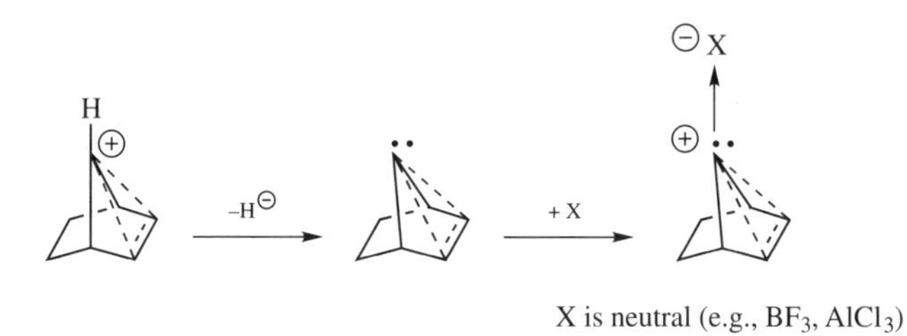

Scheme 3

3.2.3 Computational Criteria for the Viability of Zwitterionic "Neutral" Nonclassical Carbocation Analogs

Scientific progress enables problems to be investigated at higher resolution after the passage of only a few years. Although electronic structure computations date from the middle of the last century (e.g., see Ref. 18) routinely feasible ab initio investigations became common only with the advent of fast computers and sophisticated software much later. Since experiments are difficult (and sometimes even impossible), costly, and time-consuming, theoretical computations provide a valuable alternative for structure determination of stable ground-state molecules, short-lived intermediates and even transition states. High-level ab initio and DFT methods allow *general* and *systematic* investigations of structures; accurate ^{1}H, ^{13}C, and ^{11}B chemical shifts; IR frequencies; and HOMO-LUMO gaps, and much additional information also can be computed. Comparisons of experimental with computed geometries and spectroscopic details has demonstrated the reliability of the

B3LYP density functional theory (DFT) method with either the 6-31G* or the 6-311 + G** basis set as used here.[19]

The computational criteria for the existence of zwitterionic carbocation systems are the same as those for carbocation structures. Since the potential energy surfaces of carbocations can be extremely flat, the same may expected for the respective zwitterions: (1) not only should the species be minima (no imaginary vibrational frequencies), but also (2) it is desirable that the lowest positive frequency be substantial; stable multicenter bonding in nonclassical carbocations should lead to (3) large HOMO-LUMO separations whereas small HOMO-LUMO gaps result in wavefunction instability and open-shell singlet character; and (4) the zwitterionic structure should have high barriers toward dissociation or ring opening. These are important considerations for the existence of internally charge-compensated species. If its nonclassical delocalization is insufficient energetically, the "neutral" carbocation may rearrange to classical isomers.

3.3 APPLICATIONS OF THE PRINCIPLES: EXAMPLES

3.3.1 Remote H Replacement by a Highly Delocalized Negative Anion

The SO_3^- anion is one of the most effective we have found in its ability to stabilize the negative charge in designing charge-compensated carbocation analogs. Remarkably, SO_3^- substitution results in a neutral 2-norbornyl species, **13**.[2;13a–c,f,h;20] The C–C bond lengths of this zwitterion mirror the corresponding distances of 2-norbornyl cation, **13a**, also optimized at B3LYP/6-311 + G**. In particular, the critical C2–C6 (1.921 Å) and the C1–C2 (1.393 Å) lengths (distances) of **13** are similar to the corresponding distances in **13a**; 1.890 and 1.392 Å. Furthermore, the strong charge electron-attracting ability of SO_3 (the electron affinity of SO_3 itself is 1.9 ± 0.1 eV) results in the stabilization of a substantial negative charge (−0.62). An equally substantial positive charge (+0.65), essential for three-center two-electron bridging, resides on the three groups constituting the nonclassical bonding. The extent and stability of nonclassical bonding in **13** also are indicated by large HOMO-LUMO separation of 3.15 eV.

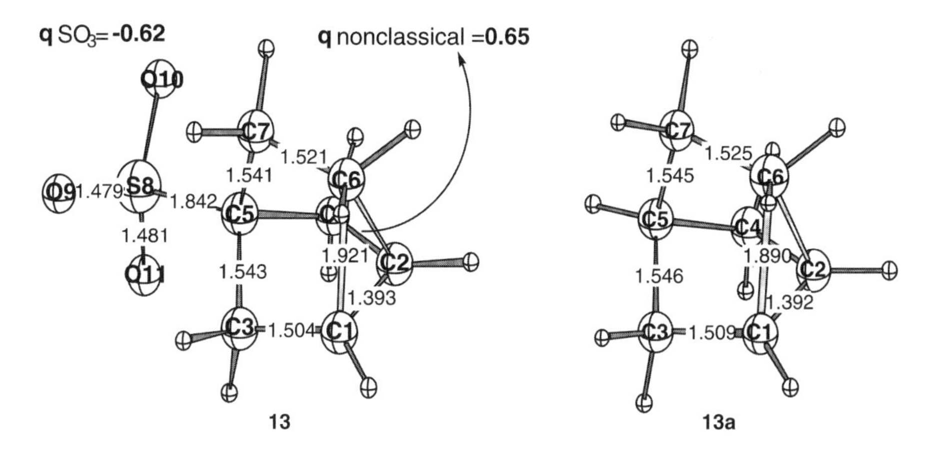

Excellent agreement is found between the proton chemical shifts, computed at GIAO B3LYP/6-311 + G**, for **13** (δ HC2 = 6.7 and δ HC6 = 3.3) and the measured[21] NMR values for **13a** (δ HC2 = 6.6 and δ HC6 = 3.1). Likewise, the computed ^{13}C NMR shifts advocate **13** as a neutral nonclassical carbocation; the computed δ^{13}Cs of C1,2 (144.3) and C6 (53.3) in **13** are close to 138.3 and 38.7 for the respective carbons in **13a**. Hence, the similar geometry of **13** and **13a** and agreement of their ^{1}H NMR and ^{13}C chemical shifts unequivocally establish **13** to be a "neutral" nonclassical ion.

3.3.2 Stabilization by a Remote Highly Delocalized Carbanion

The parent 7-norbornadienyl cation, **14a**, is one of the most stable long-lived stable nonclassical carbocations.[22] It is the parent ion of neutral **10**, discussed previously and subsequently, and can also be stabilized by the cyclopentadienyl anion appendage in **14**. The latter is a remarkable example of a nonclassical neutral hydrocarbon.

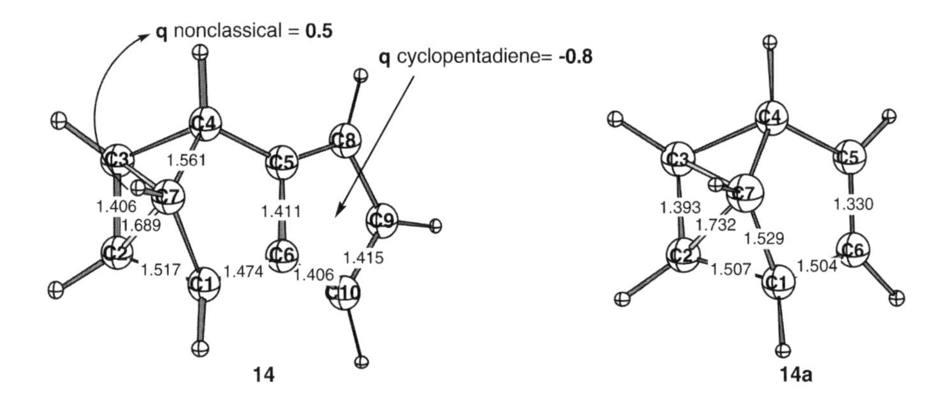

The key structural features of the charge compensated species **14** compare well with those of the 7-norbornadienyl cation, **14a**. The C2–C7 (1.689 Å) and C2–C3 (1.406 Å) bond distances of the three-center two-electron substructure of **14** actually indicate greater delocalization than in **14a** with its 1.732 and 1.393 Å separations. NBO population analysis indicates a near-unit (−0.80) negative cyclopentadienyl moiety has a significant negative charge, −0.8, on the cyclopentadienyl anion moiety. The C–C bond length equalization documents the aromatic character of this five-membered ring appendage. Much of the positive charge (+0.5) is delocalized over the nonclassical bridging entities (C2, C3, and C7 and their attached Hs). The stabilization of the positive charge and the short multicenter C–C distances in **14** also is reflected by the large HOMO-LUMO gap, 2.6 eV. However, the vertical ionization potential (IP) of **14** is only 0.1 eV, at B3LYP/6-311+G**. Although **14** is a minimum with an appreciable first positive vibrational frequency (127.6 cm^{-1}), the barrier for decomposition of **14** to the more stable classical

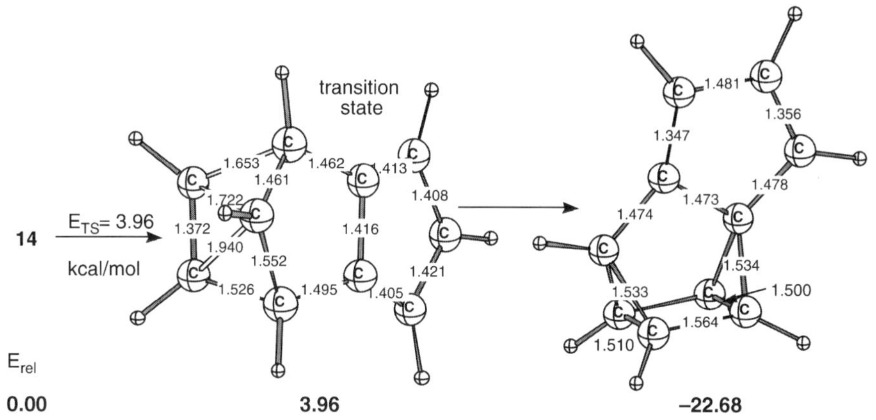

Scheme 4

isomer (Scheme 4), is small (4.0 kcal/mol). Hence, the low IP and the small isomerization barrier predict that the zwitterionic carbocation **14** will be viable only at extremely low temperatures.

3.3.3 Linking Stabilized Positive and Negative Organic Fragments

Arnett et al.[23] have investigated the heterolysis of weak (easily dissociated) C–C bonds into carbocations and carbanions. The X-ray structure of trimethylcyclopropenium (*p*-nitrophenyl)malononitrile (shown below) revealed an unusually long C1′–C1 bond distance, 1.588 Å ≫ C1–C4 (1.485 Å). In a polar solvent, this covalent compound dissociated into the stabilized carbanion and carbocation fragments:

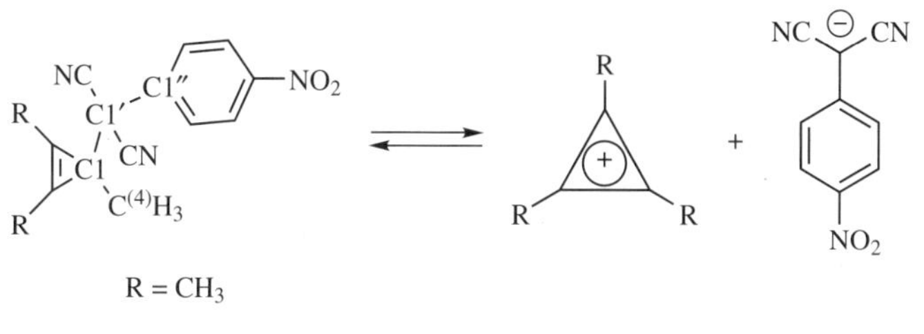

R = CH_3

We have investigated a closely related zwitterionic species, **15**, *trans*-1-dicyano-3-(2′,3′-dimethyl)cyclopropylcyclobutane, in which the stabilized positive and negative entities are linked together. The cyclobutane ring in **15** separates the cyclopropenium ion from the dicyanomethyl anion. The zwitterionic character of **15** is evident from the huge computed dipole moment (21.2 D). The lowest vibrational frequency at B3LYP/6-31G*, is small (+14 cm^{-1}) but corresponds to the noncritical out-of-plane motion of the CH_3 groups. The NBO charges are −0.76 on the dicyanomethyl group and +0.64 on the cyclopropenium group.

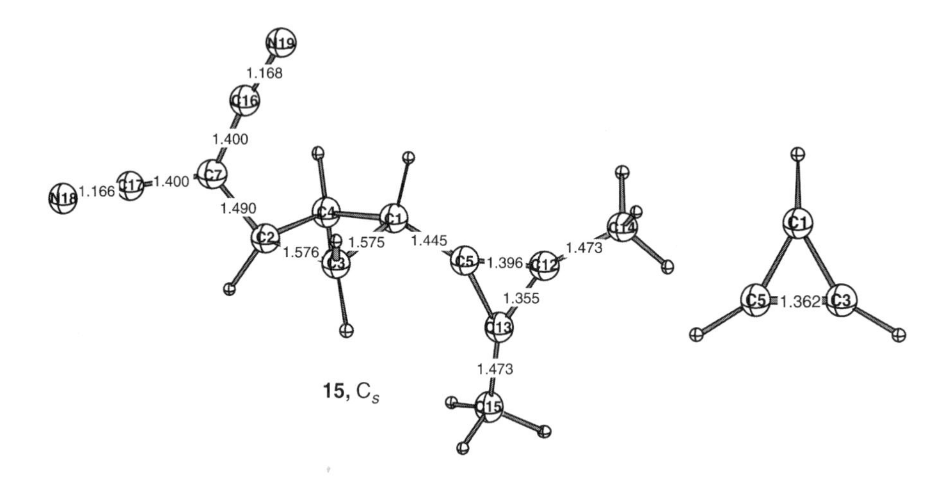

15, C_s

The cyclopropenium fragment in **15** has C12–C13 bond elongation, 1.355 Å, as compared to those in cyclopropene, 1.291 Å, optimized at the same level (B3LYP/6-311+G**). Note that all the bond lengths of the three-membered ring in **15** are much closer to those of cyclopropenium cation, 1.362 Å. Additionally, the computed ^{13}C NMR chemical shifts of C12 ($\delta = 169.5$) and C13 ($\delta = 170.0$) of the three-membered ring in **15** are much closer to those in cyclopropenium cation ($\delta = 178.6$) than those in cyclopropene ($\delta = 117.1$).

3.3.4 Replacing CH^+ by Isoelectronic BH

The isolobal substitution of CH^+ by the isoelectronic BH is a frequently investigated way to generate neutral nonclassical carbocation equivalents.[3,24] For example, the 3c–2e bonding in 7-boranorbornene (**16**) and 7-boranorbornadiene (**17**) (analogs of the 7-norbornenyl and 7-norbornadienyl cations, respectively), arise from the interaction of the electron-deficient bridging boron with a double bond. Although the parent 7-boranorbornene is not known experimentally, Fagan et al.[25] reported that the Diels–Alder reaction of a zirconium metallacycle with phenylboron dichloride gave the *endo*-1-phenyl-2,3,4,5-tetramethylborole, **18**, as the major product. The single crystal X-ray investigation of **18** showed that plane defined by B1, C1, and C4 was tilted substantially toward the C2=C3 double bond, resulting in equal B1–C2 and B1–C3 bond distances (1.864 Å). Although these structural features are very similar to those of the analogous 7-norbornenyl carbocation derivative reported by Laube[14]; (**18**) as well as **17** and **18**, are not carbocations. Our goal is to use isoelectronic boron substitution differently to devise "neutral" carbocation analogs of the 7-norbornenyl and 7-norbornadienyl cations;

the strategy is based on the Wang–Schleyer approach,[17] discussed above:

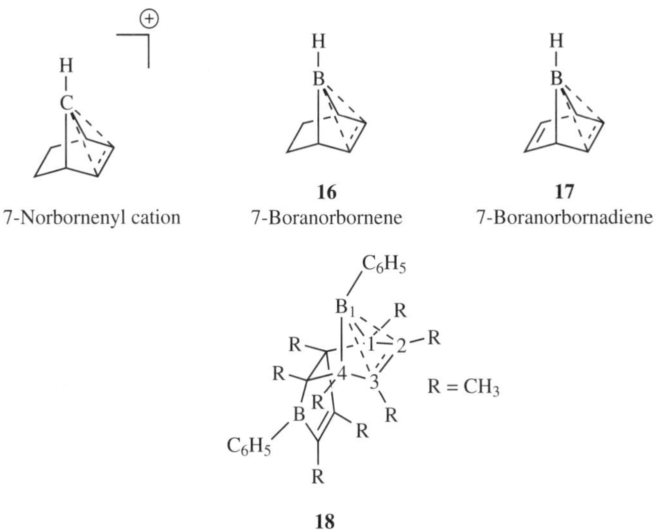

Although the replacement of apical bridgehead CH^+ in 7-norboradiene by neutral BH leads to **17**, one can in principle substitute any of the carbons by borons to generate nonclassical zwitterionic species. The neutral **17**, **19**, and **20** 7-norboradiene cation analogs are examples.[26] All their nonclassical 3c–2e bridging distances are like those of the carbocation (for 7-norbornadienyl cation, the C2–C7 distance is 1.732 Å and the C2–C3 distance is 1.393 Å).

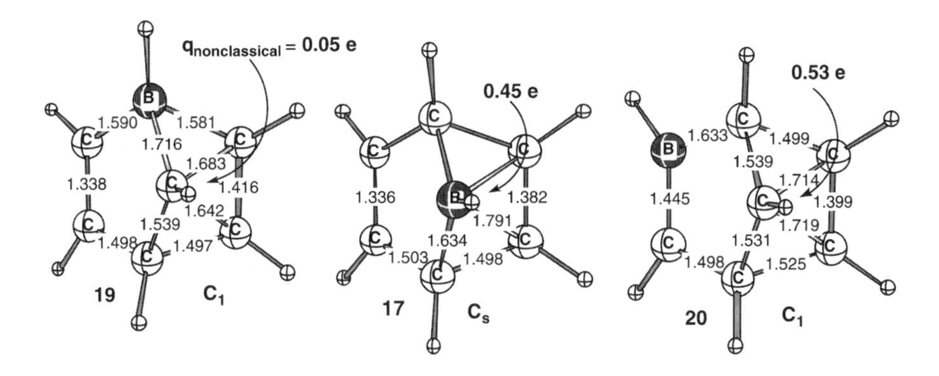

The boradiadamantyl, **21**, is an analog of the adamantyl cation, **21a**, first prepared by Olah and Schleyer as a stable bridgehead cation.[27] Three CC bonds are arranged parallel to the *p* orbital in **21**, and the optimal CC hyperconjugative interaction results in 1.642 Å bond lengths. The C1–C2 (1.451 Å) distance in **21** is similar to 1.458 Å, in **21a**. The corresponding bond lengths of Laube's X-ray[14] structure of the 3,5,7-trimethyladamantyl cation compare well those of **21a**. The smallest

B3LYP/6-311+G** computed vibrational frequency for **21** is +61 cm^{-1}, and the HOMO-LUMO separation is 2.5 eV. As expected, the dipole moment of **21** (9.1 D) is large, due to the substantial separation, 4.54 Å between the positively (C1) and negatively (B1) charged centers. The computed δ^{13}C chemical shift of C1 (270.1) in **21a** is significantly downfield and comparable to that (309.3 ppm) in 1-adamantyl cation, **21a**:

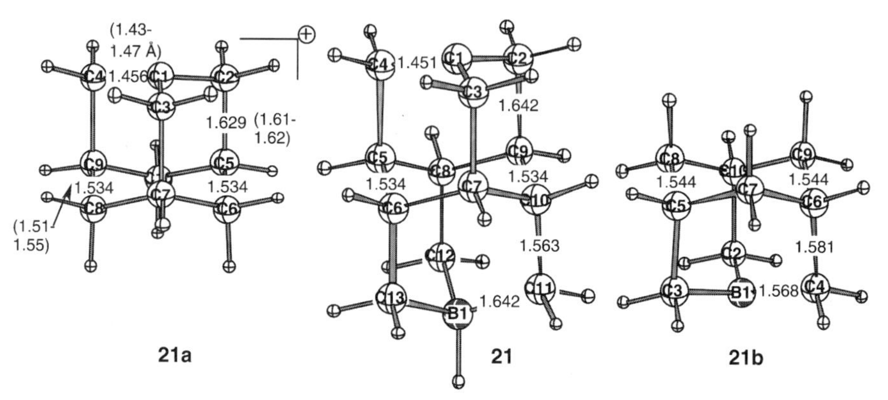

The figures in parentheses are Laube's X-ray data for 3,5,7-trimethyladamantyl cation

The adamantyl cation, **21a**, also has a known boron counterpart, 1-boraadamantane (**21b**). Both serve as precedents for the zwitterionic carbocation analog **22** (an isomer of **21b**),[26] designed by replacing the carbon of any of the remote methylene groups of **21a** by boron. The key structural features of **21a** are preserved in **22**. The C1–C3 and C1–C4 distances are short, 1.441 and 1.474 Å, respectively, due to rehybridization and planarization of C1. The C2–C5 and C4–C7 bond lengths are longer, 1.673 and 1.597 Å, respectively, owing to hyperconjugation interaction with the empty *p* orbital on C1. In addition, **22** also exhibits a large dipole moment of 5.4 D and a considerable HOMO-LUMO gap, 3.87 eV. Furthermore, the computed ^{13}C chemical shift of electron deficient C1 in **22** ($\delta = 285.9$) is close to that (309.3 ppm) in **21a**.

Olah et al. compared the 1,3-dehydro-5-adamantyl cation, **23a**, with its related zwitterionic boron analog, **23**.[28] This is one of the first examples of the application of skeletal boron incorporation to compensate the carbocation charge. Although its intermediacy can be invoked in the solvolysis of 5-bromo-1,3-dehydroadamantane, **23a** has not yet been characterized under long-lived ion conditions.[29] However, the parent homocyclopropenium ion (**23b**) was generated successfully.[30] B3LYP/6-31G* computations show **23b**, **23a**, and **23** to have similar geometric features. All have long partial CC bonds involved in the 3c–2e delocalization, but these vary progressively in length from 1.860 Å in **23b** to 1.911 Å in **23a** and to 2.067 Å in **23**. This trend can be attributed to the increased stabilization provided by the cages of **23a** and **23b**. The ^{13}C chemical shifts of the 3c–2e centers (C1) vary similarly, from

+4.9 for **23b**[30] to the computed 43.8 for **23a** and to 76.1 ppm for **23**.

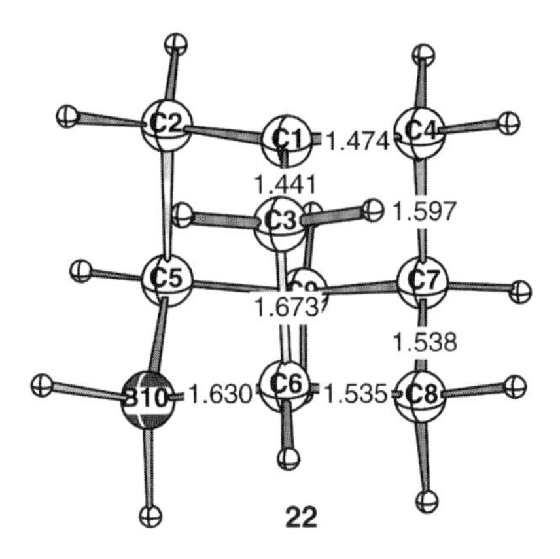

The zwitterionic boron analog (**23**) of the 1,3-dehydro-5-adamantyl cation (**23a**) has a partial positive charge (+0.26) delocalized over the trishomoaromatic unit. Moreover, the HOMO shown below indicate that the positive charge is compensated by a highly delocalized negative charge distributed among orbitals on every skeletal atom the (cf. Wang–Schleyer's planar tetracoordinate carbon[17]). Note that the computed ^{13}C NMR chemical shift of the tertiary C7 carbon, 70.4 ppm, is well downfield of that in adamantane.

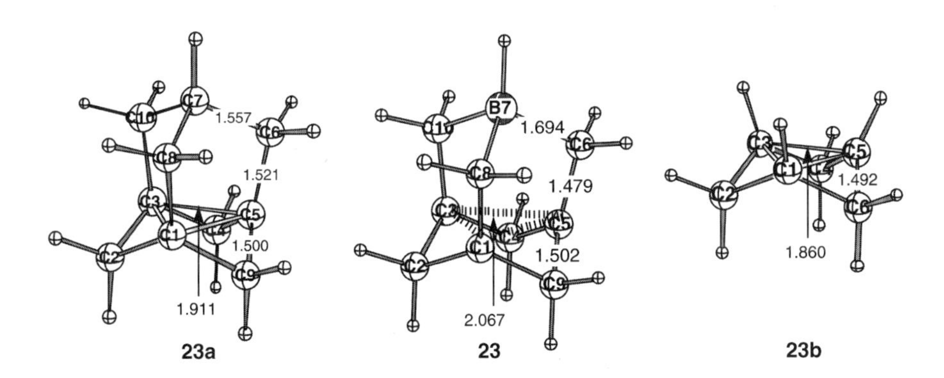

3.3.5 Symmetric Geometry States for α-Ketol Rearrangements

Schleyer and Maerker[31] recognized the analogy between Evans and Forrestal's[12] nonclassical carbocation (**10**) and the arrested bridged enediolates transition states[32] for the α-ketol rearrangement. The isomerization of α-hydroxy ketones was studied first by Ruzicka and Mehldahl in the late 1930s. The "α-ketol rearrangement" (either acid or base catalyzed) has synthetic utility and involves a 1,2-alkyl group migration.[32] Because of the considerable strain in **10a**, the nonclassical intermediate **10** actually is more stable than the alcoholate anion derivatives of **10a** (which can be regarded formally as a negatively charged nonclassical "carbocation"!):

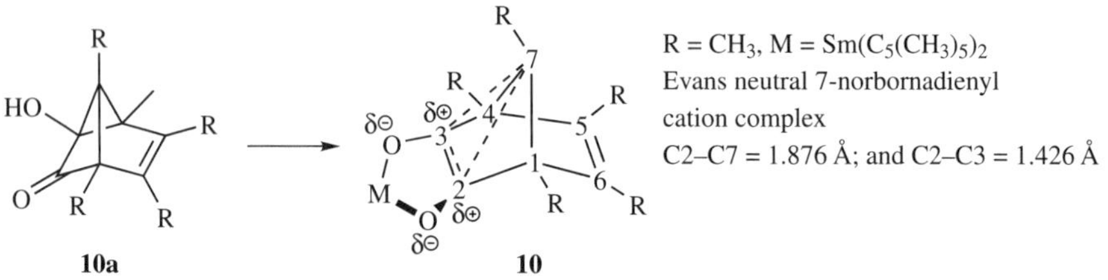

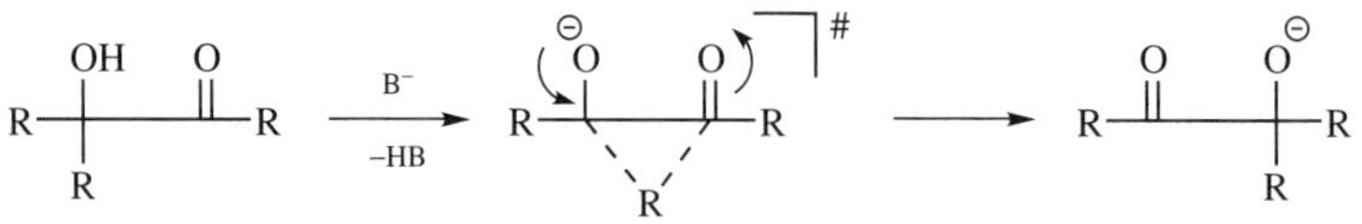

10a → **10**

$R = CH_3$, $M = Sm(C_5(CH_3)_5)_2$
Evans neutral 7-norbornadienyl
cation complex
C2–C7 = 1.876 Å; and C2–C3 = 1.426 Å

The simplified model of Evans neutral nonclassical (**10**) is a minimum **24**, in which the methyl groups on the norbornadienyl derivative are substituted by hydrogens and the samarium complex is replaced by $AlCl_2$. The optimized geometry of model **24** mirrors the details of the Evans' X-ray structure; the C2–C3 distance in **10** (1.426 Å) is similar to that in **24** (1.430 Å), while the C2–C7 distance (1.876 Å) in **10** is slightly longer than the 1.820 Å in **24**. The two negative charges in the enediolate $(RCO)_2^{2-}$ moieties are counterbalanced by the carbocationic and positively charged metal groups in **24**. The strong polarization of the C—O groups leads to substantial positive charges (~0.4) on C2 and C3, which participate in the nonclassical bridging.

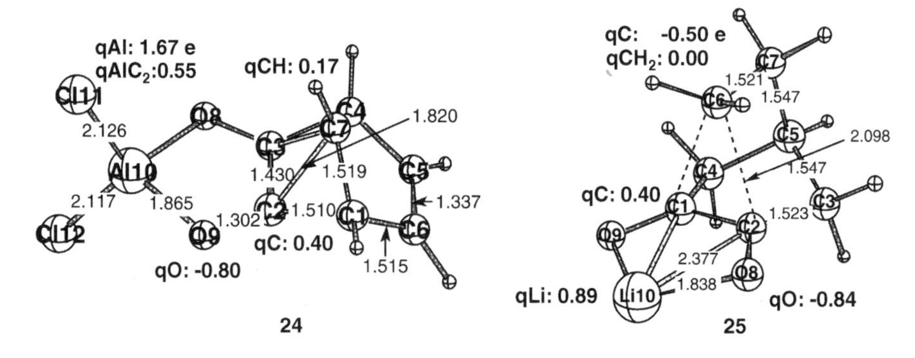

The 2-norbornyl α-ketol anion rearrangement has been investigated by Nickon and his associates and found to have a low barrier.[31] Our computions the C_s (bridged) structure, an analog of the 2-norbornyl cation(!), to be a transition state, rather than a minimum.[31] However, the barrier for the enantiomerization of the lithium salt via transition state **25** is only 13.4 kcal/mol at B3LYP/6-31G*. The 3c–2e bonding in **25** is characterized by longer C6–C1,2 (2.098 Å) and C1–C2 (1.450 Å) separations than those computed for the bridged 2-norbornyl cation at the same level (1.892 and 1.395 Å, respectively). The total charge on the C_5H_9 moiety (C3, C4, C5, C6, and C7 and the attached hydrogens) of **25**, as well as the bridging CH_2 group, is essentially zero. The positive charges on the two oxygen-substituted carbons, near +0.4 in **25**, facilitate the delocalized bonding, presumably in Evans' complex, **10**, as well.

3.3.6 Carbene–Lewis Acid Complexes

Freeman presented computational evidence that singlet carbenes, such as **26** and the examples in Scheme 5, behave like neutral counterparts of the corresponding nonclassical carbocations.[33] Indeed, 7-carbenanorbornene (**26**), was predicted much

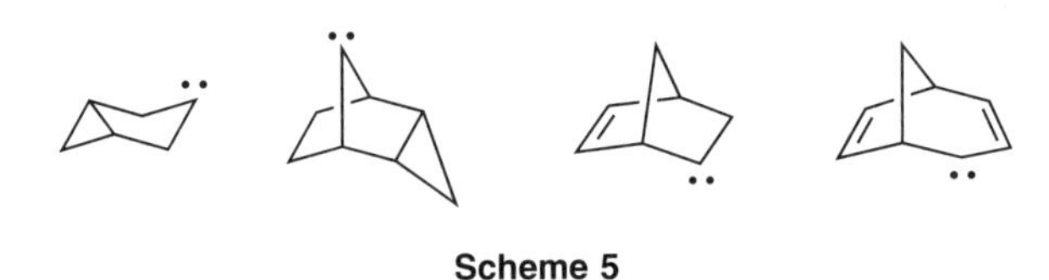

Scheme 5

earlier by Hoffmann[34] to have a nonclassical carbocation-like structure based on extended Hückel MO theory (EHT) calculations; Moss et al. reported experimental evidence based its reactivity.[35] We have gone a step further by investigating the effects on geometries and energies by complexing such carbenes with various Lewis acids, as in **27**.

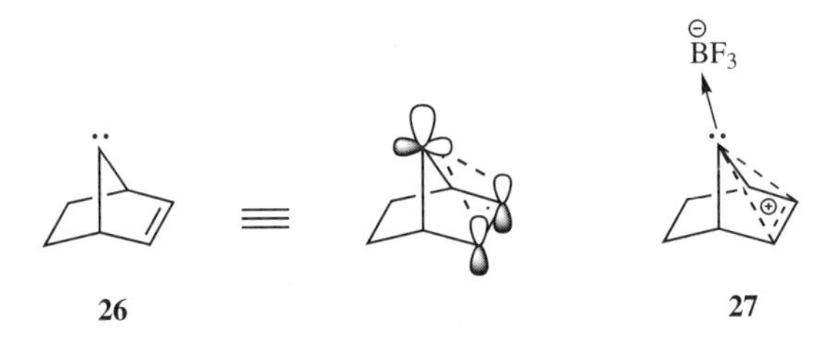

The BF_3 complexation in **27** enhances the degree of nonclassical bonding exhibited by **26**. While the B3LYP/6-311+G**-optimized geometries of **26** and of **27** resemble the bridged structure of the 7-norbornenyl cation (**26a**), the C7–C2 distance (1.897 Å) in **26** is 0.15 Å longer than that in **26a** (1.749 Å); the C7–C2 distance (1.808 Å) in the BF3 complex **27** is intermediate. The C2–C3 bond lengths these structures also show a corresponding trend: 1.372, 1.383, and 1.397 Å for **26**, **27**, and **26a**, respectively. For comparison, the C2–C7 distances reported by Laube for the 2,3-dimethyl-7-phenylnorbornen-7-yl ion is 1.86 Å[14] and by Evans et al. for

11 is 1.876 Å.[12]

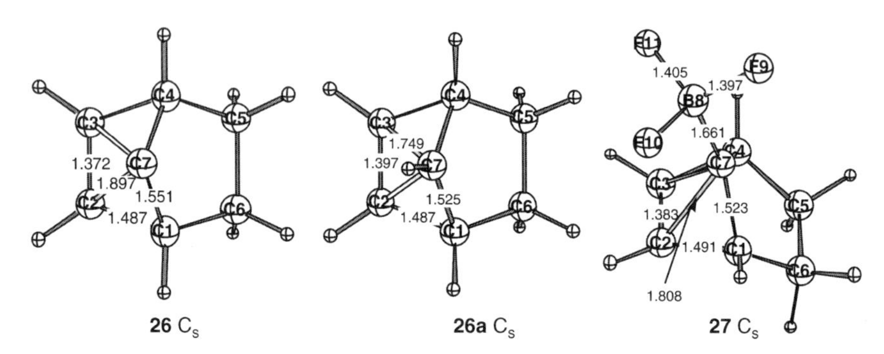

26 C_s **26a** C_s **27** C_s

Freeman[33f] used the isodesmic equation shown in Scheme 6 to evaluate a 15.6 kcal/mol stabilization energy of singlet carbene **26**. Remarkably, this is not much less than the 20.9 kcal/mol stabilization energy for the 7-norbornenyl cation (**26a**) obtained similarly. Applying the same isodesmic equation to **27** reveals that BF_3 complexation increases the stabilization energy of **26a** to 19.1 kcal/mol. The strengthening of the 3c–2e bond in **27** also is demonstrated by the computed natural charges. The strong electron withdrawing ability of BF_3 results in delocalization of substantial positive charge (+0.44) on nonclassical bridging in C_7H_8 unit. The increased stabilization energies in the complexes of **26** with various Lewis acids are listed in Table 3.1, except PF_3.

X X

X = carbene, $\Delta E = -15.6$ kcal/mol
X = carbocation, $\Delta E = -20.9$ kcal/mol
X = **:** ⟶ BF_3, $\Delta E = -19.1$ kcal/mol

Scheme 6

Table 3.1 C7–C1, C2, and C1–C2 bond lengths and the isodesmic energies ($E_{isomeric}$, Scheme 6) for various Lewis acid complexes of 7-carbenanorbornene optimized at B3LYP/6-311 + G[a]**

Lewis Acid	PG	C7–C1,C2 (Å)	C1–C2 (Å)	$E_{isomeric}$ (kcal/mol)
BF_3	C_s	1.808	1.383	−19.1
AlF_3	C_s	1.808	1.386	−16.2
$AlCl_3$	C_s	1.806	1.386	−20.1
$ScCl_3$	C_s	1.821	1.385	
ScF_3	C_1	1.828, 1.823	1.384	
PF_3	C_s	2.320	1.336	3.2
Carbene	C_s	1.897	1.372	−15.6
Carbocation	C_s	1.749	1.397	−20.9

[a]PG is the point group.

Note that the positive isodesmic energy, 3.2 kcal/mol, for 7-carbenanorbornene-PF_3 complex indicates its *destabilization* (Table 3.1). The C7–C1 distance is very large, 2.3 Å, and C7 actually is bent away from C1C2, which has a typical C=C double-bond length, 1.336 Å (ethene, 1.329 Å). Unlike the other examples in Table 3.1, this PF_3 complex is an ylide with $C^{(-)}$ rather than $C^{(+)}$ polarization.

Pyramidane C_{4v} ([3.3.3.3]fenestrane) is the simplest possible hydrocarbon carbene with a pyramidal center.[36] Minkin et al.'s original HF/STO-3G computations, predicting this singlet carbene to be a local minimum, were confirmed by Lewars' high-level QCISD(T)/6-31G*//MP2(fc)/6-31G* study, which found pyrimidane to lie "in a fairly deep well, with a barrier of 96.1 kcal/mol separating it from its kinetically most readily accessible isomer." Pyrimidane is closely related to cation $C_5H_5^+$, which Hoffmann[37] predicted to have a square pyramidal C_{4v} geometry based on extended Hückel MO theory. Although the parent $C_5H_5^+$ ion has not been observed, Masamune and coworkers succeeded in preparing methylated derivatives:[38]

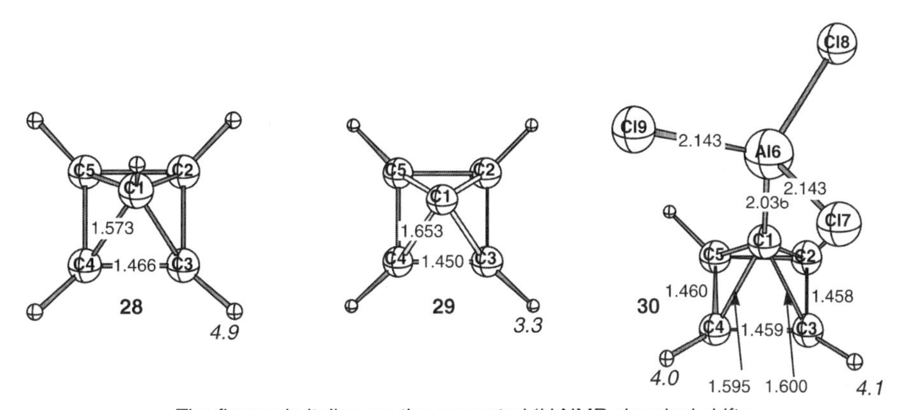

The figures in italics are the computed ^{1}H NMR chemical shifts

The computed NMR chemical shifts (4.9 ppm) of the protons on the four-membered ring (4MR) of cation **28**, (close to Masamune's measured 4.6 ppm for the dimethyl derivative) are 1.6 ppm upfield (shielded) compared to those in pyramidane carbene **29** (3.3 ppm). This indication that carbene **29** is more highly delocalized (aromatic) than **28** is confirmed by their nucleus-independent chemical shift (NICS) values (−30.4 for **29** vs. −23.8 for **28**). The 4MR proton chemical shifts (4.1) and NICS (−26.2) of the charge-compensated Lewis acid complex, **30**, are intermediate between cation **28** and carbene **29**, but are closer to the cation **28**.

The same is true for the CC bond distances (see above). In general, all the carbenes listed in Table 3.2 undergo marked changes in geometry after complexing with Lewis acids (also see **31** and **32**). All the pyramidane-Lewis acid complexes are minima characterized by positive vibrational frequencies. Although q_{C5H4} in Table 3.2 indicates that these zwitterions have small positive charge delocalized

Table 3.2 C1–C2 and C1–C3 bond lengths for various Lewis acid–pyramidane carbene complexes optimized at B3LYP/6-311+G[a]**

	PG	C1–C2	C1–C3	ν	q_{C5H4}	HOMO-LUMO	E_{assoc}
$AlCl_3$	C_s	1.600	1.595	23.4	0.22	6.7	44.3
AlF_3	C_s	1.603	1.595	18.7	0.18	8.5	47.5
$Al(CH_3)_3$	C_s	1.608	1.618	11.2	0.21	5.7	23.4
$ScCl_3$	C_s	1.606	1.611	18.0	0.20	6.1	43.5
ScF_3	C_s	1.608	1.615	12.1	0.16	8.7	38.9
$TiCl_4$	C_{2v}	1.600	1.600	12.0	0.30	4.6	16.1
TiF_4	C_s	1.604	1.613	19.7	0.19	6.2	20.9
C_5H_4	C_{4v}	1.653	1.653	405.5	—	6.7	—

[a]PG is the point group of the complex, ν is the smallest vibrational frequency (cm^{-1}), q_{C5H4} is the charge on the pyramidane ring, HOMO-LUMO separation is indicated in eV, and E_{assoc} (kcal/mol) is the B3LYP/6-311 + G** computed association energy of the complex with respect to its respective monomers.

over the pyramidane carbene moiety, the computed HOMO-LUMO gaps and association energies E_{assoc} for all the structures depicted in Table 3.2 are considerable. Hence, Lewis acid complexation stabilizes carbene **29**:

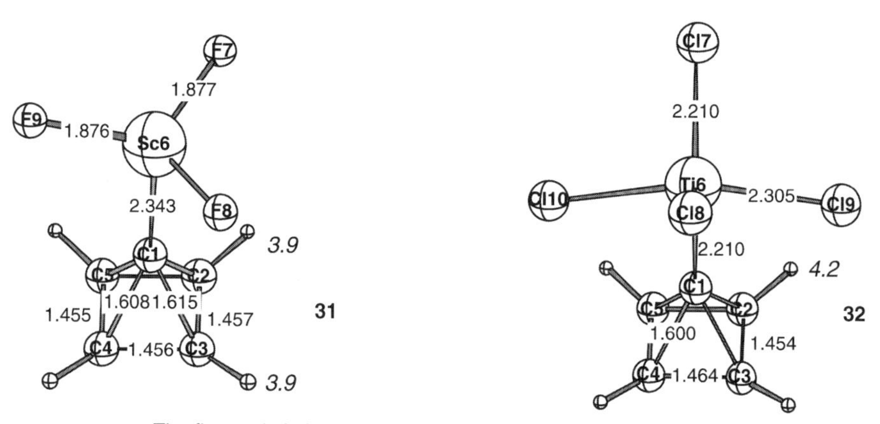

The figures in italics are the computed ¹H NMR chemical shifts

3.3.7 Other Lewis Acid Complexes

The X-ray structure of tetramethylcyclobutadiene complexed with $AlCl_3$ determined by Krüger et al.[7] is reproduced nicely computationally (see **33**). Both geometries are similar to that of the tetramethylcyclobutenyl cation (**33a**) computed at the same level; the C1–C2 and C2–C4 bond lengths in **33** are almost the same as in **33a**, and the C2-C3 distances differ by 0.06 Å. The GIAO-B3LYP/6-311+G**//B3LYP/6-311 + G** computed ^{13}C NMR shifts of C2, C3, and C4 (182.1 and

158.8, respectively) in **33** agree reasonable well with those measured for the 4-chlorotetramethyl homocyclopropenyl cation (191.5 and 174.4, respectively) and with those computed for **33a** (196.9 and 180.2, respectively). In addition, the theoretical δ ^{13}C for C_1 in **33** (73.6) is similar to Olah's experimental 76.0 for the 4-chlorotetramethylhomocyclopropenyl cation value;[39] the computed HOMO-LUMO gap of **33** is large: 4.3 eV.

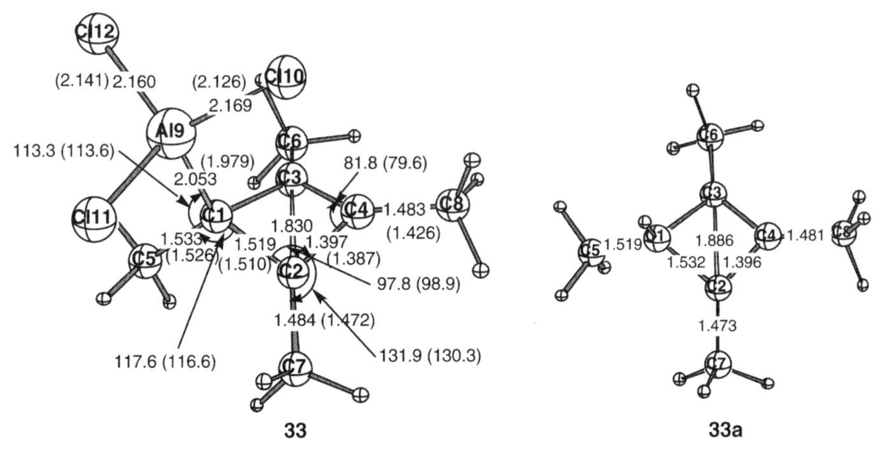

Values in parentheses are the X-ray data

Following Masamune's[30] first preparation, Olah[13e] studied the parent trishomoaromatic homocyclopropenium ion (**34a**) as a long-lived ion in superacid; **34a** also has been a subject of numerous ab initio studies.[40] Our B3LYP/6-31G*-optimized geometry of the neutral nonclassical homocyclopropenium–AlF_3 complex, **34**, is in remarkable agreement with the structure of the parent homocyclopropenium cation.

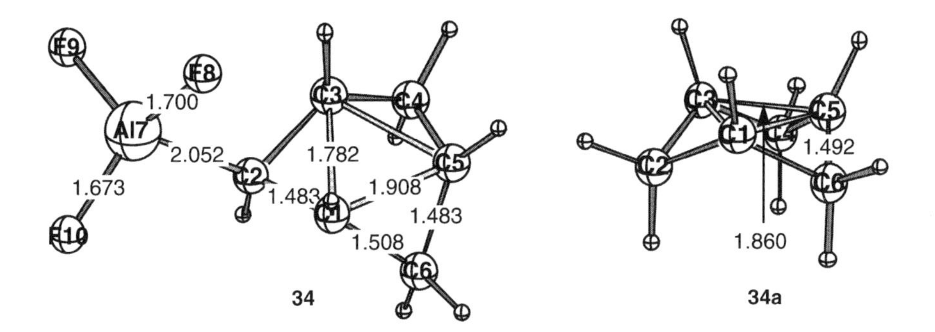

The C1–C2, C1–C6, and C5–C6 distances in both these structures match closely, as does the weighted average of the 3c–2e nonclassical bridging C1–C3 (1.782 Å) and C1–C5 (1.908 Å) bond lengths in zwitterionic structure **34** (1.866 Å) with the

1.860 Å homocyclopropenium ion **34a** distance. The key natural charges of the complex are −0.20 localized on the AlF_3 unit and +0.64 on the 3c–2e nonclassical bridging moiety. The GIAO B3LYP/6-311 + G** ^{13}C chemical shifts of the 3c–2e bridging, pentacoordinated carbons C1,3 are shielded: 6.7 in **34** and 9.2 ppm in **34a**. The $\delta^{13}C$ chemical shifts of the other C4,6 carbons are 25.4 for **34** and 21.0 for C2,4,6 of **34a**.

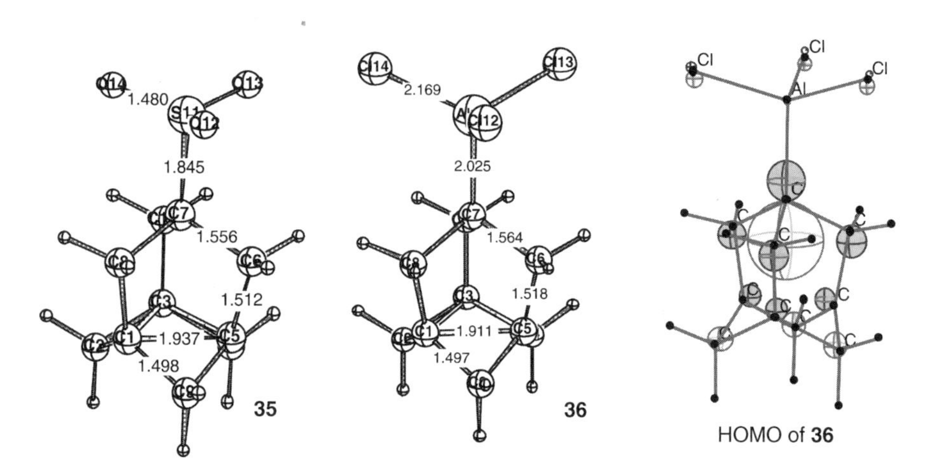

The zwitterions **35** and **36** represent zwitterion analogs of the 3c–2e 1,3-dehydro-5-adamantyl cation **23a** designed by replacing the H^+ on C7 with SO_3 and $AlCl_3$, respectively. The structural features of **35** and **36** are remarkably similar to those of both **23a** and the charged compensated **23**. The GIAO B3LYP/6-311+G** ^{13}C NMR shifts of C1 and C2 in **35**, 62.2 and 37.9, respectively, may be compared to 43.8 and 37.9, respectively, evaluated for **23a**, and the 76.1 and 45.2 δ ^{13}C for C1 and C2, respectively, of **23**.

With a 27.5 kcal/mol barrier (at B3LYP/6-31G*) toward ring opening, **35** should be observable; **35** also has a large complexation energy, −77 kcal/mol, with respect to 1,3,5,7-bisdehydroadamantane and SO_3. All the vibrational frequencies **35** and **36** are real at B3LYP/6-311 + G** (the lowest are 39.2 and 22.1 cm^{-1}, respectively) and also have relatively large HOMO-LUMO separations (3.9 and 5.3 eV, respectively). As expected, both these zwitterionic complexes exhibit large dipole moments, 16.5 and 15.7 D, respectively. Remarkably, none of the atoms comprising the framework have large positive or negative charges. Like **23**, the HOMOs of complexes **35** and **36** have coefficients distributed over the entire structures.

3.4 EXAMPLES OF ZWITTER"ANIONIC" CARBOCATION ANALOGS

Since the five-membered rings act as electron acceptors and three- membered rings act as electron donors, the [3]radialene derivative, **38**, is expected to be stabilized as a dianion. Indeed, the B3LYP/6-311 + G**-optimization of **38** leads to a D_{3h} minimum with a 57.8 cm^{-1} lowest vibrational frequency. Although **38** has not been realized experimentally, Iyoda et al.[41] reported the synthesis of its derivatives, bis(9-fluorenylidene)-9′-xanthenylidene cyclopropane. The three-membered ring in **38** has CC bond lengths, 1.401 Å, only somewhat longer than those in the cyclopropenium ion, 1.362 Å. The CC bond lengths in the five-membered rings in **38** are close to the 1.415 Å in the cyclopentadienyl anion, and are consistent with the −0.6 charges of these rings (hydrogens included). The residual charge on the 3MR is also large and positive, +1.8. The NICS(1) in the center of the 3MR (−8.2) and the 5MRs (−8.6) of **38** are close to those in cyclopropenium cation (−14.7) and cyclopentadienyl anion (−9.5), respectively; furthermore, **38** has an appreciable HOMO-LUMO gap (3.2 eV):

Although the zwitter"anionic" carbocations croconate ($C_5O_5^{2-}$) and rhodizonate ($C_6O_6^{2-}$) have been known for more than 160 years,[42] the lower analogues of oxocarbon family, squarate ($C_4O_4^{2-}$) and deltate ($C_3O_3^{2-}$), were synthesized and characterized more recently.[43] Extensive literature on dianionic oxocarbons pertaining to various aspects ranging from synthesis to physical properties is well documented.[44] Using geometric, energetic, and magnetic indices as aromaticity criteria, Schleyer et al. more recently[45] concluded that delocalization of π electrons in the $C_nO_n^{2-}$ oxocarbon dianion series (and the aromatic stabilization energy) decreases with increasing ring size, from $n = 3$ to $n = 6$. We restrict our discussion here to $C_3O_3^{2-}$, **39**:

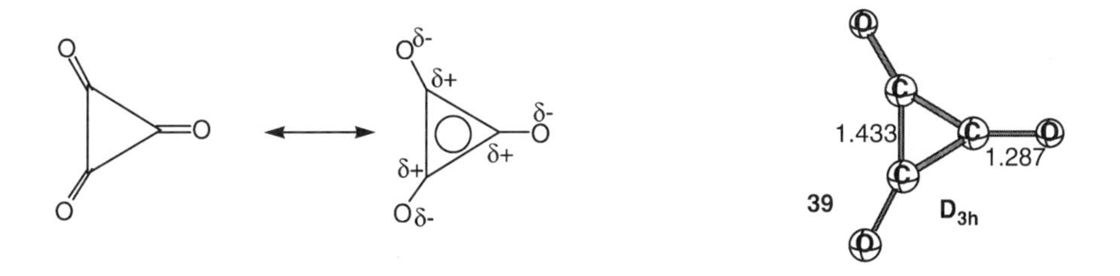

The deltate dianion, $C_3O_3^{2-}$, has D_{3h} symmetry and C–C bond lengths (1.433 Å) considerably longer than those (1.362 Å) in $C_3H_3^+$. The C–O bond length (1.287 Å) also is longer than computed for acetone, 1.212 Å. The calculated Wiberg bond indices, 1.195 for the CC and 1.312 for the CO bonds (the latter compared to 1.8 for C=O in acetone) indicate delocalization. The total natural charge on the three-membered ring in **39**, +0.69, is appreciable, but less than the +2 expected formally. The computed large diamagnetic susceptibility exaltation (Λ) of −15.6 cgs ppm, indicates that $C_3O_3^{2-}$ is highly diatropic as does the −11.0 NICS(1). The three-membered ring of **39** is clearly aromatic. Similarly, the stabilization of the cyclopropenium ring of the sulfur deltate oxocarbon analog, $C_3S_3^{2-}$, has been reported recently.[46] This is another zwitter "anionic" carbocation.

Section 3.3.5 (above) also presents examples of zwitter"anionic" carbocations. Along with **10a** (and **10**), "anionic" nonclassical carbocation equivalents of the 7-norbornadienyl and 7-norbornenyl cations are illustrated by **40**, which has two cyclopentadienyl anion appendages (cf. **14**) and **41**, which has two anionic SO_4 groups. Both these structures are minima at B3LYP/6-311+G** with lowest vibrational frequencies (70.3 and 32.5 cm^{-1}, respectively) and with 2.4 and 2.8 eV HOMO-LUMO gaps. The −0.4 charge of the 5MR appendage involved in nonclassical bridging in **40** is much less than the −0.8 charge of the other cyclopentadienyl moiety. The appreciable negative NICS(0) values in the center of five-membered rings (−11.1 and −14.3, respectively) demonstrates that **40** bears two cyclopentadienyl anion appendages. The nonclassical bridging unit of **40** features has longer CC bonds (1.803 and 1.473 Å) than do those of the 7-norbornadienyl cation (1.732 and 1.393 Å):

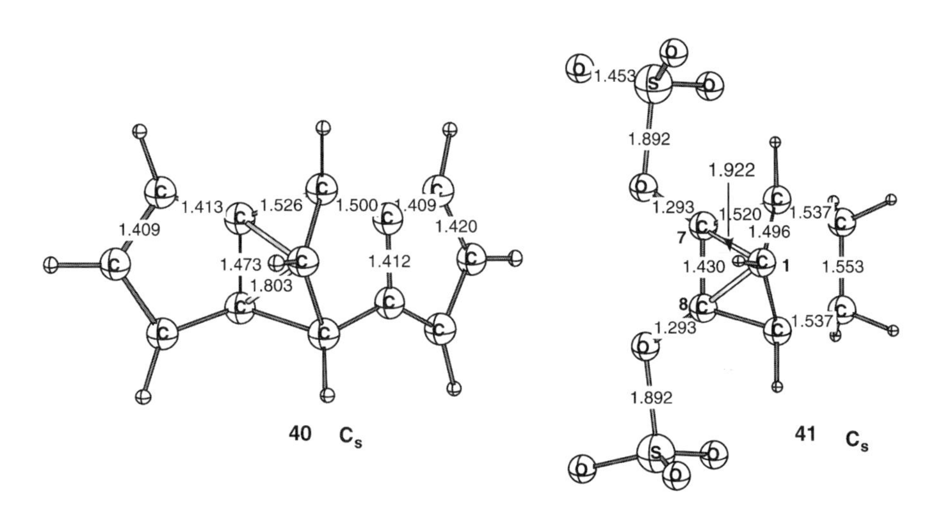

Negative natural charges (−0.7) reside on the two SO_4 groups in the zwitter-anionic carbocation, **41**; the nonclassical bridging moiety has large positive charge, 0.95. Both the C1–C7 (1.922 Å) and the C7–C8 distances (1.430 Å) in **41** are somewhat longer than those in the 7-norbornenyl cation (1.749 and 1.397 Å, respectively).

3.5 CONCLUDING REMARKS AND OUTLOOK

Neutral and negatively charged zwitter species can have the geometric, spectroscopic (e.g., NMR chemical shifts), and electronic structure features that characterize both classical and nonclassical carbocations. While a few experimental examples are known, many possibilities exist. They simply have not been explored.

The existence of "long" CC bonds well over 2 Å in length is an additional point to be stressed. If molecules only of the conventional organic type are investigated, only CC bonds of conventional lengths can be expected. The long CC separations in nonclassical carbocations are not peculiarities of an exotic class of chemical species, but represent structural features that may be found far more generally if only one looks for them.

ACKNOWLEDGMENTS

We thank George Olah and Surya Prakash most warmly for inspiring our studies, which are discussed in this chapter, and for their warm and enduring friendship. Paul Ankan (UGA) also provided for helpful discussions and suggestions. The work was supported by National Science Foundation Grant CHE-0209857.

REFERENCES

1. (a) G. A. Olah, *J. Org. Chem.* **66**, 5943 (2001); (b) G. A. Olah, *Angew. Chem., Int. Ed. Engl.* **34**, 1393 (1995); (c) G. K. S. Prakash and P. v. R. Schleyer, eds., *Stable Carbocation Chemistry*, Wiley-Interscience, New York, 1997; (d) H. C. Brown, *The Nonclassical Ion Problem*, with comments by P. v. R. Schleyer, Plenum Press, New York, 1977.
2. (a) The term "nonclassical ions" was first used by J. D. Roberts, and was defined by Paul Bartlett as containing too few electrons to allow a pair for each "bond"; see P. D. Bartlett, *Nonclassical Ions*, W. A. Benjamin, New York, 1965; (b) J. D. Roberts and R. H. Mazur, *J. Am. Chem. Soc.* **73**, 3542 (1951).
3. (a) M. Unverzagt, G. Subramanian, M. Hofmann, P. v. R. Schleyer, S. Berger, K. Harms, W. Massa, and A. Berndt, *Angew. Chem. Int. Ed. Engl.* **36**, 1469 (1997); (b) A. Berndt, D. Steiner, D. Schweikart, C. Balzereit, M. Menzel, H. J. Winkler, S. Mehle, M. Unverzagt, T. Happel, P. v. R. Schleyer, G. Subramanian, and M. Hofmann, in *Advances in Boron Chemistry*, W. Seibert, ed., Cambridge Univ. Press, Cambridge, UK, p. 61 (1997); (c) P. v. R. Schleyer, H. Jiao, M. N. Glukhovtsev, J. Chandrasekhar, and E. Kraka, *J. Am. Chem. Soc.* **116**, 10129 (1994); (d) J. Chandrasekhar, E. D. Jemmis, and P. v. R. Schleyer, *Tetrahedron Lett.* 3707 (1979); (e) R. L. Disch, M. L. Sabio, and J. M. Schulman, *Tetrahedron Lett.* **24**, 1863 (1983); (f) A. A. Fokin, B. Kiran, M. Bremer, X. Yang, H. Jiao, P. v. R. Schleyer, and P. R. Schreiner, *Chem. Eur. J.* **6**, 1615 (2000).

4. W. Schäfer and H. Hellmann, *Angew. Chem.*, **79**, 566 (1967).
5. G. M. Whitesides and W. J. Ehmann, *J. Am. Chem. Soc.* **91**, 3800 (1969).
6. J. B. Koster, G. J. Timmermans, and H. van Bekkum, *Synthesis* 139 (1971).
7. C. Krüger, P. J. Robers, Y.-H. Tsay, and J. B. Coster, *J. Organomet. Chem.* **78**, 69 (1974).
8. D. A. Applequist and J. D. Roberts, *J. Am. Chem. Soc.* **81**, 6523 (1959).
9. (a) W. H. Knoth, J. C. Sauer, H. C. Miller, and E. J. Muetterties, *J. Am. Chem. Soc.* **86**, 115 (1964); (b) W. H. Knoth, J. C. Sauer, J. H. Balthis, H. C. Miller, and E. J. Muetterties, *J. Am. Chem. Soc.* **89**, 115 (1967); (c) W. R. Hertler and M. S. Raasch, *J. Am. Chem. Soc.* **86**, 3661 (1964); (d) W. R. Hertler, *Inorg. Chem.* **3**, 1195 (1964); (e) H. C. Miller, W. R. Hertler, E. L. Muetteties, W. H. Knoth, and N. E. Miller, *Inorg. Chem.* **4**, 1216 (1965); (f) S. A. Jasper, Jr. R. B. Jones, J. Mattern, J. C. Huffmann, and L. J. Todd, *Inorg. Chem.* **33**, 5620 (1994); (g) H. C. Miller, N. E. Miller, and E. L. Muetteties, *J. Am. Chem. Soc.* **85**, 3885 (1963); (h) H. C. Miller, N. E. Miller, and E. L. Muetteties, *Inorg. Chem.* **3**, 1456 (1964); (i) W. H. Knoth, J. C. Sauer, D. C. England, W. R. Hertler, and E. J. Muetterties, *J. Am. Chem. Soc.* **86**, 3973 (1964); (j) J. Wright and A. Kaczmarczyk, *Inorg. Chem.* **12**, 1453 (1973); (k) E. J. M. Hamilton, G. T. Jordan, IV, E. A. Meyers, and S. G. Shore, *Inorg. Chem.* **35**, 5335 (1996); (l) R. G. Kultyshev, J. Liu, E. A. Meyers, and S. G. Shore, *Inorg. Chem.* **38**, 4913 (1999); (m) O. Tutusaus, F. Teixidor, R. Núñez, C. Viñas, R. Sillanpää, and R. Kivekäs, *J. Organomet. Chem.* **657**, 247 (2002) and reference cited therein.
10. (a) J. Stieglitz, *Am. Chem. J.* **21**, 110 (1899); (b) J. F. Norris and W. W. Sanders, *Am. Chem. J.* **25**, 54 (1901); (c) J. F. Norris, *Am. Chem. J.* **25**, 117 (1901); (d) A. v. Baeyer and V. Villiger, *Chem. Ber.* **35**, 1189 (1902); (e) M. Gomberg, *Chem. Ber.* **35**, 2397 (1902).
11. R. Santillan, N. Farfán, D. Castillo, A. Gutiérrez, and H. Höpfl, *Chem. Eur. J.* **4**, 1904 (1998).
12. W. J. Evans, K. J. Forrestal, and J. M. Ziller, *J. Am. Chem. Soc.* **117**, 12635 (1995).
13. (a) S. Winstein, M. Shatavsky, C. Norton, and R. B. Woodward, *J. Am. Chem. Soc.* **77**, 4183 (1955); (b) W. G. Woods, R. A. Carboni, and J. D. Roberts, *J. Am. Chem. Soc.* **78**, 5653 (1956); (c) S. Winstein and E. T. Stafford, *J. Am. Chem. Soc.* **79**, 505 (1957); (d) P. R. Story and M. Saunders, *J. Am. Chem. Soc.* **82**, 6199 (1960); (e) G. A. Olah and G. Liang, *J. Am. Chem. Soc.* **97**, 6803 (1975); (f) R. Houriet, T. Schwitzguebel, P.-A. Carrupt, and P. Vogel, *Tetrahedron Lett.* **27**, 37 (1986).
14. (a) T. Laube, *Angew. Chem. Int. Ed. Engl.* **25**, 349 (1986); (b) T. Laube, *Angew. Chem.* **99**, 580 (1987); (c) T. Laube, *J. Am. Chem. Soc.* **111**, 9224 (1989); (d) S. Hollenstein and T. Laube, *J. Am. Chem. Soc.* **115**, 7240 (1993); (e) T. Laube, E. Bannwart, and S. Hollenstein, *J. Am. Chem. Soc.* **115**, 1731 (1993); (f) T. Laube and C. Lohse, *J. Am. Chem. Soc.* **116**, 9001 (1994); (g) T. Laube, *Helv. Chim. Acta* **77**, 943 (1994); (h) T. Laube and S. Hollenstein, *Helv. Chim. Acta* **77**, 1773 (1994); (i) T. Laube and E. Schaller, *Acta Crystallogr.* **B51**, 177 (1995); (j) T. Laube, *Acc. Chem. Res.* **28**, 399 (1995).
15. (a) L. Schafer, K. Siam, J. D. Ewbank, and E Osawa, *J. Mol Struct.* (*Theochem*) **32**, 125 (1986); (b) D. A. Dougherty, C. S. Choi, G. Kaupp, A. B. Buda, J. M. Rudzinski, and E. Osawa, *J. Chem. Soc. Perkins Trans. 2* 1063 (1986); (c) E. Osawa, P. M. Ivanov, and C. Jaime, *J. Org. Chem.* **48**, 3990 (1983); (d) E. Osawa, Y. Onuki, and K. Mislow, *J. Am. Chem. Soc.* **103**, 7475 (1981); (e) K. Harano, T. Ban, M. Yasuda, and E. Osawa, *J. Am. Chem. Soc.* **103**, 2310 (1981); (f) T. Suzuki, K. Ono, J. Nishida, H. Takahashi, and T. Tsuji, *J. Org. Chem.* **69**, 4944 (2000); (g) K. K. Baldridge, T. R. Battersby, R. VernonClark, and J. S. Siegel, *J. Am. Che. Soc.* **119**, 7048 (1997); (h) J. Cioslowski, S. T. Mixon, and W. D. Edwards, *J. Am. Chem. Soc.* **113**, 1083 (1991); (i) P. Maslak, J. N. Narvaez, and M. Parvez,

J. Org. Chem. **56**, 602 (1991); (j) P. H. M. Budzelaar, E. Kraka, D. Cremer, and P. v. R. Schleyer, *J. Am. Chem. Soc.* **108**, 561 (1986).

16. F. Stahl, P. v. R. Schleyer, H. Jiao, H. F. Schaefer III, K.-H. Chen, and N. L. Allinger, *J. Org. Chem.* **67**, 6599 (2002).
17. (a) Z.-X. Wang and P. v. R. Schleyer, *J. Am. Chem. Soc.* **124**, 11979 (2002); (b) Z.-X. Wang and P. v. R. Schleyer, *J. Am. Chem. Soc.* **123**, 994 (2001).
18. E. G. N. Muller and R. S. Mulliken, *J. Am. Chem. Soc.* **80**, 3489 (1958).
19. P. v. R. Schleyer and C. Maerker, *Pure Appl. Chem.* **67**, 755 (1995) and references cited therein.
20. (a) S. Winstein and D. S. Trifan, *J. Am. Chem. Soc.* **71**, 2953 (1949); (b) J. D. Roberts and C. C. Lee, *J. Am. Chem. Soc.* **73**, 5009 (1951); (c) S. Winstein and D. S. Trifan, *J. Am. Chem. Soc.* **74**, 1147 (1952); (d) S. Winstein and D. S. Trifan, *J. Am. Chem. Soc.* **74**, 1154 (1952); (e) J. D. Roberts, C. C. Lee, and W. H. Saunders, *J. Am. Chem. Soc.* **76**, 4501 (1954); (f) R. K. Lustgarten, M. Brookhart, S. Weinstein, P. G. Gassman, D. S. Patton, H. G. Richey Jr., and J. D. Nichols, *Tetrahedron Lett.* 1699 (1970); (g) H. C. Brown, *The Nonclassical Ion Problem*, with comments by P. v. R. Schleyer, Plenum Press, New York, 1977; (h) M. Saunders and M. R. Kates, *J. Am. Chem. Soc.* **102**, 6867 (1980); (i) G. A. Olah, G. K. S. Prakash, M. Arvanaghi, and F. A. L. Anet, *J. Am. Chem. Soc.* **104**, 7105 (1982); (j) R. M. Jarret and M. Saunders, *J. Am. Chem. Soc.* **109**, 3366 (1987); (k) C. S. Yannoni, V. Macho, and P. C. Myhre, *J. Am. Chem. Soc.* **104**, 7380 (1982); (l) W. Koch, B. Liu, and D. J. DeFrees, *J. Am. Chem. Soc.* **111**, 1527 (1989); (m) W. Koch, B. Liu, D. J. DeFrees, D. E. Sunko, and H. Vancik, *Angew. Chem. Int. Ed. Engl.* **29**, 183 (1990); (n) P. C. Mhyre, G. C. Webb, and C. S. Yannoni, *J. Am. Chem. Soc.* **112**, 8991 (1990).
21. (a) M. Saunders, P. v. R. Schleyer, and G. A. Olah, *J. Am. Chem. Soc.* **86**, 5680 (1964); (b) G. A. Olah, G. D. Mateescu, L. A. Wilson, and M. H. Gross, *J. Am. Chem. Soc.* **92**, 7231 (1970); (c) G. A. Olah, G. D. Mateescu, and J. L. Riemenschneider, *J. Am. Chem. Soc.* **94**, 2529 (1972); (d) S. A. Johnson and D. T. Clark, *J. Am. Chem. Soc.* **110**, 4112 (1988).
22. (a) S. Winstein and C. Ordronneau, *J. Am. Chem. Soc.* **82**, 2084 (1960); (b) P. R. Story and M. Saunders, *J. Am. Chem. Soc.* **84**, 4876 (1962); (c) H. G. Richey Jr. and R. K. Lustgarten, *J. Am. Chem. Soc.* **88**, 3136 (1966); (d) M. Brookhart, A. Diaz, and S. Winstein, *J. Am. Chem. Soc.* **88**, 3135 (1966); (e) Brookhart, R. K. Lustgarten, and S. Winstein, *J. Am. Chem. Soc.* **89**, 6352 (1967); (f) G. A. Olah, A. M. White, J. R. DeMember, A. Commeyras, and C. Y. Lui, *J. Am. Chem. Soc.* **92**, 4627 (1970) and references cited therein; (g) M. Bremer, K. Schötz, P. v. R. Schleyer, U. Fleischer, M. Schindler, W. Kutzelnigg, W. Koch, and P. Pulay, *Angew. Chem. Int. Ed. Engl.* **28**, 1042 (1989); (h) G. A. Olah and S. Yu, *J. Org. Chem.* **40**, 3638 (1975).
23. (a) E. M. Arnett, E. B. Troughton, A. T. McPhail, and K. E. Molter, *J. Am. Chem. Soc.* **105**, 6172 (1983); (b) E. M. Arnett, *J. Chem. Ed.* **62**, 385 (1985); (c) E. M. Arnett and K. E. Molter, *Acc. Chem. Res.* **18**, 339 (1985).
24. (a) J. M. Schulman, R. L. Disch, P. v. R. Schleyer, M. Bühl, M. Bremer, and W. Koch, *J. Am. Chem. Soc.* **114**, 7897 (1992); (b) D. Scheschkewitz, A. Ghaffari, P. Amseis, M. Unverzagt, G. Subramanian, M. Hofmann, P. v. R. Schleyer, H. F. Schaefer III, G. Geiseler, W. Massa, and A. Berndt, *Angew. Chem. Int. Ed. Engl.* **39**, 1272 (2000); (c) W. Lösslein, H. Pritzkow, P. v. R. Schleyer, L. R. Schmitz, and W. Siebert, *Angew. Chem. Int. Ed. Engl.* **39**, 1276 (2000); (d) R. R. Sauers, *Tetrahedron Lett.* **42**, 6625 (2001).
25. (a) P. J. Fagan, E. G. Burns, and J. C. Calabrese, *J. Am. Chem. Soc.* **110**, 2979 (1988); (b) P. G. Fagan, W. A. Nugent, and J. C. Calabrese, *J. Am. Chem. Soc.* **116**, 1880 (1994).

26. C. S. Wannere and P. v. R. Schleyer, unpublished results.
27. P. v. R. Schleyer, R. C. Fort, W. E. Watts, M. B. Comisarow, and G. A. Olah, *J. Am. Chem. Soc.* **86**, 4195 (1964).
28. G. A. Olah, G. Rasul, and G. K. S. Prakash, *J. Org. Chem.* **65**, 5956 (2000).
29. W. B. Scott and R. E. Pincock, *J. Am. Chem. Soc.* **95**, 2040 (1973).
30. (a) S. Masamune, M. Sakai, A. V. K. Jones, and T. Nakashima, *Can. J. Chem.* **52**, 855 (1974); (b) G. A. Olah, G. K. S. Prakash, T. Rawdah, D. Whittaker, and J. C. Rees, *J. Am. Chem. Soc.* **101**, 3935 (1979).
31. (a) C. Maerker and P. v. R. Schleyer, unpublished results; (b) A. Nickon and J. L. Lambert, *J. Am. Chem. Soc.* **84**, 4604 (1964); (c) A. Nickon, J. H. Hammons, J. L. Lambert, and R. O. Williams, *J. Am. Chem. Soc.* **85**, 3713 (1965); (d) A. Nickon, J. L. Lambert, and J. E. Oliver, *J. Am. Chem. Soc.* **88**, 2787 (1966); (e) A. Nickon, J. L. Lambert, R. O. Williams, and N. H. Werstiuk, *J. Am. Chem. Soc.* **88**, 3354 (1966); (f) A. Nickon, T. Nishida, and Y. Lin, *J. Am. Chem. Soc.* **91**, 6860 (1969).
32. (a) L. Ruzicka and H. F. Meddahl, *Helv. Chim. Acta* **21**, 1760 (1938); (b) L. Ruzicka and H. F. Meddahl, *Helv. Chim. Acta* **22**, 421 (1939); (c) L. Ruzicka and H. F. Meddahl, *Helv. Chim. Acta* **23**, 364 (1940); (d) D. N. Kirk and M. P. Hartshorn, *Steroid Reaction Mechanism*, Elsevier, Amsterdam, 1968, p. 388; (e) C. D. Gutsche and D. Redmore, *Carbocyclic Ring Expansion Reactions*, Academic Press, New York, 1968; (f) L. F. Fieser and M. Fieser, *Steroids*, Reinhold, New York, 1959, p. 577; (g) C. J. Collins and J. F. Eastham, in *The Chemistry of the Carbonyl Group*, S. Patai, ed., Wiley, London, 1966, p. 761.
33. (a) P. K. Freeman and D. G. Kuper, *J. Org. Chem.* **30**, 1047 (1965); (b) J. M. Wheeler, R. H. Chung, Y. N. Vaishnav, and C. C. Shroff, *J. Org. Chem.* **34**, 545 (1969); (c) J. E. Baldwin and H. C. Krauss, Jr., *J. Org. Chem.* **35**, 2426 (1970); (d) P. K. Freeman, T. A. Hardy, R. S. Raghavan, and D. G. Kuper, *J. Org. Chem.* **42**, 3882 (1977); (e) P. K. Freeman, R. S. Raghavan, and D. G. Kuper, *J. Am. Chem. Soc.* **93**, 5288 (1971); (f) P. K. Freeman, *J. Am. Chem. Soc.* **120**, 1619 (1998); (g) P. K. Freeman and J. K. Pugh, *J. Org. Chem.* **65**, 6107 (2000).
34. (a) F. Geliter and R. Hoffmann, *J. Am. Chem. Soc.* **90**, 5457 (1968); (b) M. W. Wang and C. Wentrup, *J. Org. Chem.* **61**, 7022 (1996); (c) R. A. Moss, U.-H. Dolling, and J. R. Whittle, *Tetrahedron Lett.* 931 (1971).
35. (a) R. A. Moss and U.-H. Dolling, *Tetrahedron Lett.* 5117 (1972); (b) R. A. Moss and C.-T. Ho, *Tetrahedron Lett.* 1651 (1976); (c) W. Kirmse and T. Meinert, *J. Chem. Soc. Chem. Commun.* 1065 (1994); (d) U. H. Brinker and J. Ritzer, *J. Am. Chem. Soc.* **103**, 2116 (1981).
36. (a) V. I. Minkin, R. M. Minyaev, I. I. Zacharov, and V. I. Avdeev, *Zh. Org. Khim.* **3**, 14 (1978); (b) E. D. Jemmis and P. v. R. Schleyer, *J. Am. Chem. Soc.* **104**, 4781 (1982); (c) V. I. Minkin, R. M. Minyaev, and G. V. Orlova, *J. Mol. Struct.* (*Theochem*) **110**, 241 (1984); (d) V. Balaji and J. Michl, *Pure Appl. Chem.* **60**, 189 (1988); (e) E. Lewars, *J. Mol. Struct.* (*Theochem*) **423**, 173 (1998); (f) E. Lewars, *J. Mol. Struct.* (*Theochem*) **507**, 165 (2000); (g) J. P. Kenny, K. M. Krueger, J. C. Rienstra-Kiracofe, and H. F. Schaefer III, *J. Phys. Chem. A* **105**, 7745 (2001); (h) V. I. Minkin, R. M. Minyaev, and R. Hoffmann, *USP Khim.* **71**, 989 (2002).
37. W. D. Stohrer and R. Hoffmann, *J. Am. Chem. Soc.* **94**, 1661 (1972); this species was predicted independently by R. E. Williams, *Inorg. Chem.* **10**, 210 (1971).

38. (a) S. Masamune, M. Sakai, H. Ona, and A. J. Jones, *J. Am. Chem. Soc.* **94**, 8956 (1972); (b) S, Masamune, *Pure Appl. Chem.* **44**, 861 (1975).
39. (a) G. A. Olah, J. S. Staral, and G. Liang, *J. Am. Chem. Soc.* **96**, 6233 (1974); (b) G. A. Olah and J. S. Staral, *J. Am. Chem. Soc.* **98**, 6290 (1976).
40. See, for example, R. V. Williams, *Chem. Rev.* **101**, 1185 (2001).
41. (a) M. Iyoda, H. Kurata, M. Oda, C. Okubo, and K. Nishimoto, *Angew. Chem. Int. Ed. Engl.* **32**, 89 (1993); (b) for aromaticity and electron affinity of various [3]radialenes, see C. Lepetit, M. B. Nielsen, F. Diederich, and R. Chauvin, *Chem. Eur. J.* **9**, 5056 (2003).
42. (a) L. Gmelin, *Ann. Phys.* (Leipzig) **4**(2), 31 (1825); (b) J. F. Heller, *Justus Liegigs Ann. Chem.* **24**, 1 (1837).
43. (a) S. Cohen, J. R. Lacher, and J. D. Park, *J. Am. Chem. Soc.* **81**, 3480 (1959); (b) D. Eggerding and R. West, *J. Am. Chem. Soc.* **97**, 207 (1975); (c) D. Eggerding and R. West, *J. Am. Chem. Soc.* **98**, 3641 (1976).
44. (a) G. Seitz and P. Imming, *Chem. Rev.* **92**, 1227 (1992); (b) S. Fukuzumi and T. Yorisue, *J. Am. Chem. Soc.* **113**, 7764 (1991); (c) B. Zhao and M. H. Back, *Can. J. Chem.* **70**, 135 (1992).
45. P. v. R. Schleyer, K. Najafian, B. Kiran, and H. Jiao, *J. Org. Chem.* **65**, 426 (2000).
46. Z. Chen, L. R. Sutton, D. Moran, A. Hirsch, W. Thiel, and P. v. R. Schleyer, *J. Org. Chem.* **68**, 8808 (2003).

4

RECENT STUDIES OF LONG-LIVED CARBOCATIONS AND CARBODICATIONS

G. K. Surya Prakash

Loker Hydrocarbon Research Institute and Department of Chemistry
University of Southern California
Los Angeles, California

V. Prakash Reddy

Department of Chemistry
University of Missouri—Rolla
Rolla, Missouri

Carbocation Chemistry, Edited by George A. Olah and G. K. Surya Prakash
ISBN 0-471-28490-4

4.1 INTRODUCTION

The field of carbocation chemistry has come a long way since its discovery more than 100 years ago. The significance of the carbocations as true reaction intermediates in acid catalyzed processes has been highlighted with the award of Nobel Prize to Professor George A. Olah in 1994.[1] The primary focus of the current review is on the more recent work in the area of long-lived carbocations and carbodications, carried out in collaboration with Professor George Olah and other investigators. In this review, we have discussed a broad variety of carbocations, such as sterically crowded carbocations, nonclassical and classical cyclopropylmethyl cations (including Cram's phenonium ions) and dications, cyclobutylmethyl cations, dienylic and allylic dications, cyclobutylmethyl cations, and hexaaryl(trimethylene)methane dications.

We have included discussions on a nonclasical triaxane-2-methyl cation, a novel cyclopropylmethyl cation. Also discussed in depth is the formation of the otherwise highly unstable cyclobutylmethyl cation as a long-lived carbocation that has been established by its preparation at low temperatures in superacidic media. The cyclobutylmethyl cation has been shown to exist preferentially in the bisected conformation from its ^{13}C NMR data and the measured rotational barrier. The rotational barriers have also been measured for the highly hindered 2-(adamantylidenemethyl)-2-adamantyl allyl cation by dynamic NMR spectroscopy. Studies on long-lived dienylic-allylic dications, nonclassical 2,10-*para*[$3^2.5^6$]octahedrane dication, a cyclopropyl-group-stabilized 1,3 carbodication and aryl-substituted trimethylenemethane dications are covered in some detail. Our collaborative studies with Prinzbach on polycyclic [1.1.1.1] and [2.2.1.1]isopagodyl dications, however, are not covered.[2,3] Cyclopropyl cation stabilized by ferrocenyl group, and β-silyl carbocations have also been reviewed earlier,[4] and will not be discussed in the current review.

For clarity of description, we have emphasized on the ^{13}C NMR data in the assignment of the structures to the carbocations. The total chemical shift difference criterion, originally proposed by Shleyer, Prakash, and Olah,[5] is applied to many of these carbocationic and carbodicationic species to probe the nature of these cations and to establish the extent of the neighboring group participation. Thus, the summation of the ^{13}C NMR chemical shifts of the classical carbocations, as compared with those of their corresponding neutral hydrocarbons, are typically very large, and their difference corresponds to about 350 ppm (or more). The difference in the summation of the chemical shifts for the nonclassical carbocations, on the other hand, is mush smaller, usually less than 200 ppm. Using this crieterion, the classical trivalent nature of a 1,3-carbodication and a 1,4-carbodication, both involving neighboring cyclopropyl group stabilization is confirmed.

4.2 STERICALLY CROWDED CARBOCATIONS

The study of sterically crowded carbocations is challenging, as the carbocationic center is far removed from the counteranion in such structures. In addition, the solvation of the cationic centers is minimized because of the unfavorable steric interactions between the carbocation and the solvent species. Among the most hindered carbocations, tris(1-adamantyl)methyl cation (**1**) has been prepared by the ionization of the corresponding alcohol in superacidic media at low temperatures, and characterized by ^{13}C NMR spectroscopy. The carbocation shows a deshielded carbocationic center at δ^{13}C 327.1.[6] Whereas the less crowded secondary carbocation, bis(1-adamantly)methyl cation, undergoes rearrangement partially to the 4-(1-adamantyl)-3-homoadamantyl cation,[7] such ring expansion rearrangement is not possible in the tris(1-adamantyl)methyl cation because of the steric hindrance. Below we discuss some of the recently prepared sterically hindered carbocations.

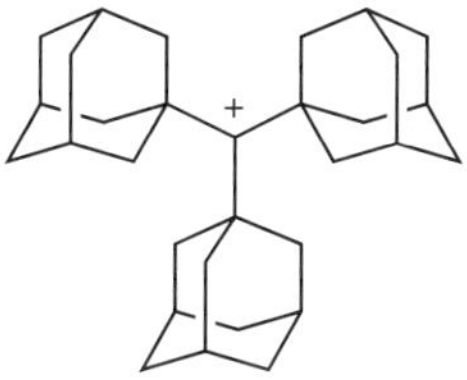

δ ^{13}C 327.1
1

4.2.1 1,1′-Diadamantylbenzyl Cations

Steric hindrance in cumyl-type carbocations (1,1-dialkylarylmethylium cations; e.g., **2**) causes decreased resonance stabilization from the aromatic ring, as the twisting of the ring supresses conjugation with the empty *p* orbital of the carbocationic center. In order to study the extent of π delocalization in such sterically crowded carbocations, a series of *para*-substituted 1,1′-diadamantylbenzyl cations (**3**) were prepared in FSO_3H/SO_2ClF at low temperatures, and the substituent effects on the δ^{13}C of the carbocationic centers were systematically probed using the Hammett-type plot: the 1,1′-diadamantylbenzyl cations showed negligible slope as compared to a similar plot for the 1-aryl-1-cyclopentyl cations, indicating virtual absence of the π-resonance stabilization in the sterically crowded 1,1′-diadamantylbenzyl cations, **3**. The relative absence of the π-delocalization is also indicated by the ^{13}C NMR chemical shifts for the carbocationic centers. A δ^{13}C (C+) value of 286.5 was observed for the 1,1′-diadamantylbenzyl cation (**3**, R = H), which is much deshielded as compared to that of the cumyl cation, **2** (δ^{13}C 255). In agreement with this conclusion the *para*-carbon chemical shifts are virtually unaltered with respect to those of the corresponding alcohol in case of the 1,1′-diadamantylbenzyl cation (**3**, R = H). To compensate for the lack of resonance stabilization, significant hyperconjugative interactions through the adamantly moiety are involved, as evidenced by the ^{13}C NMR chemical shifts of the adamantyl ring carbons.[8] The (*o*-tolyl)di-1-adamantylmethylium cation (**3**; R = *o*-Me), a highly crowded benzylic carbocation, in particular, shows the most deshielded carbocationic center for a tertiary benzylic cation (δ^{13}C 300.7).[9]

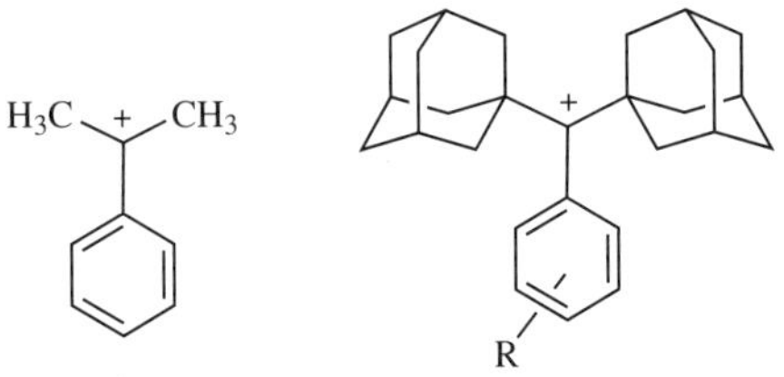

H_3C + CH_3

R

2
δ^{13}C = 255.0

3 (R = H, *o*-Me, *p*-OMe, etc.)
δ^{13}C (R = H) = 286.5
δ^{13}C (R = *o*-Me) = 300.7

4.2.2 2-(Adamantylidenemethyl)-2-adamantyl Cation

A highly hindered allyl cation involving the adamantly rings, the 2-(adamantylidenemethyl)-2-adamantyl cation (**5**), has been prepared by the ionization of the corresponding allyl alcohol (**4**) in either neat FSO_3H or FSO_3H in SO_2ClF as the solvent. The allyl cation **5** is shown to be stable up to +80°C.[10]

OH
FSO_3H
1 2 3 4 8 9 10 +

4 **5**

In the allyl cation **5**, there is a restricted rotation around the allylic bonds resulting in the distinct absorptions for the α-bridgehead allylic carbons (δ^{13}C 53.0 and 47.1 for C1 and C3, respectively), and the β-methylene carbons [δ^{13}C 45.1 (C8, C9) and 44.9 (C4, C10)]. The ^{13}C NMR spectrum of the carbocation is unchanged even upon warming the carbocation solution to room temperature. On further heating to 50°C (neat FSO_3H), the *cis*- and *trans*-β-methylene signals [(C4, C10) and (C8, C9), respectively] coalesced to give a single sharp signal, and the intensity of the

C1 and C3 signals was significantly reduced. On further heating of the solution in the NMR spectrometer, the C1 and C3 signals continued to decrease in intensity and merged into the baseline at 80°C. On the basis of the coalescence of the β-methylene signals at 50°C, a rate constant of 28.8 s^{-1} and a $\Delta G^{\#}$ of 16.8 kcal/mol (at 50°C) was estimated for the rotation around the allylic bonds. The estimated barrier is in accordance with the theoretical calculations (at semiempirical levels), and is only slightly higher than that of related 1,1,3,3-tetramethylallyl cation.

Of more interest, the allyl cation (**5**), despite its high inherent steric strain, could be prepared under superacidic conditions, and is a highly stable species even at elevated temperatures.

4.2.3 Tris(1-naphthyl)- and Tris(2-naphthyl)methyl Cations

The highly crowded triarylmethyl cations tris(1-naphthyl)methyl cation (**7**) and tris(2-napthyl)methyl cation (**9**) have been prepared under long-lived conditions and characterized by ^{13}C NMR spectroscopy at low temperatures. The tris(2-napthyl) methyl cation (**9**) can abstract hydride ion from the cycloheptatriene to give the tris (2-naphthyl)methane (**10**), and tropylium ion (**11**), while the tris(1-naphthyl)methyl cation (**7**) does not react with the cycloheptatriene, indicating the more crowded nature of the carbocation **7** as compared to the carbocation **9**.[11] In order to better understand the nature of the hydride abstraction, a primary kinetic deuterium isotope effect was measured by treating **9** with an equimolar mixture of cycloheptatriene and cycloheptatriene-d_8. Analysis of the product mixture by ^{1}H NMR revealed $k_H/k_D = 7.1 + 0.5$, indicating a collinear nature of hydride transfer.[12]

FSO₃H/SO₂ClF, −78 to −20°C: **6** (COH) → **7** (C+)

FSO₃H/SO₂ClF, −78 to −20°C: **8** (COH) → **9** (C+)

9 + cycloheptatriene → **11** + **10** (CH)

4.3 CYCLOPROPYLMETHYL CATIONS

The cyclopropylmethyl cations have been among the most extensively investigated species. Thus, the solvolysis of cyclopropylmethyl and cyclobutyl substrates (e.g., **12**) gave similar composition of the cyclopropylmethanol (**16**), cyclobutanol (**17**), and homoallyl alcohol (**18**) products, irrespective of which substrate was used for the solvolysis. The enormous rate accelerations, combined with other labeling experiments, suggested the involvement of equilibrating bicyclobutonium structures (**13**–**15**) to the reaction intermediates. Historically, such cations were first named as "nonclassical" by J. D. Roberts.[13] Extensive kinetic, stereochemical, theoretical, and NMR studies confirmed the nonclassical nature (i.e., stabilization of the cationic centers through σ-delocalization) for the primary cyclopropylmethyl cations.[14]

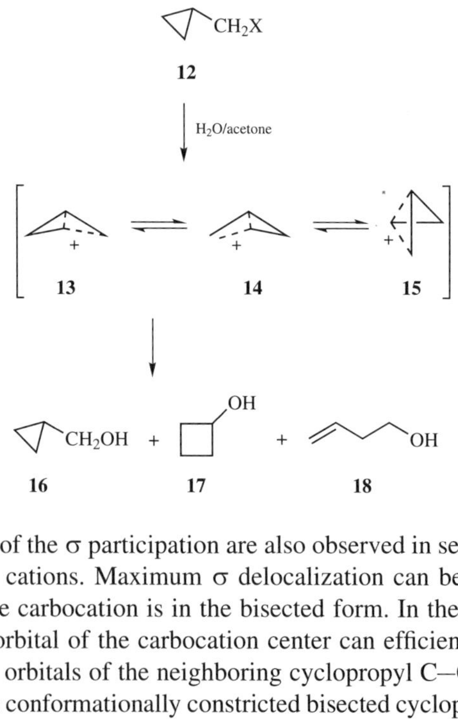

CH_2X

12

H_2O/acetone

13 **14** **15**

CH_2OH + OH + OH

16 **17** **18**

Varying extents of the σ participation are also observed in secondary and tertiary cyclopropylmethyl cations. Maximum σ delocalization can be involved when the conformation of the carbocation is in the bisected form. In the bisected conformation, the vacant *p* orbital of the carbocation center can efficiently overlap with the bonding molecular orbitals of the neighboring cyclopropyl C—C σ bonds. Some of the examples of the conformationally constricted bisected cyclopropylmethyl cations include 2-phenyl-8,9-dehydro-2-adamantyl cation (**19**), 2-cyclopropyl-8,9-dehydro-2-adamantyl cation (**20**), and 3-nortricyclyl cation (**21**). They have been prepared in superacidic media and thoroughly characterized by ^{1}H and ^{13}C NMR spectroscopy.[15]

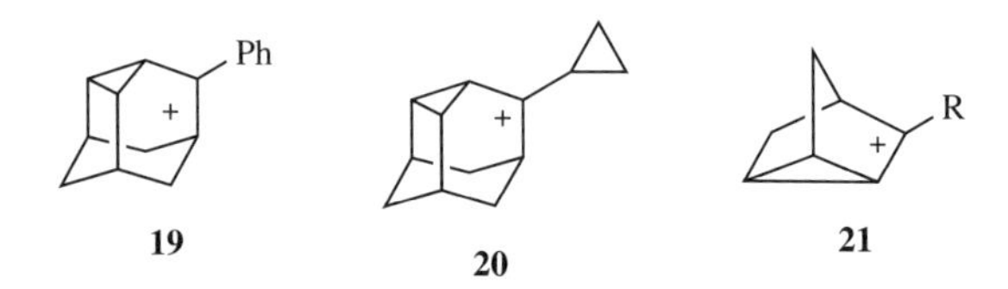

4.3.1 Triaxane-2-methyl Cation

Sorensen and coworkers prepared the primary, secondary, and tertiary nortricyclylmethyl cations (**22–24**) by ionizing the corresponding alcohols using SbF_5-SO_2ClF.[16] Due to the extensive charge delocalization these conformationally fixed bisected cyclopropylmethyl cations are highly stable; the primary carbocation (**22**) could be observed up to −20°C, the secondary carbocation (**23**) at room temperature, and the tertiary carbocation (**24**) was stable up to 100°C.

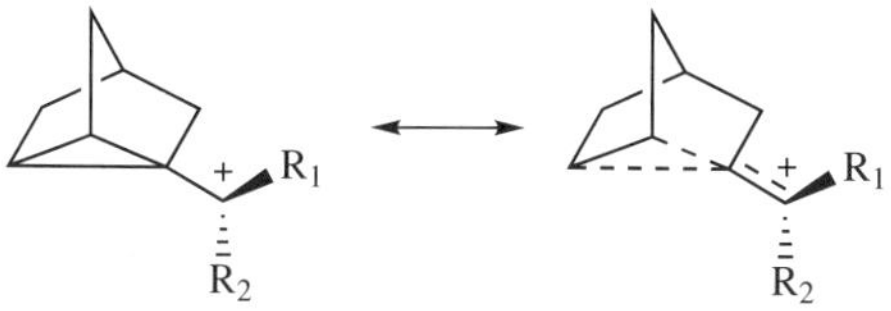

22 (R_1/R_2 = H)
23 (R_1 = H, R_2 = CH_3)
24 (R_1/R_2 = CH_3)

In the cyclopropylmethyl alcohol derivative, triaxane-2-methyl alcohol **25**, the cyclopropyl moiety, is also locked into a rigid hydrocarbon framework. Therefore, ionization of **25** under superacidic conditions has been explored.[17]

Triaxane-2-methyl alcohol **25** was ionized at −78°C in SO_2ClF with SbF_5, resulting in an orange solution. The 75 MHz proton-decoupled ^{13}C NMR spectrum of this species at −80°C shows only seven resonances (assigned on the basis of multiplicities and coupling constants): $\delta^{13}C$ 147.9 (s), 97.6 (d, J_{CH} = 187.1 Hz), 52.7 (t, J_{CH} = 134.7 Hz), 47.3 (t, J_{CH} = 168.3 Hz), 43.4 (d, J_{CH} = 151.4 Hz), 38.7 (t, J_{CH} = 136.5 Hz), and 36.6 (d, J_{CH} = 149.9 Hz) in an approximate ratio of 1 : 2 : 1 : 1 : 1 : 2 : 2, respectively. The chemical shifts remain constant over the temperature range of −100 to −40°C. This is in accordance with either a single ground-state structure for the ion or a set of rapidly equilibrating unsymmetrical bicyclobutonium ion structures. By similar studies on the parent cyclopropylmethyl cation, the fast temperature-dependent degenerate equilibration of the cyclopropylmethyl cation and bicyclobutonium ion is well established.[14] Since the ion has apparent C_s symmetry, possible structures for **26** include the static bisected cyclopropylmethyl cation **27**, a fast equilibrium of the nonclassical unsymmetrically bridged bicyclobutonium ions **28a** and **28b**, or a fast equilibrium of the classical 2,5-dehydro-3-protoadamantyl cations **29a** and **29b**.

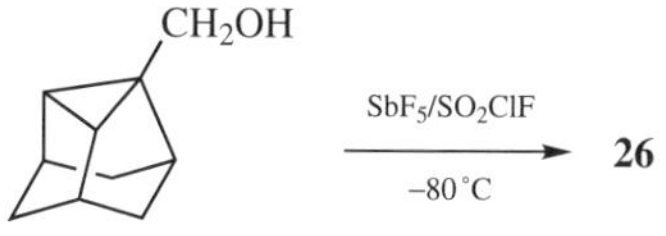

CH_2OH — SbF_5/SO_2ClF, −80°C → **26**

25

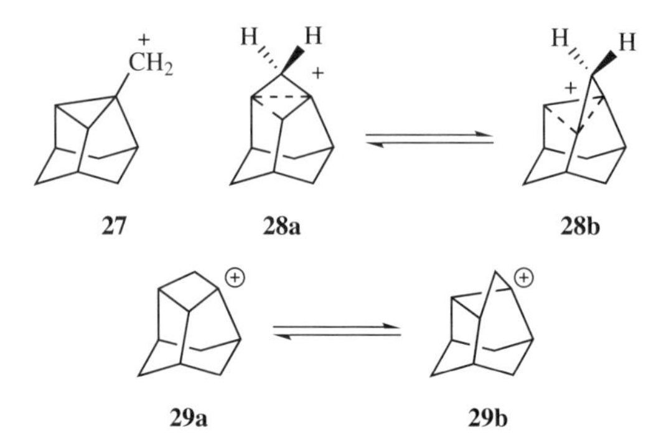

The geometries of these possible ions were optimized by DFT B3LYP/6-31G* level. The ^{13}C NMR chemical shifts were calculated by IGLO (individual gauge for localized orbitals).[17] By comparison of the calculated chemical shifts with the experimentally determined values the C_s symmetric structure **27** for **26** was ruled out. IGLO calculations at the DZ//B3LYP/6-31G* level gives a ^{13}C chemical shift of δ^{13}C 200.7 for the carbocationic center of the static bisected ion **27**, which is analogous to Sorensen's nortricyclylmethyl cation **24** (δ^{13}C 191.4). The difference of about 153 ppm between the IGLO calculated ^{13}C chemical shift for the carbocationic center in **27** and the experimentally observed shift of δ^{13}C 47.3 excludes **27** as the major species in the equilibrium. Applying the empirical ^{13}C NMR chemical shift additivity concept[5] the difference of $\Delta\delta^{13}\mathrm{C} = 226$ between the total sum of the ^{13}C NMR chemical shifts of the ion (638 ppm) and the total sum of the ^{13}C NMR chemical shifts of the corresponding hydrocarbon (412 ppm) further supports a nonclassical structure. Typically, $\Delta\delta$ of 350 ppm or greater indicate classical carbocationic nature, whereas $\Delta\delta$ values for nonclassical carbocations are significantly smaller. The most likely structure for **26** is probably a set of rapidly equilibrating bicyclobutonium ions (**28a/b**) with apparent C_s symmetry. This conclusion is also based on IGLO calculations. The protoadamantyl cations (**29a/b**) should show much more classical character.

4.3.2 3-Spirocyclopropyl-2-norbornyl Cations

The 3-spirocyclopropyl-2-norbornyl cation, **31**, a bisected cyclopropylmethyl cation, could not be obtained as a stable entity by ionization of the corresponding alcohol (**30**) under stable ion conditions, as it spontaneously rearranged to the allylic cation, 2-methylbicyclo[3.2.1]oct-3-en-2-yl cation (**32**).[18] The bisected nature of the carbocation **31** results in enhanced σ participation from the cyclopropyl group resulting in its rearrangement to the more stable allylic cation, **32**. However, tertiary 2-methyl-, 2-phenyl-, and 2-cyclopropyl-3-spirocyclopropyl-2-norbornyl cations (**33–35**) have been prepared through the ionization of their corresponding alcohols at low temperatures.[19] These tertiary carbocations are indeed observed as stable entities at relatively low temperatures. From the observed ^{13}C NMR chemical shifts of these cations it is evident that the spirocyclopropyl group exerts dominant neighboring group participation in stabilizing the carbocationic centers. The extent of the spirocyclopropyl group participation is directly related to the electron-releasing characteristics of the substitutents attached to the carbocationic center. For example, the 2-methyl derived carbocation **33** rearranges to the corresponding allyl cation, **36**, at −90°C, whereas the 2-phenyl derived carbocation, (**34**) is quite stable at −90°C, and it only rearranges to the allyl cation **37** at above −80°C. The highly stabilized 2-cyclopropyl- derived carbocation **35** is stable up to −20°C, above which it decomposes to as yet unidentified products. Although stabilization due to the spirocyclopropyl group in the carbocations **33–35** virtually eliminates the necessity for the C1,C6 σ-bond participation in the 2-norbornyl framework, such nonclassical σ participation still exists to a significant extent in the carbocations **33** and **34**, as evidenced by their ^{13}C chemical shift deshieldings.

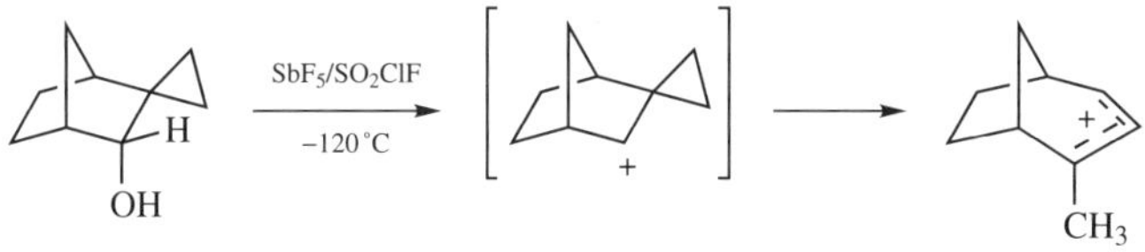

SbF_5/SO_2ClF, −120°C

30 **31** **32**

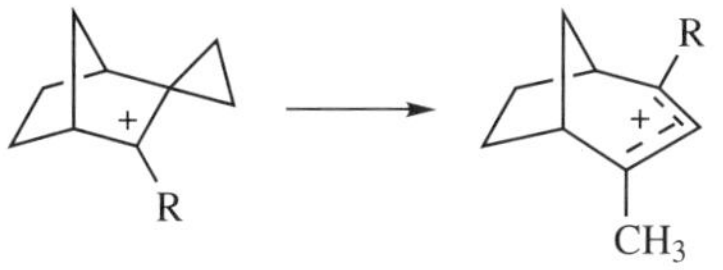

33 (R = Me) **36** (R = Me)
34 (R = Ph) **37** (R = Ph)
35 (R = cyclopropyl)

4.3.3 3-Spirocyclopropyl-2-bicyclo[2.2.2]octyl Cation

In contrast to the well-known 2-norbornyl cation, the analogous long-lived secondary 2-bicyclo[2.2.2]octyl cation (**39**) is not observed under stable ion conditions. All attempts of its preparation resulted in its spontaneous rearrangement to the thermodynamically more stable bicyclo[3.3.0]oct-1-yl cation (**40**), which further rearranges to 2-methyl-2-norbornyl cation (**41**) at above −10°C.[20]

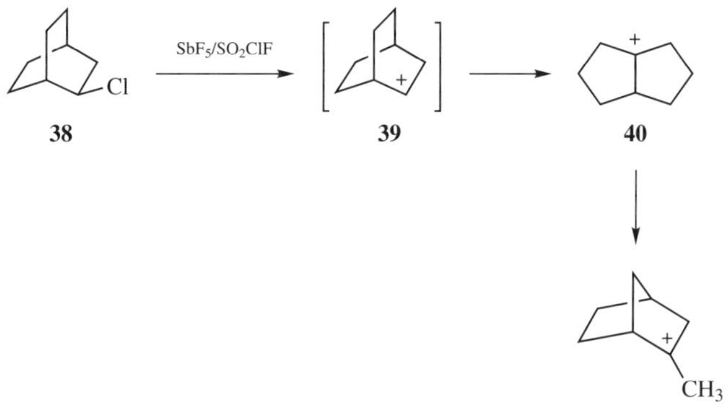

SbF_5/SO_2ClF

38 **39** **40**

41

The ionization of 1,3′-spirocyclopropylbicyclo[2.2.2]octan-2′-ol (**42**) also did not result in the formation of the expected 1,3′-spirocyclopropylbicyclo[2.2.2] oct-2′-yl cation (**43**), but instead it gave a rearranged allylic carbocation, the 2-methylbicyclo[3.2.2]-non-3-en-2-yl cation (**44**).[21] The rearrangement occurs through the ring opening of the neighboring spirocyclopropyl group. The allylic cation **44** shows ^{13}C NMR signals similar to those of 2-methylbicyclo[3.2.1]oct-3-en-2-yl cation (**32**). The formation of the carbocation **44** implies the involvement of the 1,3′-spirocyclopropylbicyclo[2.2.2]oct-2′-yl cation, **43**, as a highly energetic intermediate, insufficiently stabilized by the neighboring spirocyclopropyl group.

SbF_5/SO_2ClF

OH

CH_3

42 **43** **44**

4.3.4 Cram's Phenonium Ions

Ionization of β-phenylethyl systems involves neighboring π-orbital participation to give Cram's phenonium ions (e.g., **46**).[22] The phenonium ions were originally postulated as the reaction intermediates from extensive kinetic and stereochemical studies of Cram, and later thoroughly characterized in superacidic media by Olah and coworkers.[23,24] The proton-proton coupling constants observed for the methylene carbons in a derivative unsymmetrically substituted in the aromatic ring established the bisected structure of this phenonium ion. Olah, Prakash, and coworkers also obtained[25a] the phenonium ion **46** by protonation of benzocyclobutene (**47**) under stable ion conditions. On the basis of the observed ^{13}C NMR chemical shift data, total chemical shift difference criteria, theoretical calculations, and comparison with relevant models, the phenonium ion was shown to exist as the bisected classical cyclopropylcarbinyl cation (i.e., spirocyclopropylbenzenium ion, **46**), rather than the Freudenberg type ion (**48**), which was favored by Sieber, Schleyer, and Gauss' ab initio calculations (GIAO//MP2(fu)/6-31G*).[25b] The latter structure (**48**) can be considered to be a true nonclassical structure containing a hypercoordinate sp^2 carbon and displaying 6π-aromatic nature. In an attempt to prepare such a nonclassical phenonium ion 2-fluoro(tetrafluoro)benzonorbornene (**49**) containing strongly deactivated aromatic ring, was ionized under stable ion conditions, but it gave instead the corresponding tetrafluorobenzonortricyclyl cation (**50**). The observed ^{13}C NMR data for this carbocation strongly favored the classical (bisected) spirocyclopropylbenzenium ion, as shown in the structure for **50**.

CH_2CH_2Cl

HF/SbF_5 or SbF_5/SO_2ClF −80°C

HF/SbF_5–SO_2ClF −80°C

45 **46** **47**

48

F F F F F

SbF_5/SO_2ClF −80°C

F F F F

49 **50**

4.4 CYCLOBUTYLMETHYL CATIONS

The anchimeric assistance provided by the strained cyclobutyl ring in the solvolysis of the cyclobutylmethyl substrates was probed extensively through kinetic studies, and characterization of the products derived from the ring expansion rearrangements. Winstein and Holness proposed that a nonclassical cyclobutylmethyl cation was involved as the reaction intermediate in these solvolysis reactions.[26] Solvolytic studies by Dauben et al. and Wiberg and coworkers on the cyclobutylmethyl substrates such as bicyclo[2.2.0]hexane-1-methyl *p*-nitrobenzoate and bicyclo[2.1.1]-hexane-1-methyl tosylate further gave support for the intermediacy of a nonclassical carbocation.[27]

The cyclobutylmethyl cations could not be prepared as long-lived (persistent) species in superacidic media. In fact, all earlier attempts of preparation of the primary, secondary, and tertiary cyclobutylmethyl cations through the ionization of their corresponding alcohols or halides were not successful. The ionization of the cyclobutylmethyl chloride (**51**)[28] in SbF_5/SO_2ClF at −78°C, for example, resulted in the clean formation of the equilibrating fivefold degenerate 1-cyclopentyl cation (**52**). Similarly, the ionization of the 1-cyclobutylethanol (**53**), cyclobutylphenylmethanol (**54**), and cyclobutylcyclopropylmethanol (**55**) using SbF_5/SO_2ClF gave only the corresponding ring-expanded cations, 1-methylcyclopentylium (**59**), 1-phenylcyclopentylium (**60**), and 1-cyclopropylcyclopentylium (**61**) carbocations. Persistent secondary cyclobutylmethyl cations (**56–58**) could not be observed under the reaction conditions, apparently due to the large driving force for ring expansion rearrangements associated with the neighboring group participation of the cyclobutyl ring.[28] The ionization of the tertiary alcohol, cyclobutyldimethylmethanol (**62**), in superacidic media also resulted in the formation of the rearranged carbocation, the 1,2-dimethylcyclopentylium cation (**64**).[28]

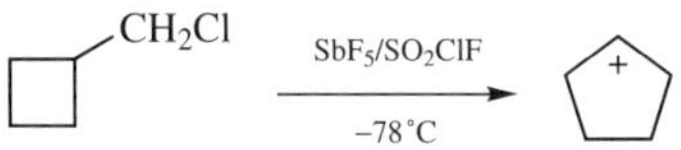

CH_2Cl —SbF_5/SO_2ClF, −78°C→

51 **52**

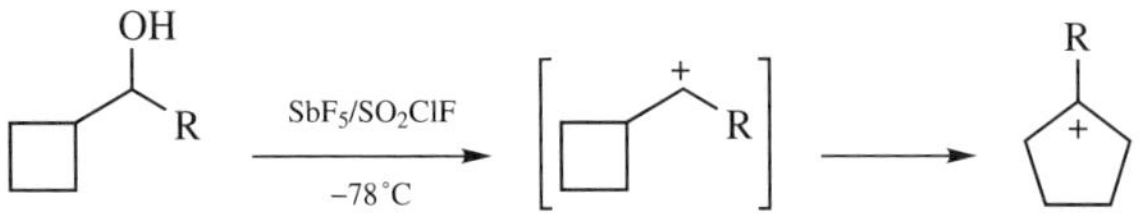

OH R —SbF_5/SO_2ClF, −78°C→ [+ R] → R +

53 (R = CH_3) **56** (R = CH_3) **59** (R = CH_3)
54 (R = Ph) **57** (R = Ph) **60** (R = Ph)
55 (R = cyclopropyl) **58** (R = cyclopropyl) **61** (R = cyclopropyl)

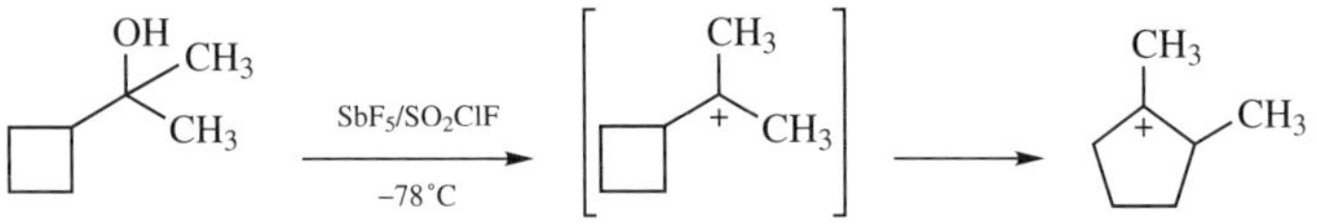

OH CH_3 CH_3 —SbF_5/SO_2ClF, −78°C→ [CH_3 + CH_3] → CH_3 + CH_3

62 **63** **64**

It is noteworthy that ionization of the cyclobutylmethanol (**65**), using SbF_5/SO_2ClF gave the 1-bicyclo[4.4.0]dec-1-yl cation (**67**), on warming to −60°C from −78°C. The formation of the later cation involves deep-seated rearrangements, probably through formation of the 1-(cyclobutylethyl)cyclobutyl cation (**69**) involving the dehydrative dimerization of the cyclobutylmethanol. The 1-(cyclobutylethyl)cyclobutyl cation (**69**) undergoes a series of Wagner–Meerwein rearrangements and 1,2-hydride shifts to give the bicyclo[4.4.0]octan-1-yl cation (**67**).[28]

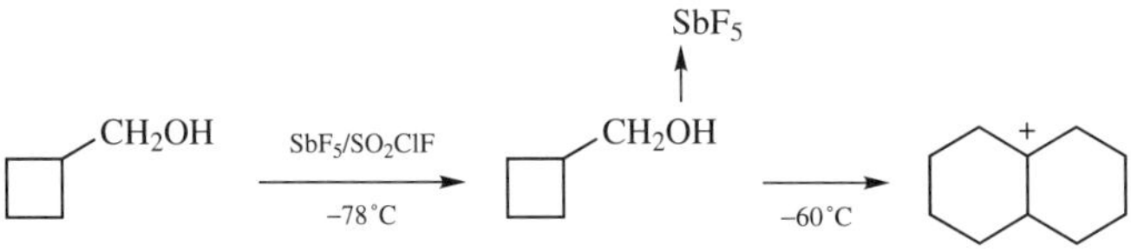

CH_2OH —SbF_5/SO_2ClF, −78°C→ SbF_5 ↑ CH_2OH —−60°C→ +

65 **66** **67**

On the other hand, cyclobutyldicyclopropylmethanol (**74**), on ionization using FSO_3H/SO_2ClF at −90°C, cleanly gave the cyclobutyldicyclopropylmethyl cation (**75**).[28] The ^{13}C NMR spectrum of the cation at −80°C shows the following absorptions: $\delta^{13}C$ 275.9 (C^+), 44.3 (d, $J = 139.2$ Hz, C1), 39.6 (d, $J = 177$ Hz, C_α), 36.6 (t, $J = 170.8$ Hz, C_β), 32.2 (d, $J = 172.0$ Hz, C_α'), 31.5 (t, $J = 171.8$ Hz, C_β'), 29.3 (t, $J = 142.0$ Hz, C2, C4), 17.1 (t, $J = 137.4$ Hz, C3). The carbocationic center of **75** ($\delta^{13}C$ 275.9) is strongly deshielded and is very similar to that of the tricyclopropylmethylium cation ($\delta^{13}C$ 280.5). The cyclobutyldicyclopropylmethylium cation therefore is substantially trivalent (classical) in nature, with the expected delocalization into the cycloalkyl rings. The classical nature of the carbocation is also inferred from the ^{13}C NMR chemical shift additivity criterion.[5] The summation of the chemical shifts for all the carbons of the carbocation is 611.3, which is 470 ppm higher than that for the corresponding neutral hydrocaron, cyclobutyldicyclopropylmethane (estimated value = 161.8, based on the observed chemical shifts of the alcohol **74**). The difference in chemical shift, 449.5 ppm, is similar in magnitude to that of the static 3-nortricyclyl cation (**21**, R = H; $\Delta\delta = 470$) and is substantially greater than that for the σ-delocalized cyclopropylmethyl cation (**13–15**; $\Delta\delta = 283$).

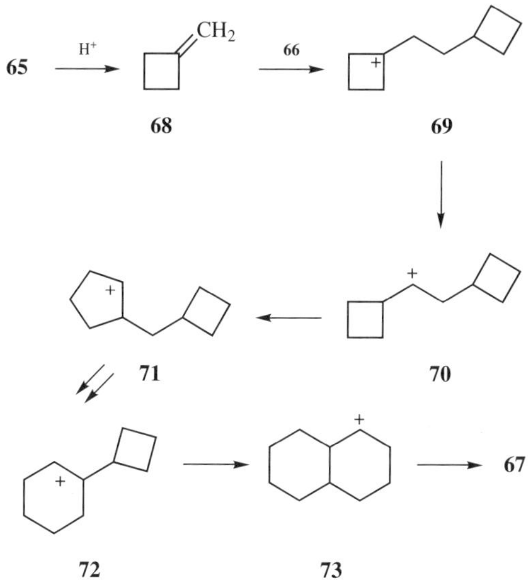

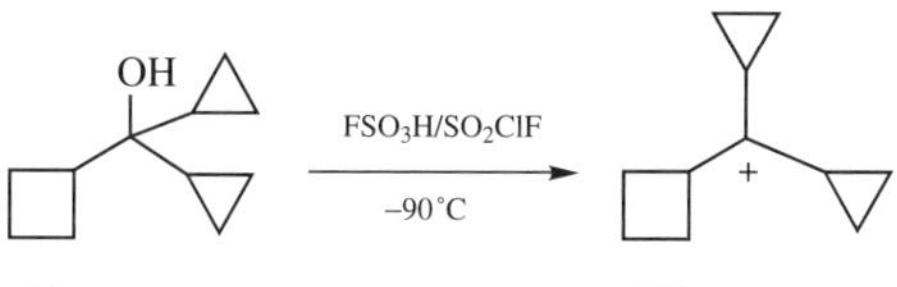

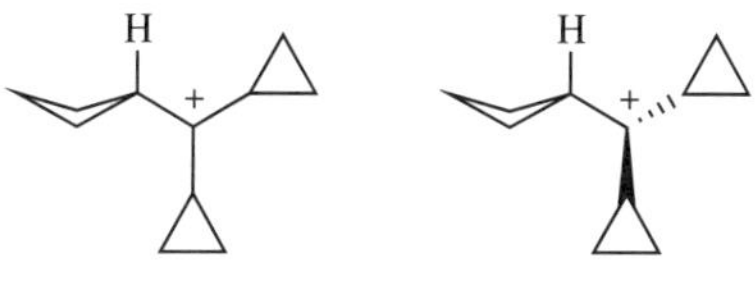

The ^{13}C NMR data also show that the carbocation **75** exists in solution mainly as the C_s symmetric bisected conformer, **76**. The perpendicular conformation of the cation **77** would show identical chemical shifts for the C_α and C_α' carbons, as well as the C_β and C_β' carbons of the cyclopropyl ring, whereas the bisected conformer, which is C_s symmetric, would show distinct chemical shifts for the C_α and C_α' carbons (C_β and C_β' carbons), as is observed. The distinct chemical shifts for the cyclopropyl methine and methylene carbons at $-80°C$ is indicative of the absence of the C1–C^+ free rotation on the NMR time scale. When the carbocation solution in superacid was warmed from -80 to $-40°C$, the signals for the cyclopropyl methine (C_α and C_α') and methylene (C_β and C_β') carbons merged into the baseline, while other signals are relatively unaffected. On the basis of this evidence the free energy of activation ($\Delta G^{\#}$) for the rotational barrier around the C1–C^+ bond was calculated as 11 ± 0.5 kcal/mol at $-40°C$. The barrier is comparable to that of the α,α–dimethylcyclopropylmethyl cation (13.7 kcal/mol at $-21°C$),[14] and suggests that although the bisected cyclobutyldicylopropylmethylium cation (**76**) is trivalent (classical) ion, it involves significant σ participation from the adjacent cycloalkyl groups. The theoretical calculations at the B3LYP/6-31G* level also supported the bisected nature of the cation. Interestingly, the perpendicular conformation of the carbocation **77** is not even a stationary point on the potential energy surface at this theoretical level.[28]

4.5 CARBODICATIONS

Long-lived carbodications can be generated in superacidic media if the carbocationic centers are tertiary and the cationic centers are well separated from each other. The charge–charge repulsion leads to increasing destabilization of the dications when the charge centers are proximally located. In simple alkyl-substituted carbodications, the carbocationic centers should be separated from each other by at least two carbon atoms, in order for them to have reasonable stability. A variety of 1,4 and 1,5 carbodications involving tertiary carbocationic centers were prepared and well characterized in superacidic media at low temperatures. In general, the strong electron-releasing groups such as phenyl and cyclopropyl substituents help to stabilize the carbodications. The extent of the stabilization of the cyclopropyl group is comparable, and in some cases superior, to those of the phenyl group. Cycloalkyl carbodications stabilized by cyclopropyl groups, such as 2,6-dicyclopropyl-2,6-adamantanediyl dication (**78**),[29] and *anti*-tricyclo[5.1.0.0^{3,5}]octa-2,6-diyl dications (**79**)[30] are examples of stable carbodications stabilized by the delocalization involving the cyclopropyl groups. Interestingly, the carbodications show enhanced charge dispersal as compared to their related carbomonocations, as reflected by their relatively shielded $\delta^{13}C$ absorptions for the cationic centers.

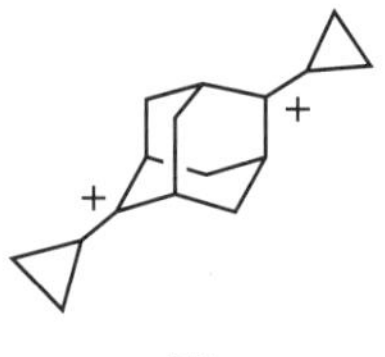

78 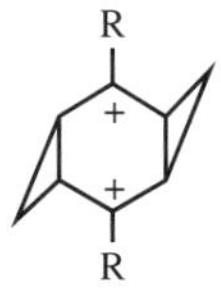 **79**

Below, we describe a number of carbodications including a bisallylic benzene dication, a nonclassical 2,10-*para*[$3^2.5^6$]octahedrane dication, 1,3- and 1,4-carbodications stabilized by neighboring cyclopropyl groups, and related systems.

4.5.1 2,6-Dimethylmesitylene-2,6-diyl Dication

Whereas a series of ring-substituted benzylic monocations (arylcarbenium ions) have been observed,[31] long-lived phenyl-1,3-dimethyldiyl dication wherein the carbocation centers are both primary are presently few. Using methods developed for the preparation of monocarbocationic intermediates under long-lived, stable ion conditions,[32] have been utilized for the preparation of mesityl-2,6-dimethyldiyl dication **81** with unique structural features.

Ionization of 2,6-bis(chloromethyl)mesitylene (**80**) in fivefold excess of SbF_5 in SO_2ClF at $-78°C$ resulted in a deep-red-colored solution. The 75 MHz ^{13}C NMR spectrum exhibited seven well-resolved, peaks at $\delta^{13}C$ 218.5 (s), 198.2 (t, $J_{C,H} = 170.3$ Hz) 195.4 (s), 143.9 (s), 140.0 (d, $J_{C,H} = 177.3$ Hz), 25.6 (q, $J_{C,H} = 131.9$ Hz), 23.9(q, $J_{C,H} = 132.3$ Hz). The 300-MHz ^1H NMR showed absorptions at δ^1H 8.77 (br, 2H), 8.51 (br, 2H), 7.10 (singlet, 1H), 2.62 (singlet, 3H) and 2.10 (singlet, 6H). The ion is remarkably stable even at $-10°C$. The observed NMR data indicate that a dienylic allylic dication system **81a** is the major contributor to the structure **81**.

The ^1H (in brackets) and ^{13}C NMR chemical shift assignments are shown on the structure **81a**. The remarkable stability of the dication can be attributed to highly stabilized dienyl–allyl dication nature (i.e., **81a**).

CH3 ClH2C CH2Cl CH3 H CH3 **80** — SbF5/SO2ClF, −78 °C → CH3 H2C CH2 CH3 H CH3 **81** 2+

CH3 25.9 (2.62); 218.5; CH2 198.2 (8.77, 8.51); H2C; 143.9; 195.4; CH3; 140.0; CH3 25.6 (2.10); H (7.10)

81a

Dicationic structure **81a** is reminescent of the bisallylic benzene dication **82**. The benzene dication **82** is still experimentally elusive, although di- and polycyclic analogs were obtained by two-electron oxidation of the corresponding arenes by SbF_5.[33] MINDO/3 calculations by Dewar and Holloway[34a] showed that the benzene

82

dication **3** favors a C_{2h} chair conformation as the most stable form, with essentially isolated allyl cation units. Schleyer and Lammertsma subsequently found[34b] that **82** is subject to Jahn–Teller distortion upon optimization forming a double allylic cation. According to Schleyer et al.,[34c] the uncoupling of the allyl units in **82** tends to keep the pairs of π electrons as far apart from one another as possible, thus minimizing the repulsions between them. This leads to unequal bond lengths in the ring and forces the ring to be distorted from planarity.

The predominant contribution of **81a** to the structure of the dication **81** as observed by NMR in superacids is further supported by density functional theory (DFT)[35]/IGLO[36] calculations. The geometry was fully optimized at the DFT B3LYP/6-31G* level in its C_1 symmetry. However, final geometry **81a** was found to be very close to C_s symmetry (Fig. 4.1). C6–C11 bond distance is 1.377 Å, is slightly longer than that of a double bond (1.34 Å). On the other hand, the C1–C2 bond distance is 1.451 Å, in between those of single (1.54 Å) and double bond. Thus, one of the positive charge of the dication is asymmetrically delocalized over C8–C2–C1–C6–C11 atoms (i.e., dienyl cation). The second positive charge is delocalized among C5–C4–C3 atoms (i.e. allyl cation) as the bond distance of C3–C4 (1.393 Å) is between those of single and double bond. Consequently, the

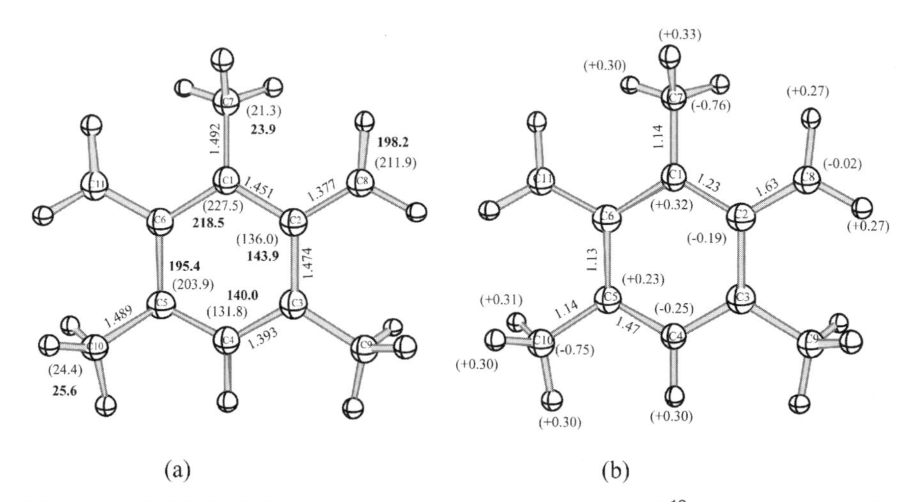

Figure 4.1 B3LYP/6-31G* calculated (a) bond distance and IGLO/DZ ^{13}C NMR chemical shifts (in parentheses); experimental shifts are given in bold (b) Löwdin bond order and NBO charges (in parentheses) of **81a**.

dication **81a** can be described as dienyl-allyl dication. The structure can also be considered as that of a substituted benzene dication. **81a** is slightly distorted from planarity as the C1–C2–C3–C4 dihedral angle was found to be 3.0°. There seems to be very little interaction between dienyl and allyl units of the dication as the C2–C3 bond distance (1.474 Å) is close to that of a single bond. These results are also consistent with calculated Löwdin bond orders[37] and natural bond orbital (NBO) charges[38] of **81a** (Fig. 4.1). Thus, the bond orders of C6–C11 and C1–C6 are 1.63 and 1.23, respectively, and the atomic charge of C1, C2, and C3 are +0.32, −0.19, and −0.02 au (atomic unit) are again indicating asymmetrical charge delocalization over C8–C2–C1–C6–C11 atoms. The C2–C3 bond order of 1.13, indicating only little interaction between dienyl and allyl units of the dication. IGLO-calculated ^{13}C NMR chemical shifts of **81a** are also correlate well with the experimentally obtained data (see Fig. 4.1).

The calculated ^{13}C NMR chemical shifts of carbocationic center (CH_2 carbon) of **81a** is 13.7 ppm more deshielded than the experimentally observed results. The agreement between experimental and calculated values may be improved by using correlated level calculations such as the GIAO MP2 method.[39] However, GIAO-MP2 calculations using ACES II program[40] at the time were limited to only small molecules.

To substantiate the structural nature of **81a** and **82**, we also carried out calculations on the benzene dication system at the same DFT B3LYP/6-31G* level. Optimized structure of the dication is a C_{2h} symmetric **82** (chair conformation) with substantial distortion from planarity (C3–C4–C5–C6 dihedral angle is 31.6°). Comparison of the geometries calculated for **81a** (Fig. 4.1) and those of calculated for **3** (Fig. 4.2) shows considerable similarities, although **81a** involves a dienyl–allyl dication the later encompasses allyl–allyl dication.

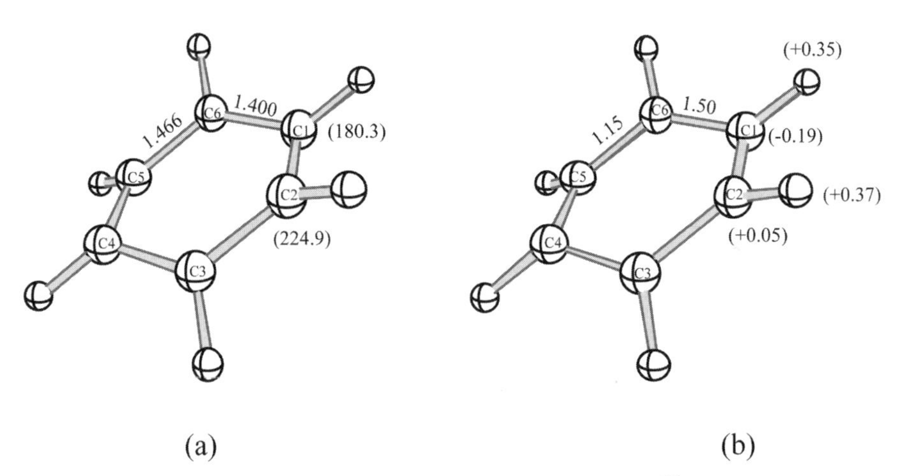

Figure 4.2 B3LYP/6-31G* calculated (a) bond distance and IGLO/DZ ^{13}C NMR chemical shifts (in parentheses) (b) Löwdin bond order and NBO charges (in parentheses) of **82**.

Similarly, dienylic oxoallylic dication **83** was also generated and studied.[41] All attempts generate a trication from 1,3,5-tris(chloromethyl)trimethylbenzene led to dicationic dienylic–allylic donor–acceptor complex **84**.

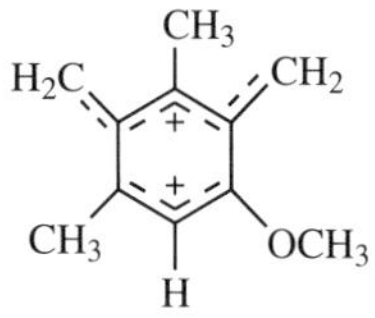

83 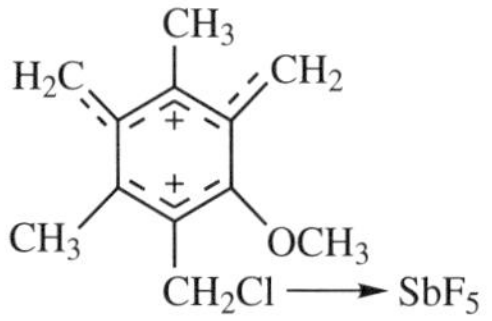 **84**

4.5.2 2,10-*para*[$3^2.5^6$]Octahedrane Dication

Structurally similar to alcohol **25** is the diol ditriaxane-2,10-dimethyl alcohol **85**. In **85** two cyclopropylmethyl alcohol units are also locked into a bisnordiadamantane framework. Because its geometry is similar to that of **25**, comparable spectroscopic features to **28a/b** for the ionization product (dication **86**) can be expected. On the other hand, formation of a distonic[17] primary cyclopropylmethyl dication gives a considerably higher charge/atom ratio compared to **26**, which could lead to relatively lower extent of σ delocalization into the strained cyclopropyl moieties.

Ionization of **85** with antimony pentafluoride (SbF_5) in SO_2ClF at $-78°C$ resulted in a dark yellow solution. The 75-MHz proton-decoupled ^{13}C NMR spectrum of the ion at $-80°C$ shows only five absorptions, indicating the formation of a highly symmetric dication or a mixture of rapidly equilibrating dications with an apparent high symmetry: $\delta^{13}C$ 135.7 (s, quaternary carbon), 118.9 (t, $J_{CH} = 174$ Hz, methylene), 105.0 (d, $J_{CH} = 193.2$, four b CHs), 66.4 (d, $J_{CH} = 148.8$, two belt CHs), and 53.3 (d, $J_{CH} = 156.1$, four belt CHs) in a ratio of 1 : 1 : 2 : 1 : 2, respectively. Considering the similar geometry and a correlation between charge density and ^{13}C chemical shifts,[42] the downfield shift of the methylenes by 71 ppm compared to that of **28a/b** clearly indicates higher positive charge density on the methylene moieties in **86** and a lower degree of charge delocalization into the cyclopropylmethyl/cyclobutyl framework. On the other hand, this value is still relatively shielded when compared to Sorenson's nortricyclylmethyl cation **22**[16] by 72.5 ppm, indicating still substantial positive charge delocalization into the cyclopropane moiety, clearly ruling out the static bisected structure **87** for the ionization product of **85**. A temperature dependent NMR spectroscopic investigation (-100 to $-30°C$) showed no significant temperature dependence of chemical shifts.[43] Application of the empirical ^{13}C NMR chemical shift additivity rule also supports the assignment of this ion as being bridged. The difference between the total sum of the ^{13}C NMR chemical shifts for this ion (1275.6 ppm) and the total sum of ^{13}C NMR chemical shifts for the structurally closely related parent hydrocarbon 2,10-dimethylditriaxane (774.6 ppm)[5,44] is 250 ppm per unit of positive charge. Comparison of this value with related monocharged systems such as the highly delocalized cyclopropylmethyl cation (283 ppm) and the static bisected nortricyclylmethyl cation (470 ppm) supports the bicyclobutonium-type nonclassical structures **88ac/b**[45] for the ionization product of **85**.

$$\mathbf{85} \xrightarrow[-80\,°C]{SbF_5/SO_2ClF} \mathbf{86}$$

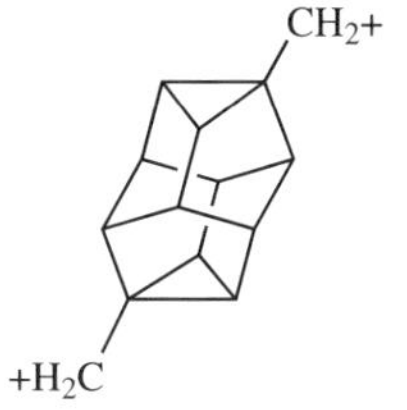

CH_2+ +H_2C

87 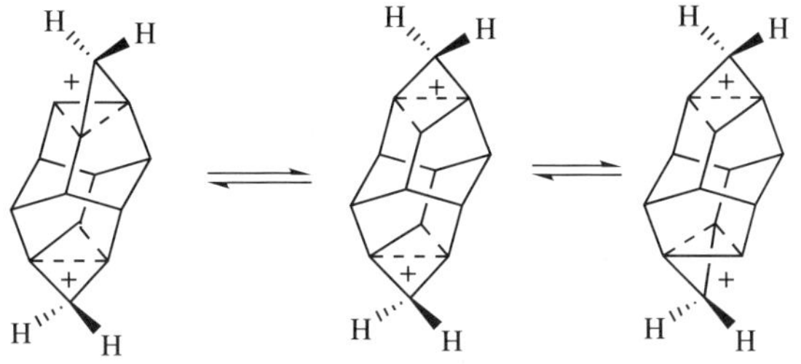 **88a** **88b** **88c**

4.5.3 *trans*-Cyclopropane-1,2-diylbis(dicyclopropylmethylium) Dication

An unique cyclopropyl stabilized carbodication, *trans*-cyclopropane-1,2-diylbis(dicyclopropylmethylium) dication (**89**), has been prepared by the ionization of the corresponding diol using $FSO_3H–SbF_5$ in SO_2ClF at $-78°C$.[46] The dication showed the following ^{13}C NMR absorptions at $-73°C$: $\delta^{13}C$ 39.3 (C_α, C_α'), 37.6, 37.3, 35.9, 34.9, 33.7, 32.5 (C3), and 264.1 (C^+). The five absorptions from 37.6 to 33.7 most probably arise from the four distinct cyclopropane methylene carbons and the C1, C2 carbons. On warming to $-50°C$, the absorptions for the C^+, C_α, C_α', and C3 remained relatively unaltered, appearing at $\delta^{13}C$ 264.5 (s), 39.9 (d, $J = 175$ Hz), and 33.0 (t, $J = 178$ Hz). The remaining absorptions coalesced to give a broad signal at $\delta^{13}C$ 37.3. Cooling the cation solution back to $-73°C$ resulted in the regeneration of the original multiline spectrum, showing that the cation exhibits reversible temperature-dependent behavior of its ^{13}C NMR spectrum.

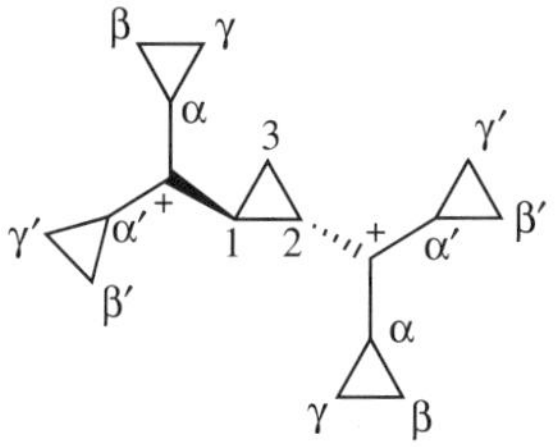

89
$\delta^{13}C(C^+) = 264.5$

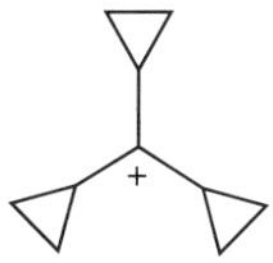

90
$\delta^{13}C(C^+) = 280.5$

The equivalence of the C_α and $C_{\alpha'}$ signals at $-73°C$ indicates an extremely low barrier to rotation around the C^+–C1 or C^+–C2 bond. The barrier to rotation around the C^+–C_α or C^+–C_α' bond is also very small, on the order of 10–12 kcal/mol (an estimate), as all absorptions due to the cyclopropane methylenes (C_β, C_β', C_γ, and C_γ') coalesced at only -50°C. It is also interesting to note that the carbocationic centers of the dication are shielded by about 16 ppm, as compared to that of the analogous carbomonocation, the tricyclopropylcarbinyl cation, **90** (δ^{13}C 280.5).[14] The carbodication, therefore, is significantly stabilized by delocalization involving the four peripheral cyclopropane rings. The delocalization from the central cyclopropane ring is relatively weak, indicated by the extremely low barrier for the rotation around the C^+–C1 or C^+–C2 bonds, as shown by the equivalence of the C_α and C_α' methine carbons.

The classical carbocations may be distinguished from the bridged species using the chemical shift additivity criterion (i.e., on the basis of total chemical shift difference) as suggested by Prakash, Schleyer, Olah, and coworkers (see below).[5] The summation of the chemical shifts for the carbodication **89** is larger than that of the corresponding hydrocarbon by 893 ppm. This value is nearly double the value expected for a typical classical carbomonocation, and is similar to those of the other static dications, 1,5-bicyclo[3.3.3]undecyl dication (**91**) ($\Delta\Sigma\delta = 830$) and 2,5-dimethyl-2,5-hexyl dication (**92**) ($\Delta\Sigma\delta = 758$).[5] Thus the dication shows "classical" behavior, with substantial charge delocalization into the adjacent cyclopropane carbons.

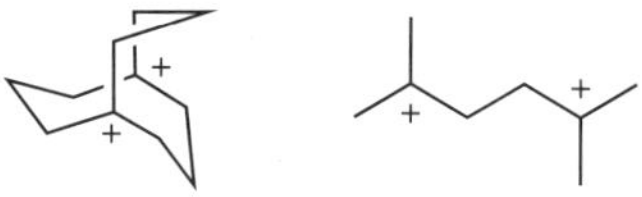

91 **92**

Tha analogous dication stabilized by the neighboring phenyl groups, the *trans*-cyclopropane-1,2-ylbis(diphenylmethylium) dication (**79**), also shows significantly enhanced delocalization of the positive charge into the aromatic rings as compared to the analogous monocation, the diphenylcyclopropylmethyl cation.[47] The dication also derives significant stabilization from the cyclopropane ring.

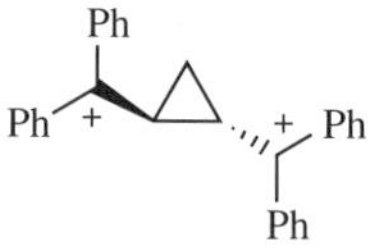

Ph Ph + + Ph Ph

93

Attempts to prepare the analogous carbodication, **94**, stabilized by the neighboring methyl groups, afforded only the ring-opened product, 2,6-dimethylhepta-3,5-dien-2-yl dication (**95**), showing the ineffectiveness of the methyl groups in the stabilization of this carbodication:[46]

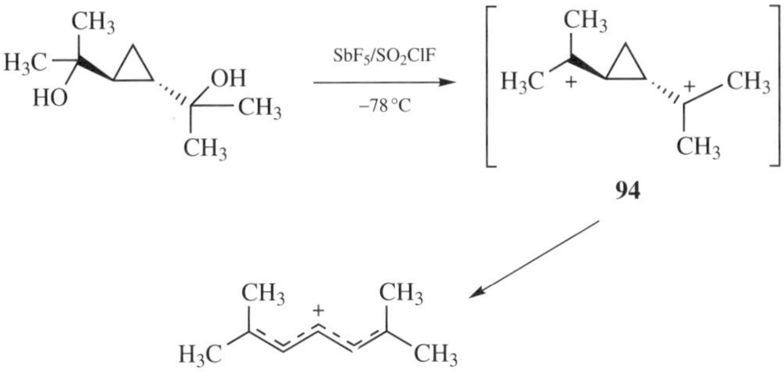
CH3
H3C
HO
OH
CH3
CH3
SbF5/SO2ClF
−78 °C
CH3
H3C +
+ CH3
CH3
94
CH3 + CH3
H3C CH3

95

4.5.4 1,1,3,3-Tetracyclopropyl-1,3-propanediyl Dication

Generation of 1,3-carbodications, in which the charge centers are separated by only one methylene group, were mostly unsuccessful. The ionization of the 2,4-dichloropentane (**96**) and 2,3,3,4-tetramethyl-2,4-pentanediol (**99**) failed to give the expected 1,3-carbodications. The former reaction gave the 3-penten-2-yl cation (**98**), whereas the latter gave only the products of disproportionation, the rapidly equilibrating 2,3-dimethyl-2-butyl cation (**103**) and the *O*-protonated acetone (**102**).

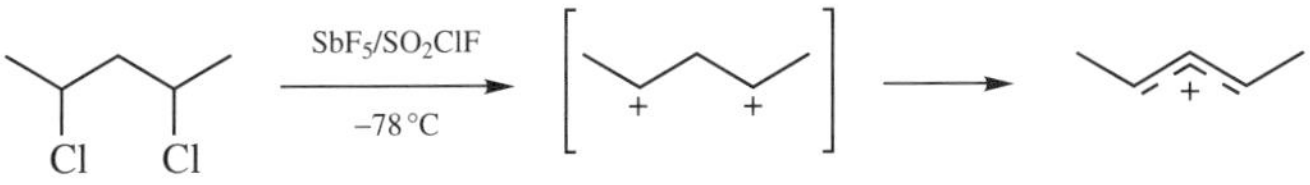
Cl Cl
SbF5/SO2ClF
−78 °C
+ +
+

96 **97** **98**

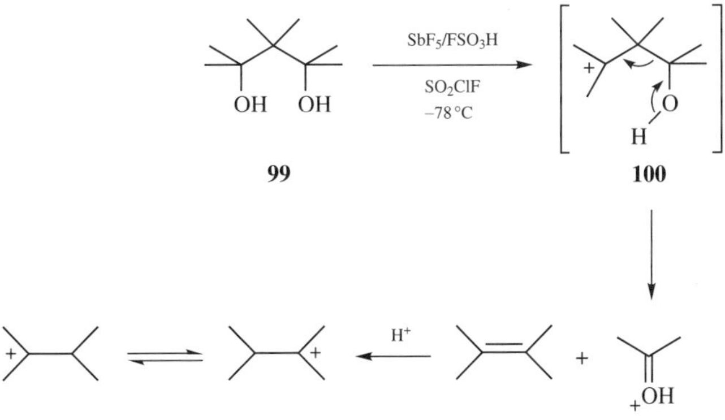
OH OH
99
SbF5/FSO3H
SO2ClF
−78 °C
+
O
H
100
H+
+
+
+
+OH

103 **101** **102**

High-level theoretical calculations [up to Gaussian-2 (G2) theory] predict that even the parent 1,3-propanediyl dication (i.e., the protonated allyl cation) is a minimum, with significant hyperconjugative interactions.[48] Thus the failure to obtain the 1,3-carbodications from the 1,3-dichloro- and 1,3-dihydroxyalkanes may be, in part, due to the availability of other kinetically readily accessible pathways for formation of the more stable carbocations.

Schleyer and coworkers were able to observe a σ-resonance-stabilzed carbodication, the 1,3-dehydro-5,7-adamantanedilyl dication (**105**), formed through the ionization of the 1,3-dehydro-5,7-difluoroadamantane (**104**).[49] The dication **92** owes its exceptional stability (up to 0°C) to its unique three-dimensional aromaticity; specifically, the four p-orbitals overlap in a tetrahedral fashion. The ^{13}C NMR spectrum of the carbodication shows uniquely shielded absorption (δ^{13}C 6.6) for the bridgehead carbons, even though they have formal positive charges. The attempted preparation of the 1,3-adamantanediyl dication (**107**) by the ionization of 1,3-difluoroadamantane (**106**) was, however, unsuccessful, and resulted only in the formation of the monocation monodonor–acceptor complex (**108**).[50]

F F SbF_5/SO_2ClF 2+

104 **105**

F F SbF_5/SO_2ClF + F SbF_5 + +

106 **108** **107**

In principle, the neighboring charge-stabilizing phenyl or cyclopropyl groups at the cationic centers would make it possible to observe the otherwise unstable acyclic 1,3-carbodications. The ionization of the 1,1,3-triphenyl-1,3-propanediol (**109**), and 1,1,3,3-tetraphenyl-1,3-propanediol (**112**), in FSO_3H/SbF_5-SO_2ClF at −78°C gave only the disproportionated products: the former giving 1,1-diphenylethyl cation (**110**) and *O*-protonated benzaldehyde (**111**), and the latter giving the 1,1-diphenylethyl cation (**110**) and *O*-protonated benzophenone (**113**).[51] The latter ionization also resulted in formation of the minor amount of the allyl cation, the 1,1,3,3-tetraphenylallyl cation, **116**. The ionization of the latter substrate using FSO_3H/SO_2ClF at extremely low temperature (−130°C), however, gave the allyl cation (**116**) as the only product. The bistrifluoroacetate **114** derived from the 1,1,3,3-tetraphenyl-1,3-propanediol (**112**) also provided the allyl cation (**116**) as the only product. The latter observations indicate the intermediate formation of the 1,3-carbodication, the 1,1,3,3-tetraphenyl-1,3-propanediyl dication (**115**) in these reactions.

Ph Ph OH OH Ph H SbF_5/FSO_3H SO2ClF −78 °C Ph + Ph CH_3 + Ph H $^+$OH

109 **110** **111**

The ionization of 1,1-dicyclopropyl-2-(2-hydroxy-2-adamantyl)ethanol (**117**) similarly gave the allyl cation, the 2-adamantylidene-1,1-dicyclopropylethyl cation (**119**). In this case, it was possible to observe the hydroxyl-protonated dication (**120**) when the ionization was carried out below −100°C. Presumably the 1,3-carbodication (**118**) may exist as the transient species that undergoes deprotonation to give the observed allyl cation (**107**).

Ionization of the spirocyclopropyl-derived 1,1,3,3-tetracyclopropyl-1,3-propanediol (**121**) in $FSO_3H/SbF_5–SO_2ClF$ at −78°C also resulted in formation of the disproportionated cationic products, the tricyclopropylmethylium cation (**90**) and the *O*-protonated dicyclopropylketone (**122**). The spirocyclopropyl group, in fact, acts as a powerful neighboring group σ donor and usually gives the ring-opened rearranged carbocations.[21]

$\xrightarrow[\text{SO}_2\text{ClF}, -78\,^\circ\text{C}]{\text{FSO}_3\text{H/SbF}_5}$ **90** +

121 **122**

Ionization of the nonspirocyclopropyl analog, the 1,1,3,3-propane-1,3-diol (**123**) in FSO_3H/SbF_5–SO_2ClF at $-78°C$, on the other hand, gave the expected 1,3-carbodication, the 1,1,3,3-tetracycloprpyl-1,3-propanediyl dication (**124**) [$\delta^{13}C$ 262.8 (s, C^+), 47.1 (t, $J = 134$ Hz, C2), 39.6 (d, $J = 181$ Hz), 48.7 (d, $J = 179$ Hz, cyclopropyl CH), 38.2 (t, $J = 182$ Hz), and 46.6 (t, $J = 171$ Hz, cyclopropyl CH_2)]. The structure of the carbodication was further confirmed by matching the experimentally observed ^{13}C NMR chemical shifts with those obtained from theoretically calculated values (IGLO method using B3LYP/6-31G*-optimized geometries).[51]

$\xrightarrow[\text{SO}_2\text{ClF}, -78\,^\circ\text{C}]{\text{FSO}_3\text{H/SbF}_5}$

123 **124**

125

The cationic center of the carbodication **124** ($\delta^{13}C$ 262.8) is shielded by 12 ppm as compared to that of the 1,1-dicyclopropylethyl cation (**125**).[14] The enhanced shielding of the cationic centers in the carbodication (**112**) signifies increased delocalization of the charge from the cationic centers into the neighboring cyclopropyl groups. The enhanced charge delocalization into the cyclopropyl groups, as compared to that of the monocation **125**, is also reflected in the relatively deshielded absorptions for the cyclopropyl methine ($\Delta\delta_{average} = 13.8$) and methylene carbons ($\Delta\delta_{average} = 15.1$). In spite of the significant charge delocalization into the neighboring cyclopropyl groups, the carbodication **124** is a classical species as shown by the chemical shift additivity criterion. The total chemical shift difference between the carbodication (**124**) and the corresponding hydrocarbon, 1,1,3,3-tetracyclopropylpropane (estimated value based on the diol), is 901 ppm {$[\Sigma\delta^{13}C\ (R^{2+})] - [\Sigma\delta^{13}C\ (RH)] = 901$}, which is in fact of the same order of magnitude as the *trans*-cyclopropane-1,2-diylbis(dicyclopropylium) dication, **89** ($\Delta\delta = 893$).[46] The successful preparation of the first acyclic 1,3-carbodication thus proves the exceptional stabilization of the carbocation centers by the cyclopropyl groups.

4.5.5 (Hexaaryltrimethylene)methane Dications

The (trimethylene)methane dication (**126**), a 2π-electron system, is still unknown, although highly stabilized versions, such as sulfur- and oxygen-stabilized derivatives of the cation, have been reported.[52–56] In a study to probe the Y aromaticity, formally 1,3-carbodications involving Y conjugation, the hexaphenyl(trimethylene) methane dication (**128**) and its substituted versions (**130**) were obtained by ionization of their corresponding diols (**127** and **129**) in FSO_3H/SO_2ClF.

2+

126

Ph Ph Ph Ph HO Ph Ph OH

127

FSO_3H/SO_2ClF −78°C

Ph Ph 2+ Ph Ph Ph Ph

128

The (hexaphenyltrimethylene)methane dication (**128**) is a highly symmetric species as shown by its ^1H NMR spectrum [δ^1H 208.5 (s, C2), 145.4 (s, C1), 144.1 (d, *ortho*), 142.4 (d, *ortho′*), 141.5 (s, *ipso*), 137.5 (d, *para*), 131.21 (d, *meta*), 131.18 (d, *meta′*). The steric congestion in this cation results in the nonequivalence of both the *ortho* and the *meta* carbons of the phenyl rings at the *para* position. This nonequivalence can also be seen in its ^{13}C NMR spectrum. The carbodication is stable up to −20°C, although quenching of the NMR sample with $NaHCO_3$-buffered methanol at −40°C afforded a complex mixture of products. Of interest is the fact that the $\delta^{13}C$ for the cationic center of this carbodication (208.5) and that of the triphenylmethyl cation are remarkably similar, ruling out the importance of Y aromaticity in this system.[57] In order to decrease the amount of aryl stabilization in this system electron-withdrawing trifluoromethyl groups were attached to the phenyl rings. Ionization of the respective diol (**129**) in FSO_3H/SO_2ClF at −120°C cleanly provided the carbodication **130**, which again showed high degree of symmetry having only nine peaks in its ^{13}C NMR spectrum [$\delta^{13}C$ 209.8 (s, C2), 146.7 (s, C1), 145.6 (d, *ortho*), 143.1 (s, *ipso*), 142.4 (q, *para*), 138.8 (d, *ortho′*), 128.4 (d, *meta*), 128.1 (d, *meta′*), 121.4 (q, CF_3)] and nonequivalence of the *ortho*- and *meta*-aryl carbons. The *p*-CF_3 substituent was found to have very little influence on the $\delta^{13}C$ of the benzylic carbocationic centers relative to that of the carbodication **128**, despite the strongly electron-withdrawing nature of the CF_3 group. The carbodication **130** is stable only up to −80°C, above which it rearranged to the corresponding indenyl cation, **131**; similar observations were made with other aryl derivatives.[57]

FSO_3H/SO_2ClF, −120°C

129

Ar = *p*-$CF_3C_6H_4$

130 $^{2+}$

−80°C

F_3C

131

Disruption of the three-fold symmetry and reduction of the number of stabilizing phenyl rings even to five also resulted in intramolecular allylation (intramolecular Friedel–Crafts reaction) and subsequent formation of the corresponding indenyl cations.[57] In all the cases studied, there was no evidence of the Y-aromatic stabilization. Although the carbodication **128** is exceptionally stabilized, this tendency may be steric (kinetic) as well as conjugative in nature rather than due to the Y aromaticity. These results are in agreement with earlier findings of Schleyer and coworkers that Y-conjugation dications do not possess any enhanced thermodynamic stability.[52]

ACKNOWLEDGMENTS

We would like to thank our senior collaborators (G. A. Olah, G. Rasul, J. Casanova, R. K. Murray. Jr., A. de Meijere, and J. S. Lomas), students and postdocs who made much of the discussed work possible. We are also grateful to the Loker Hydrocarbon Research Institute at University of Southern California for support of our work.

REFERENCES

1. (a) G. A. Olah, Nobel Lecture, *Angew. Chem. Int. Ed. Engl.* **34**, 1393 (1995); (b) G. A. Olah, in *Stable Carbocation Chemistry*, G. K. S. Prakash and P. v. R. Schleyer, eds., Wiley, New York, 1997.
2. G. K. S. Prakash, *Pure Appl. Chem.* **70**, 2001 (1998).
3. J. Reinbold, M. Bertau, T. Voss, D. Hunkler, L. Knothe, H. Prinzbach, D. Neschadin, G. Gesheidt, B. Mayer, H.-D. Martin, J. Heinze, G. K. S. Prakash, and G. A. Olah, *Helv. Chim. Acta.* **84**, 1518 (2001).

4. G. K. S. Prakash, in *Stable Carbocation Chemistry*, G. K. S. Prakash and P. v. R. Schleyer, eds., Wiley, New York, 1997.
5. P. v. R. Schleyer, P. Lenoir, P. Mison, G. Liang, G. K. S. Prakash, and G. A. Olah, *J. Am. Chem. Soc.* **102**, 683 (1980).
6. G. A. Olah, G. K. S. Prakash, and R. Krishnamurti, *J. Am. Chem. Soc.* **112**, 6422 (1990).
7. G. A. Olah, G. K. S. Prakash, G. Liang, P. v. R. Schleyer, and W. D. Graham, *J. Org. Chem.* **47**, 1040 (1982).
8. G. A. Olah, M. D. Heagy, and G. K. S. Prakash, *J. Org. Chem.* **58**, 4851 (1993).
9. M. Heagy, G. A. Olah, G. K. S. Prakash, and J. S. Lomas, *J. Org. Chem.* **60**, 7355 (1995).
10. G. A. Olah, V. P. Reddy, J. Casanova, and G. K. S. Prakash, *J. Org. Chem.* **57**, 6431 (1992).
11. G. A. Olah, Q. Liao, J Casanova, R. Bau, G. Rasul, and G. K. S. Prakash, *J. Chem. Soc. Perkin Trans. 2* 2239 (1998).
12. E. J. Karabatsos and M. Tornaritis, *Tetrahedron Lett.* **30**, 5733 (1989).
13. For leading references, see P. D. Bartlett, *Nonclassical Ions*, W. A. Benjamin, New York, 1965, p. 272 and P. Vogel, *Carbocation Chemistry*, Elsevier, Amsterdam, 1985, pp. 350–355.
14. G. A. Olah, V. P. Reddy, and G. K. S. Prakash, *Chem. Rev.* **92**, 69 (1992) and references cited therein.
15. G. A. Olah, G. Liang, K. A. Babiak, T. M. Ford, D. L. Goff, T. K. Morgan, Jr., and R. K. Murray, Jr., *J. Am. Chem. Soc.* **98**, 576 (1976).
16. (a) L. R. Shmitz and T. S. Sorensen, *J. Am. Chem. Soc.* **104**, 2600 (1982); (b) L. R. Schmitz and T. S. Sorensen, *J. Am. Chem. Soc.* **104**, 2605 (1982).
17. G. A. Olah, H. A. Buchholz, G. K. S. Prakash, G. Rasul, J. J. Sosnowski, R. K. Murray Jr., M. A. Kusnetsov, S. Liang, and A. de Meijere, *Angew. Chem. Int. Ed. Engl.* **117**, 1499 (1996).
18. G. K. S. Prakash, A. P. Fung, G. A. Olah, and T. N. Rawdah, *Proc. Natl. Acad. Sci. USA.* **84**, 5092 (1987).
19. G. A. Olah, V. P. Reddy, G. Rasul, and G. K. S. Prakash, *J. Org. Chem.* **57**, 1118 (1992).
20. G. A. Olah and G. Liang, *J. Am. Chem. Soc.* **93**, 6873 (1971) and references cited therein.
21. V. P. Reddy, G. A. Olah, and G. K. S. Prakash, *J. Org. Chem.* **58**, 7622, (1993).
22. For review, see C. J. Lancelot, D. J. Cram, and P. v. R. Schleyer, in *Carbonium Ions*, G. A. Olah and P. v. R. Schleyer, eds., Wiley, New York, 1972, Vol. III, Chapter 27.
23. G. A. Olah and R. D. Porter, *J. Am. Chem. Soc.* **92**, 7627 (1970).
24. (a) G. A. Olah, and R. D. Porter, *J. Am. Chem. Soc.* **93**, 6877 (1971); (b) G. A. Olah, R. J. Spear, and D. A. Forsyth, *J. Am. Chem. Soc.* **98**, 6284 (1976).
25. (a) G. A. Olah, N. J. Head, G. Rasul, and G. K. S. Prakash, *J. Am. Chem. Soc.* **117**, 875 (1995); (b) S. Sieber, P. v. R. Schleyer, and J. Gass, *J. Am. Chem. Soc.* **115**, 6987 (1993).
26. S. Winstein and N. J. Holness, *J. Am. Chem. Soc.* **77**, 3054 (1955).
27. (a) W. G. Dauben, J. L. Chitwood, and V. Scherer, *J. Am. Chem. Soc.* **90**, 1014 (1968); (b) K. B. Wiberg and B. R. Lorry, *J. Am. Chem. Soc.* **85**, 3188 (1963).
28. G. K. S. Prakash, V. P. Reddy, G. Rasul, J. Casanova, and G. A. Olah, *J. Am. Chem. Soc.* **120**, 13362 (1998).
29. G. K. S. Prakash, V. V. Krishnamurthy, M. Arvanaghi, and G. A. Olah, *J. Org. Chem.* **50**, 3985 (1985).

30. G. K. S. Prakash, A. P. Fung, T. N. Rawdah, and G. A. Olah, *J. Am. Chem. Soc.* **107**, 2920 (1985).
31. (a) G. A. Olah, C. A. Cupas, and M. B. Comisarow, *J. Am. Chem Soc.* **88**, 361 (1966); (b) G. A. Olah, C. A. Cupas, M. B. Comisarow, and J. M. Bollinger, *J. Am. Chem. Soc.* **89**, 5687 (1967).
32. G. A. Olah, T. Shamma, A. Burrichter, G. Rasul, and G. K. S. Prakash, *J. Am. Chem. Soc.* **119**, 3407 (1997).
33. (a) G. A. Olah and D. A. Forsyth, *J. Am. Chem. Soc.* **98**, 4086 (1976); (b) G. A. Olah and B. P. Singh, *J. Org. Chem.* **48**, 4830 (1983).
34. (a) M. J. S. Dewar and M. K. Holloway, *J. Am. Chem. Soc.* **106**, 6619 (1984); (b) P. v. R. Schleyer and K. Lammertsma, *J. Am. Chem. Soc.* **105**, 1049 (1983); (c) P. v. R. Schleyer, K. Lammertsma, and H. Schwarz, *Angew. Chem. Int. Ed. Engl.* **28**, 1321 (1989).
35. (a) T. Ziegler, *Chem. Rev.* **91**, 651 (1991); (b) M. J. Frisch, G. W. Trucks, H. B. Schlegel, P. M. W. Gill, B. G. Johnson, M. A. Robb, J. R. Cheeseman, T. A. Keith, G. A. Peterson, J. A. Montgomery, K. Raghavachari, M. A. Al-Laham, V. G. Zakrzewski, J. V. Ortiz, J. B. Foresman, J. Cioslowski, B. B. Stefanov, A. Nanayakkara, M. Challacombe, C. Y. Peng, P. Y. Ayala, W. Chen, M. W. Wong, J. L. Andres, E. S. Replogle, R. Gomperts, R. L. Martin, D. J. Fox, J. S. Binkley, D. J. Defrees, J. Baker, J. J. P. Stewart, M. Head-Gordon, C. Gonzalez, and J. A. Pople, *Gaussian 94* (revision A.1), Gaussian, Inc., Pittsburgh, PA, 1995.
36. (a) M. Schindler, *J. Am. Chem. Soc.* **109**, 1020 (1987); (b) W. Kutzelnigg, U. Fleischer, and M. Schindler, M. *NMR Basic Princ. Prog.* **23**, 165 (1991); (c) ^{13}C NMR chemical shifts were calculated by IGLO (DZ basis set; C, 7*s* 3*p* contracted to [4111,21]; H, 3*s* contracted to [21]) methods using B3LYP/6-31G* geometries (i.e., at the IGLO/DZ//B3LYP/6-31G* level); ^{13}C NMR chemical shifts were referenced to TMS [calculated absolute shift, i.e., $\sigma(C) = 218.13$].
37. P. O. Löwdin, *Phys. Rev.* **97**, 1474 (1955).
38. (a) A. E. Reed, R. B. Weinstock, and F. Weinhold, *J. Chem. Phys.* **83**, 735 (1985); (b) A. E. Reed, L. A. Curtiss, and F. Weinhold, *Chem. Rev.* **88**, 899 (1988).
39. J. Gauss, *J. Chem. Phys. Lett.* **191**, 614 (1992); J. Gauss, *J. Chem. Phys.* **99**, 3629 (1993).
40. J. F. Stanton, J. Gauss, J. D. Watts, W. J. Lauderdale, R. Bartlett, ACES II, an ab initio program system, *Quantum Theory Project*, Univ. Florida, 1991, 1992.
41. G. A. Olah, T. Shamma, A. Burrichter, G. Rasul, and G. K. S. Prakash, *J. Am. Chem Soc.* **119**, 12923 (1997).
42. (a) G. J. Ray, R. J. Kurland, and A. K. Colter, *Tetrahedron* **27**, 735 (1971); (b) P. C. Lauterbur, *J. Am. Chem. Soc.* **83**, 1838 (1961); (c) H. Spiesecke and W. G. Schneider, *Tetrahedron Lett.* 468 (1961); (d) E. W. Lalancette and R. E. Benson, *J. Am. Chem. Soc.* **85**, 1941 (1965); (e) G. A. Olah, M. Bollinger, and A. M. White, *J. Am. Chem. Soc.* **91**, 3667 (1969); (f) G. A. Olah and D. Mateescu, *J. Am. Chem. Soc.* **92** 1430 (1970); (g) R. Ditchfield, D. P. Miller, and J. A. Pople, *Chem. Phys. Lett.* **6**, 573 (1970); (h) R. Ditchfield, D. P. Miller, and J. A. Pople, *J. Chem. Phys.* **54**, 4186 (1971); (i) D. G. Farum, *Adv. Phys. Org. Chem.* **2**, 123 (1964).
43. Only small changes of up to 2 ppm in certain peaks were observed in the ^{13}C NMR spectra of the ions over the studied temperature range.
44. (a) K.-I. Hirao, Y. Ohuchi, and O. Yonemitsu, *J. Chem. Soc. Chem. Commun.* 99 (1982); (b) K.-I. Hirao, H. Takahashi, Y. Ohuchi, and O. Yonemitsu, *J. Chem. Res. (S)* 319 (1992); *(M)* 2601 (1992).

45. Compounds **88a** and **88c** are enatiomers.
46. G. A. Olah, V. P. Reddy, G. Lee, J. Casanova, and G. K. S. Prakash, *J. Org. Chem.* **58**, 1639 (1993).
47. G. A. Olah, J. L. Grant, R.-J. Spear, J. M. Bollinger, A. Seroamz, and G Sipos, *J. Am. Chem. Soc.* **98**, 2501 (1976).
48. P. M. Mayer and L. Radom, *Chem. Phys. Lett.* 244 (1997).
49. M. Bremer, P. v. R. Schleyer, K. Schoetz, M. Kausch, and M. Schindler, *Angew. Chem. Int. Ed. Engl.* **26**, 761 (1987).
50. G. A. Olah, G. K. S. Prakash, J. G. Shi, V. V. Krishnamurthy, G. D. Mateescu, G. Liang, G. Sipos, V. Buss, J. M. Gund, and P. v. R. Schleyer, *J. Am. Chem. Soc.* **107**, 2764 (1985).
51. G. A. Olah, V. P. Reddy, G. Rasul, and G. K. S. Prakash, *J. Am. Chem. Soc.* **121**, 9994 (1999).
52. K. Schotz, T. Clark, H. Schaller, and P. v. R. Schleyer, *J. Org. Chem.* **49**, 733 (1984).
53. R. Schwesinger, M. Mibfeldt, K. Peters, and H. Schnering, *Angew. Chem.* **99**, 1210 (1987).
54. (a) T. Sugimoto, K. Ikeda, and J. Yamauchi, *Chem. Lett.* 29 (1991); (b) F. Adams, R. Gompper, A. Hohenester, and H.-U. Wagner, *Tetrahedron Lett.* **29**, 6921 (1988).
55. K. Mizumoto, H. Kawai, K. Okada, and M. Oda, *Angew. Chem.* **98**, 930 (1986).
56. T. Kawase, C. Wei, N. Ueno, and M. Oda, *Chem. Lett.* 1901 (1994).
57. N. J. Head, G. A. Olah, G. K. S. Prakash, *J. Am. Chem. Soc.* **117**, 11205 (1995).

5

ANTIAROMATICITY EFFECTS IN CYCLOPENTADIENYL CARBOCATIONS AND FREE RADICALS

Annette D. Allen and Thomas T. Tidwell
Department of Chemistry
University of Toronto
Toronto, Ontario, Canada

5.1 INTRODUCTION

The "centennial of carbocations" was celebrated in 2001,[1a] and marked the discovery of the first of these species, the triphenylmethyl cation (Ph_3C+, **1**), reported in 1901 in independent studies by Norris[1b,c] and by Kehrmann and Wentzel,[1d] who

Carbocation Chemistry, Edited by George A. Olah and G. K. Surya Prakash
ISBN 0-471-28490-4

observed the formation of colored solutions from reaction of triphenylmethyl alcohol or chloride in concentrated H_2SO_4, or with $AlCl_3$ or $SnCl_4$. In 1902 Baeyer interpreted these reactions as forming salts, termed "carbonium" salts.[1e,f] Moses Gomberg, who had in 1900 reported $Ph_3C\bullet$, the first persistent free radical, also studied Ph_3C+,[1g] and the development of the chemistry of the electron-deficient intermediate carbocations and free radicals has been intertwined ever since:

$$Ph_3CCl \xrightarrow{AlCl_3} \underset{\mathbf{1}}{Ph_3C+} \xrightarrow{H_2SO_4} Ph_3COH$$

Interest in cyclopentadienyl cations began in 1925 when Ziegler and Schnell,[2a] in the course of generating the stable pentaphenylcyclopentadienyl radical, prepared pentaphenylcyclopentadienol (**2**), which, on reaction with concentrated H_2SO_4, gave an intense violet color implicating formation of the pentaphenylcyclopentadienyl cation **3**, while reaction of **2** with HCl or HBr gave the cyclopentadienyl halides **4** [Eq. (5.1)]. The reactivity of the bromide **4b** toward hydrolysis was found to be less than that of triphenylmethyl bromide,[2a] and quantitative measurements of the ionization in liquid sulfur dioxide confirmed the much lower tendency for cation formation from **4b** compared to Ph_3CBr.[2b] At that time even the existence of carbocations as discrete intermediates was still open to question, but these results already suggested that an intrinsic destabilizing factor was present in the cation **3** (this was one of the few forays into the study of carbocations by Karl Ziegler, a master of radical and carbanion chemistry, later to win the Nobel prize for studies of polymerization):

$$\underset{\mathbf{2}}{C_5Ph_5OH} \xrightarrow{H_2SO_4} \underset{\mathbf{3}}{C_5Ph_5^+} \underset{}{\overset{HHal}{\rightleftharpoons}} \underset{\mathbf{4a,b}\ (Hal = Cl, Br)}{C_5Ph_5Hal} \quad (5.1)$$

Hückel proposed in 1931 the $4n + 2$ rule for the aromaticity of benzene,[3a] and in 1952 Roberts et al.[3b] reported the delocalization energies calculated by simple Hückel molecular orbital theory of the cyclopentadienyl cation, radical, and anion as 1.24, 1.85, and 2.47β, respectively, and predicted that the cation would be a ground-state triplet. By contrast, the corresponding cyclopropenyl and cycloheptatrienyl species showed the opposite order of relative delocalization energy, namely, cation > radical > anion. Experimental studies were revived in 1959 when Bloom and Krapcho reported UV spectra for **3** and for the 1-anisyltetraphenylcyclopentadienyl cation.[3c]

A major new impetus for the study of cyclopentadienyl cations was provided by findings of Breslow and coworkers showing the spectra previously assigned[3c] to **3** were due to other products,[4a] but that the cation **3** is long-lived at −40 °C.[4b,c] The ion was characterized by UV and ESR, and was shown to have a low-lying excited triplet state.[4b,c] These investigations prompted Breslow to introduce the concept of *antiaromaticity*,[4d,e] which is exemplified by the anti-Hückel 4π-electron cyclopentadienyl cation and cyclopropenyl anion, as well as by the neutral species cyclobutadiene. The concept[4f,g] of antiaromaticity has proven to be of enormous value in the teaching of organic chemistry, and is cited in almost all basic organic textbooks as a clinching example of the importance of the 6π-electron structure providing the aromatic stabilization of benzene.[5] The cyclopentadienyl cation is a prototypical example of a *destabilized* carbocation, a family that includes cations with electron-withdrawing substituents, or with nonoptimum geometries.[4h]

Many other examples of cyclopentadienyl cations were examined subsequently, including the pentachlorocyclopentadienyl cation, which was shown to be a ground-state triplet.[4c] Such ground-state triplets not only are of theoretical interest but also have been pursued because of their possible applications as ferromagnets.[6a]

Once the concept of antiaromaticity was recognized the parent cyclopentadienyl cation **5** became an attractive object for study, and proved accessible to experimental studies. Gas-phase ionization potentials of the cyclopentenyl and cyclopentadienyl radicals were measured, and provide hydride affinities of the corresponding cations, which lead to a calculated ΔH of 31.2 kcal/mol for the reaction of Eq. (5.2).[6b] This thermodynamic criterion provides a quantitative measure of the enormous destabilization of the cyclopentadienyl cation **5** compared to the cyclopentenyl cation **6**, which is highly stabilized by delocalization.

$$\Delta H = 31.2\ \text{kcal/mol} \qquad (5.2)$$

6 5

The cyclopentadienyl cation (**5**) was moreover generated in a matrix and found by ESR spectroscopy to have a triplet ground state,[7a] as had been predicted by simple Hückel theory.[3b] A symmetric D_{5h} geometry for the triplet was predicted by ab initio calculations,[7b] and the calculations indicate the C_{2v} singlet structure with a planar *cis*-butadienyl moiety is 8.7 kcal/mol higher in energy than the triplet.[7b] It has been suggested that triplet structures with $4n$ π-electron systems may have aromatic character, as indicated by magnetic, energetic, and geometric criteria,[7h,i] and this factor may contribute to the greater stability of the triplet. Preparation of the cyclopentadienyl cation in the gas phase from the loss of HF from $C_5H_6F^+$ formed from cyclopentenone and CFO^+ was reported, and the cation was proposed to be generated as the singlet.[7j,k]

A square pyramidal structure has also been considered for the cyclopentadienyl cation (Fig. 5.1), but was found to be less stable.[7c,d] The energies and structures of a variety of alkyl-substituted cyclopentadienyl cations have also been calculated.[7e]

Further computational studies of the cyclopentadienyl cation indicated destabilization of 42.0 kcal/mol relative to the pentadienyl cation.[7b] Other criteria applied

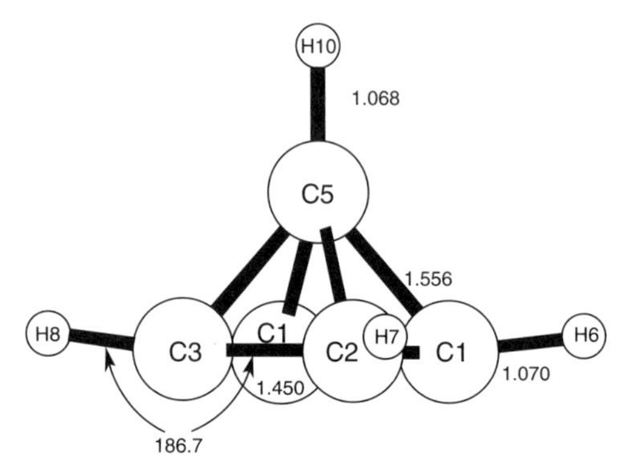

Figure 5.1 Square pyramidal cyclopentadienyl cation (reprinted from Ref. 7c with the permission of the American Chemical Society).

to the cyclopentadienyl cation are the nucleus-independent chemical shift (NICS), and the IGLO-calculated magnetic susceptibility exaltation λ_{tot}.[7b] For NICS absolute magnetic shieldings are computed at the ring centers, with negative NICS values indicating aromaticity and positive values, antiaromaticity.[7f] Magnetic susceptibility exaltations are obtained from the differences between calculated magnetic susceptibilities χ_{tot} and nonaromatic values derived from increments for hypothetical polyene systems without cyclic delocalization.[7g] For **5** the computed values of 32.6 (ppm cgs) for λ and and 54.1 for NICS are indicative of strong antiaromatic destabilization.[7b]

The pentamethylcyclopentadienyl cation **7** has been generated by reaction of pentamethylcyclopentadienyl bromide with $AgBF_4$ in CH_2Cl_2 and captured by nucleophiles, including methanol and methylamine leading to substitution products [Eq. (5.3a)].[8a] Tetramethylfulvene (**8**) was also formed from this cation by an elimination reaction:[8a]

$$C_5(CH_3)_5Br \xrightarrow[CH_2Cl_2]{AgBF_4} C_5(CH_3)_5^+ \; (\mathbf{7}) \qquad (5.3a)$$

$$\mathbf{7} \xrightarrow{CH_3OH} C_5(CH_3)_5OCH_3 + (CH_3)_4C_5{=}CH_2 \; (\mathbf{8}) \qquad (5.3b)$$

The first kinetic studies of the generation of a cyclopentadienyl cation, namely the reaction of **9** forming **10** (Scheme 1), were reported in 1997.[8b] Solvolysis in

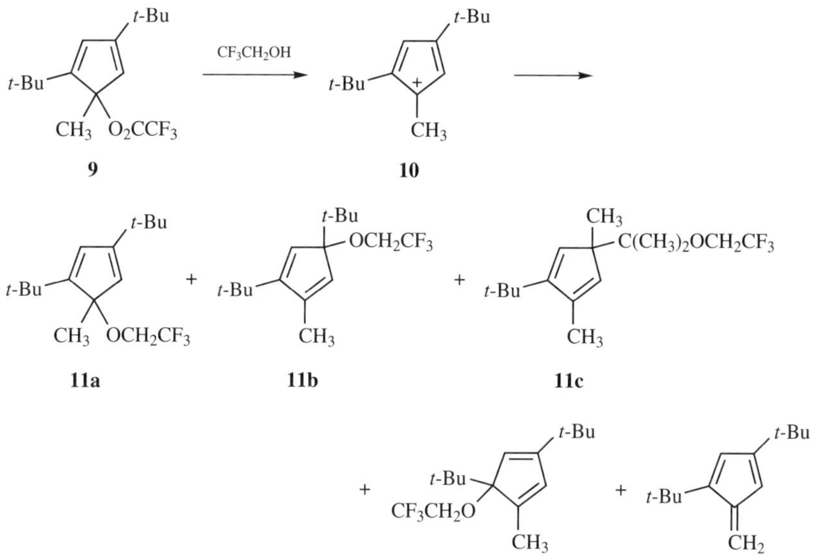

Scheme 1

CF_3CH_2OH (TFE) led to the products **11**, and the variation of the rate constants log k for **9** in different solvents gave a dependence m on the solvent parameters for ionization of trifluoroacetate Y_{OTFA} of 0.97. The diverse group of products formed and the strong dependence of the reactivity on the solvent ionizing power both gave convincing evidence for the formation of the cation **10**.

The rate constant for solvolysis of the cyclopentenyl trifluoroacetate **12** forming **13** was estimated to exceed that of **9** forming **10** by a factor of 10^{14}, and the enormous rate retardation due to the additional double bond in **9** was attributed to antiaromatic destabilization in the $4\pi e^-$ anti-Hückel system.[8b] The large magnitude of this destabilization is consistent with major destabilization due to antiaromaticity, as also indicated by the 31.2 kcal/mol destabilization of the parent cyclopentadienyl cation **3** noted in Eq. (5.2),[6b] the 41.2 kcal/mol destabilization computed relative to pentadienyl cation,[7b] and the calculated NICS and magnetic susceptibility exaltations:[7b]

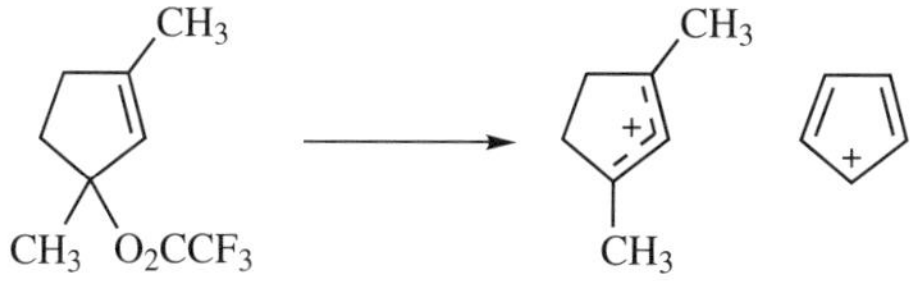

(5.4)

The 5-phenyl-5-cyclopentadienyl trifluoroacetate corresponding to **9** also showed evidence for strong antiaromatic destabilization of the derived carbocation, and was even less reactive than **9**. This effect was attributed to the presence of a twisted phenyl group that was less stabilizing than CH_3.[8b]

The secondary isotope effect $k(CH_3)/k(CD_3)$ is a useful diagnostic tool for the elucidation of the nature of carbocations, with smaller values for rather stabilized carbocations, and larger values for less stable cations with higher electron demand. For **9** this isotope effect was 1.30, consistent with the formation of a highly destabilized carbocation.[8c]

Electron-withdrawing substituents destabilize carbocations,[4h,9] and it was of interest to test for the combined effect of such a substituent in conjunction with the destabilization due to antiaromaticity in the cyclopentadienyl cation. Rate ratios for the substituents $k(H)/k(CF_3)$ ranging from 2.4 to 10^6 had been found for solvolysis of substrates **14–16** [Eq. (5.5)],[9e–g] but the cyclopentadienyl system is much more destabilized, and the fluoroalkyl substituent would be compared to CH_3. However, despite the greatly depressed reactivity of **9** compared to **12**, it was nevertheless notable that a rate constant for **9** could still be measured in hexafluoroisopropanol (HFIP) at 25°C,[8b] and this suggested that such a doubly destabilized carbocation might be experimentally accessible:

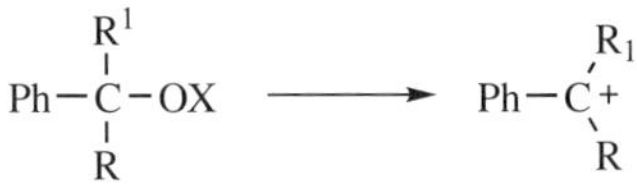

$$Ph-C(R^1)(R)-OX \longrightarrow Ph-C^+(R_1)(R) \quad (5.5)$$

14 R^1 = H, R = H or CF_3
15 R^1 = CH_3, R = H or CF_3
16 R^1 = CF_3, R = H or CF_3

The heptafluoropropylcyclopentadienyl ester **17** reacted in trifluoroethanol with initial rearrangement forming the isolable trifluoroacetate **18** [Eq. (5.6)].[8c] The structural assignment as **18** was favored over an isomeric structure **18a**.[8c] In a subsequent step **18** formed the rearranged fulvene **20**, in a process implicating the intermediate carbocation **19** [Eq. (5.7)]. The reaction of **17** forming the cation **19** was estimated to be retarded by a factor of at least 10^{20} compared to the ionization of **12**, showing the enormous destabilization of the doubly destabilized ion **19**, which nevertheless was generated in solution.[8c] These rate factors are commensurate with the thermodynamic destabilization found for the cyclopentadienyl cation,[6b] and show a cumulative effect of the destabilization due to the antiaromaticity of the cyclopentadienyl cation and the electron withdrawal by the perfluoropropyl group:

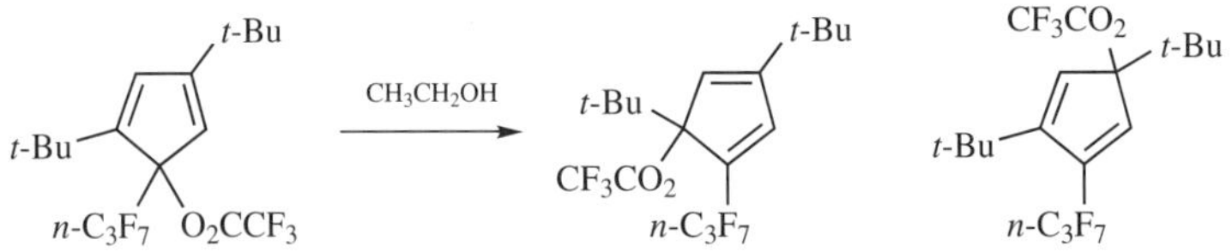

t-Bu, t-Bu, n-C_3F_7, O_2CCF_3 — **17** $\xrightarrow{CH_3CH_2OH}$ t-Bu, t-Bu, CF_3CO_2, n-C_3F_7 — **18**; CF_3CO_2, t-Bu, t-Bu, n-C_3F_7 — **18a** (5.6)

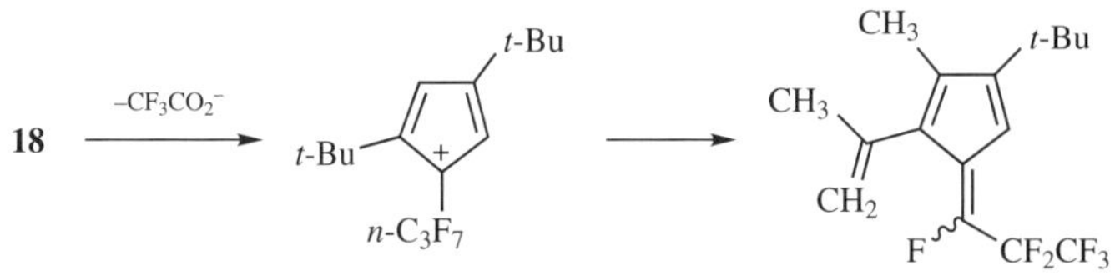

(5.7)

Two possible pathways for the rearrangement of **17** forming **18** may be envisaged. One is formation of a carbocation/trifluoroacetate ion pair, and a second is an electrocyclic rearrangement. To differentiate these possibilities, the carbonyl ^{18}O-labeled derivative **17**-^{18}O was prepared by reaction of the alcohol with ^{18}O-labeled $(CF_3CO)_2O$ [Eq. (5.8)].[8c] The ^{18}O content in **17**-^{18}O was determined by the ^{18}O-induced isotope shift of the ^{13}C NMR of the carbonyl carbon, and within the limits of detectability the rearrangement occurred exclusively with migration of the O originally bonded to the cyclopentadienyl moiety, and the reaction may be represented as a [1,5]-sigmatropic rearrangement through the transition state **21** [Eq. (5.9)]:

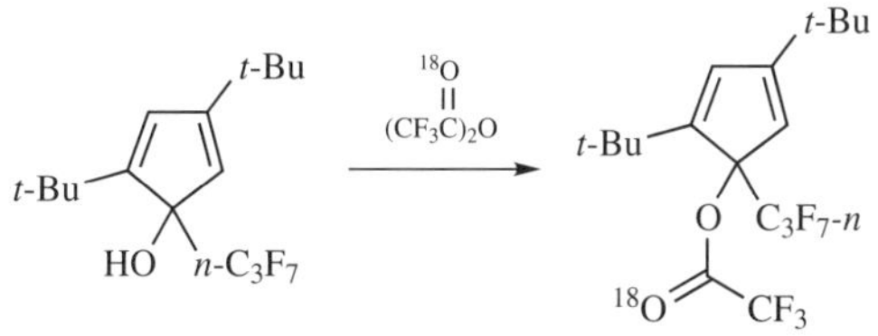

(5.8)

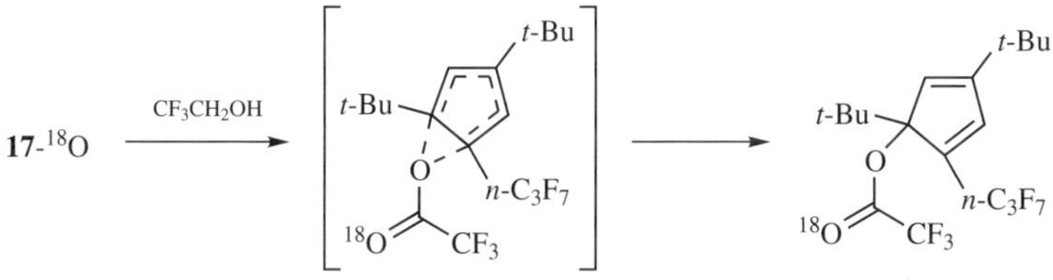

(5.9)

The occurrence of the rearrangement by a [1,5]-sigmatropic shift also provides further evidence against structure **18a**, as this could only be formed by a series of such processes, and would lead through a secondary trifluoroacetate expected to be less reactive than either **18** or **18a**. Rearrangements of trifluoroacetoxy groups in penta(aryl)cyclopentadienyl systems have been observed before, and were proposed to occur by [3,3]-sigmatropic pathways,[8d] in contrast to these results. The [3,3]-sigmatropic pathway was favored over the [1,5] pathway by analogy to another study of the rearrangement of an amidinyl species using NMR identification of the aryl groups.[8d] However, a [1,5] pathway was favored for the migration of arylthio groups in methyl(tetramethoxycarbonyl)cyclopentadienes.[8d] [1,5] Rearrangements are well documented in cyclopentadienyl derivatives.[8e,f]

The formation of the extensively rearranged fulvene **20** on further reaction of **18** provides strong evidence for carbocation formation, and this could be either **19** or the rearranged cation **22** formed directly. A possible pathway for the formation of **20** is shown in Scheme 2, and involves initial formation of **19** and then methyl migration forming **22** followed by proton loss and then ionization of fluoride to give cation **23**, which after methyl rearrangement and proton loss gives **20**. Loss of fluoride from perfluoroalkyl groups under ionizing conditions is well known.[9i]

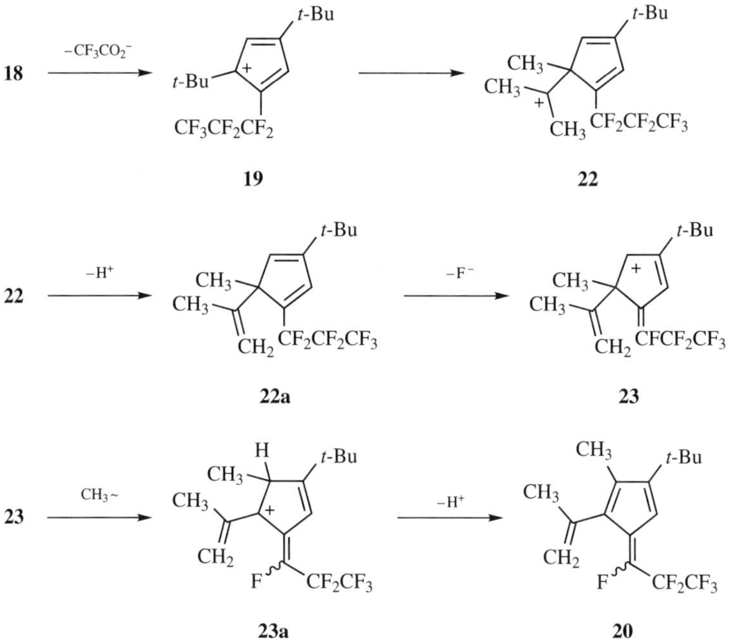

Scheme 2

In 2002 it was reported that pentamethylcyclopentadiene reacts with triphenylmethyl tetrakis(pentafluorophenyl)borate to rapidly form a crystalline salt of the pentamethylcyclopentadienyl cation **7**, which was characterized by ^{13}C NMR and X-ray crystallography [Eq. (5.13)].[10] The experimental evidence initially appeared convincing, even though the structure and the spectroscopic properties were highly unusual, and this report attracted wide attention. However, it was quickly found[11a–c] that the structure isolated was the salt of pentamethylcyclopentenyl cation **24**, which had been characterized by NMR in solution in 1969.[11d] The mechanism by which the reduced product **24** forms is still awaiting elucidation. Inasmuch as crystalline salts of the pentaphenylcyclopentadienyl cation were reported in 1963,[4b] and the pentamethylcyclopentadienyl cation was generated in solution

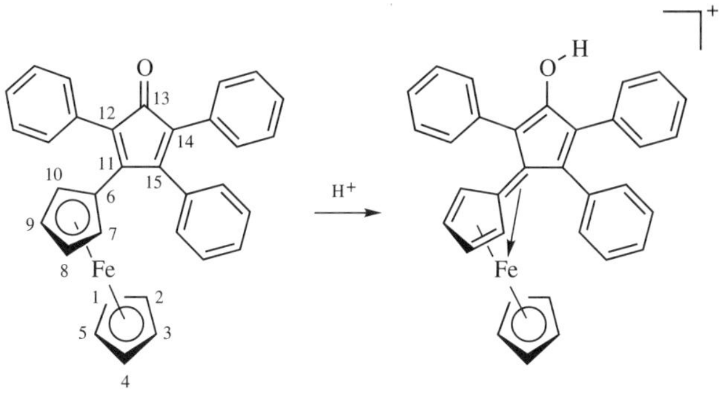

24a 24b

Figure 5.2 Superstabilized cyclopentadienyl cation (reprinted from Ref. 11e with the permission of the American Chemical Society).

and captured by nucleophiles,[8a] further efforts to characterize cyclopentadienyl cations may be expected:

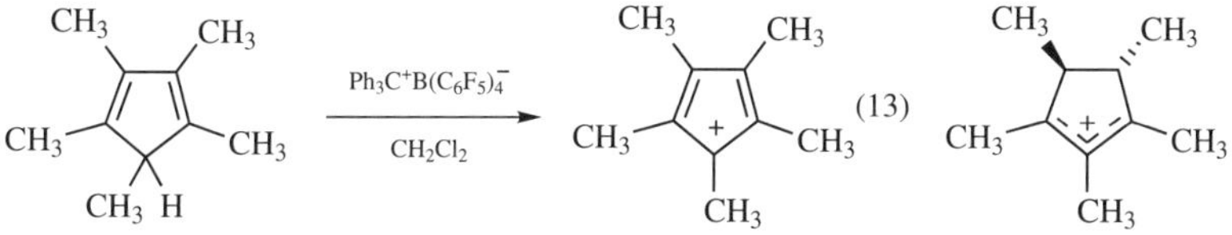

$$\text{C}_5(\text{CH}_3)_5\text{H} \xrightarrow[\text{CH}_2\text{Cl}_2]{\text{Ph}_3\text{C}^+\text{B}(\text{C}_6\text{F}_5)_4^-} \mathbf{7} \quad \mathbf{24} \tag{13}$$

An example of a superstabilized cyclopentadienyl cation substituted with 3 phenyl groups, an OH group, and a ferrocenyl group is formed by protonation of the cyclopentadienone **24a** giving **24b** (Fig. 5.2).[11e] The cation **24b** was characterized by NMR, IR, and computational studies, but an X-ray structure could not be obtained because of slow decomposition.[11e] It was thought this reaction might yield a stabilized cyclopentadienyl cation, but because of the extensive delocalization of positive charge into the ferrocenyl moiety, it was described as a hydroxyfulvalene complex.[11e]

5.2 INDENYL AND FLUORENYL CATIONS

Indenyl cation **25** and fluorenyl cation **26** are benzannelated derivatives of cyclopentadienyl cation, and are expected to have attenuated antiaromaticity compared to the cyclopentadienyl cation, and while **25** is unambiguously antiaromatic,[7b,12] there has been controversy regarding the fluorenyl cation **26**[7b,13–16] Neither of these carbocations has been generated and observed as a long-lived species in solution. 3-Indenyl 3,5-dinitrobenzoate **27** has a rate constant for solvolysis less than that for the saturated analog **27a** by a factor of 5×10^7, showing a major effect of antiaromaticity on reactivity.[12b] Reaction of 9-fluorenyl derivatives **28** show rate retardations of 10^3 compared to the benzhydryl analogues Ph_2CHX:[13]

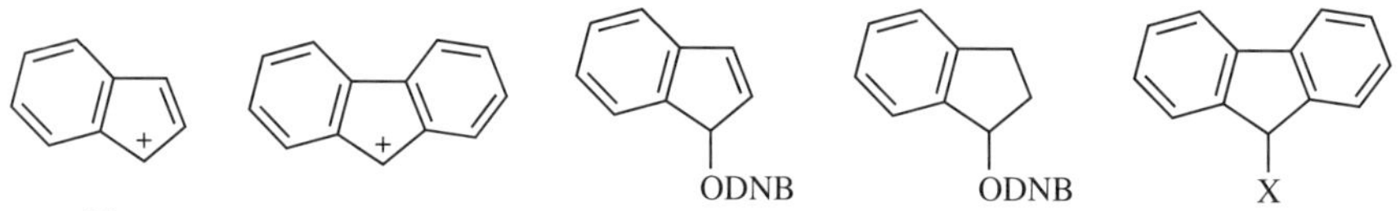

ODNB ODNB X

25 **26** **27** **27a** **28**

Solvolytic studies of the doubly destabilized 3-trifluoromethylindenyl tosylate **29a** indicated a diminution in reactivity of 10^9 relative to a comparable 1-indanyl analog, and factors of 10^6 and 10^3 were attributed to antiaromatic destabilization and to the rate-retarding effect of the electron-withdrawing CF_3 group, respectively.[9a] This reaction proceeded by internal return forming the rearranged tosylate **29c**, and an experiment with ^{18}O-labeled tosylate showed that there was no oxygen scrambling in unreacted **29a**, and extensive but not complete scrambling in **29c**. This result, the normal salt effect, and the significant dependence of the rate of reaction on solvent polarity indicated that the rearrangement occurred by an ionization/recombination pathway, and not by a sigmatropic rearrangement:

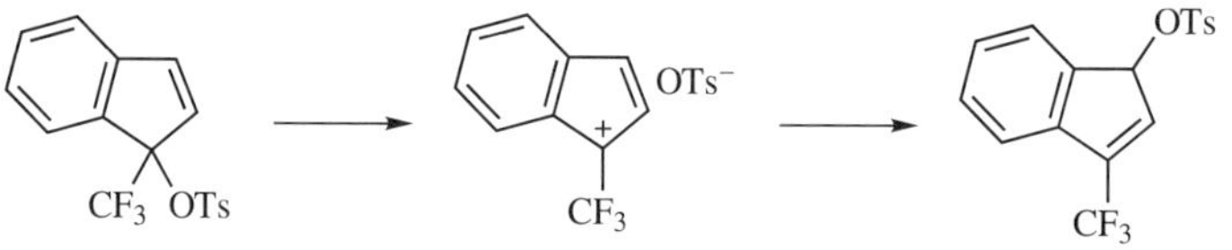

CF_3 OTs → OTs^- CF_3 → OTs CF_3 (5.10a)

29a **29b** **29c**

The reactivity of 9-CF_3-substituted fluorenyl tosylate **30a** was strongly depressed compared to **28**, and a rate factor of 10^6 due to antiaromatic destabilization of the cation **30b** was estimated [Eq. (5.10b)].[9b] In contrast to **29a**, the fluorenyl substrate showed a large special salt effect, in that addition of small amounts of NaO_2CCF_3 caused large rate accelerations attributed to a "special" salt effect, in which the salt scavenges solvent separated ion pairs and prevents their return reforming the reactant. Also **30a** showed a very strong enhancement of reactivity in more polar solvents compared to **29a**. The differences in solvent and salt effects were attributed to freer solvation of the developing carbocation center for reaction of **29a** and the presence of the allyic position in **29b** permitting internal return:

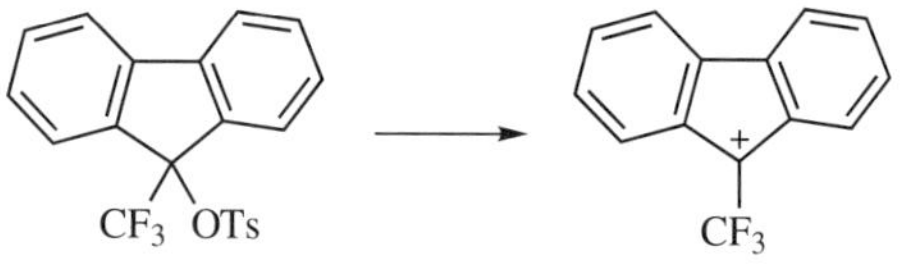

CF_3 OTs → CF_3 (5.10b)

30a **30b**

Solvolytic studies of formation of 9-fluorenyl carbocations with CO_2R,[14a,b] $CONMe_2$,[14c] and $CR{=}NOCH_3$[14d] substituents have also been reported. On the basis of the calculated geometry Creary et al.[14c,d] suggested that such cations avoid antiaromatic structures with cyclopentadienyl cation moieties, and resemble bis (dienyl) cations **31**:

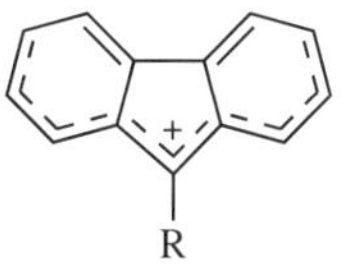

R

31

The difficulty of forming fluorenyl carbocations is reflected by pK_R values that are 4 units less than for benzhydryl derivatives,[15a] and calculations indicate that the fluorenyl cation is 8–10 kcal/mol less stable than benzhydryl analogues.[7b,16] Amyes et al.,[16a] however, judged these energy differences to be small and not indicative of antiaromatic destabilization. This view is somewhat surprising, as in many contexts an energy difference of 5 kcal/mol is considered quite large, especially in a comparison of the benzhydryl cation, which suffers from strain in attaining coplanarity and the fluorenyl cation constrained to coplanarity.

Fluorenyl cations are, however, quite readily formed and observed in flash photolysis experiments,[17] and the ease of formation of fluorenyl cations from triplet precursors is suggestive[17a,b] of ground-state destabilization of the singlet fluorenyl cation due to antiaromaticity. 9-Arylfluorenyl cations react with nucleophilic solvents with rate constants approximately two orders of magnitude greater than those for the corresponding monosubstituted triaryl cations.[17h] Such kinetic instability was one of the early criteria for antiaromaticity, and the parent fluorenyl cation **26** has been directly observed only on a nanosecond timescale in the very weakly nucleophilic hexafluoroisopropanol,[17d] or in zeolites.[17i]

Comparisons of the computed magnetic susceptibilities and nucleus-independent chemical shifts of cyclopentadienyl (**5**), indenyl (**25**), and fluorenyl (**26**) cations together with those of the corresponding aromatic anions are quite informative regarding the nature of the antiaromatic effects in these species.[7b] The same differences for more positive magnetic susceptibilities of the cations relative to the anions occur as benzene rings are incorporated, while the positive NICS values are strongly indicative of antiaromaticity for the five-membered rings of all three cations, while the six-membered ring for the indenyl cation is also antiaromatic, and even the values for the fluorenyl cation show much less aromaticity than those of the corresponding anions (-12.5 and -12.4, respectively), or of benzene (-11.5).[7b] Thus, by some criteria fluorenyl cation is nonaromatic, while by others there is appreciable residual antiaromaticity.

5.3 BISFLUORENYL DICATIONS

The tetrabenzo[5.5]fulvalene dication **32** and derivatives have been generated experimentally[18a–d] and studied computationally.[18e–h] The ^{1}H NMR spectrum of **32** has signals at δ 5.07 to 5.87 with a center of gravity of δ 5.41, and the upfield shifts from the neutral precursor, with a center of gravity of δ 7.75 ppm, indicate a substantial paratropic ring current suggesting the species is antiaromatic.[18a] The NICS values and magnetic susceptibility exaltation of **32** also indicate antiaromaticity.[18e] It was suggested that for 9-substituted fluorenyl cations **31**, compared to the parent fluorenyl cation **26**, the benzene aromaticity predominates, and that this tendency is increased by electron donor groups.[18e] However, with electron demanding substituents at C9, there is extra electron delocalization to this position from the benzene rings, and this in turn results in enhanced antiaromaticity effects. It is suggested that the fluorenyl rings in **32** are orthogonal, and that σ–π conjugation occurs from the C–C bond of one ring to the empty *p* orbital of the other as shown in **33**.[18a,e] This effect would explain the upfield ^{13}C shift at the carbocation centers in **32**:

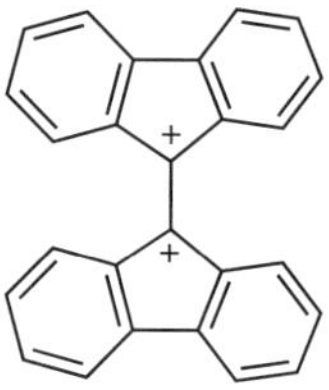

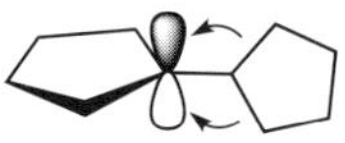

33

32

Both 3,6- and 2,7-disubstituted tetrabenzo[5.5]fulvene dications **34** and **35**, respectively, are generated by SbF_5 oxidation of the corresponding fulvenes at −78°C in SO_2ClF and characterized by ^{1}H and ^{13}C NMR [Eq. (5.11)].[18f–h] The structures of the dications and their NICS values were also computed, and it was concluded that substituents in the 3,6 positions favored exomethylene-2,5-cyclohexadienyl-type resonance structures **34a**, which decreased the cyclopentadienyl cation character and the antiaromaticity of the cations:

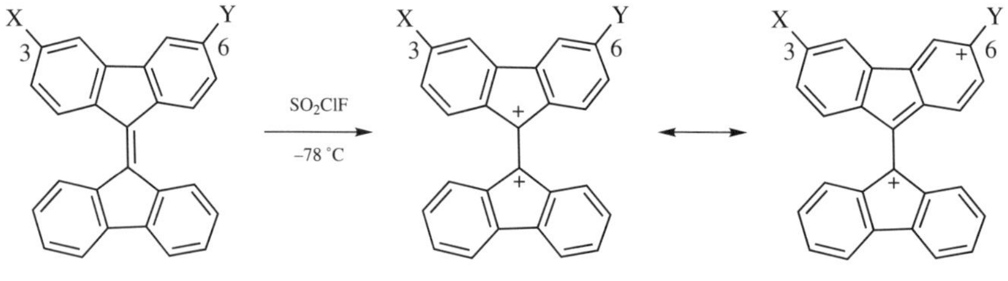

X Y 3 6 $\xrightarrow[-78\ ^{\circ}C]{SO_2ClF}$ **34** ⟷ **34a**

(5.11)

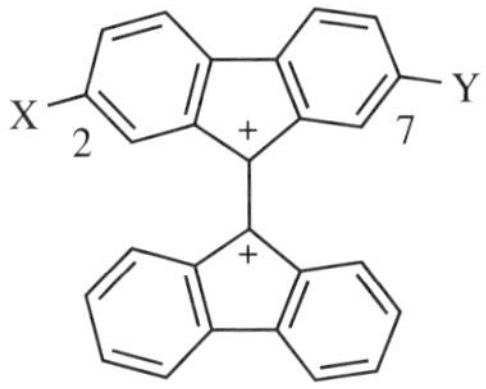

35

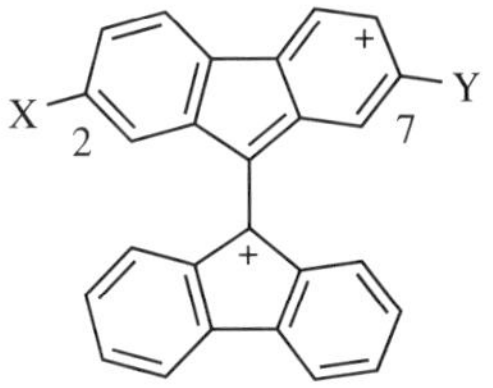

35a

5.4 AROMATICITY AND ANTIAROMATICITY IN CYCLOPENTADIENYL RADICALS

Free radicals are electron-deficient species, and many of the criteria used for assessing antiaromaticity in cations may also be applied to radicals, and so study of radicals may shed light on the analogous carbocations. Cyclopentadienyl radicals[19] have been known as long-lived species since Ziegler and Schnell reported the pentaphenyl derivative in 1925,[2a] and others include aryl substituted derivatives[19b] as well as pentaisopropyl,[19c] pentafluoro,[19d] pentachloro,[19e] pentamethyl,[19f–h] and the parent.[19i–p] The singlet and triplet states of these radicals are usually close in energy.

The ESR spectra of the parent cyclopentadienyl radical **36** and of a variety of monosubstituted derivatives $RC_5H_4\bullet$ (R = D, Me, Et, *i*-Pr, and *t*-Bu)[19e,f] were measured in solution. The spectrum of **36** favors a planar π-radical with average D_{5h} symmetry.[19e] This radical is expected to exist as an equilibrium mixture of five pairs of C_{2v} structures due to Jahn–Teller distortion, with a very low barrier for rearrangement.

Resonance stabilization energies (RSEs)[19q] of radicals provide a criterion to evaluate the effects of electron delocalization, and are obtained as the difference of the bond dissociation energies (BDEs) of precursors R_sX, which form delocalized radicals, and the BDE of substrates $R_{ns}X$, which form comparable nondelocalized radicals:

$$\mathrm{RSE}(R_s\bullet) = \mathrm{BDE}(R_{ns}X) - \mathrm{BDE}(R_sX) \tag{5.12}$$

Estimates of the RSE of cyclopentadienyl (**36**), indenyl (**37**), and fluorenyl radicals (**38**)[19p] and C—H bond dissociation energies (BDE) of precursors[19b,d,e,h] indicate that these radicals are less stabilized than are nonaromatic cyclic analog; typically the C—H BDE of cyclopentadiene, indene, and fluorene are 5–7 kcal/mol greater than those of the corresponding cyclohexadienes, and are the same as that of cyclopentene:

36 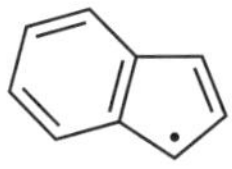 37 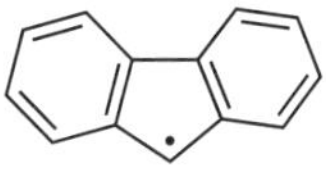 38

Calculated nucleus-independent chemical shifts (NICS) and magnetic susceptibilities for the cyclopentadienyl radical **36**[7i] are intermediate between those of the singlet cation **5** and the cyclopentadienyl anion.[7i] The calculated ^{1}H chemical shift of **36** (δ 6.5) is intermediate between those calculated for singlet and triplet **5** (δ 5.2 and 8.0, respectively), and upfield from that for benzene (δ 7.8).[7i] Further insight into the possible antiaromatic character of the cyclopentadienyl radical is provided by the magnetic properties, which have been studied computationally.[7i] The nucleus independent chemical shift (NICS), which is a shift calculated for the ring center, is 2.6 for C_5H_5•, compared to the values of 27.6 for the cyclobutadiene singlet (antiaromatic) and -5.3 for the triplet (aromatic), and 49.2 for cyclopentadienyl cation singlet (antiaromatic) and -4.5 for the triplet (aromatic). Benzene has a NICS value of -9.7, and negative values are diagnostic for aromatic species. By this criterion the triplet cyclopentadienyl radical with a positive NICS of low magnitude is nonaromatic, or modestly antiaromatic. This is consistent with the experimental result that this radical has neither strong aromatic stabilization nor antiaromatic destabilization.

Kinetic studies of reactions forming cyclopentadienyl radicals had not been investigated. Radical additions to fulvenes provide a potential means for such measurements, and the fulvenones **39**–**41** are attractive substrates for such studies. The photochemical generation of substituted derivatives of indenylideneketene (**40**) formed by Wolff rearrangement of substituted derivatives of **42** [Eq. (5.13)] and the subsequent ketene hydration to the corresponding acids play a decisive role in the manufacture of integrated circuits by microlithography,[21] and so the photochemical generation and ionic chemistry of **39**[22a], **40**[22b,c] and **41**[22d,e] has been the subject of intensive investigation. The antiaromatic character of **39** has also been examined computationally:[22f,g]

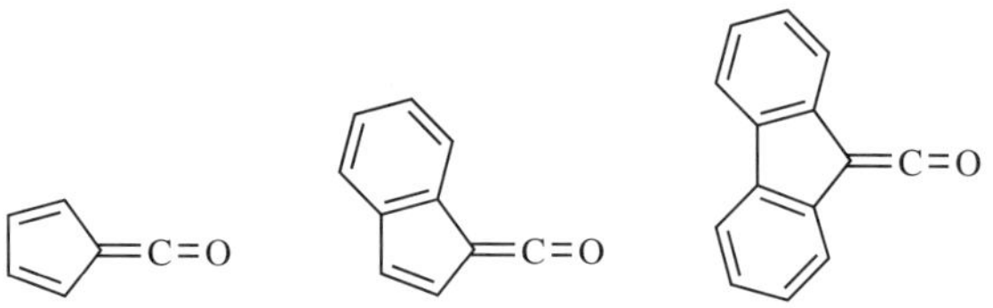

=C=O =C=O =C=O

39 **40** **41** (5.13)

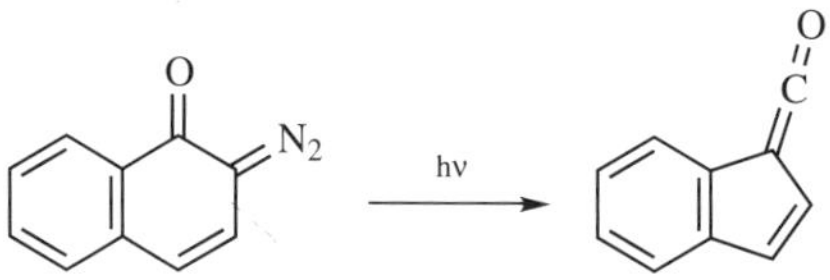

O N2 hν O C

42 **40**

We have been studying radical reactions of ketenes,[23] and in computational studies of the reaction of $CH_2{=}C{=}O$ with the aminoxyl TEMPO (TO•) using density functional methods at the B3LYP/6-31++G**//B3LYP/6-31++G** level have shown that while addition at C_β is predicted to be endothermic by 23.6 kcal/mol, the addition at the carbonyl carbon is more favorable by 18.8 kcal/mol [Eq. (5.14)].[23b,h] Substituents at C_β are expected to stabilize radical **44** and experimentally it was observed that TEMPO reacts with ketenes with a first-order dependence on [TEMPO] forming products derived from initial addition of TEMPO to the carbonyl carbon.[23d,e] As illustrated for $CH_2{=}C{=}O$, this results in an intermediate radical that may add a second TEMPO at C_β [Eq. (5.14)], while the fate of other such initial radicals depends on the particular structure involved:[23b–e]

$$\underset{\mathbf{43}}{\mathrm{TOC\dot{H}C{=}O}} \xleftarrow[23.6\ \mathrm{kcal/mol}]{\Delta E=} \mathrm{CH_2{=}C{=}O} + \mathrm{TO\cdot} \xrightarrow[4.8\ \mathrm{kcal/mol}]{\Delta E=} \underset{\mathbf{44}}{\mathrm{\dot{C}H_2C(=O)OT}} \xrightarrow{\mathrm{TO\cdot}} \underset{\mathbf{45}}{\mathrm{TOCH_2C(=O)OT}} \quad (5.14)$$

Previous investigations indicate that cyclopentadienyl radicals are quite reactive unless heavily substituted, but there has been little explicit consideration of whether these species more closely resemble the highly stabilized aromatic cyclopentadienyl anions[6b,d] or the highly destabilized antiaromatic cyclopentadienyl cations.

Calculations at the B3LYP/6-31G*//B3LYP/6-31G* level were carried out of the structures and energies of **39** and the adduct **46** formed by reaction with aminoxyl ($H_2NO\cdot$) and the reaction was predicted to proceed by attack in the molecular plane at the carbonyl carbon and to be exothermic by 24.6 kcal/mol [Eq. (5.15)].[23f] This reaction is more favorable than that of $CH_2{=}C{=}O$ [Eq. (5.14)] by 5.9 kcal/mol at the same level:[23f]

$$\underset{\mathbf{39}}{\mathrm{C_5H_4{=}C{=}O}} + \mathrm{H_2NO\cdot} \xrightarrow[-24.6\ \mathrm{kcal/mol}]{\Delta E=} \underset{\mathbf{46}}{\mathrm{C_5H_4^{\bullet}C(=O)ONH_2}} \quad (5.15)$$

Fulvenones **39–41** were generated by Wolff rearrangement, their IR spectra were measured, and the products and kinetics of their reactions with TEMPO were determined.[23f] Generation of **39** in the presence of TEMPO gave the dimeric material **47** (66%) and the new products **48** and **49**, in 8 and 12% yields of the soluble product [Eq. (5.16)]. The structure of **48** was established by 2D NMR techniques, and the structure of **49** was established by X-ray crystallography.[23f] The same products were observed when the reaction was conducted by first generating **39** by photolysis and then adding TEMPO. Evidently formation of **48** and **49** occurs via initial radical attack on the carbonyl carbon to form **50**:[23f]

$$\underset{\mathbf{39}}{\mathrm{C_5H_4{=}C{=}O}} \xrightarrow{\mathrm{TO^\bullet}} \underset{\mathbf{50}}{\mathrm{C_5H_4^{\bullet}{-}CO_2T}} \xrightarrow{\mathrm{TO^\bullet}} \underset{\mathbf{48}}{\text{(H)(TO)-cyclopentenone-}\mathrm{CO_2T}} + \underset{\mathbf{49}}{\text{TO-cyclopentenone-}\mathrm{C(CO_2T)(OH)}} \quad (5.16)$$

47

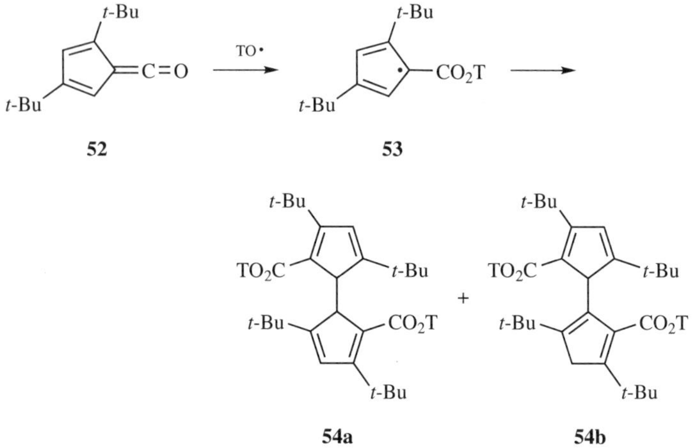

t-Bu C=O TO• t-Bu CO_2T t-Bu t-Bu

52 53

t-Bu TO_2C t-Bu t-Bu CO_2T t-Bu + t-Bu TO_2C t-Bu t-Bu CO_2T t-Bu

54a 54b

Scheme 3

Confirmation of this reaction pathway came from generation of the fulvenone **52** followed by the addition of TEMPO leading to the isomeric dimers **54a,b** in a 37/63 ratio (Scheme 3).[23f]

Reaction of the fulvenone **40** with TEMPO and examination of the product by ^{1}H NMR showed the formation of a mixture of **56–58** derived from intermediate **55**, and on chromatographic separation complete conversion to ketone **58** was observed (Scheme 4).[23f]

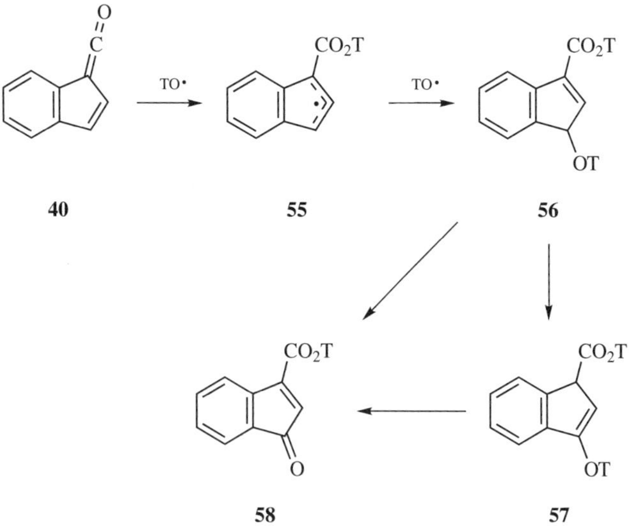

O C TO• CO_2T TO• CO_2T OT

40 55 56

CO_2T O CO_2T OT

58 57

Scheme 4

Fulvenone **41** reacted with TEMPO to give the dimer **60** of the radical **59** in 32% yield:

$$\text{41} \xrightarrow{\text{TO}\bullet} \text{59 } (\bullet\text{—CO}_2\text{T}) \longrightarrow \text{60 } (\text{CO}_2\text{T}, \text{T}_2\text{OC}) \qquad (5.17)$$

41 **59** **60**

The kinetics of the reactions of **39**–**41** and **52** with TEMPO were measured, and compared to the rates of hydration.[23f] 2,4-Di-*tert*-butylfulvenone **52** is significantly less reactive than **39** in both the reaction with H_2O and with TEMPO, by factors of 180 and 170, respectively, so the rate ratio $k(H_2O)/k_2(\text{TEMPO})$ is essentially constant. The indenylidene and fluorenylidene ketenes **40** and **41** are significantly more reactive toward TEMPO than **39**, by factors of 21 and 24, respectively.

A qualitative correlation found[23c,d] between the rate constants for reaction of ketenes with TEMPO and with H_2O has been extended as shown in Eq. (5.18) (Fig. 5.3).[23f,h] For the indenyl and fluorenyl ketenes **40** and **41**, the ratios of the rate constants predicted by this equation and those observed for reaction with

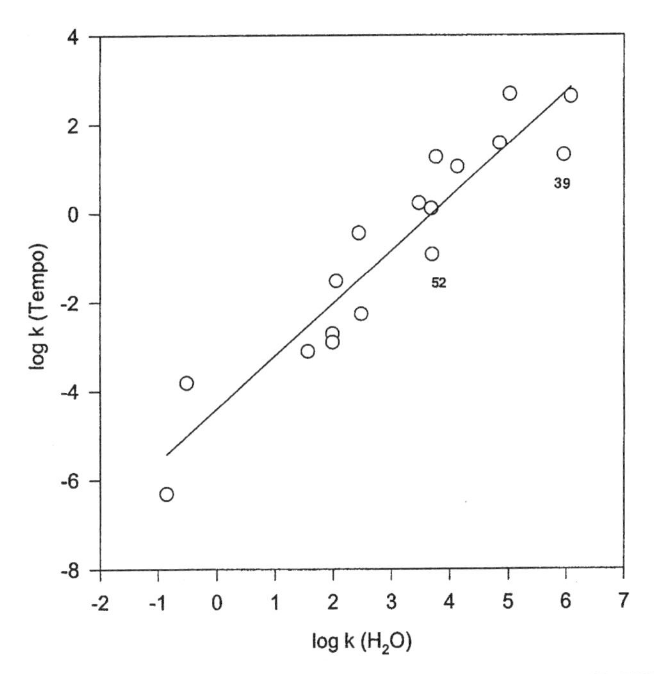

Figure 5.3 Correlation between rate constants for reactions of ketenes with TEMPO versus water.

TEMPO are k_2(obs)/k_2(pred) of 0.70 and 14, respectively (Fig. 5.3). For the cyclopentadienyl derivatives **39** and **52**, these rate constant ratios are 0.046 and 0.13, respectively, indicating significantly slower rates for **39** and **52** with TEMPO than predicted by the correlation:

$$\log k_2(\text{TEMPO}) = 1.20 \log k(\text{H}_2\text{O}) - 4.45, (r = 0.94) \qquad (5.18)$$

The reactivities of the ketenes may also be compared by the k_2(TEMPO)/k(H_2O) rate ratios, which are 35 and 450 for the indenyl and fluorenyl ketenes **40** and **41**, and 2.2 and 2.4 for **39** and **52**. This comparison is also suggestive of rate retardations of the reactions of **39** and **52** with TEMPO, which are attenuated in the benzannelated derivatives **40** and **41**.

The rate retardations of the pentafulvenones **39** and **52** with TEMPO are relevant to the question of the possible destabilization of the radical species formed in these reactions due to the presence of 5π-electron antiaromatic effects. These effects echo in greatly diminished form the destabilization of the cyclopentadienyl, indenyl, and fluorenyl cations, which decreases progressively through this series.[6b]

In summary, the chemistry of cyclopentadienyl cations and radicals continues to provide new insights into the phenomena of aromaticity and antiaromaticity more than 75 years after they were first investigated by Ziegler. These species by no means have yielded all their secrets, and further revelations may be expected.

ACKNOWLEDGMENT

Financial support by the Natural Sciences and Engineering Research Council of Canada, the Petroleum Research Fund, and the Canada Council for the Arts for a Killam Fellowship to T. T. T. is gratefully acknowledged.

REFERENCES

1. (a) G. A. Olah, *J. Org. Chem.* **66**, 5943–5957 (2001); (b) J. F. Norris, and W. W. Sanders, *Am. Chem. J.* **25**, 54–62 (1901); (c) J. F. Norris, *Am. Chem. J.* **25**, 117–122 (1901); (d) F. Kehrmann and F. Wentzel, *Chem. Ber.* **34**, 3815–3819 (1901); (e) A. Baeyer and V. Villiger, *Chem. Ber.* **35**, 1189–1201 (1902); (f) A. Baeyer and V. Villiger, *Chem. Ber.* **35**, 5943–5957 (1902); (g) M. Gomberg, *Chem. Ber.* **35**, 2397–2408 (1902).
2. (a) K. Ziegler and B. Schnell, *Liebigs Ann.* **445**, 266–282 (1925); (b) K. Ziegler and H. Wollschitt, *Liebigs Ann.* **479**, 90–110 (1930).
3. (a) E. Hückel, *Z. Physik* **70**, 204–286 (1931); (b) J. D. Roberts, A. Streitwieser, Jr., and C. M. Regan, *J. Am. Chem. Soc.* **74**, 4579–4582 (1952); (c) S. M. Bloom and A. P. Krapcho, *Chem. Ind.* 882 (1959).
4. (a) R. Breslow and H. W. Chang, *J. Am. Chem. Soc.* **83**, 3727–3728 (1961); (b) R. Breslow, H. W. Chang, and W. A. Yager, *J. Am. Chem. Soc.* **85**, 2033–2034 (1963); (c) R. Breslow, H. W. Chang, R. Hill, and E. Wasserman, *J. Am. Chem. Soc.* **89**, 1112–1119 (1967); (d) R. Breslow, *Chem. Eng. News* 90–99 (June 28, 1965); (e) R. Breslow, *Acc. Chem. Res.*

6, 393–398 (1973); (f) K. B. Wiberg, *Chem. Rev.* **101**, 1317–1332 (2001); (g) A. D. Allen and T. T. Tidwell, *Chem. Rev.* **101**, 1333–1348 (2001); (h) T. T. Tidwell, *Angew. Chem. Int. Ed. Engl.* **23**, 20–32 (1984).

5. P. Y. Bruice, *Organic Chemistry*, 3rd ed., Prentice-Hall, New York, 2001; T. W. G. Solomons, *Organic Chemistry*, 5th ed., Wiley, New York, 1992; S. N. Ege, *Organic Chemistry*, 4th ed., Wiley, New York, 1999; F. A. Carey, *Organic Chemistry*, 5th ed., McGraw-Hill, New York, 2003.
6. (a) R. Breslow, in *Magnetic Properties Organic Materials*, P. Lathi, ed., Marcel Dekker, New York, 1999, Chapter 3, pp. 27–40; (b) F. P. Lossing and J. C. Traeger, *J. Am. Chem. Soc.* **97**, 1579–1580 (1975).
7. (a) M. Saunders, R. Berger, A. Jaffe, J. M. McBride, J. O'Neill, R. Breslow, J. M. Hoffman, Jr., C. Perchonock, E. Wasserman, R. S. Hutton, and V. J. Kuck, *J. Am. Chem. Soc.* **95**, 3017–3018 (1973); (b) H. Jiao, P. v. R. Schleyer, Y. Mo, M. A. McAllister, and T. T. Tidwell, *J. Am. Chem. Soc.* **119**, 7075–7083 (1997); (c) J. Feng, J. Leszczynski, B. Weiner, and M. C. Zerner, *J. Am. Chem. Soc.* **111**, 4648–4655 (1989); (d) M. N. Glukhovtsev, R. D. Bach, and S. Laiter, *J. Phys. Chem.* **100**, 10952–10955 (1996); (e) B. Reindl and P. v. R. Schleyer, *J. Comput. Chem.* **19**, 1402–1420 (1998); (f) P. v. R. Schleyer, C. Maerker, A. Dransfeld, H. Jiao, and N. J. R. v. E. Hommes, *J. Am. Chem. Soc.* **118**, 6317–6318 (1996); (g) H. J. Dauben, Jr., J. D. Wilson, and J. L. Laity in *Non-benznoid Aromatics*, J. Snyder, ed., Academic Press, 1971, Vol 2; (h) N. C. Baird, *J. Am. Chem. Soc.* **94**, 4941–4948 (1972); (i) V. Gogonea, P. v. R. Schleyer, and P. R. Schreiner, *Angew. Chem. Int. Ed. Engl.* **37**, 1945–1948 (1998); (j) D. Leblanc, J. Kong, P. S. Mayer, and T. H. Morton, *Int. J. Mass Spectrom.* **222**, 451–463 (2003); (k) T. H. Morton, *Abstracts of Papers*, 225th National Meeting of the American Chemical Society, March 23–27, 2003, New Orleans, LA, 2003; ORGN 38.
8. (a) P. Jutzi and A. Mix, *Chem. Ber.* **125**, 951–954 (1992); (b) A. D. Allen, M. Sumonja, and T. T. Tidwell, *J. Am. Chem. Soc.* **119**, 2371–2375 (1997); (c) A. D. Allen and T. T. Tidwell, *J. Org. Chem.* **66**, 7696–7699 (2001); (d) G. A. Dushenko, I. E. Mikhailov, I. A. Kamenetskaya, R. V. Skachkov, A. Zhunke, K. Myugge, and V. I. Minkin, *Russ. J. Org. Chem.* **30**, 1559–1564 (1994); (e) S. McLean, C. J. Webster, and R. J. Rutherford, *Can. J. Chem.* **47**, 1555–1559 (1969); (f) H. Jiao and P. v. R. Schleyer, *J. Chem. Soc. Faraday Trans.* **90**, 1559–1567 (1994).
9. (a) A. D. Allen, N. Mohammed, and T. T. Tidwell, *J. Org. Chem.* **62**, 246–252 (1997); (b) A. D. Allen, J. D. Colomvakos, O. S. Tee, and T. T. Tidwell, *J. Org. Chem.* **59**, 7185–7187 (1994). (c) A. D. Allen, M. Fujio, O. S. Tee, T. T. Tidwell, Y. Tsuji, Y. Tsuno, and K. Yatsugi, *J. Am. Chem. Soc.* **117**, 8974–8981 (1995); (e) W. Kirmse, A. Wonner, A. D. Allen, and T. T. Tidwell, *J. Am. Chem. Soc.* **114**, 8828–8835 (1992); (f) A. D. Allen, V. M. Kanagasabapathy, and T. T. Tidwell, *J. Am. Chem. Soc.* **108**, 3470–3474 (1986); (g) A. D. Allen, V. M. Kanagasabapathy, and T. T. Tidwell, *J. Am. Chem. Soc.* **105** 5961–5962 (1983); (h) A. D. Allen, I. C. Ambidge, C. Che, H. Micheal, R. J. Muir, and T. T. Tidwell, *J. Am. Chem. Soc.* **105**, 2343–2350 (1983); (i) A. D. Allen, M. P. Jansen, K. M. Koshy, N. N. Mangru, and T. T. Tidwell, *J. Am. Chem. Soc.* **104**, 207–211 (1982); (j) K.-T. Liu, M.-Y. Kuo, and C.-F. Sheu, *J. Am. Chem. Soc.* **104**, 211–215 (1982); (k) K. M. Koshy, D. Roy, and T. T. Tidwell *J. Am. Chem. Soc.* **101**, 357–363 (1979).
10. J. B. Lambert, L. Lin, and V. Rassolov, *Angew. Chem. Int. Ed.* **41**, 1429–1431 (2002).
11. (a) M. Otto, D. Scheschkewitz, T. Kato, M. M. Midland, J. B. Lambert, and G. Bertrand, *Angew. Chem. Int. Ed.* **41**, 2275–2276 (2002); (b) T. Müller, *Angew. Chem. Int. Ed.* **41**,

2276–2277 (2002); (c) J. B. Lambert, *Angew. Chem. Int. Ed.* **41**, 2278 (2002); (d) P. H. Campbell, N. W. K. Chiu, I. J. Miller, and T. S. Sorensen, *J. Am. Chem. Soc.* **91**, 6404–6410 (1969); (e) L. E. Harrington, I. Vargas-Baca, N. Reginato, and M. J. McGlinchey, *Organometallics* **22**, 663–669 (2003).

12. (a) E. C. Friedrich and D. B. Taggart, *J. Org. Chem.* **43**, 805–808 (1978); (b) E. C. Friedrich and T. M. Tam, *J. Org. Chem.* **47**, 315–319 (1982); (c) E. C. Friedrich, D. B. Taggart, and M. A. Saleh, *J. Org. Chem.* **42**, 1437–1443 (1977).
13. (a) A. Ledwith and D. G. Morris, *J. Chem. Soc.* 508–509 (1964); (b) G. W. Cowell, T. D. George, A. Ledwith, and D. G. Morris, *J. Chem. Soc. B* 1169–1172 (1966); (c) G. W. Cowell and A. Ledwith, *J. Chem. Soc. B* 695–697 (1967).
14. (a) L. J. Johnston, P. Kwong, A. Shelemay, and E. Lee-Ruff, *J. Am. Chem. Soc.* **115**, 1664–1669 (1993); (b) C. S. Q. Lew, B. D. Wagner, M. P. Angelini, E. Lee-Ruff, J. Lusztyk, and L. J. Johnston, *J. Am. Chem. Soc.* **118**, 12066–12073 (1996); (c) X. Creary and J. Tricker, *J. Org. Chem.* **63**, 4907–4911 (1998); (d) X. Creary and A. Wolf, *J. Phys. Org. Chem.* **13**, 337–343 (2000).
15. (a) N. Deno, J. Jaruzelski, and A. Schriesheim, *J. Am. Chem. Soc.* **77** 3044–3051 (1955); (b) G. A. Olah, G. K. S. Prakash, G. Liang, P. W. Westerman, K. Kunde, J. Chandrasekhar, and P. v. R. Schleyer, *J. Am. Chem. Soc.* **102**, 4485–4492 (1980); (c) Y. Hou and C. Y. Meyers, *J. Org. Chem.* **69**, 1186–1195 (2004).
16. (a) T. L. Amyes, J. P. Richard, and M. Novak, *J. Am. Chem. Soc.* **114**, 8032–8041 (1992); (b) C. F. Rodriquez, D. L. Vuković, and A. C. Hopkinson, *J. Mol. Struct. (Theochem)* **363**, 131–138 (1996).
17. (a) P. Wan and E. Krogh, *J. Chem. Soc. Chem. Commun.*, 1207–1208 (1985); (b) P. Wan and E. Krogh, *J. Am. Chem. Soc.* **111**, 4887–4895 (1989); (c) S. L. Mecklenburg and E. F. Hilinski, *J. Am. Chem. Soc.* **111**, 5471–5472 (1989); (d) R. A. McClelland, N. Mathivanan, and S. Steenken, *J. Am. Chem. Soc.* **112**, 4857–4861 (1990); (e) W. Kirmse, J. Kilian, and S. Steenken, *J. Am. Chem. Soc.* **112**, 6399–6400 (1990); (f) F. Cozens, J. Li, R. A. McClelland, and S. Steenken, *Angew. Chem. Int. Ed. Engl.* **31**, 743–745 (1992); (g) C. S. Q. Lew, R. A. McClelland, L. J. Johnston, and N. P. Schepp, *J. Chem. Soc. Perkin Trans. 2* 395–397 (1994); (h) F. L. Cozens, N. Mathivanan, R. A. McClelland, and S. Steenken, *J. Chem. Soc. Perkin Trans. 2* 2083–2090 (1992); (i) M. A. O'Neill, F. L. Cozens, and N. P. Schepp, *Tetrahedron* **56**, 6969–6977 (2000).
18. (a) J. L. Melandra, N. S. Mills, D. E. Kadlecek, and J. A. Lowery, *J. Am. Chem. Soc.* **116**, 11622–11623 (1994); (b) N. S. Mills, J. L. Malandra, E. E. Burns, A. Green, K. E. Unruh, D. E. Kadlecek, and J. A. Lowery, *J. Org. Chem.* **62**, 9318–9322 (1997); (c) N. S. Mills, E. E. Burns, J. Hodges, J. Gibbs, E. Esparza, J. L. Malandra, and J. Koch, *J. Org. Chem.* **63**, 3017–3022 (1998); (d) N. S. Mills, T. Malinky, J. L. Malandra, E. E. Burns, and P. Crossno, *J. Org. Chem.* **64**, 511–517 (1999); (e) N. S. Mills, *J. Am. Chem. Soc.* **121**, 11690–11696 (1999). (f) N. S. Mills, M. M. Benish, C. Ybarra, *J. Org. Chem. Soc.* **67**, 2003–2012 (2002); (g) N. S. Mills, *J. Org. Chem.* **67**, 7029–7036 (2002); (h) A. Levy, A. Rakowitz, and N. S. Mills, *J. Org. Chem. Soc.* **68**, 3990–3998 (2003).
19. (a) W. Kieslich and H. Kurreck, *J. Am. Chem. Soc.* **106**, 4328–4335 (1984); (b) H. Sitzmann, H. Bock, R. Boese, T. Dezember, Z. Havlas, W. Kaim, M. Moscherosch, and L. Zanathy, *J. Am. Chem. Soc.* **115**, 12003–12009 (1993); (c) T. Chen and Hs. H. Günthard, *Chem. Phys.* **97**, 187–203 (1985); (d) P. Bachmann, F. Graf, and Hs. H. Günthard, *Chem. Phys.* **9**, 41–56 (1975); (e) P. J. Barker, A. G. Davies, and M.-W. Tse, *J. Chem. Soc. Perkin Trans. 2* 692–696 (1981); (g) P. N. Culshaw, J. C. Walton, L. Hughes, and K. U. Ingold,

J. Chem. Soc. Perkin Trans. 2 879–886 (1993); (h) W. R. Roth and F. Hunold, *Liebigs Ann.* 1119–1122 (1995); (i) L. Yu, J. M. Williamson, and T. A. Miller, *Chem. Phys. Lett.* **162**, 431–436 (1989); (j) L. Yu, S. C. Foster, J. M. Williamson, M. C. Heaven, and T. A. Miller, *J. Phys. Chem.* **92**, 4263–4266 (1988); (k) S. Furuyama, D. M. Golden, and S. W. Benson, *Int. J. Chem. Kinet.* **3**, 237–248 (1971); (l) D. J. DeFrees, R. T. McIver, Jr., and W. J. Hehre, *J. Am. Chem. Soc.* **102**, 3334–3338 (1980); (m) F. G. Bordwell, J.-P. Cheng, and J. A. Harrelson, Jr., *J. Am. Chem. Soc.* **110**, 1229–1231 (1988); (n) E. M. Arnett, T. C. Moriarity, L. E. Small, J. P. Rudolph, and R. P. Quirk, *J. Am. Chem. Soc.* **95**, 1492–1495 (1973); (o) W. Yi, A. Chattopadhyay, and R. Bersohn, *J. Chem. Phys.* **94**, 5994–5998 (1991); (p) J. P. Puttemans, G. P. Smith, and D. M. Golden, *J. Phys. Chem.* **94**, 3226–3227 (1990); (q) J. J. Brocks, H.-D. Beckhaus, A. L. J. Beckwith, and C. Rüchardt, *J. Org. Chem.* **63** 1935–1943 (1998).

20. (a) D. A. Robaugh and S. E. Stein, *J. Am. Chem. Soc.* **108**, 3224–3229 (1986); (b) S. E. Stein and R. L. Brown, *J. Am. Chem. Soc.* **113**, 787–793 (1991); (c) F. G. Bordwell and A. V. Satish, *J. Am. Chem. Soc.* **114** 10173–10176 (1992); (d) B. Römer G. A. Janaway, and J. I. Brauman, *J. Am. Chem. Soc.* **119**, 2249–2254 (1997); (e) I. W. C. E. Arends, P. Mulder, K. B. Clark, and D. D. M. Wayner, *J. Phys. Chem.* **99**, 8182–8189 (1995); (f) I. A. Opeida, A. G. Matvienko, and O. Z. Ostrovskaya, *Russ. J. Org. Chem.* **30**, 681 (1994); (g) A. Atto, A. Hudson, R. A. Jackson, and N. P. C. Simmons, *Chem. Phys. Lett.* **33**, 477–478 (1975); (h) D. F. McMillen and D. M. Golden, *Ann. Rev. Phys. Chem.* **33**, 493–532 (1983).
21. (a) G. M. Wallraff and W. D. Hinsberg, *Chem. Rev.* **99**, 1801–1821 (1999); (b) A. Reiser and H.-Y. Shih, *Angew. Chem. Int. Ed. Engl.* **35**, 2429–2440 (1996); (c) A. Reiser, *Photoreactive Polymers: The Science and Technology of Resists*, Wiley, New York, 1989; (d) V. V. Ershov, G. A. Nikiforov, and C. R. H. I. de Jonge, *Quinone Azides*, Elsevier, New York, 1981; (e) L. F. Thompson, C. G. Willson, and M. J. S. Birven, *Introduction to Microlithography.* Wiley, New York, 1984; (f) W. J. DeForest, *Photoresists Material and Processes*, McGraw-Hill, New York, 1975.
22. (a) B. Urwyler and J. Wirz, *Angew. Chem. Int. Ed. Engl.* **29**, 790–792 (1990); (b) M. Barra, T. A. Fischer, G. J. Cernigliaro, R. Sinta, and J. C. Scaiano, *J. Am. Chem. Soc.* **114**, 2630–2634 (1992); (c) J. Andraos, A. J. Kresge, and V. V. Popic, *J. Am. Chem. Soc.* **116**, 961–967 (1994); (d) J. Andraos, Y. Chiang, A. J. Kresge, and V. V. Popic, *J. Am. Chem. Soc.* **119**, 8417–8424 (1997); (e) D. B. Wagner, B. R. Arnold, G. W. Brown, and J. Lusztyk, *J. Am. Chem. Soc.* **120**, 1827–1834 (1998); (f) M. A. McAllister and T. T. Tidwell, *J. Am. Chem. Soc.* **114**, 5362–5368 (1992); (g) K. Najafian, P. v. R. Schleyer, and T. T. Tidwell, *Org. Biomol. Chem.* **1**, 3410–3417 (2003).
23. (a) A. D. Allen, M. H. Fenwick, H. Henry-Riyad, and T. T. Tidwell, *J. Org. Chem.* **66**, 5759–5765 (2001); (b) W. Huang, H. Henry-Riyad, and T. T. Tidwell, *J. Am. Chem. Soc.* **121**, 3939–3943 (1999); (c) A. D. Allen, B. Cheng, M. H. Fenwick, W. Huang, S. Missiha, D. Tahmassebi, and T. T. Tidwell, *Org. Lett.* **1**, 693–696 (1999); (d) A. D. Allen, B. Cheng, M. H. Fenwick, B. Givehchi, H. Henry-Riyad, V. A. Nikolaev, E. A. Shikova, D. Tahmassebi, T. T. Tidwell, and S. Wang, *J. Org. Chem.* **66**, 2611–2617 (2001); (e) J. Carter, M. H. Fenwick, W. Huang, V. V. Popik, and T. T. Tidwell, *Can. J. Chem.* **77**, 806–809 (1999); (f) A. D. Allen, J. Porter, D. Tahmassebi, and T. T. Tidwell, *J. Org. Chem.* **66**, 7420–7426 (2001); (g) H. Henry-Riyad and T. T. Tidwell, *J. Phys. Org. Chem.* **16**, 559–563 (2003); (h) A. D. Allen, H. Henry-Riyad, and T. T. Tidwell, *Arkivoc* **xii**, 63–74 (2002).

6

LONG-LIVED CARBOCATIONS IN COLD SIBERIA

Vyacheslav G. Shubin and Gennady I. Borodkin
N. N. Vorozhtsov Novosibirsk Institute of Organic Chemistry
Novosibirsk, Russia

In memory of our encouraging preceptor Prof. V. A. Koptyug

6.1 INTRODUCTION

Some words about the title of this chapter. The term *long-lived carbocations* is well known, widely used, and hardly requires any comment. The term *cold Siberia* has become a stock phrase. Speaking in jest, this cold Siberia is probably a more suitable place for generating the long-lived carbocations with respect to California,

Carbocation Chemistry, Edited by George A. Olah and G. K. Surya Prakash
ISBN 0-471-28490-4

where, as one of us had a chance to learn, the air temperature exceeds +20°C even in February.

The basic contribution to chemistry of the long-lived carbocations was brought by Professor G. A. Olah and his school at the University of Southern California in Los Angeles. It was the school of Professor V. A. Koptyug that developed the chemistry of such species in Novosibirsk, the informal capital of Siberia in Russia. The main part of the works on structure and reactivity of the long-lived carbocations has been published in Russian. They are most likely inaccessible to the majority of foreign researchers. The invitation of the editors of this collective volume, Professors G. A. Olah and G. K. S. Prakash, opens an excellent opportunity to present a summary of these works in English.

First, let us give a word on origin and development of the long-lived chemistry in Russia. Being a postgraduate student of Professor N. N. Vorozhtsov junior, V. A. Koptyug was involved in research of acid-catalyzed isomerization reactions of aromatic compounds with the use of those labeled with ^{14}C. As it was established for methyl- and chloronaphthalenes, the mechanism of isomerization consisted of the intramolecular 1,2 shift. In addition, acid-catalyzed sixfold degenerate rearrangement of toluene was found. It proceeds through the 1,2-methyl shift.

In 1963, V. A. Koptyug published his monograph *Isomerization of Aromatic Compounds* in Russian.[1]

The isomerization mechanism was considered to include the formation of short-lived arenium intermediates (Scheme 1).

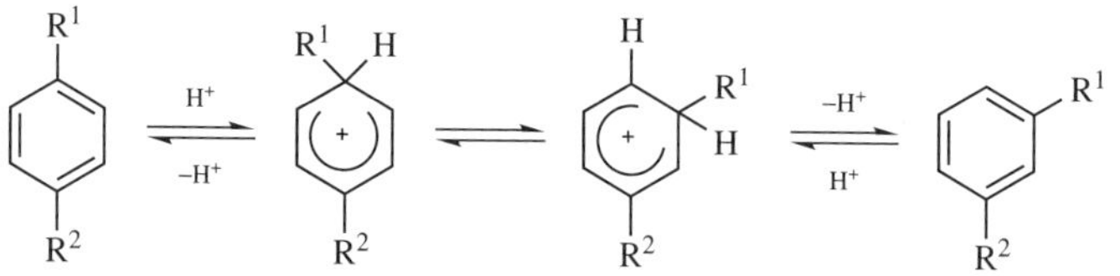

Scheme 1

In order to confirm the mechanism, it was necessary to somehow prolong the lifetime of the intermediates in solutions. The two methods were used for this: systems of high acidity (such as $HBr–Al_2Br_6$) and low temperatures. A series of long-lived arenium ions was generated. IR spectroscopy was used for studying thereof.[2]

NMR was, however, much more informative. Once, having returned from summer vacation, V. A. Koptyug enthusiastically told us that in his long train journey he had read a remarkable monograph by J. D. Roberts titled *Nuclear Magnetic Resonance.* In Koptyug's opinion, the NMR method should be highly applicable to carbocation chemistry, since it allowed one to "watch" each hydrogen atom separately. Having burrowed into the literature, we found that the method had already been used for studying long-lived carbocations.[3] Especially attractive results were obtained by Professor W. von E. Doering et al.,[4] who managed to generate long-lived heptamethylbenzenium ion by protonation of 4-methylene-1,1,2,3,5,6-hexamethylcyclohexa-2,5-diene. That cation was a modeling compound for studying key stage of isomerization, namely, the 1,2 shift of the methyl group.

In Doering's paper[4] there was no information on the degenerate rearrangement of heptamethylbenzenium cation and it seemed that the process might be slow enough. To detect the rearrangement, Koptyug and coworkers attempted to prepare that cation, containing the deuterium label in one methyl group[*] (Scheme 2).

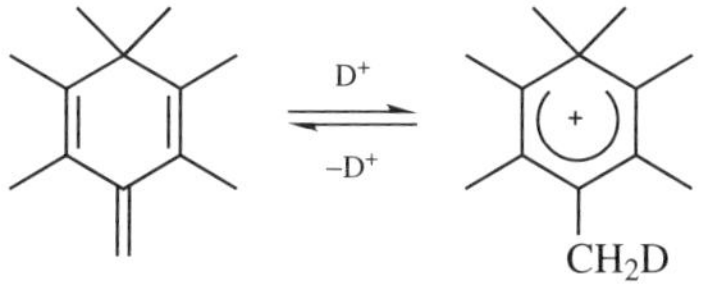

D^+
$-D^+$
CH_2D

Scheme 2

To the researchers' surprise, each methyl group of this cation proved to be enriched with deuterium. As it turned out, the rate of the degenerate rearrangement was high enough. This was corroborated by dynamic magnetic resonance.[5] The mechanism consisted in 1,2-methyl shift (Scheme 3) and was confirmed by the method of spin saturation transfer.[6]

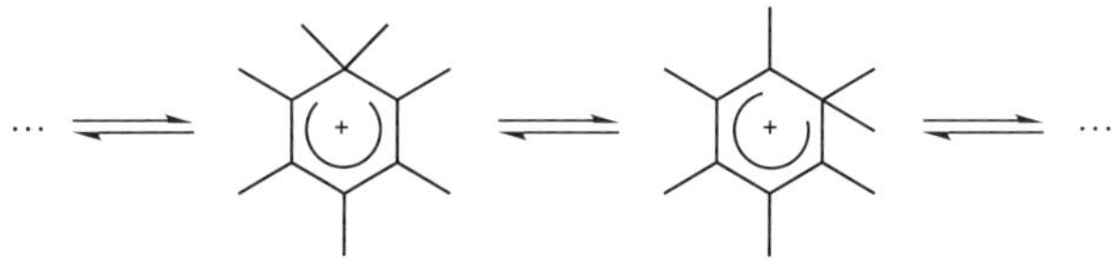

Scheme 3

The participation of all the methyl groups of heptamethylbenzenium ion in the isotope exchange was used for elaboration of a new method of preparation of fully deuteriated hexamethylbenzene[7] (Scheme 4).

The works cited in the References at the end of this chapter began a long series of studies in the field of long-lived carbocations by Koptyug and his associates.

[*] Here and below the free-valence dash in structural formulas denotes the CH_3 group.

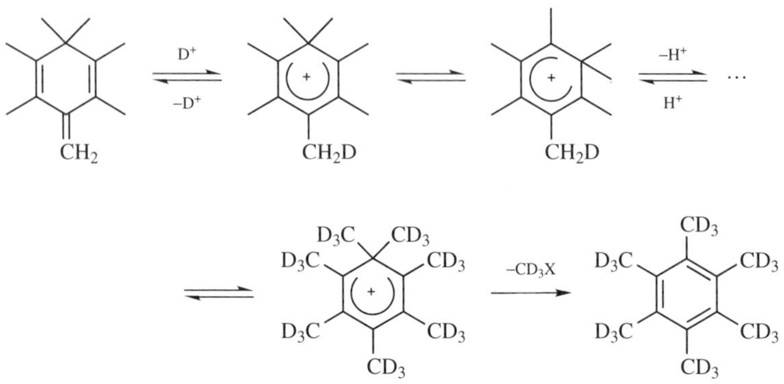

Scheme 4

6.2 ARENIUM IONS: STRUCTURE AND REACTIVITY

The next challenge was to penetrate into the world of the acid-catalyzed molecular rearrangements especially at the quantitative level. Koptyug constructed the research plan based on the following principles: (1) to develop the long-lived carbocation models of key intermediates and (2) to employ nuclear magnetic resonance for "direct observation" of such "frozen" intermediates. In the first stage, the structure of arenium ions was scrutinized by means of NMR, infrared, and Raman spectroscopy as well as X-ray diffraction.

In the sense of the structure of the protonated arenes, it was important to answer the question on hybridization of protonated carbon atom: whether change in hybridization takes place from sp^2 to sp^3. The data obtained by ^{13}C NMR testified that such transition of hybridization did take place.[8]

The transformation of aromatic compounds into arenium ions is accompanied by loss of aromaticity. Koptyug assumed that some compensation effect was borne by the ring CH_2 fragment hyperconjugation with the electron-deficient part of the arenium ion. This assumption was confirmed by IR spectroscopy.[9] It was found that intensity of the CH_2 absorption band was reduced when a positive charge at the carbocation center was diminished and the expected shift of this band to the high-frequency area was observed. The computational analysis[10] of the C—H stretching vibrations of the CH_2 group in arenium ions showed that the increase in the force constant by passing from the benzenium ion to the 1-H-naphthalenium and 9-H-anthracenium ions could be explained by a decreasing interactions of the 1*s* orbital of the CH_2 group hydrogen with the $2p_z$ orbitals of the adjacent carbon atoms. This was related to the suggestion[9] that the hyperconjugation between the CH_2 group and the electron-deficient π system decreases in the abovementioned series of arenium ions.

As has been noted above, long-lived arenium ions were usually generated by protonation of aromatic compounds by superacids, such as $HBr–Al_2Br_6$. However, in some cases Koptyug and coworkers managed to generate the long-lived arenium

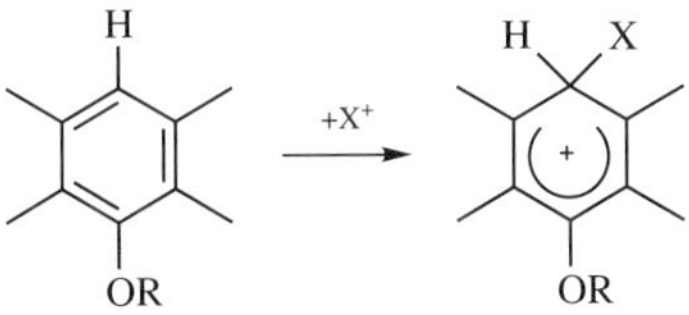

$X = SO_3H$, R = H, Me; X = Br, R = H

Scheme 5

ions by addition of an electrophile X^+ other than proton ($X = Br$, SO_3H) to an aromatic compound[11] (Scheme 5).

The main results obtained at that stage of the survey were as follows:

- Protonation of an aromatic compound results in formation of tetragonal carbon fragment.
- New C—H bond is in hyperconjugation with the electron-deficient part of the arenium ion.
- Positive charges are located mainly in *ortho* and *para* positions of benzenium cations.

The indebatable experimental data on geometry of arenium ions were obtained by X-ray diffraction. The following structures were determined: 1-R-1,2,3,4,5,6-hexamethylbenzenium (**1**) ($R = Me$, $X^- = BF_4^-$, $SbCl_6^-$,[12] $R = CH_2Cl$, $X^- = AlCl_4^-$[13], and $R = Ph$, $X^- = AlCl_4^-$.[14] The ions enumerated have structural features close to C_{2v} symmetry, with the similar carbon–carbon distances regardless of the R substituent:

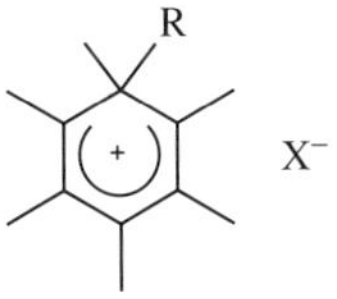

X^-

1

A geometric peculiarity of ions **1** ($R = Me$, CH_2Cl, Ph, $X^- = AlCl_4^-$) is that carbon atoms of skeleton (C2–C6) are located practically in the same plane. Atom C1 is deviated from this plane and the corresponding dihedral angles are 5.4°, 8.0°, and 7.4°, respectively.[13] The deviation increases with an increase in the electron acceptor character of group R. For all the arenium ions of type **1**, lengths of C2–C3 and C5–C6 bonds are less than those of C3–C4 and C4–C5 bonds. Lengths of C2,4,6–CH_3 bonds are markedly shorter than C3–CH_3, C5–CH_3 bonds, obviously because of the hyperconjugation effect. Angle C2–C1–C6 significantly differs from typically tetrahedral angles and comes nearer to the trigonal angle.

Studies on the structure and chemical behavior of the long-lived arenium ions have eventually led to the quantitative description of the isomerization reactions. Before Koptyug's works, there was no such description. Despite a vast number of papers published, no general relationships between structure of rearranging carbocations and rates of their rearrangements were revealed. Professor Koptyug's school studies were based on the approach consisting in experimental determination of the rates of 1,2 shifts of different migrants in long-lived carbocations of a various structures; the main focus was on the carbocations undergoing degenerate rearrangements. These are the processes in which "intrinsic" migration ability of different atoms and groups can be revealed most clearly.

Arenium ions were proved to be perspective models for studying carbocation rearrangements. Degenerate rearrangements of many long-lived arenium ions have been studied. Those were ions (**1–10**) with various R migrants of the σ type (H[15–22], CH_3,[5,6,23–33] C_2H_5,[34–36] $CH_2CH{=}CH_2$,[37] CH_2Ar,[38–41] CH_2Cl,[36,42,43] $CHCl_2$,[36] CH_2CHO,[36] of the π type (XC_6H_4,[30,44–50] NO_2,[17,51,52] SO_2R[51]), and of the *n* type (OR,[36,53,54] Cl,[17,22,55,56] Br):[17,57]

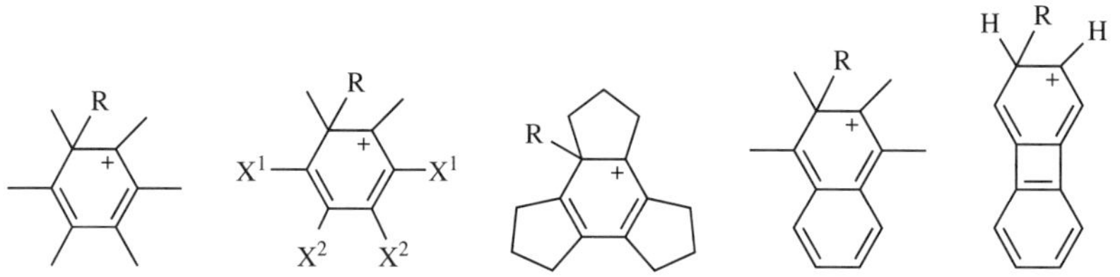

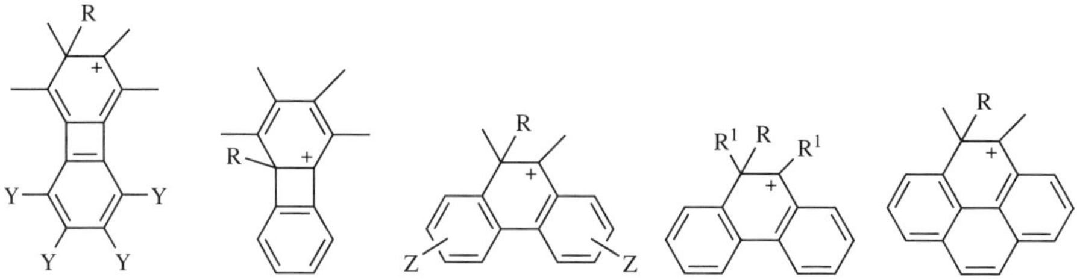

$X^1 = Cl$, $X^2 = Me$; $X^1 = X^2 = Cl$; $X^1 = X^2 = F$; Y = H, Y = Me; Z = 2,7-Br, 3,6-Br, 3,6-Me, 4,5-Me, 3,6 = CF_3; R^1 = Me, R^1 = Et

Studies on mechanisms of the rearrangements of the long-lived 1-R-1,2,3,4,5,6-hexamethylbenzenium ions (**1**, R = CH_2Ph,[38] Br,[57] NO_2,[49] and SO_3H[51]) using the "spin saturation transfer" method showed that these migrants were transferred by 1,2 shift. Such a mechanism was also established by the method of isotope labeling in the case of degenerate rearrangement of the 1-(X-phenyl)-1,2,3,4,5,6-hexamethylbenzenium ions.[43,44]

An alternative mechanism of degenerate rearrangement of ions **9** involving the intermediary formation of the 9-R$(CH_3)_2$C-9-fluorenyl cations (Scheme 6, pathway b)

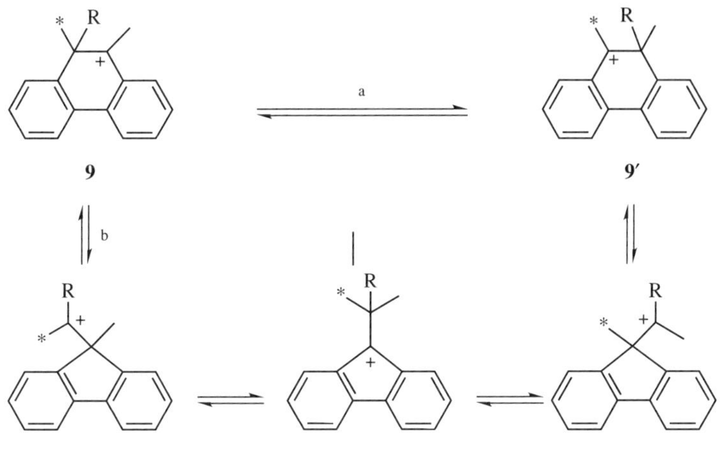

9

9′

11

Scheme 6

was rejected; the lifetime of the 9-*tert*-butyl-9-fluorenyl cation (**11**, $R = CH_3$) proved to be much longer than that of "fixed" structure of the ion **9** ($R = CH_3$).[58]

It is interesting to note that the 9-methyl-9-fluorenyl- (**11a**) and 9-ethyl-9-fluorenyl (**11b**) cations generated under the long-life conditions[58] do not undergo degenerate rearrangement according to Scheme 7[59] at least up to −10°C.

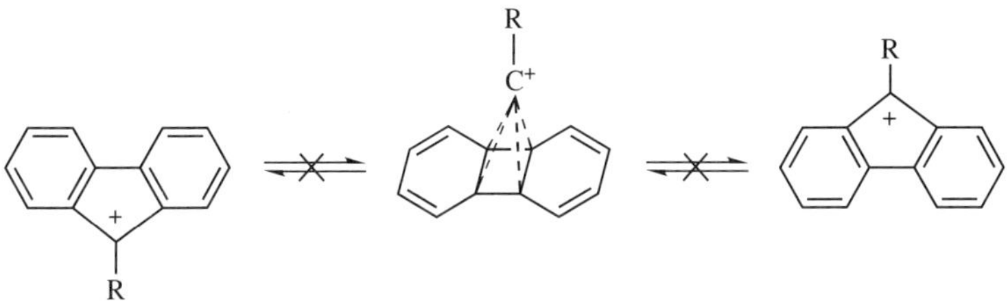

11a,b

R = Me (**11a**), Et (**11b**)

11′a,b

Scheme 7

The carbocation rearrangement rates are strongly dependent on the nature of the migrating group. The rate of the degenerate rearrangement of nitrohexamethylbenzenium ion seemed to be unexpectedly high.[51,60] At the same time, it is known that nitroaromatic compounds are seldom involved in acid-catalyzed isomerizations. This cast in doubts the early proposed structure of the cation formed during the nitronium addition to hexamethylbenzene. Special research was undertaken[61] to confirm the structure of the cation. Comparison of $J_{^{13}C-^{15}N}$ constant (8 Hz) with those for isopropyl nitrite (J $\leq$ 2 Hz) and 4-hydroxy-1-nitro-1,2,3,5,6-pentamethylbenzenium ion ($J = 8$ Hz) did not leave any doubt that the generated cation was really nitrohexamethylbenzenium. Hence, the ability of the nitro group to migrate is really high. The inability of nitroaromatic compounds to participate in acid-catalyzed isomerization reactions is probably due to the very low concentration of the cation **12** in its equilibrium with the isomeric cation **13** (Scheme 8).

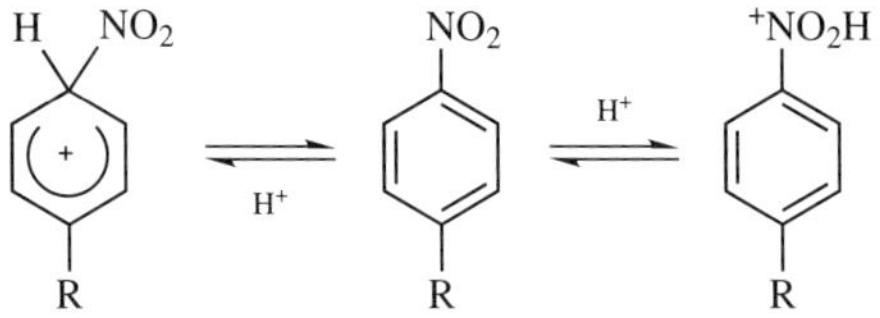

Scheme 8

Predicting the ability of different groups to migrate in cationic sigmatropic rearrangements is one of the most important problems. For *p*- and *m*-XC_6H_4 groups in the 9-(X-phenyl)-9,10-dimethylphenanthrenium,[48] 1-(X-phenyl)-1,2,3,4,5,6-hexamethylbenzenium,[46] and 1-(X-phenyl)-2-hydroxy-1,3,4,5,6-pentamethylbenzenium[62] ions, the rates of 1,2 shift obey the following relationship:

$$\log \frac{k_X}{k_H} = \rho^+ \sigma_x^+ \tag{6.1}$$

It is interesting to note that the values of ρ^+ are not constant. At -50°C they are -4.5, -4.47, and -3.37, respectively. Therefore, the relative migrating abilities of structurally similar migrants are not equal. Analogous relationships hold for 1,2-shifts of $XC_6H_4CH_2$ groups in 1-(X-benzyl)-1,2,3,4,5,6-hexamethylbenzenium ($\rho^+_{-50°C} = -4.42$)[39] and 9-(X-benzyl)-9,10-dimethylphenanthrenium ($\rho^+_{-50°C} = -1.08$)[40] ions. Comparison of the selectivity (ρ^+) and activity ($\log k_{Ph}$) parameters for 1,2-aryl shifts in the arenium ions shows that the reactivity–selectivity principle is not obeyed even for isosteric migrants.[46] This violation of the reactivity–selectivity principle may be associated with the simultaneous influence of inductive and resonance effects of X substituent. The reactivity–selectivity principle is obeyed in the case of 1,2 shifts of σ migrants CH_2R (R = H, Me, Cl) in arenium ions (**1**,**8**), because the migrating ability of these migrants is determined mainly by inductive effect of the substituent R.[43]

A more complicated problem arises in developing approaches to predicting the rates of rearrangements of carbocations with different types of migrants. A solution was presented in Ref. 63. For 1,2 shifts of different migrants (Me, Et, XC_6H_4, Cl, Br) in the arenium ions **1**, **4**, **8**, and **10**, the values of activation free energy related to parameters characterizing the nature of the migrant and the cation skeleton according to the following equation:

$$\Delta G^{\neq}_{25^\circ C} = 0.9 + 0.329\,E_b - 0.00129\,PA(\delta_{C^+} - 130) \quad (r = 0.987, s = 0.7, n = 29) \tag{6.2}$$

Here E_b is C–R splitting energy in MeR compounds, δ_{C^+} is the chemical shift of carbocationic center for ions **1**, **4**, **8**, and **10** with the standard migrant (Me) and PA is the proton affinity of the MeR compounds. The values of activation free energy for 1,2 shifts of the hydrogen atom in ions **1** and **8** (R = H) do not obey this relationship. However, the use of the deformation vibration force constant of the fragment C–C–R (K_{def}) instead of the E_b parameter leads to a more generalized correlation that also includes ions **1** and **8** with R = H:[63]

$$\Delta G^{\neq}_{25^\circ C} = 17.1 + 14.5\,K_{def} - 0.00127\,PA(\delta_{C^+} - 130) \quad (r = 0.963, s = 1.1, n = 33) \tag{6.3}$$

An equation of type (6.2) was then extended to include nondegenerate rearrangements of the 1-R-2-hydroxy-1,3,4,5,6-pentamethylbenzenium ions (Scheme 9).[64]

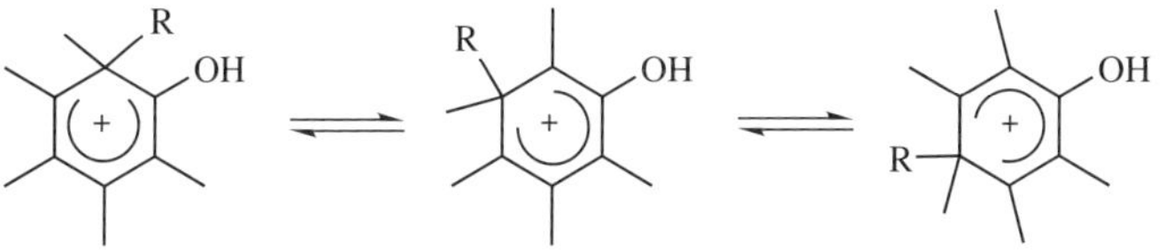

R = Me, Et, XC_6H_4, Cl

Scheme 9

Structural models of arenium ions (**1**, **4**, **8**, and **10**) allow correlation of the rates of 1,2 shifts linearly and the estimations of electron deficiency at the carbocation centers for R = H,[16,19] Me,[28,31] Ph,[49] and Cl:[55]

$$y = a_i + b_i x \tag{6.4}$$

where $y = \log k$, $\Delta G^{\neq}$, or E_a; x is the ^{13}C chemical shift of the carbocationic center (δ_{C^+}), σ^+ constants of substituents Z,[16,31,49] and the stretching frequencies of the C=O bond of ketones that correspond to substituting C=O for the $C^+–CH_3$ fragment in ions **1**, **8**, and **10**.[49,65]

The series of relative migrating tendencies of different types of migrants (H, CH_3, C_6H_5, Cl, and NO_2) in the rearrangements of the arenium ions are not universal,[16,52,55] and the rearrangements do not comply with the reactivity–selectivity principle.[66]

Some restrictions are encountered when one applies Eq. (6.4) for the quantitative description of arenium ion rearrangements in the cases of cations **5** and **6**. Morozov et al.[67] showed that the data obtained could be satisfactorily kept within a two-parameter correlation. The parameters used were q_π^+ (π charge calculated by CNDO/2 method) and the basicity constant for arenes obtained as a result of the formal detachment of a migrant as a cation R^+ (K_i) from both benzoid (**1**, **4**, **8**, R = Me) and nonbenzoid (**6**, Y = H, R = Me) arenium ions:

$$\Delta G^{\neq}_{25^\circ C} = 25.9 - 30.6 q_\pi^+ + 0.38 \log K_i \qquad (r = 0.995, s = 0.56, n = 7) \qquad (6.5)$$

It is necessary to emphasize the very wide range of the rearrangement rates covered by this equation. The difference in the values obtained for free energy of activation is 13 kcal/mol.

Nondegenerate rearrangements of arenium ions and the equilibria of isomeric ions are described in numerous papers published by V. A. Koptyug and his associates.[68] References 69 and 70 describe an approach to estimation of relative stability of substituted benzenium ions. The approach was based on using known σ_p^+ and σ_m^+ constants. Besides, σ_{ortho}^+ and σ_{ipso}^+ were used. The latter were calculated from the ratios of the isomeric ion contents. An agreement between the predicted and observed ratios was obtained for all the ions considered except for the isomeric ions formed from polymethylphenols.

It has been shown[71] that in conditions of practically full protonation by superacids it is possible to shift the equilibrium of isomeric arenium ions to those with electron donor substituents in *ortho* and *para* positions in relation to the protonated carbon atom of the ring. For example, the mixture of 1,2- and 1,4-dichlorobenzenes in $HF–SbF_5$ at room temperature transforms totally into 1,3 isomer.

The phenanthrenium ion model was used to compare the 1,2-shift activation barriers in their dependence on the orientation of the vacant orbital of carbocationic center relative to the carbon–migrant bond.[33,72] The orbital orientation was changed by introducing CH_3 groups at positions 4 and 5. The spatial distortion of skeleton (like a "spring washer"; Scheme 10) leads to different orientations of two 9-CH_3 groups in the 4,5,9,9,10-pentamethylphenanthrenium ion (**14**) with respect to the vacant *p*-AO at C10. A theoretical and experimental analysis of this ion rearrangement made it necessary to account for the mutual orientation of the carbon–migrant bond and the vacant *p*-AO not only in the initial state of the rearranging carbocation but also en route to the transition state.

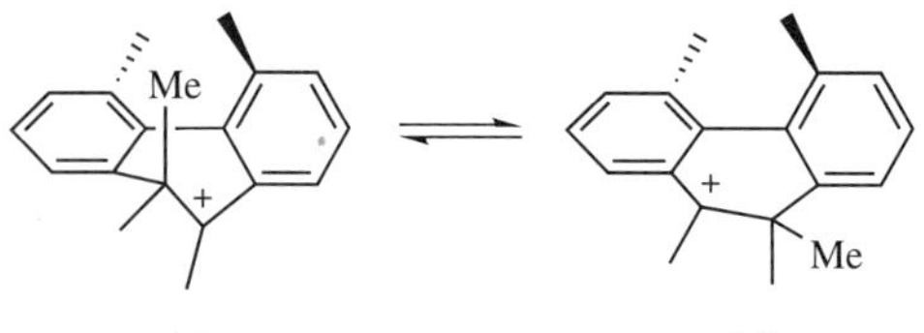

Me

14

Me

14′

Scheme 10

For the quantitative description of molecular rearrangements covering not only arenium ions but also cations of other types and neutral structures, it is possible to use rate of the rearrangement for a certain migrant.[73] It allows us to describe structural effects on the rates. For instance, a linear relationship does exist between free activation energies of 1,2-hydrogen and methyl shifts in the arenium ions (**1,4,6,8**), 1-R-1,2-acenaphthylenium ions, norbornyl cations, and 5-R-1,2,3,4,5-pentamethylcyclopentadienes.[74]

One of the remarkable features of carbocation rearrangements, at least those that proceed by 1,2 shifts of migrants, not subject to specific solvation, is independence of the rates of the medium.[75,76] The most wide medium variation was tested in the case of the long-lived heptametnylbenzenium cation. The latter undergoes sixfold degenerate rearrangement by 1,2-methyl shift (Scheme 3). This cation was generated in various acids: superacids such as FSO_3H or its mixture with SbF_5, ordinary acids such as trifluoroacetic and formic acids as well as in neutral solvents (hexafluorobenzene, nitrobenzene, methylene chloride, and $SOCl_2$). Anions were also different.[75,76] Such an unusual anion as the carbanion was also used.[77] No substantial medium effect has been found. On the contrary, rates of 1,2 shifts of such migrants as OR ($R = H$, CH_3)[54,78,79] and NO_2,[52] subject to specific solvation, are dependent on the medium acidity.

Data on the strong influence of an ordered medium on the rates of 1,2 shifts in carbocations were obtained for the first time for the crystalline tetrachloroaluminate of the heptamethylbenzenium ion.[12,24] Variation of the counterions in crystalline heptamethylbenzenium salts ($AlCl_4^-$, BF_4^-, BPh_4^-) results in strong inhibition of migration of the methyl group (at 25°C k_{cryst}/k_{solv} ratios are 0.0032, 0.041, and 0.035, respectively).[12] Analogous data were obtained for degenerate 1,2-phenyl shifts in 1-phenyl-1,2,3,4,5,6-hexamethylbenzenium ion (**15**).[45] Unexpectedly, the intramolecular rearrangement of the $SbCl_6^-$ salt of heptamethylbenzenium ion accelerates on going from solution to the crystalline state (at 25°C $k_{cryst}/k_{solv} = 3.2$).[12]

It is noteworthy that nondegenerate rearrangement of 1-phenyl-1,2,3,4,5,6-hexamethylbenzenium tetrachloroaluminate (Scheme 11) does not take place in the crystalline state even at 70°C for 2 h. However, in solution at room temperature the equilibrium between ions **15** and **16** is achieved within a few minutes.[45]

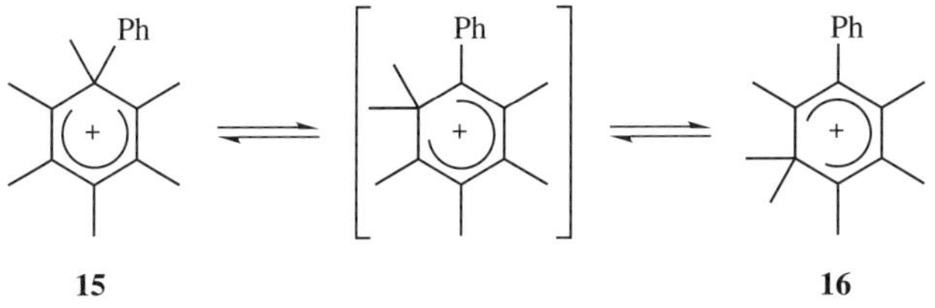

15 **16**

Scheme 11

These results show that the transfer of the reacting system from solution to crystalline state can serve as an effective tool to retard nondegenerate carbocation rearrangments. The efficiency of such tuning was demonstrated by the reaction between tetrachloroaluminate of cation (**15**) and Grignard's reagent (MeMgI)[80] or between 1-chloromethyl-1,2,3,4,5,6-hexamethylbenzenium (**17**) tetrachloroaluminate and secondary amines (Scheme 12).[81]

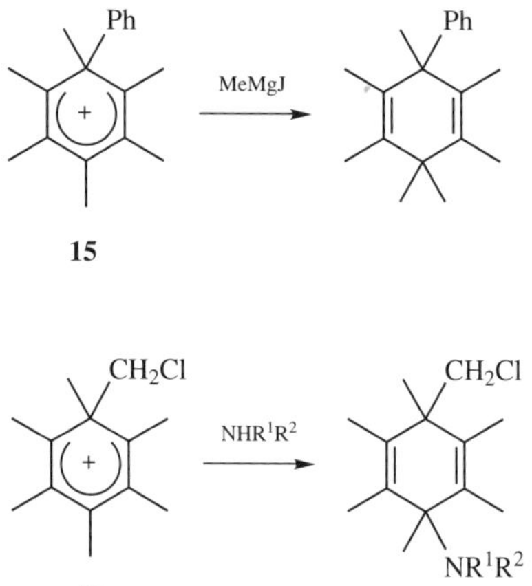

Ph MeMgJ Ph

15

CH_2Cl NHR^1R^2 CH_2Cl

NR^1R^2

17

Scheme 12

The phenomenon of acceleration of carbocation rearrangements is observed in a "partially ordered" (surface of Al_2O_3) medium.[27] When heptamethylbenzenium tetrachloroaluminate is deposited on alumina, the the Arrhenius parameters of the rearrangement are the following: $E_a = 15.0 \pm 0.2$ kcal/mol, $\log A = 10.3 \pm 0.2$).[27] Interestingly, such values are intermediate between those found for the reactions in solution ($E_a = 16.5 \pm 0.1$ kcal/mol, $\log A = 11.5 \pm 0.1$)[12] and in the crystalline phase ($E_a = 10.5 \pm 0.2$ kcal/mol, $\log A = 4.8 \pm 0.2$).[12,24] The fact that the $\log A$ value is larger than that found for the rearrangement of the cation in the crystalline state may be due to its relatively higher mobility on the Al_2O_3 surface.

One interesting aspect of the theory of sigmatropic rearrangements proceeding by 1,2 shift of tetragonal migrants is stereochemistry at the migrating carbon atom. Diastereotopy of protons of the migrating group attached to chiral center and dynamic NMR was used to solve the problem.[41,82,83] It was found[41] that the rate constants obtained from lineshape analysis of the signal of CH_3 and CH_2Ar groups of the long-lived 9-(3′,5′-dichlorobenzyl)-3,6,9,10-tetramethyphenanthrenium ion (**18**) were the same. It means that the configuration at the tetragonal C atom is retained. This is in accordance with the Woodward–Hoffmann rules (Scheme 13).[84]

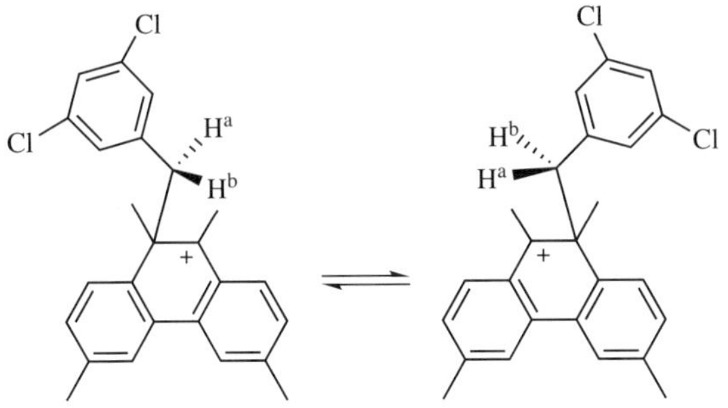

18 ⇌ 18′

Scheme 13

Analogous results were obtained for degenerate rearrangements of the long-lived 1-(*p*-chlorbenzyl)-1,2-dimethylacenaphthylenium and 1-ethyl-1,2-dimethylacenaphthylenium ions.[82,83]

One intriguing problem of detailed mechanism of aryl group migration is to determine whether a classical phenonium ion **19** or a nonclassical cation **20** is formed (Scheme 14, pathways a and b, respectively).

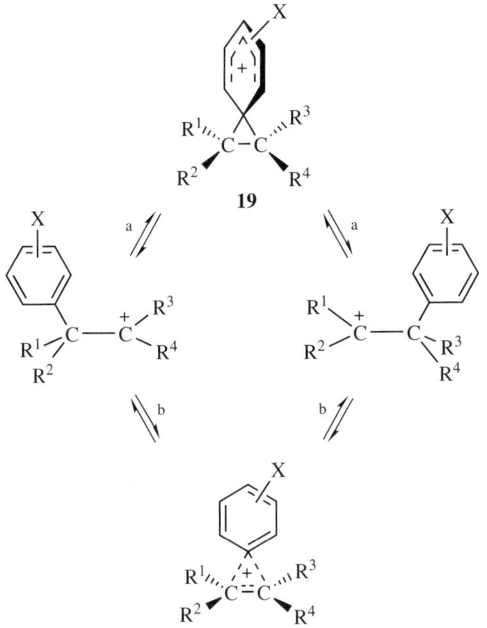

Scheme 14

Studies of the rearrangements of 9-(X-phenyl)-9,10-dimethylphenanthrenium (**8**, $R = XC_6H_4$, $X = CH_3$, p-CH_3O, $Z = H$)[48] and 2,7-, 3,6-di-Z-9-phenyl-9,10-dimethylphenanthrenium (**8**, R = Ph, Z = 2,7-Br, 3,6-CF_3) ions [49] by dynamic NMR revealed that the rate of rotation of XC_6H_4 groups around the C9–C_{Ar} bond was much lower than the rates of rearrangements proceeding by 1,2 shift of these groups and, therefore, the mechanism involving the formation of nonclassical phenonium ion of type **20** can be ruled out.

The results obtained at studying rearrangements of long-lived arenium ions have opened ways to description of the isomerization reactions, due to the kinetics of automerization reactions (degenerate rearrangements). This kinetics characterizes an "internal barrier" of the reaction. Thermodynamics of the reaction is defined with respect to the difference in relative stability of the initial and rearranged ion, the extent of aromatic protonation, and the content of rearranging ions in equilibrium with isomeric cations. Taking into account such thermodynamic factors, Koptyug et al. developed an approach to quantitative description of the isomerization reactions.[68]

One characteristic reaction of arenium ions is their transformation into aromatic compounds on interaction with nucleophilic reagents. As it turned out, not only a proton but also some functional groups could be abstracted from an arenium ion. For example, 3-phenyl-1,1,2,4,5,6-hexamethylbenzenium ion transforms into 1,2,3,4,5-pentamethylbiphenyl[85] (Scheme 15).

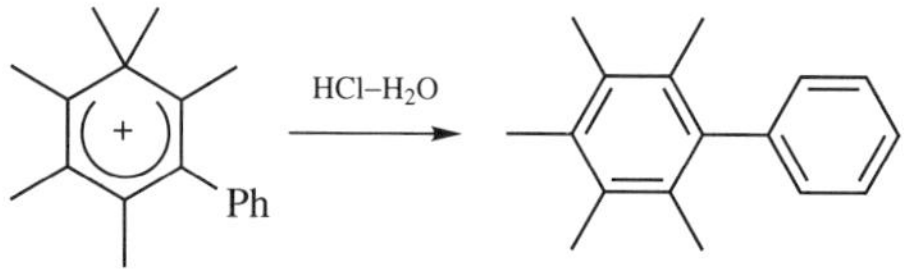

HCl–H_2O

Ph

Scheme 15

However, attempts to obtain pentamethylphenol by demethylation of 4-hydroxy-1,2,3,4,5,6,6-hexamethylbenzenium ion have failed.[68] It is obvious that this is due to substantial localization of the positive charge on oxygen atom of the ion.

Quenching **1** (R = Br) and **1** (R = SO_3H) ion salts with Et_2O/Na_2CO_3, MeOH/Et_2NH, or MeOH/Na_2CO_3 leads to hexamethylbenzene.[57,86,87] Reversibility of aromatic sulfonation and bromination make the product formation understandable.

4-Hydroxy-1-X-1,2,3,5,6-pentamethylbenzenium ions (**21**) bearing electronegative groups X are thermally transformed into 4-hydroxy-2,3,5,6-tetramethylbenzyl cation (**22**)[88–92] (Scheme 16). The substituents X shown below allow the reaction to be completed in FSO_3H within 5–10 min.[68]

A similar reaction was found for 4-methoxy-1-X-1,2,3,5,6-pentamethylbenzenium ions with X = NO_2, SO_3H.[88] The reaction mechanism is not clear. Scheme 17 indicates the most likely routes.[68]

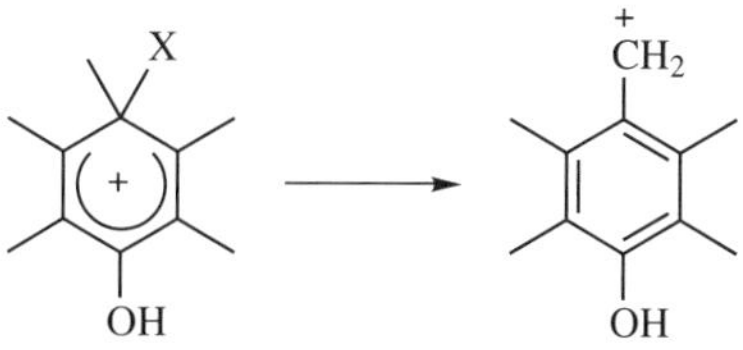

21 22

X = $OCOCH_3$ (–10° C), SO_3H (–10° C), OH (0° C), OCH_3 (0° C), NO_2 (10° C), F (80° C), Cl (>80° C), Br (>100° C)

Scheme 16

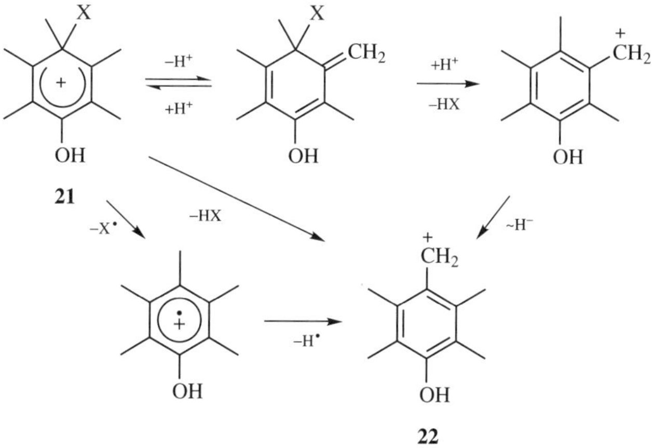

Scheme 17

Considering the ion **21** conversions, one should also take into account the possibility of the acid-catalyzed removal of the X group as X^- anion with the formation of antiaromatic dication **23**. The latter then loses a proton[68] (Scheme 18).

The 1-X-1,2,3,4,5,6-hexamethylbenzenium ions are not rearranged into benzyl cation in strongly acid media, such as in FSO_3H. "Heating" evokes rearrangement

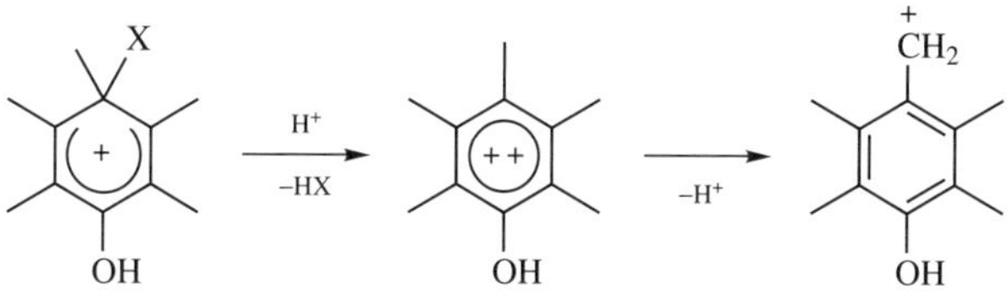

Scheme 18

of the ions generated by adding SO_3 and NO_2^+ to hexamethylbenzene at −25°C and 0°C, respectively. The reaction leads to ions with oxygen function at the ring C_{sp^3} carbon atom.[86,93] The latter rearrange by the 1,2-methyl shift according to Scheme 19.[68,86,93] Similar reactions were observed in the case of 1-methoxy- and 1-hydroxy-1,2,3,4,5,6-hexamethylbenzenium ions.[53]

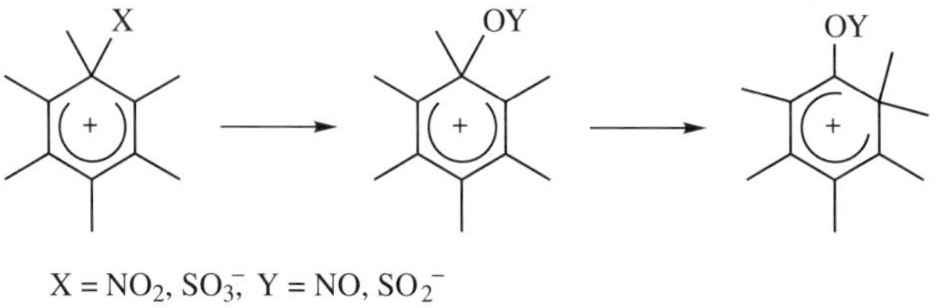

X = NO_2, SO_3^-, Y = NO, SO_2^-

Scheme 19

The 1,3-dinitro-1,2,4,5,6-, 1-nitro-4-halogeno-1,2,3,5,6-pentamethylbenzenium[94,95] and 1-H-sulfo-4-methoxy-2,3,5,6-tetramethylbenzenium[11] ions undergo rearrangements of the same type.

During studies of the rearrangement mechanism of 1-nitro-^{18}O-1,2,3,4,5,6-hexamethylbenzenium ion, it was shown that in the product, 2,3,4,5,6,6-hexamethylcyclohexa-2,4-dienone, the carbonyl group is formed at the expense of the nitro group.[95] The authors of this paper concluded that the rearrangement proceeded either by homolytic scission of the C–N bond with the subsequent recombination

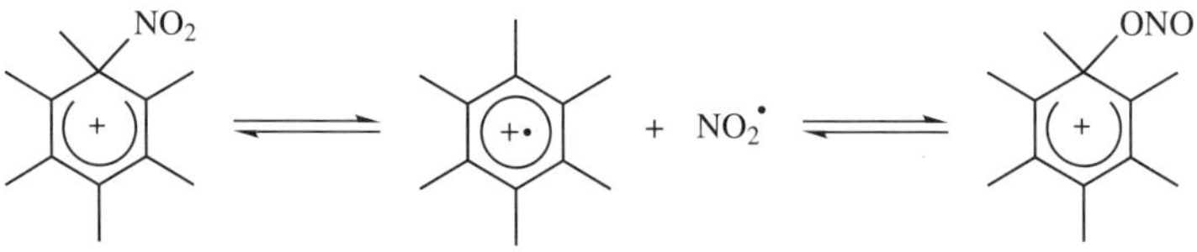

NO_2 ⇌ + $NO_2^{\bullet}$ ⇌ ONO

Scheme 20

of the hexamethylbenzene radical cation and $NO_2^{\bullet}$ (Scheme 20) or via a cyclic intermediate (or transition state) with a bond between oxygen of the nitro group and the carbon atom carrying a positive charge.

Other papers[53,86] discuss the mechanism of the polymethylbenzenium transformation into cations of the benzyl type.

6.3 COMPLEXES OF PHENOLS WITH LEWIS ACIDS

Professor Koptyug and his school also studied complexes of phenols with the Lewis acids.[96–115] It is well known that phenols exist predominantly in the hydroxy forms. Complexation with Lewis acid provokes isomerization into the keto forms; that was established for the complexes formed from phenols, naphthols, and aluminum halides. It was established by ^{13}C NMR that the distribution of π-electronic density in complexes of keto forms of phenols is close to that in related hydroxybenzenium ions. The bipolar structure depicted in Scheme 21 describes the electronic structure of these complexes.[102,103]

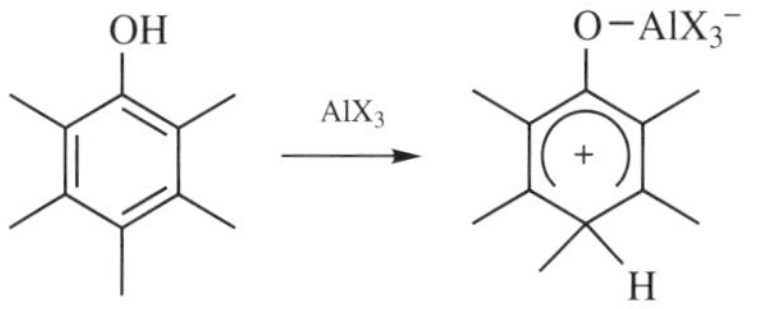

Scheme 21

Reaction of 2-naphthol-aluminum chloride complex with benzene leads to 4-phenyl-2-tetralone.[104] Thus, 2-naphthol behaves as a nonsaturated ketone in conditions of Friedel–Crafts reaction.[116,117] Under the same conditions, 1-naphthol yields 4-phenyl-1-tetralone. 3-Phenyl-1-tetralone does not rearrange into 4-phenyl-1-tetralone under the conditions of this reaction. Accordingly, the authors proposed Scheme 22 as a special route to 4-phenyl-1-tetralone.[104]

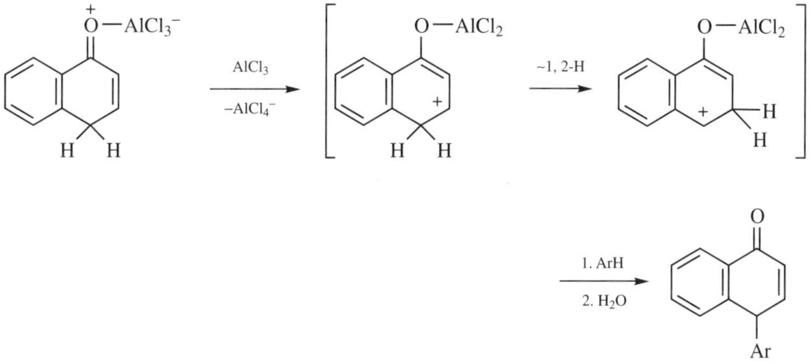

Scheme 22

Complexes of the keto forms of 1- and 2-naphthols and 4-methyl- and 3-phenyl-1-naphthols with aluminum chloride react with triethylsilane under mild conditions (0–20°C) and give rise to corresponding tetralones. In the authors' opinion, this reaction can be considered as a new, rather simple route to tetralones.[105]

Complexes of phenols with the Lewis acids are promising synthons. For example, the *m*-cresol complex reacts with benzene, giving the corresponding cyclic ketone, 3-methyl-5-phenylcyclohexenone-2.[107] It opens a new way to derivatives of phenylcyclohexene. Complexes of keto forms of phenols and naphthols with Lewis acids form dications. These "superelectrophiles" enable reaction with very weak nucleophiles such as arenes and even alkanes.[108–112]

6.4 NONARENIUM CARBOCATIONS: STRUCTURE AND REACTIVITY

Alongside arenium ions, such long-lived carbocations as acenaphthylenium, bicyclo[3.1.0]hexenyl, bicyclo[2.1.1]hexenyl, benzobicycloheptenyl, and cyclobutenyl cations were studied.

1,1,2-Trimethylacenaphthylenium tetrachloroaluminate was the first one whose X-ray diffraction pattern was recorded and interpreted.[118] One feature of the cation (**25**) is alteration of the lengths of C–C bonds in the naphthalene fragment. The length of C2–CH_3 bond [1.46(2)] is significantly shorter than the sum of covalent radii of C_{sp^2} and C_{sp^3} atoms. It is apparently due to the hyperconjugation effect. There is substantial reduction of the values of C1C2C2a, C2C2aC8b and C1C8aC8b angles [110(1), 106(1) and 107(1)°, respectively] in comparison with typical values for trigonal carbon atom fragments:

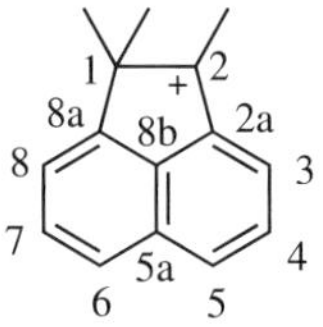

25

Acenaphthylenium ions proved to be perspective models for studies on structure–reactivity relationships in carbocation rearrangements. These ions are topochemically similar to cyclopentyl cation **26** and cyclopentadiene derivatives (**28**). The latter are prone to undergo degenerate rearrangements by 1,2 shifts (sigmatropic 1,5 shifts).[119] This similarity has allowed us to determine the feasibility of relationships such as (6.4) (X = δ_{C^+} at a very wide range of δ_{C^+} (formal change of π charge from 0 to 1):[120]

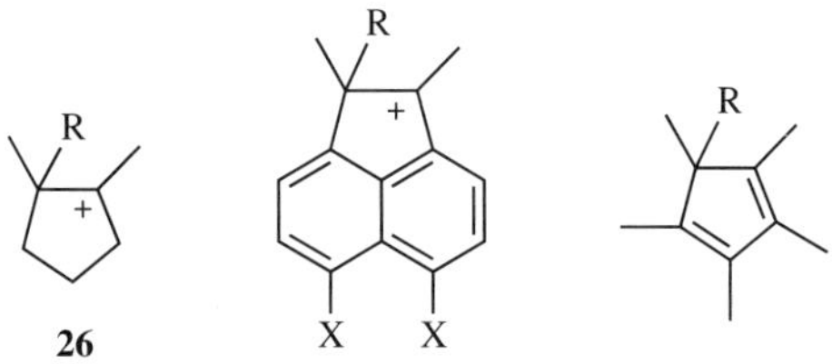

26

X = H, Br, Me
27

28

For 1,2-hydrogen shifts in cations **26** and **27** and pentamethylcyclopentadiene (**28**), comparison between $\Delta G^{\#}$ and δ_{C^+} values leads to a satisfactory correlation:[120]

$$\Delta G^{\#}_{25^\circ C} = 41.3 - 0.49\,\delta_{C^+} \qquad (r = 0.986, s = 1.6, n = 5) \tag{6.6}$$

Such correlations for rearrangements of the carbocations and neutral compounds may have common origins.

1,2-Aryl shifts in degenerate rearrangements of 1-(X-phenyl)-1,2-dimethylacenaphthylenium ions obey relationship (6.1).[121] Comparison of the parameters ρ^+ for rearrangements of arenium ions (**1**, **8**, $R = XC_6H_4$).[46,48] and 1-(X-phenyl)-1,2-dimethylacenaphthylenium ions ($\rho^+_{-50°C} = -6.64$) shows that the relative migration abilities of the substituted phenyl groups are different.[121,122] Infringement of the reactivity–selectivity principle takes place. The O'Ferrall–Jenks mode is, however, applicable to correlation of ρ^+ sensitivity constant for 1,2 shifts of XC_6H_4 groups.[46]

For a series of 1-R-1,2-dimethylacenaphthylenium ions, the relationships of type (6.2) and (6.3) are transformed into simpler dependences.[64,123]

$$\Delta G^{\#} = a + bD + c\text{PA} \tag{6.7}$$

where $D = E_b$ (MeR) or K_{def}.

Comparison of the activation free energies of the 1,2 shifts of different migrants (Me, Et, Ph, p-MeC_6H_4, p-ClC_6H_4, m-ClC_6H_4, p-$CF_3C_6H_4$, Cl, Br) with the corresponding E_b (MeR) and PA parameters gives the relationship.[123]

$$\Delta G^{\#}_{-50°C} = 20.7 + 0.413\,E_b - 0.273\,\text{PA} \qquad (r = 0.98, s = 1.1, n = 9) \tag{6.8}$$

The activation free energies of the 1,2-hydrogen shift do not obey correlation (6.8). However, when the parameter E_b replaced by K_{def}, the correlation becomes more general.[64]

Relationship (6.8) has been used to construct a "kinetic scale" of relative stability of cationic σ and π complexes.[124]

The acenaphthylenium ion is a unique model for direct studies on interconversion of cationic σ and π complexes. The interconversion of the episulfonium and episelenonium ions (weak π complexes according to Dewar's classification[125] with the corresponding σ complexes has been studied[126–130] (Scheme 23).[131]

Extremely high interconversion rates of ions **30d** $\leftrightarrows$ **29d** and **30i** $\leftrightarrows$ **29i** ($k_{-120°C} \geq 10^8$ and $k_{-80°C} > 10^7$ s^{-1}, respectively) imply easy rupture of the C–S or C–Se bonds in the corresponding π complexes.[127,128] The nature of the substituents R, R^1, and X and the element E affects the equilibrium in the cases of **29** $\leftrightarrows$ **30** σ and π complexes.[127–129] Analysis of the data on equilibrium resorting to those reported for 1-OR-1,2-dimethylacenaphthylenium ions[78,79] led to the conclusion[127–129] that the content of the cationic π-complexes **30** in the equilibrium (Scheme 23) increases with an increase in the electron-donating properties of substituents R and R^1, with an increase in the atomic number of the central element E, and with a diminution of the electron-donating properties of the substituent X. The equilibrium **30d** $\leftrightarrows$ **29d** is not sensitive to medium effects.[129,130]

Medium ordering changes the rates of 1,2-methyl shifts in the crystalline 1,2,2-trimethylacenaphthylenium tetrachloroaluminate.[118] The 1,2 shift of the methyl

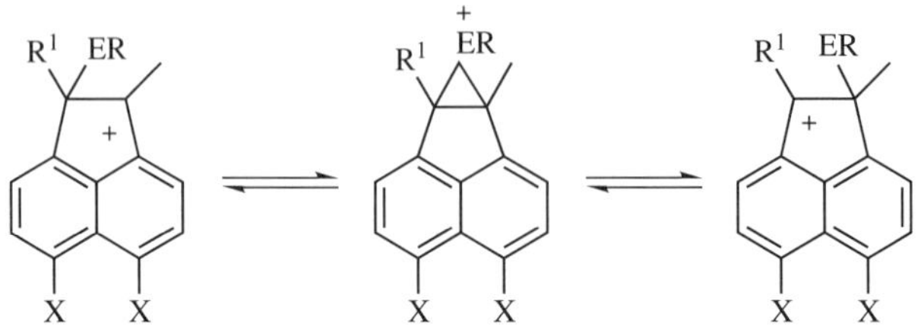

29 ⇌ 30 ⇌ 31

	E	R	R^1	X		E	R	R^1	X
a	S	Me	Me	H	**h**	Se	Ph	Me	H
b	S	Ph	Me	H	**i**	Se	C_6F_5	Me	H
c	S	Ph	CH_2Cl	H	**j**	Se	Ph	CH_2Cl	H
d	S	C_6F_5	Me	H	**k**	Se	C_6F_5	CH_2Cl	H
e	S	C_6H_5	CH_2Cl	H					
f	S	C_6H_5	Me	Br					
g	S	C_6F_5	Me	Me					

Scheme 23

group in this ion leads to a structure that is formally identical to the original one. However, this shift is nondegenerate in regard to the environment. This is a "topo-nondegenerate" rearrangement; see Scheme 24.*

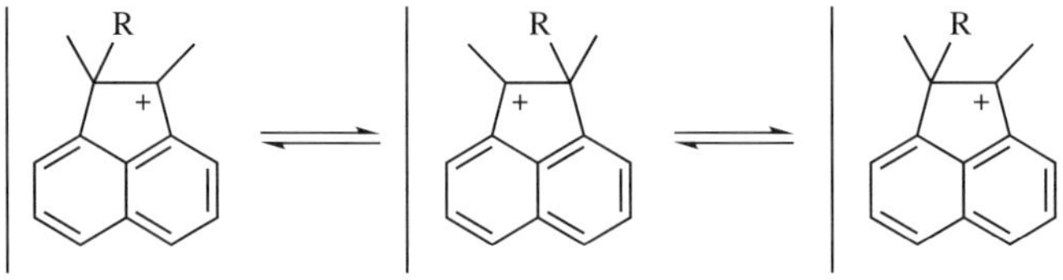

25 ⇌ ⇌ 25′

$R = CH_3, CD_3$

Scheme 24

In the crystalline state, the rearrangement of cation **25** proceeds much slower than that in solution (at 0°C $k_{cryst}/k_{solv} = 3 \times 10^{-3}$).[118] This is obviously due to the fact that the structure of ion **25′** does not correspond to the unit cell.

Koptyug et al. devoted a number of publications to bicyclo[3.1.0]hexenyl cations. The presence of the cyclopropane fragment in these cations has been

*A vertical line in the Scheme 24 denotes a "fixed" environment of cations **25, 25′** in a crystal.

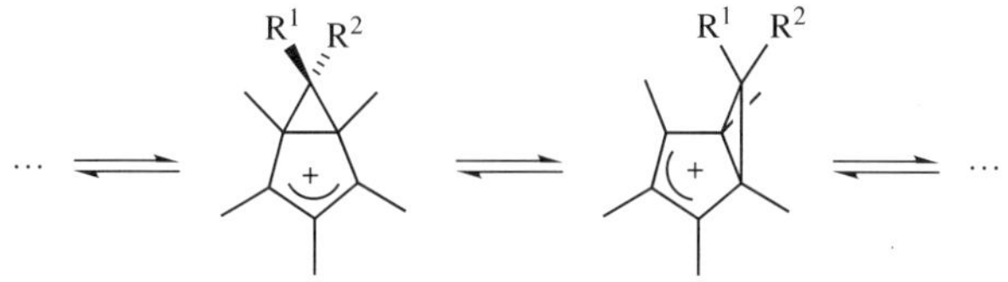

$R^1 = H, R^2 = Me,$[134–136] $R^1 = R^2 = Me,$[133] $R^1 = Et, R_2 = Me,$[133] $R^1 = Me, R^2 = Et,$[133]

$R^1 = Me, R^2 = CH_2Cl,$[133] $R^1 = CH_2Cl, R^2 = Me$[133]

Scheme 25

confirmed by ^{13}C NMR.[132] The characteristic property of these cations is fivefold degenerate rearrangement[133–136] (Scheme 25).[137]

This rearrangement is stereoselective, and it accelerates with enhancement of electron donor ability of fragments R^1 and R^2 in geminal units. The lifetime of bicyclo[2.1.1]hexenyl cation **33** is much more than that of the "fixed" structure of the respective bicyclo[3.1.0]hexenyl cation **32**. From these data it follows that the cation **33** cannot be an intermediate of the rearrangement. The latter proceeds by the mechanism of sigmatropic 1,4 shift[133,137] (Scheme 26).

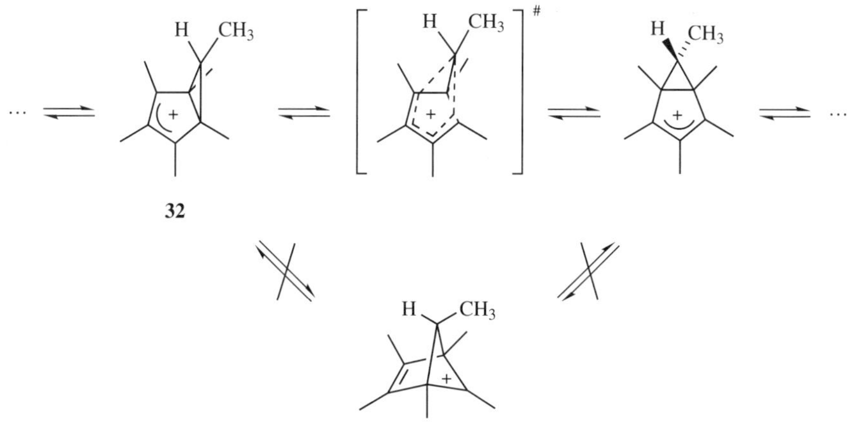

32

33

Scheme 26

Bicyclo[3.1.0]hexenyl cations undergo retrorearrangements giving rise to benzenium ions.[133,134,136]

Interaction of halogens with Dewar hexamethylbenzene in the presence of AlX_3 ($X = Cl$, Br) in CH_2Cl_2 at low temperatures results in the formation of the long-lived 5-*endo*-X-1,2,3,4,5,6-hexamethylbicyclo[2.1.1]hexenyl cations (**34a**, $X = Cl$; **b**, $X = Br$).[87,138–140] By ^{13}C NMR it has been established that substantial part of the positive charge in the 5-*endo*- and 5-*exo*-H-hexamethylbicyclo[2.1.1]hexenyl cations centered on C2 and C3 atoms.[141] These cations are nonclassical ones.[141] Koptyug et al.[87] studied degenerate rearrangements of **34** class of cations:

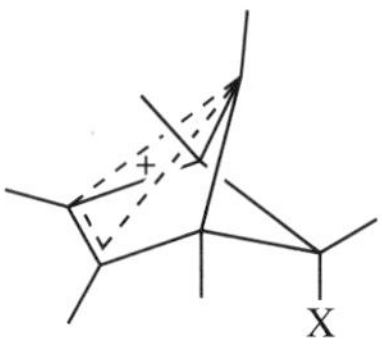

34

Koptyug's group also performed wide studies on the long-lived cyclobutenyl cations. They observed that the pentamethylcyclobutenyl cation does not undergo degenerate rearrangement by 1,2-methyl shift. Even under rather harsh conditions (FSO_3H, 70°C, 20 h), no redistribution of the deuterium label occurs;[142] comparitively similar data were obtained by dynamic NMR.[143] It has been found[144] that the 4-phenyl-1,2,3,4-tetramethylcyclobutenyl cation does not undergo degenerate rearrangement by 1,2-phenyl shift. Even at 100°C, there are no changes in ^1H NMR spectrum of the cation. These data are in accordance with the theory of pericyclic

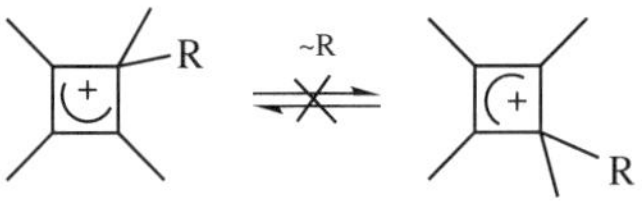

35 ~R **35′**

Scheme 27

reactions,[184] which qualifies 1,2 shifts in cyclobutenyl cations as forbidden processes (Scheme 27).

But the intrinsic instability of the cationic structures provokes other rearrangements, either keeping the four-membered cycle intact[145] (Scheme 28) or expanding the cycle (Scheme 29).[142]

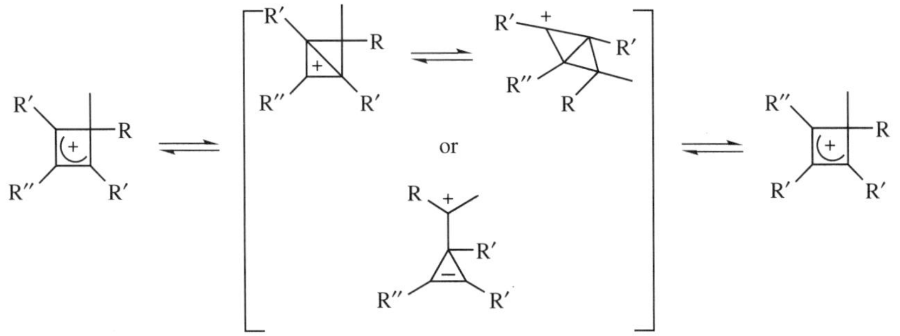

Scheme 28

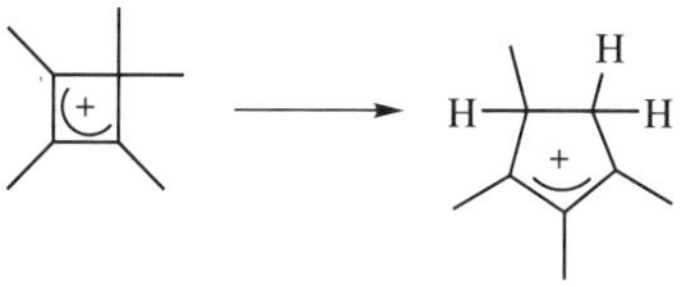

Scheme 29

In the authors' opinion nonclassical structure **36** may be formed as an intermediate or transition state during the rearrangement proceeding with the cycle safety:[145]

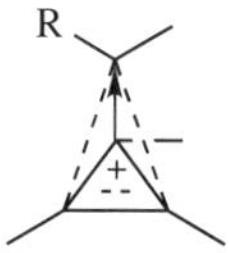

36

Section 6.5 describes cyclobutenyl cations as model structures for studying cationic cyclization reactions.

A study[146] considered the kinetics and mechanism of the degenerate rearrangement of the long-lived 2,3,3-trimethyl-2-benzobicyclo[2.2.1]heptenyl cation (**37**) (Scheme 30). The rearrangement mechanism was shown to be stereoselective. Namely, only the 3-*exo*-CH_3 group migrates to the carbocationic center; the rate of the rearrangement is insensitive to medium variation over a wide range of acidity, in FSO_3H binary mixtures with CF_3COOH or SbF_5.

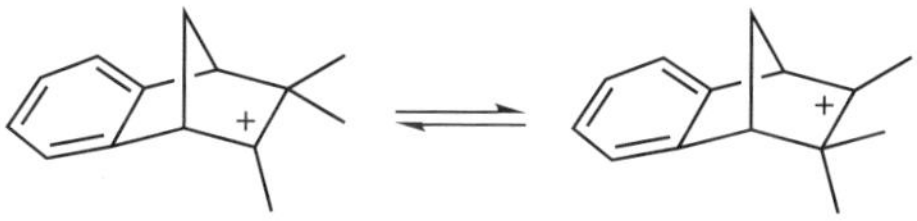

37 **37′**

Scheme 30

This result may mean that the effect of changing bond alignment suggested for the rearrangement of related carbocations[147] operates only in much more nucleophilic media such as acetic acid. Alternatively, this effect may be absent and other factors may be responsible for the change in the rate ratio observed. Perhaps, in nucleophilic media, not carbocations, but some related species are formed and some kind of ion pair effect takes place.[146]

Data obtained by ^{13}C NMR spectroscopy with isotopic perturbation suggest that the 2,3,3-trimethyl-2-benzobicyclo[2.2.1]heptenyl cation is in fact a mixture of cations differing in π-bridging degree (Scheme 31).[148,149]

Scheme 31

Deuterium perturbation of the ^{13}C NMR spectra suggests that ionization of 2-*exo*-chloro-2-methylbenzonorbornene in FSO_3H-SbF_5/SO_2ClF results in formation of similar mixture of the respective σ and π complexes. The latter are involved in ultrafast reciprocal conversion even at −127°C.[150] The authors of the paper[151] tried to stop this process by performing the reaction at the liquid nitrogen temperature. FTIR spectra of the long-lived carbocation of the 2-methylbenzotricyclyl type formed on the ionization of 2-*exo*-chloro-2-methylbenzonorbornene in SbF_5 solid matrix at −196°C show that this ion (**38**) rearranges at −123°C into another carbocation, apparently the π-bridged 2-methyl-2-benzonorbornenyl cation (**39**) (Scheme 32).

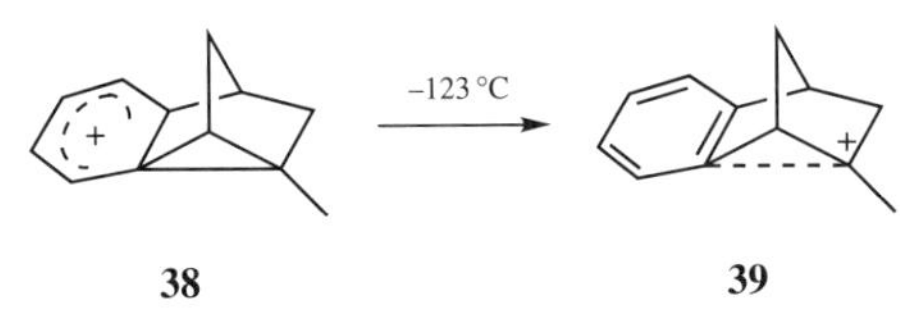

Scheme 32

6.5 CATIONIC π-CYCLIZATION REACTIONS

It has been found that long-lived cyclobutenyl cations **35** are fruitful model compounds for studying cationic π-cyclization reactions. Cyclization of long-lived

R

35

a	R = CH_2-CH = CH_2[152]
b	R = CH_2-C ≡ CH[153]
c	R = $C(CH_3)_2C$ ≡ CH[154]
d	R = $C(CH_3)_2C$ ≡ CCH_3[154]
e	R = $CH_2C_6H_5$[155]
f	R = α-$C_{10}H_7$[156]
g	R = β-$CH_2C_{10}H_7$[157]
h	R = *ortho*-biphenylyl[158]

cations **35 a–h** have hitherto been studied. Depending on R, different final products of rearrangements steps of initial cyclization take place. In the cases of **35a,e–g** cations the final products are carbocations bearing cyclopropane fragment (**40**, **41a,b**, **42a,b**, and **43a,b**, respectively).

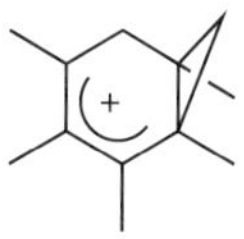

40 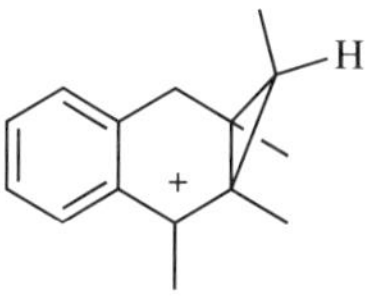 41a 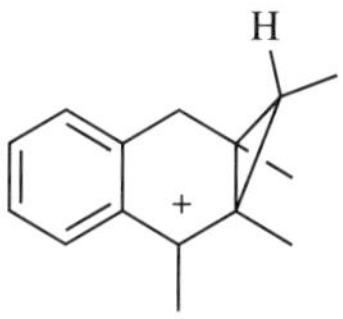 41b 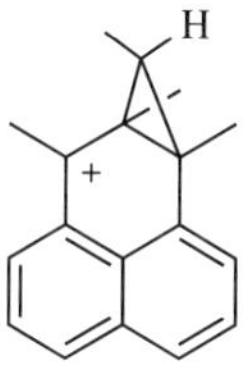 42a

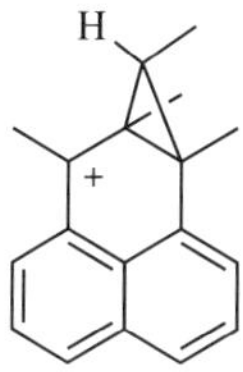

42b 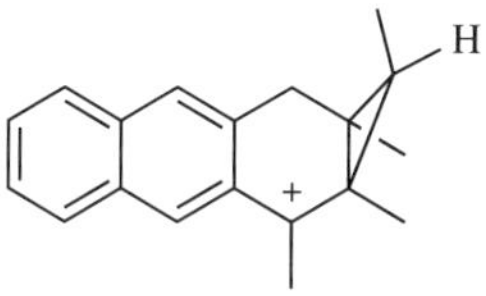 43a 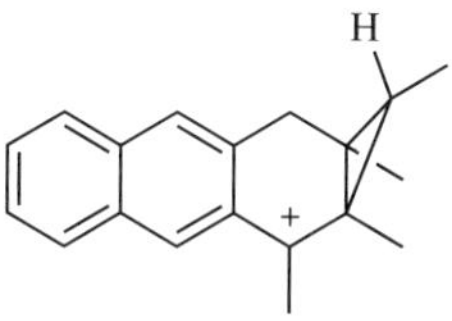 43b

It is interesting that initially only isomers having *endo*-methyl group (**41a**, **42a**, and **43a**) are formed. It shows that proton attacks the endocyclic double bond from the *exo* side only (Scheme 33).

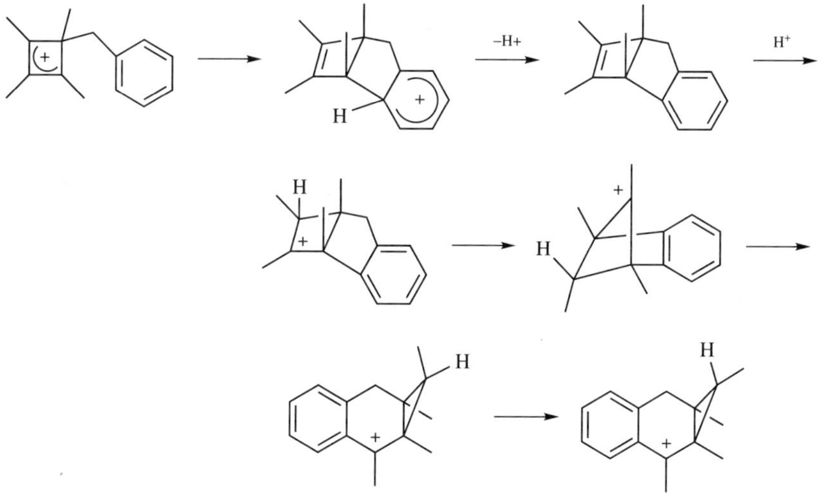

−H+ H⁺

endo *exo*

Scheme 33

Carbocations **35c,d** rearrange into polymethylbenzyl cations (**44a,b**, respectively):

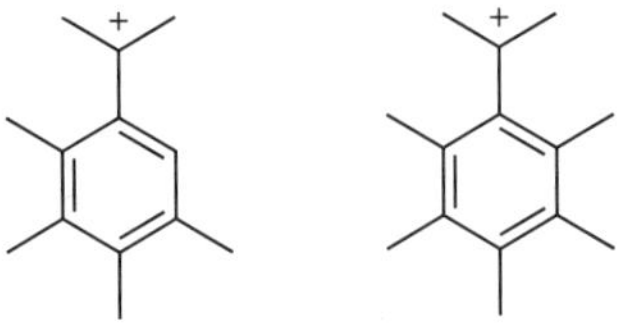

44a 44b

Cation **35h** rearranges into cations (**45a,b**)[158] Scheme 34).

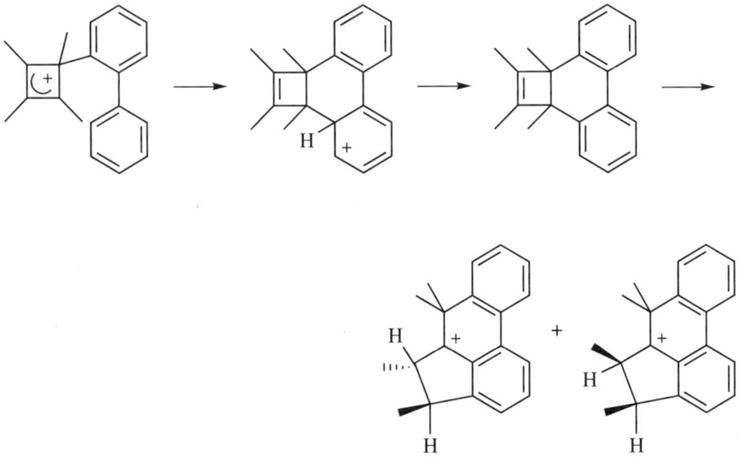

45a 45b

Scheme 34

These data demonstrate a diversity in opportunities that carbocations of the cyclobutenyl type open for organic synthesis.

In conclusion, it should be noted that π cyclization of carbocation **35a** involving double carbon–carbon bond refers to *disfavored* 5-*endo-trig* processes according the Baldwin rules.[159] This example illustrates the restricted character of the rules; the reaction disfavored by the rules, however, proceeds even at low temperatures, as the rate is rather high ($t_{1/2} \sim 10$ min at −90°C).[152]

The carbocation cyclization of adamantane derivatives proceeds unusually. Interaction of 3,7-dimethylenebicyclo[3.3.1]nonane (**46**) with triflic acid (TfOH) in $SO_2ClF–CD_2Cl_2$ at −90 to − 10°C leads to the equilibrium mixture of 3-methyl-1-adamantyl cation (**47**) and 3-methyladamantyl-1-triflate (**48**). The reciprocal transformation of cation **47** and triflate **48** (Scheme 35) and also the circumambulatory rearrangement of the ion **47** (Scheme 36) have been found; both processes are fast on the NMR timescale.[160]

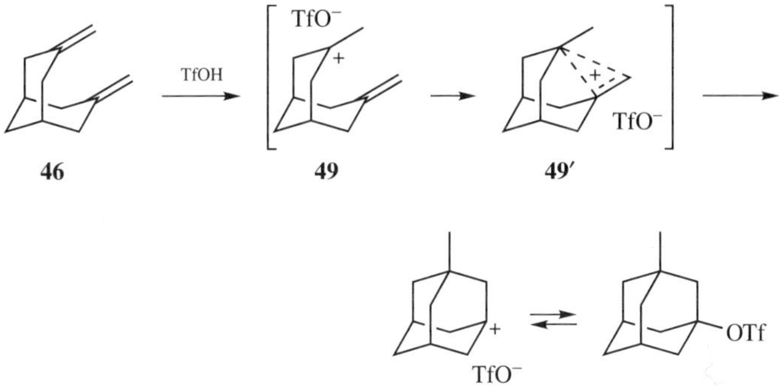

TfOH

TfO⁻ TfO⁻ TfO⁻ OTf

46 49 49′ 47 48

Scheme 35

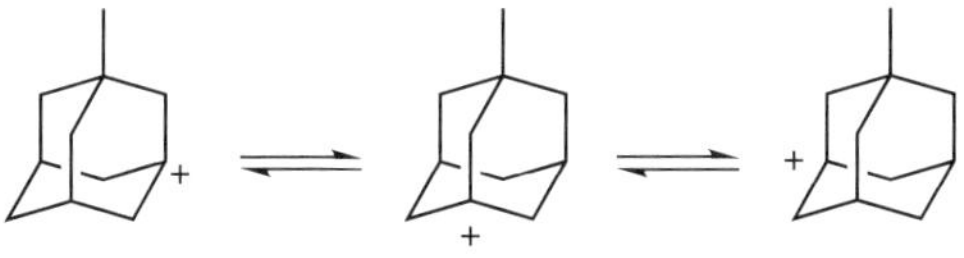

47

Scheme 36

No primary product of protonation of diene **46** or cation **49** (or **49′**) has been observed by NMR. According to the results of quantum chemistry calculations (MP2/6-31G*), these cations are not minima on PES.[160]

Professor Koptyug paid significant attention to organizing and developing researches in the field of chemistry of the carbocations generated from natural compounds—terpenes and their analogs—mainly due to (1) an important role of renewed natural raw materials as a source of basic compounds for fine organic synthesis and (2) attractivness of terpenes and their analogs as objects for development of the structural kinetic theory of molecular rearrangements.

It is important to note that the method developed by Koptyug and coworkers for quantitative description of carbocationic rearrangements of terpenoids was the continuation of works in chemistry of the carbocations generated from aromatic compounds. For the first time comparison has been carried out between the routes of the terpenoid rearrangements predicted by molecular mechanics and observed by NMR. A good agreement between experimental and calculating data was achieved when the orbital and thermodynamic factors were taken into account in addition to the magnitudes of intrinsic barrier of rearrangement.[161]. As a result of studies carried out on Koptyug's initiative and under his and Prof. V. A. Barkhash's leadership, regularities were established for chemical transformations of the complex polycyclic carbocations generated from terpenoids, creating new ways to use these compounds in organic syntheses.

A detailed review of these researches and those by Professor V. A. Barkhash in the field of nonclassical carbocations is beyond the scope of this chapter. The main results of these studies are published in a review article[162] and in a chapter of the Topics in Current Chemistry series.[163] Results of studies in the field of fluorine containing carbocations are presented as a chapter of this collective volume written by Professor V. D. Shteingarts.

ACKNOWLEDGMENT

We thank the Russian Foundation for Basic Research (Grant 03-02-32881) for financial support. Helpful remarks by Dr. Z. V. Todres are greatly appreciated.

REFERENCES

1. V. A. Koptyug, *Isomerizatsiya Aromaticheskikh Soedinenii*, Novosibirsk, 1963.
2. V. A. Koptyug, V. G. Shubin, I. K. Baeva, D. V. Korchagina, A. M. Komagorov, and A. I. Rezvukhin, *Zh. Obshch. Khim.* **35**, 1111 (1965).
3. C. MacLean and E. L. Mackor, *Mol. Phys.* **4**, 241 (1961).
4. W. von E. Doering, M. Saunders, H. G. Boyton, H. W. Earhart, E. F. Wadley, W. R. Edwards, and G. Laber, *Tetrahedron* **4**, 178 (1958).
5. V. A. Koptyug, V. G. Shubin, A. I. Rezvukhin, D. V. Korchagina, V. P. Tret'yakov, and E. S. Rudakov, *Dokl. Akad. Nauk SSSR* **171**, 1109 (1966).
6. B. G. Derendyaev, V. I. Mamatyuk, and V. A. Koptyug, *Tetrahedron Lett.* 5 (1969).
7. V. A. Koptyug, V. G. Shubin, and D. V. Korchagina, *Tetrahedron Lett.* 1535 (1965).
8. V. A. Koptyug, I. S. Isaev, and A. I. Rezvukhin, *Tetrahedron Lett.* 823 (1967).
9. V. A. Koptyug, I. K. Korobeinicheva, T. P. Andreeva, and V. A. Bushmelev, *Zh. Obshch. Khim.* **38**, 1979 (1968).
10. I. M. Sycheva, L. N. Shchegoleva, M. M. Mitasov, and B. G. Derendyaev, *Izv. Sib. Otd. Akad. Nauk SSSR, Ser. Khim. Nauk* **12**(5), 143 (1981).
11. L. P. Kamshii, V. I. Mamatyuk, and V. A. Koptyug, *Zh. Org. Khim.* **13**, 810 (1977).
12. G. I. Borodkin, S. M. Nagy, Yu. V. Gatilov, M. M. Shakirov, T. V. Rybalova, and V. G. Shubin, *Zh. Org. Khim.* **28**, 1806 (1992).
13. G. I. Borodkin, Yu. V. Gatilov, S. M. Nagy, and V. G. Shubin, *Zh. Strukt. Khim.* **37**, 534 (1996).
14. G. I. Borodkin, S. M. Nagy, I. Yu. Bagryanskaya, and Yu. V. Gatilov, *Zh. Strukt. Khim.* **25**, 114 (1984).
15. N. V. Bodoev, V. I. Mamatyuk, A. P. Krysin, and V. A. Koptyug, *Izv. Akad. Nauk SSSR, Ser. Khim.* 1199 (1978).
16. G. I. Borodkin, M. M. Shakirov, V. G. Shubin, and V. A. Koptyug, *Zh. Org. Khim.* **14**, 989 (1978).
17. V. F. Loktev, M. M. Shakirov, and V. G. Shubin, *Izv. Sib. Otd. Akad. Nauk SSSR, Ser. Khim. Nauk* **5**(2), 104 (1985).

18. V. F. Loktev and V. G. Shubin, *Izv. Sib. Otd. Akad. Nauk SSSR, Ser. Khim. Nauk*, **2**(1), 108 (1981).
19. S. V. Morozov, M. M. Shakirov, and V. G. Shubin, *Zh. Org. Khim.* **16**, 2103 (1980).
20. S. V. Morozov, M. M. Shakirov, and V. G. Shubin, *Zh. Org. Khim.* **17**, 154 (1981).
21. S. V. Morozov, M. M. Shakirov, and V. G. Shubin, *Izv. Akad. Nauk SSSR, Ser. Khim.*, 204 (1983).
22. V. F. Loktev, M. M. Shakirov, and V. G. Shubin, *Izv. Akad. Nauk SSSR, Ser. Khim.*, 696 (1986).
23. B. G. Derendyaev, V. I. Mamatyuk, and V. A. Koptyug, *Izv. Akad. Nauk SSSR, Ser. Khim.* 972 (1971).
24. G. I. Borodkin, S. M. Nagy, V. I. Mamatyuk, M. M. Shakirov, and V. G. Shubin, *J. Chem. Soc. Chem. Commun.* 1533 (1983).
25. G. I. Borodkin, S. M. Nagy, M. M. Shakirov, and V. G. Shubin, *Izv. Akad. Nauk SSSR, Ser. Khim.* 952 (1987).
26. G. I. Borodkin, B. G. Derendyaev, V. G. Shubin, and V. A. Koptyug, *Izv. Akad. Nauk SSSR, Ser. Khim.* 235 (1984).
27. G. I. Borodkin, S. M. Nagy, M. M. Shakirov, and V. G. Shubin, *Izv. Akad. Nauk SSSR, Ser. Khim.* 133 (1989).
28. S. V. Morozov, M. M. Shakirov, V. G. Shubin, and V. A. Koptyug, *Zh. Org. Khim.* **15**, 770 (1979).
29. S. V. Morozov, M. M. Shakirov, and V. G. Shubin, *Zh. Org. Khim.* **19**, 1011 (1983).
30. V. G. Shubin, D. V. Korchagina, A. I. Rezvukhin, and V. A. Koptyug, *Dokl. Akad. Nauk SSSR* **179**, 119 (1968).
31. G. I. Borodkin, M. M. Shakirov, V. G. Shubin, and V. A. Koptyug, *Zh. Org. Khim.* **12**, 1297 (1976).
32. G. I. Borodkin, M. M. Shakirov, V. G. Shubin, and V. A. Koptyug, *Zh. Org. Khim.* **12**, 1303 (1976).
33. V. A. Bushmelev, M. M. Shakirov, B. G. Derendyaev, and V. A. Koptyug, *Zh. Org. Khim.* **15**, 1934 (1979).
34. D. V. Korchagina, B. G. Derendyaev, V. G. Shubin, and V. A. Koptyug, *Zh. Org. Khim.* **12**, 384 (1976).
35. D. V. Korchagina, B. G. Derendyaev, and V. G. Shubin, *Izv. Akad. Nauk SSSR, Ser. Khim.* 441 (1971).
36. V. G. Shubin and V. A. Koptyug, *Izv. Sib. Otd. Akad Nauk SSSR, Ser. Khim. Nauk* **4**(2), 131 (1976).
37. G. I. Borodkin, E. B. Panova, M. M. Shakirov, and V. G. Shubin, in *Khimiya Karbokationov*, Nauka, Novosibirsk, 1979, p. 84.
38. L. I. Sazonova, M. M. Shakirov, and V. G. Shubin, *Zh. Org. Khim.* **13**, 2456 (1977).
39. G. I. Borodkin, E. B. Panova, M. M. Shakirov, and V. G. Shubin, *Zh. Org. Khim.* **18**, 2312 (1982).
40. G. I. Borodkin, E. B. Belikova, M. M. Shakirov, and V. G. Shubin, *Zh. Org. Khim.* **27**, 2508 (1991).
41. G. I. Borodkin, E. B. Belikova, M. M. Shakirov, and V. G. Shubin, *Zh. Org. Khim.* **25**, 1831 (1989).

42. V. G. Shubin, A. A. Tabatskaya, B. G. Derendyaev, and V. A. Koptyug, *Izv. Akad. Nauk SSSR, Ser. Khim.* 2417 (1968).
43. G. I. Borodkin, M. M. Shakirov, and V. G. Shubin, *Zh. Org. Khim.* **26**, 2254 (1990).
44. G. I. Borodkin, M. M. Shakirov, and V. G. Shubin, *Zh. Org. Khim.* **13**, 2152 (1977).
45. G. I. Borodkin, S. M. Nagy, V. I. Mamatyuk, M. M. Shakirov, and V. G. Shubin, *Zh. Org. Khim.* **20**, 552 (1984).
46. G. I. Borodkin, M. M. Shakirov, and V. G. Shubin, *Zh. Org. Khim.* **27**, 455 (1991).
47. G. I. Borodkin, L. M. Pletneva, M. M. Shakirov, and V. G. Shubin, *Zh. Org. Khim.* **15**, 652 (1979).
48. V. G. Shubin, D. V. Korchagina, B. G. Derendyaev, G. I. Borodkin, and V. A. Koptyug, *Zh. Org. Khim.* **9**, 1041 (1973).
49. G. I. Borodkin, M. M. Shakirov, V. G. Shubin, and V. A. Koptyug, *Zh. Org. Khim.* **14**, 321 (1978).
50. G. I. Borodkin, M. M. Shakirov, and V. G. Shubin, *Zh. Org. Khim.* **14**, 374 (1978).
51. V. I. Mamatyuk, B. G. Derendyaev, A. N. Detsina, and V. A. Koptyug, *Zh. Org. Khim.* **10**, 2487 (1974).
52. V. F. Loktev, D. V. Korchagina, and V. G. Shubin, *Izv. Sib. Otd. Akad. Nauk SSSR, Ser. Khim. Nauk* **14**(6), 86 (1980).
53. L. M. Mozulenko and V. A. Koptyug, *Zh. Org. Khim.* **8**, 2152 (1972).
54. A. V. Chikinev, V. A. Bushmelev, M. M. Shakirov, and V. G. Shubin, *Zh. Org. Khim.* **22**, 1454 (1986).
55. V. F. Loktev, D. V. Korchagina, M. M. Shakirov, and V. G. Shubin, *Izv. Sib. Otd. Akad. Nauk SSSR, Ser. Khim. Nauk* **12**(5), 146 (1978).
56. V. F. Loktev, D. V. Korchagina, and V. G. Shubin, *Izv. Sib. Otd. Akad. Nauk SSSR, Ser. Khim. Nauk* **2**(1), 112 (1981).
57. A. N. Detsina, V. I. Mamatyuk, B. G. Derendyaev, and V. A. Koptyug, *Zh. Org. Khim.* **12**, 610 (1976).
58. D. V. Korchagina, B. G. Derendyaev, V. G. Shubin, and V. A. Koptyug, *Zh. Org. Khim.* **7**, 2582 (1971).
59. G. A. Olah, G. K. S. Prakash, G. Liang, P. W. Westerman, K. Kunde, J. Chandrasekhar, and P. v. R. Schleyer, *J. Am. Chem. Soc.* **102**, 4485 (1980).
60. G. A. Olah, H. C. Lin, and Y. K. Mo, *J. Am. Chem. Soc.* **94**, 3667 (1972).
61. A. N. Detsina and V. I. Mamatyuk, *Izv. Akad. Nauk SSSR, Ser. Khim.* 2337 (1973).
62. R. N. Berezina, L. E. Ugryumova, D. V. Korchagina, and V. G. Shubin, *Zh. Org. Khim.* **18**, 592 (1982).
63. G. I. Borodkin, V. A. Koptyug, and V. G. Shubin, *Dokl. Akad. Nauk SSSR* **255**, 587 (1980).
64. G. I. Borodkin and V. G. Shubin, *Izv. Akad. Nauk SSSR, Ser. Khim.* 998 (1985).
65. G. I. Borodkin, I. K. Korobeinicheva, and V. G. Shubin, *Zh. Org. Khim.* **12**, 2381 (1976).
66. G. I. Borodkin and V. G. Shubin, *Voprosy Fizicheskoi Organicheskoi Khimii*, LGU, Leningrad, 1984, Vol. 2, p. 3.
67. S. V. Morozov, V. G. Shubin, and V. A. Koptyug, *Zh. Org. Khim.* **25**, 889 (1989).
68. V. A. Koptyug, *Top. Curr. Chem.* **122**, 1 (1984).
69. V. A. Koptyug and V. I. Buraev, *Zh. Org. Khim.* **16**, 1882 (1980).

70. V. A. Koptyug and V. I. Buraev, *Zh. Org. Khim.* **14**, 18 (1978).
71. Yu. G. Erykalov, A. P. Belokurova, I. S. Isaev, and V. A. Koptyug, *Zh. Org. Khim.* **9**, 343 (1973).
72. V. A. Bushmelev, M. M. Shakirov, D. M. Brouwer, and V. A. Koptyug, *Zh. Org. Khim.* **15**, 1579 (1979).
73. G. I. Borodkin, M. M. Shakirov, V. G. Shubin, and V. A. Koptyug, *Zh. Org. Khim.* **12**, 2033 (1976).
74. G. I. Borodkin, E. R. Susharin, M. M. Shakirov, and V. G. Shubin, *Zh. Org. Khim.* **21**, 1809 (1985).
75. V. G. Shubin and G. I. Borodkin, in *Stable Carbocation Chemistry*, G. K. S. Prakash and P. v. R. Schleyer, eds., Wiley, New York, 1997, p. 231.
76. G. I. Borodkin and V. G. Shubin, *Chemistry Rev.* **24** (part 2), 1 (1999).
77. E. B. Belikova, G. I. Borodkin, T. I. Savina, M. M. Shakirov, V. G. Shubin, and Yu. L. Yagupol'skii, *Izv. Akad. Nauk SSSR, Ser. Khim.* 1231 (1991).
78. A. V. Chikinev, M. M. Shakirov, and V. A. Bushmelev, *Izv. Akad. Nauk SSSR, Ser. Khim.* 815 (1988).
79. A. V. Chikinev, V. A. Bushmelev, and V. G. Shubin, *Izv. Akad. Nauk SSSR, Ser. Khim.* 1315 (1992).
80. G. I. Borodkin, S. M. Nagy, T. N. Trushchenko, and V. G. Shubin, *Zh. Org. Khim.* **25**, 217 (1989).
81. G. I. Borodkin and V. G. Shubin, *Tez. Dokl. IV Vses. Konf. Khim. Nizkikh Temperatur*, MGU, Moscow, 1988, p. 123.
82. G. I. Borodkin, Ye. B. Panova, M. M. Shakirov, and V. G. Shubin, *J. Chem. Soc. Chem. Commun.* 354 (1979).
83. G. I. Borodkin, E. B. Panova, M. M. Shakirov, and V. G. Shubin, *Zh. Org. Khim.* **19**, 114 (1983).
84. R. B. Woodward and R. Hoffmann, *The Conservation of Orbital Symmetry*, Academic Press, New York, 1970.
85. V. A. Koptyug and L. M. Mozulenko, *Zh. Org. Khim.* **6**, 102 (1970).
86. A. N. Detsina and V. A. Koptyug, *Zh. Org. Khim.* **8**, 2158 (1972).
87. I. S. Isaev, V. I. Buraev, G. P. Novikov, L. M. Sosnovskaya, and V. A. Koptyug, *Zh. Org. Khim.* **10**, 567 (1974).
88. L. A. Ostashevskaya, A. N. Detsina, V. I. Mamatyuk, I. S. Isaev, and V. A. Koptyug, *Zh. Org. Khim.* **10**, 2374 (1974).
89. A. N. Detsina and V. A. Koptyug, *Zh. Org. Khim.* **7**, 2575 (1971).
90. V. A. Koptyug and L. M. Mozulenko, *Zh. Org. Khim.* **7**, 1419 (1971).
91. L. M. Mozulenko and V. A. Koptyug, *Zh. Org. Khim.* **8**, 2531 (1972).
92. L. M. Mozulenko, A. I. Rezvukhin, and V. A. Koptyug, *Zh. Org. Khim.* **8**, 2535 (1972).
93. A. N. Detsina and V. A. Koptyug, *Zh. Org. Khim.* **8**, 2215 (1972).
94. A. N. Detsina, V. I. Mamatyuk, and V. A. Koptyug, *Izv. Akad. Nauk SSSR, Ser. Khim.* 2163 (1973).
95. A. N. Detsina, N. V. Sidorova, E. B. Panova, E. V. Malykhin, and M. M. Shakirov, *Zh. Org. Khim.* **15**, 1887 (1979).

96. V. A. Koptuyg, T. P. Andreeva, and V. I. Mamatyuk, *Izv. AN SSSR, Ser. Khim.* 2844 (1968).
97. V. A. Koptuyg, T. P. Andreeva, and V. I. Mamatyuk, *Zh. Org. Khim.* **6**, 1848 (1970).
98. V. A. Koptyug and A. V. Golounin, *Zh. Org. Khim.* **8**, 607 (1972).
99. T. P. Andreeva, G. P. Tregub, V. I. Mamatyuk, and V. A. Koptyug, *Zh. Org. Khim.* **8**, 1271 (1972).
100. L. P. Kamshii, V. I. Mamatyuk, and V. A. Koptyug, *Zh. Org. Khim.* **11**, 344 (1975).
101. A. V. Golounin and V. A. Koptyug, *Zh. Org. Khim.* **8**, 2555 (1972).
102. V. I. Mamatyuk, A. I. Rezvukhin, A. V. Golounin, and V. A. Koptyug, *Zh. Org. Khim.* **9**, 2359 (1973).
103. V. I. Mamatyuk, A. I. Rezvukhin, A. N. Detsina, V. I. Buraev, I. S. Isaev, and V. A. Koptyug, *Zh. Org. Khim.* **9**, 2429 (1973).
104. V. A. Koptyug and T. P. Andreeva, *Zh. Org. Khim.* **7**, 2398 (1971).
105. V. A. Koptyug and T. P. Andreeva, *Zh. Org. Khim.* **8**, 1265 (1972).
106. V. A. Koptyug, *Zh. Vsesouzn. Khim. Obshch.* **21**, 315 (1976).
107. V. A. Koptyug and A. V. Golounin, *Zh. Org. Khim.* **9**, 2158 (1973).
108. K. Yu. Koltunov and I. B. Repinskaya, *Zh. Org. Khim.* **31**, 1579 (1995).
109. K. Yu. Koltunov, M. M. Shakirov, and I. B. Repinskaya, *Zh. Org. Khim.* **34**, 630 (1998).
110. K. Yu. Koltunov, L. A. Ostashevskaya, and I. B. Repinskaya, *Zh. Org. Khim.* **34**, 1870 (1998).
111. K. Yu. Koltunov, E. N. Subbotina, and I. B. Repinskaya, *Zh. Org. Khim.* **33**, 750 (1997).
112. L. A. Ostashevskaya, K. Yu. Koltunov, and I. B. Repinskaya, *Zh. Org. Khim.* **36**, 1511 (2000).
113. A. V. Golounin, Yu. S. Shchedrin, and V. A. Fedorov, *Zh. Org. Khim.* **37**, 1170 (2001).
114. A. V. Golounin, Yu. S. Shchedrin, and V. A. Fedorov, *Zh. Org. Khim.* **37**, 1334 (2001).
115. A. V. Golounin, Yu. S. Shchedrin, and E. A. Ivanova, *Zh. Org. Khim.* **37**, 1338 (2001).
116. J. F. J. Dippy and A. L. L. Palluel, *J. Chem. Soc.* 1415 (1951).
117. J. E. Hofmann and A. Schriesheim, in *Friedel-Crafts and Related Reactions*, Vol. 2, G. A. Olah, ed., Wiley-Interscience, New York, 1964, p. 597.
118. G. I. Borodkin, S. M. Nagy, Yu. V. Gatilov, and V. G. Shubin, *Dokl. Akad. Nauk SSSR* **280**, 881 (1985).
119. G. I. Borodkin, E. R. Susharin, M. M. Shakirov, and V. G. Shubin, *Izv. Akad. Nauk SSSR, Ser. Khim.* 2622 (1985).
120. G. I. Borodkin, E. R. Susharin, I. R. Elanov, M. M. Shakirov, and V. G. Shubin, *Izv. Akad. Nauk SSSR, Ser. Khim.* 1352 (1991).
121. G. I. Borodkin, S. M. Nagy, M. M. Shakirov, and V. G. Shubin, *Zh. Org. Khim.* **17**, 202 (1981).
122. G. I. Borodkin, V. A. Bushmelev, S. M. Nagy, M. I. Rudnev, M. M. Shakirov, and V. G. Shubin, *Zh. Org. Khim.* **27**, 468 (1991).
123. G. I. Borodkin, E. I. Chernyak, M. M. Shakirov, and V. G. Shubin, *Zh. Org. Khim.* **22**, 330 (1986).
124. G. I. Borodkin, I. R. Elanov, M. M. Shakirov, and V. G. Shubin, *Zh. Org. Khim.* **27**, 889 (1991).

125. M. J. S. Dewar and G. P. Ford, *J. Am. Chem. Soc.* **101**, 783 (1979).
126. G. I. Borodkin, E. I. Chernyak, M. M. Shakirov, and V. G. Shubin, *Zh. Org. Khim.* **22**, 663 (1986).
127. G. I. Borodkin, E. I. Chernyak, M. M. Shakirov, and V. G. Shubin, *Zh, Org. Khim.* **26**, 785 (1990).
128. G. I. Borodkin, E. I. Chernyak, M. M. Shakirov, Yu. V. Gatilov, T. V. Rybalova, and V. G. Shubin, *Zh. Org. Khim.* **26**, 1163 (1990).
129. G. I. Borodkin, E. I. Chernyak, M. M. Shakirov, and V. G. Shubin, *Metalloorg. Khim.* **3**, 1017 (1990).
130. G. I. Borodkin, E. I. Chernyak, M. M. Shakirov, and V. G. Shubin, *Izv. Akad. Nauk, Ser. Khim.* 1057 (1993).
131. G. I. Borodkin, E. I. Chernyak, M. M. Shakirov, and V. G. Shubin, *Zh. Org. Khim.* **30**, 397 (1994).
132. I. S. Isaev, V. I. Mamatyuk, T. G. Egorova, L. I. Kuzubova, and V. A. Koptyug, *Izv. Akad. Nauk SSSR, Ser. Khim.* 2089 (1969).
133. I. S. Isaev, V. I. Mamatyuk, L. I. Kuzubova, T. A. Gordymova, and V. A. Koptyug, *Zh. Org. Khim.* **6**, 2482 (1970).
134. V. A. Koptyug, L. I. Kuzubova, I. S. Isaev, and V. I. Mamatyuk, *J. Chem. Soc., Chem. Commun.* 389 (1969).
135. V. A. Koptyug, V. I. Mamatyuk, L. I. Kuzubova, and I. S. Isaev, *Izv. Akad. Nauk SSSR, Ser. Khim.* 1635 (1969).
136. V. A. Koptyug, L. I. Kuzubova, I. S. Isaev, and V. I. Mamatyuk, *Zh. Org. Khim.* **6**, 2258 (1970).
137. R. F. Childs and S. Winstein, *J. Am. Chem. Soc.* **96**, 6409 (1974).
138. R. Hüttel, P. Tauchner, and H. Forkl, *Chem. Ber.* **105**, 1 (1972).
139. U. Burger and A. Delay, *Helv. Chim. Acta* **56**, 1345 (1973).
140. H. Hogeveen, P. W. Kwant, E. P. Schudde, and P. A. Wade, *J. Am. Chem. Soc.* **96**, 7518 (1974).
141. V. I. Mamatyuk, A. I. Rezvukhin, I. S. Isaev, V. I. Buraev, and V. A. Koptyug, *Zh. Org. Khim.* **10**, 662 (1974).
142. V. A. Koptyug, I. A. Shleider, I. S. Isaev, L. V. Vasil'eva, and A. I. Rezvukhin, *Zh. Org. Khim.* **7**, 1089 (1971).
143. T. J. Katz and E. H. Gold, *J. Am. Chem. Soc.* **86**, 1600 (1964).
144. I. A. Shleider, *Dissertation*, Novosibirsk, 1972.
145. I. A. Shleider, I. S. Isaev, and V. A. Koptyug, *Zh. Org. Khim.* **8**, 1337 (1972).
146. V. A. Bushmelev, A. M. Genaev, V. I. Mamatyuk, and V. G. Shubin, *Zh. Org. Khim.* **30**, 1761 (1994).
147. R. P. Haseltine and T. S. Sorensen, *Canad. J. Chem.* **53**, 1067 (1975).
148. A. M. Genaev, V. A. Bushmelev, and V. G. Shubin, *Zh. Org. Khim.* **31**, 1159 (1995).
149. G. A. Olah, N. J. Head, G. Rasul, and G. K. Surya Prakash, *J. Am. Chem. Soc.* **117**, 875 (1995).
150. V. A. Bushmelev, V. P. Vetchinov, A. M. Genaev, V. I. Mamatyuk, and V. G. Shubin, *Mendeleev Commun.* 187 (1996).
151. H. Vančik, A. M. Genaev, and V. G. Shubin, *J. Chem. Soc., Perkin Trans.* **2**, 667 (1997).

152. S. A. Osadchii, V. A. Drobysh, M. M. Shakirov, V. I. Mamatyuk, and V. G. Shubin, *Zh. Org. Khim.* **24**, 267 (1988).
153. S. A. Osadchii, V. A. Drobysh, M. M. Shakirov, V. I. Mamatyuk, and V. G. Shubin, *Zh. Org. Khim.* **24**, 1417 (1988).
154. S. A. Osadchii, M. M. Shakirov, and V. G. Shubin, *Zh. Org. Khim.* **29**, 930 (1993).
155. S. A. Osadchii, V. A. Drobysh, N. V. Mikushova, and V. G. Shubin, *Zh. Org. Khim.* **25**, 1838 (1989).
156. S. A. Osadchii, N. V. Mikushova, and V. G. Shubin, *Zh. Org. Khim.* **35**, 1813 (1999).
157. N. V. Mikushova, V. I. Mamatyuk, and V. G. Shubin, *The 13th IUPAC Conference on Physical Organic Chemistry*, Inchon, 1996, *Abstracts*, p. 117.
158. V. A. Bushmelev, A. M. Genaev, S. A. Osadchii, M. M. Shakirov, and V. G. Shubin, *Zh. Org. Khim.* **39**, 1374 (2003).
159. J. E. Baldwin, *J. Chem. Soc. Chem. Commun.* 734 (1976).
160. A. M. Genaev, W. A. Sokolenko, G. E. Salnikov, V. I. Mamatyuk, and V. G. Shubin, *Zh. Org. Khim.* **38**, 1051 (2002).
161. Yu. V. Gatilov and V. A. Barkhash, *Uspekhi Khim.* **61**, 1969 (1992).
162. V. A. Barkhash and M. P. Polovinka, *Uspekhi Khim.* **68**, 430 (1999).
163. V. A. Barkhash, *Top. Curr. Chem.* **116/117**, 1 (1984).

7

POLYFLUORINATED CARBOCATIONS

Vitalij D. Shteingarts
N. N. Vorozhtsov Novosibirsk Institute of Organic Chemistry
Siberian Division of Russian Academy of Sciences
Novosibirsk, Russia

7.1 INTRODUCTION

Since the era of long-lived carbocation chemistry in superacids started, the carbocations of diverse structural types have become objects for the direct observation and investigation in their own right.[1] Among them are polyfluorinated, including

Carbocation Chemistry, Edited by George A. Olah and G. K. Surya Prakash
ISBN 0-471-28490-4

perfluorinated carbocations. At the outset, an interest in this type of species was apparently provoked by the seemingly unnatural character of a combination of the electron deficiency of carbocation and the high electronegativity of fluorine. However, this interest was greatly strengthened and became more concrete with discovery and study of the acid initiated and electrophilic reactions of polyfluoroorganics. In this regard, the generation and, by so doing, the evidence for the existence of fluorinated carbocations as discrete species, would have a paramount significance for the possibility to rationalize the mechanism of acid-catalyzed and electrophilic reactions of polyfluorinated compounds in the terms of intermediacy of these cations. Moreover, knowledge of regularities of structure and reactivity of polyfluorinated carbocations stipulates the possibility of molecular design in fluoroorganic chemistry based on reactions proceeding via these species as intermediates. At last, the possibilities of obtaining fluoroorganic compounds are realized with the usage of long-lived polyfluorinated carbocations as synthons.

The diversity of issues arising in the connection with carbocation fluorination can be reduced to three principal questions:

- What is a limiting extent of fluorine substitution, including perfluorination, allowing a carbocation to exist as a long-lived species?
- What is the influence of poly(per)fluorination on the carbocation electronic structure and relative stability?
- What is the possible chemistry of polyfluorinated carbocations and the scope of these carbocations as synthons?

The early stage of a development of this area has been reviewed,[2,3] as were subsequently polyfluorinated arenium ions, which are the most significant carbocations of polyfluoroarene chemistry.[4] However, until now (2004) there has been no adequate review available on long-lived polyfluorinated carbocations. Recent reviews,[5] in all of their extensiveness, are concerned with cationic reactions of polyfluoroorganic compounds, predominantly of aliphatic ones, rather than with polyfluorinated carbocations themselves. The present chapter is aimed in filling this gap.

7.2 STRUCTURAL TYPES OF POLYFLUORINATED CARBOCATIONS AND THEIR GENERATION

Basically, two types of fluorine substitution are significant for this review: (1) substitution only in the carbocationic center, called the α or *resonance position*, since this location provides fluorine with an opportunity to delocalize the positive charge by resonance back-donation effect; and (2) substitution at neighboring carbons, which are called β or *nonresonance positions*. Sequentially, three basic types of multiple fluorination should be considered: at the same or different (in case of a delocalized π system) carbocationic centers or resonance positions, at the same or different nonresonance positions, and mixed fluorine substitution at the resonance and nonresonance positions.

Historically, the source of fluorinated carbocations was afforded by the series of studies by Olah and his coworkers in which the general methodology developed for the generation of long-lived carbocations was applied to fluorinated carbocations.[2] This methodology is based on the heterolysis by superacid of a σ-bond C_{sp^3}-X, where X is a halogen, hydroxy, or other group, or a π-bond (C=C or C=O) of neutral organic molecules, NMR spectroscopy was employed as a tool for the direct observation of a long-lived species, using the characteristics of NMR spectroscopy for qualifying carbocations as a long-lived species. Pioneering the studies of long-lived fluorinated carbocations in general and *α-fluorinated carbocations* in particular was the application of this methodology to the generation of 2-fluoroisopropyl cation **1**, 1-fluoro-1-phenylethyl cation **2** and α-fluorodiphenylmethyl cation **3**.[6] Subsequently, cyclic cations of the same type were also generated and studied[7] (Scheme 1).

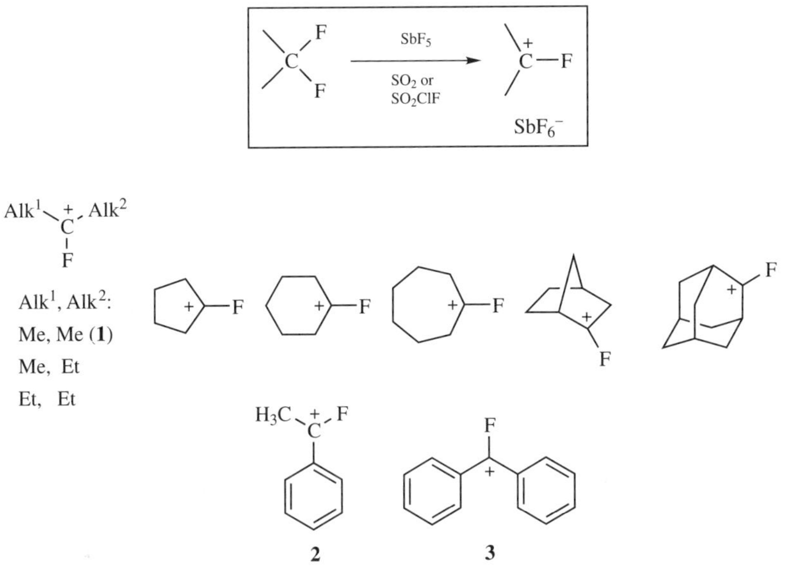

Scheme 1

These results showed that the presence of at least two alkyl groups or a phenyl substituent to be necessary at the same carbocationic center for persistent long-lived α-monofluorinated carbocations. The attempt to remove one of the alkyl groups and prepare long-lived α-fluoroethyl cation failed.[8] However, the important finding was made that substitution of an alkyl group by the second fluorine atom at the same carbocationic center provides a long-lived 1,1-difluoroethyl cation **4**,[8] α,α-difluorobenzyl cation **5** (X = H), and a series of its related halogen-substituted analogs,[6a,9] thereby opening up the chemistry of polyfluorinated carbocations in general and *α-polyfluorinated alkyl* and *arylmethyl* cations in particular (Scheme 2).

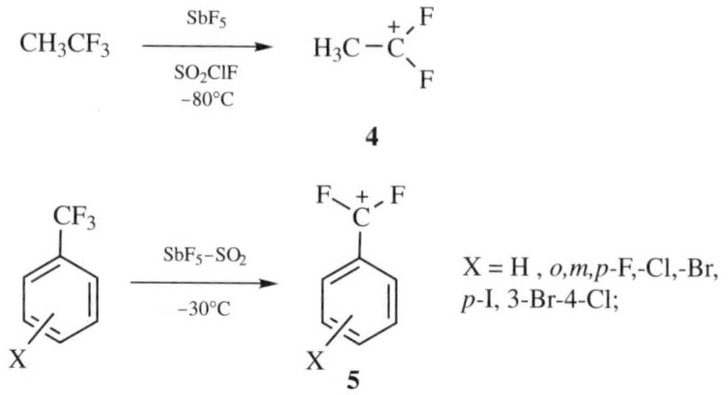

$$CH_3CF_3 \xrightarrow[\substack{SO_2ClF \\ -80°C}]{SbF_5} H_3C-\overset{+}{C}F_2$$

4

X = H , *o,m,p*-F,-Cl,-Br, *p*-I, 3-Br-4-Cl;

5

Scheme 2

The analogous substitution of the last methyl or phenyl group, however, at the same carbocationic center to generate a long lived carbocation was unsuccessful; the trifluoromethyl cation CF_3^+ could not been obtained in a superacid solution because of its high electrophilicity and, consequently, instantaneous quenching of the ensuing cation by any nucleophile present in these media.[10] This parent of perfluorinated carbocations was subsequently generated together with CF_2Hlg^+ cations by photolysis of trifluorohalogenomethanes CF_3Hlg (Hlg = Cl, Br, I) in the argon matrix and characterized by IR spectroscopy[11] (Scheme 3).

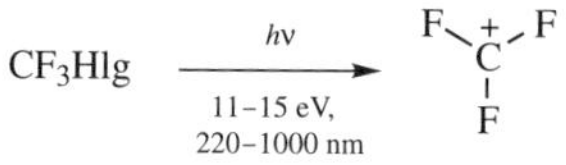

$$CF_3Hlg \xrightarrow[\substack{11-15\ eV, \\ 220-1000\ nm}]{h\nu} CF_3^+$$

Scheme 3

Sargeant and Krespan first demonstrated the possibility of multiple fluorine substitution at cyclically arranged and directly linked carbocationic sites by preparing the perfluorinated *cyclopropenyl* cation **6**,[12] as well as its analogs containing other halogens[13] from appropriate cyclopropenes (Scheme 4). Similarly, perfluoro*cycloheptatrienyl* cation **7** was generated by the action of $BF_3 \cdot Et_2O$ on perfluorocycloheptatriene.[14]

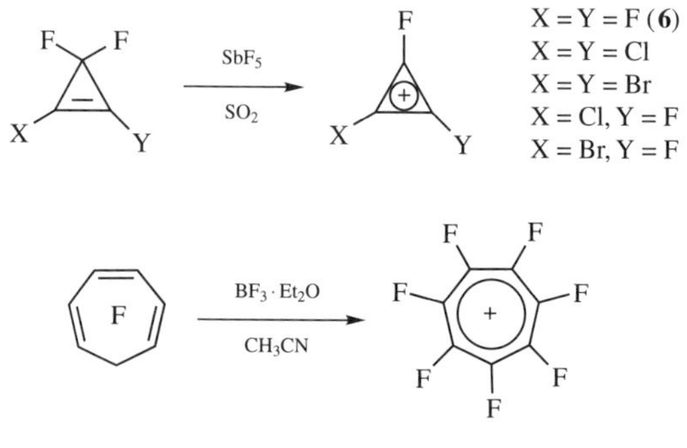

X = Y = F (**6**)
X = Y = Cl
X = Y = Br
X = Cl, Y = F
X = Br, Y = F

SbF_5, SO_2

$BF_3 \cdot Et_2O$, CH_3CN

7

Scheme 4

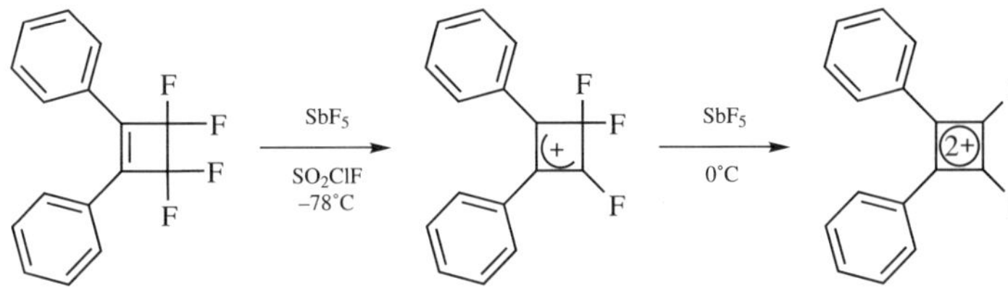

$\xrightarrow[\text{SO}_2\text{ClF, } -78^\circ\text{C}]{\text{SbF}_5}$ $\xrightarrow[0^\circ\text{C}]{\text{SbF}_5}$ **8**

Scheme 5

Even the long-lived cyclobutadiene dication **8** was obtained by Olah and Staral[15] (Scheme 5).

In all these cases the aromaticity of the cations is important. This is illustrated by the failure to generate perfluorocyclopentadienyl cation in a similar way.[16]

Subsequently, Chambers and coworkers prepared allyl and highly delocalized alkapolyenyl cations with fluorines at all resonance positions as well as the dication **9**, which is composed of two head-to-head linked allylic cations[17] (Scheme 6).

$$(CF_3)_2CF(CH_2CF_2)_{n+1}CH_2CF_3 \xrightarrow{SbF_5} (CF_3)_2CFCH_2-\overbrace{(CF-CH)_n-CF}^{+}-CH_2CF_3$$

$n = 1,2,3$

$$(CF_3)_2CFCH_2-\overbrace{CF-CH-CF}^{+}-\overbrace{CF-CH-CF}^{+}-CH_2CF(CF_3)_2$$

9

Scheme 6

For the existence of long-lived β-*polyfluorinated* carbocations, such as those with a trifluoromethyl group at a carbocationic center, Olah and coworkers have demonstrated that two phenyl groups, or a phenyl or hydroxy group together with alkyl or functionalized alkyl group must be present simultaneously at the same carbocationic center (Scheme 7).[18]

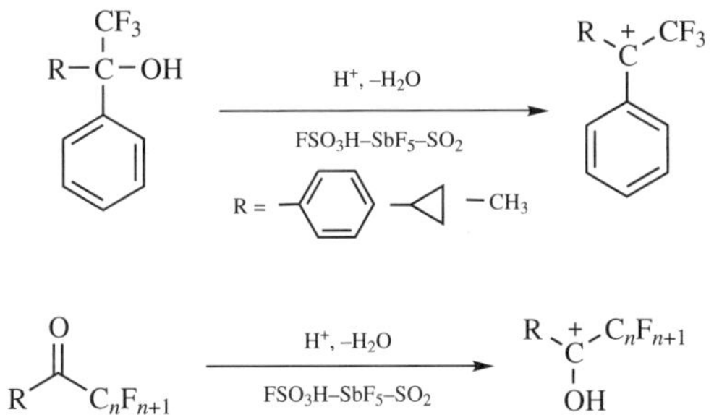

$$R-C(CF_3)(OH)-C_6H_5 \xrightarrow[\text{FSO}_3\text{H-SbF}_5\text{-SO}_2]{\text{H}^+, -\text{H}_2\text{O}} R\overset{+}{C}(CF_3)-C_6H_5$$

R = phenyl, cyclopropyl, $-CH_3$

$$RC(=O)C_nF_{n+1} \xrightarrow[\text{FSO}_3\text{H-SbF}_5\text{-SO}_2]{\text{H}^+, -\text{H}_2\text{O}} R\overset{+}{C}(C_nF_{n+1})OH$$

$n = 1$, R = Ph, Me, CH_2COCH_3, CH_2COCF_3

$n = 2$, R = Et

Scheme 7

Attempts to remove other substituent from α carbon or insert a second perfluoroalkyl group[18] failed to give long-lived carbocation.

The most widespread are the *mixed multiple fluorinated* carbocations, which were first realized by Olah and coworkers by generation of the mono- **10**, di- **11**, and tripentafluorophenyl **12** methyl cations,[19] the cation **12** was the first *long-lived perfluorinated* carbocation, and the pentafluorophenylfluoromethyl cation **13** was the first monohydroperfluoro carbocation (Scheme 8). Pentafluorobenzyl cation **9** was obtained also by protolysis of C—H bond of 2,3,4,5,6-pentafluorotoluene.[20]

SbF_5–SO_2ClF or net SbF_5

X = H (**10**), F (**13**)

11 **12**

Scheme 8

In the context of nonclassical carbocation problem benzyl-type cations **14**,[21a,b] **15**, **16**,[22] and dication **17**[21b] with carbocationic center incorporated into bicyclic hydrocarbon skeletons were reported. Interestingly, the isomers **15** and **16** were derived from the epimeric tosylates, thus indicating the different σ-bond participation in the charge delocalization in a transition state and, accordingly, the primary formation of two different nonclassical cations in the course of ionization[22] (Scheme 9).

Of special stability are polyfluorinated arylmethyl cations stabilized by oxygen backdonation and generated by protonation or ionization of the respective carbonyl precursors[23] (Scheme 10).

Further long-lived perfluorinated arylmethyl cations include perfluorinated benzyl **18** ($R^{1\div4}$, X = F), diphenylmethyl **19** (X,Y = F) and α-methyldiphenylmethyl **20** cations.[24] Subsequently, a series of polyfluorinated α,α-difluorobenzyl cations was prepared by ionizing benzylic C—F bond with SbF_5.[24b] The latter cations were used for alkylation of polyfluoroarenes in neat SbF_5 to yield the polyfluorinated α-fluorodiarylmethyl cations, which can also be prepared from appropriate polyfluorinated α,α-difluorodiarylmethanes with SbF_5. In dissolving perfluorinated *p*- and *o*-xylenes in SbF_5, the appropriate perfluorinated *p*- and *o*-methylbenzyl cations are formed in amounts that are insufficient for NMR studies. However, at least in the first case, the cation is involved in the alkylation of

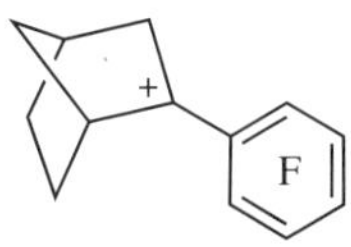
F

14

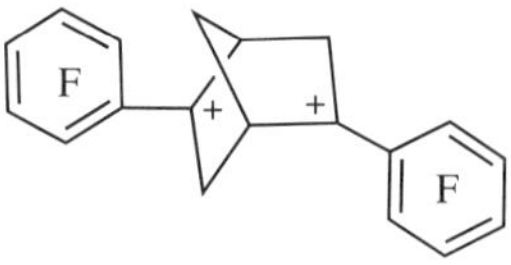
F
F

17

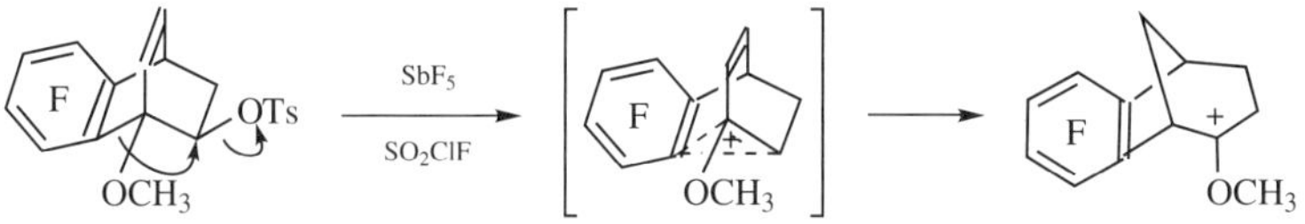
F
OTs
OCH$_3$
SbF$_5$
SO$_2$ClF
F
OCH$_3$
F
OCH$_3$

15

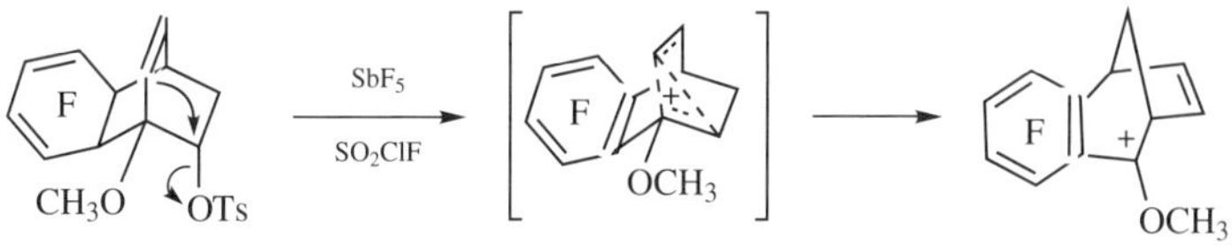
F
CH$_3$O
OTs
SbF$_5$
SO$_2$ClF
F
OCH$_3$
F
OCH$_3$

16

Scheme 9

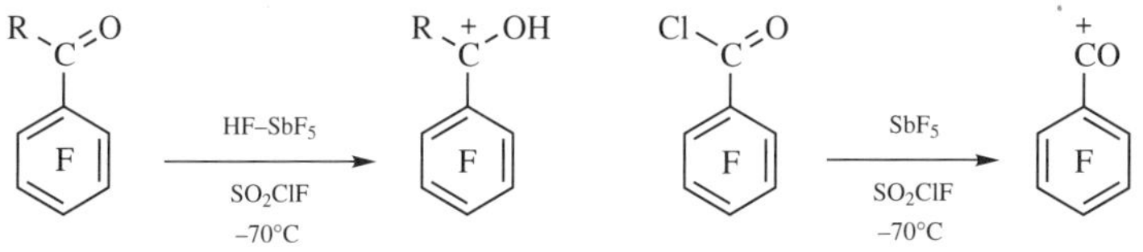
R
O
F
HF–SbF$_5$
SO$_2$ClF
–70°C
R
OH
F
Cl
O
F
SbF$_5$
SO$_2$ClF
–70°C
CO
F

R = H, C_6F_5, OH

Scheme 10

polyfluorobenzene and eventually leads to the respective perfluorodiphenylmethyl cation[25] (Scheme 11).

Similarly, the long-lived dication **21** has been obtained from the reaction of perfluoro-4,4′-dimethylbiphenyl with 2 equiv of pentafluorobenzene in SbF_5[25] (Scheme 12).

Polyfluorinated indenyl cations, including the perfluorinated indenyl **22** and 2-methylindenyl **23** cations, were prepared from corresponding indenes with SbF_5[26] (Scheme 13).

Intramolecular polyfluoroarene alkylation was used to obtain the perfluorinated fluorenyl **24** and 9-phenylfluorenyl **25** cations[27] (Scheme 14).

Of further substantial significance were cations with more contracted π systems such as polyfluorinated *arenium* cations, which are of paramount importance in electrophilic chemistry of polyfluoroarenes. In the traditional method of arene protonation, which in the case of polyfluorinated arenes necessitated a superacid $FSO_3H–SbF_5$ or $HF–SbF_5$, Olah and coworkers have demonstrated that these

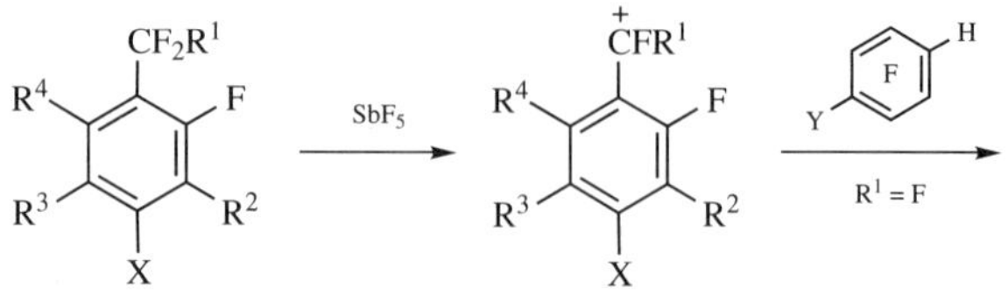

CF_2R^1 $\xrightarrow{SbF_5}$ $\overset{+}{C}FR^1$ $\xrightarrow{R^1 = F}$

18

R^{1-4} = F, X = F, Cl, Br; CH_3, H
R^{1-3}, X = F; R^4 = H;
R^1 = H; R^{2-4}, X = F

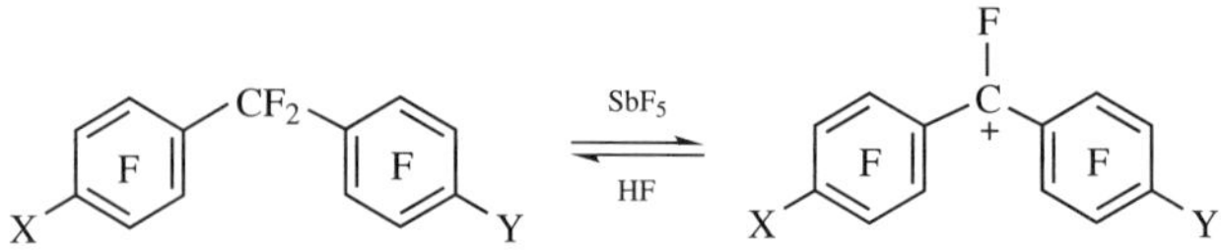

$\underset{HF}{\overset{SbF_5}{\rightleftharpoons}}$

19

X = F, Cl, Br, CH_3, H, CF_3
Y = F, Cl, Br, H, CH_3

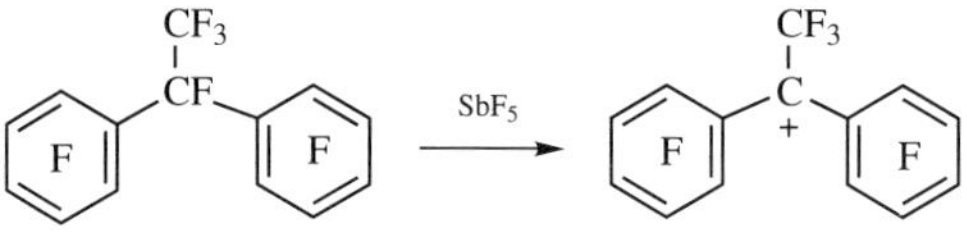

$\xrightarrow{SbF_5}$

20

Scheme 11

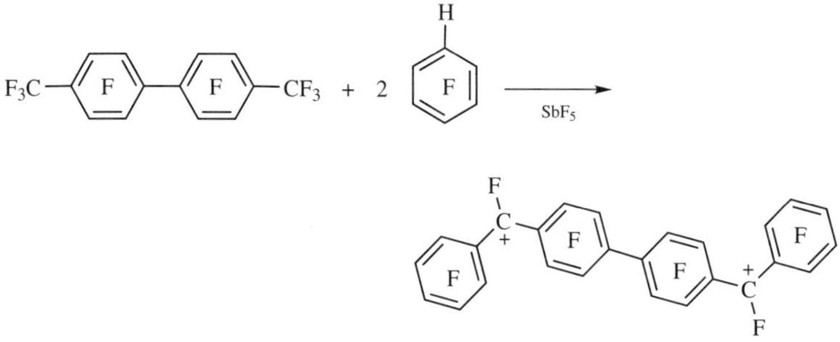

F_3C ... CF_3 + 2 $\xrightarrow{SbF_5}$

21

Scheme 12

$\xrightarrow{SbF_5}$

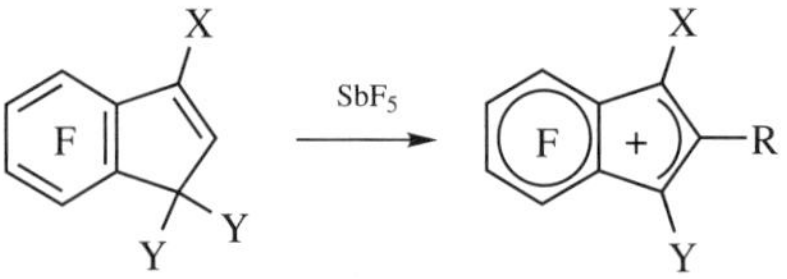

R = F; X, Y = F (**22**), Cl; X = H, Y = F
R = CF_3; X, Y = F (**23**)

Scheme 13

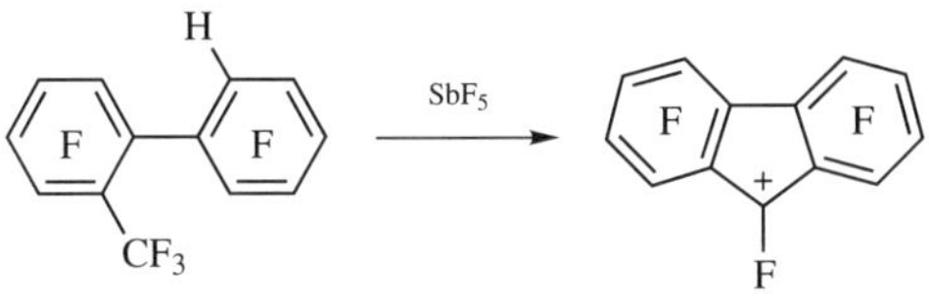

24

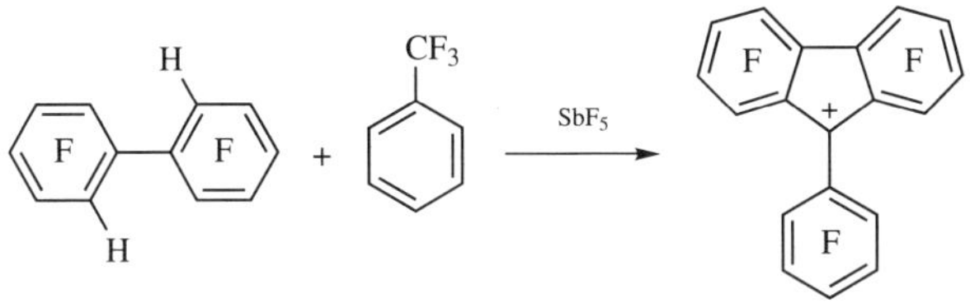

25

Scheme 14

cations can be prepared by proton addition to either the unsubstituted or *ipso* (to methyl group) positions of a polyfluorinated benzene ring, but never *ipso* to fluorine[28] (Scheme 15).

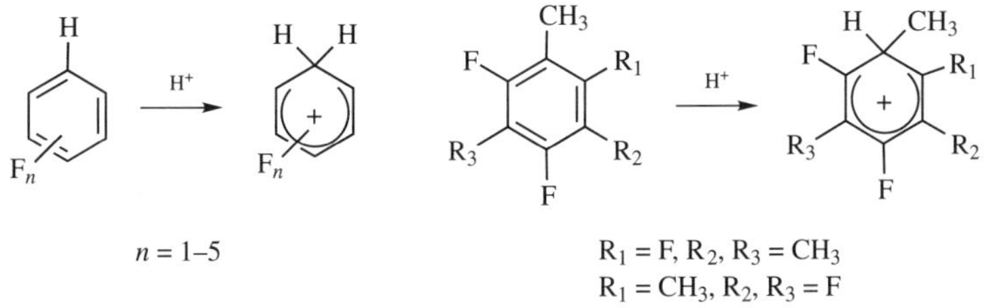

$n = 1–5$

$R_1 = F, R_2, R_3 = CH_3$
$R_1 = CH_3, R_2, R_3 = F$

Scheme 15

An interesting development resulted in the protonation of compounds containing directly nonconnected and transannularly interacting cyclic fragment, [2.2] *para*-cyclophane and its derivatives with complete ring fluorine substitution in one or both rings. Protonation was found to occur at the bridgehead position. Fluorination of one benzene ring of [2.2]*para*-cyclophane decreases basicity of both phenyl rings. Thus, whereas [2.2]*para*-cyclophane was monoprotonated by dissolving in $HSO_3F–SO_2ClF$ and double protonation takes place in $HSO_3F–SbF_5$ ("magic acid"), tetrafluoro- and octafluoro[2.2]*para*-cyclophanes were only monoprotonated in "magic acid"–SO_2ClF solution, at the nonfluorinated ring. However, the latter compound appeared more basic than the parent 2,3,5,6-tetrafluoro-*para*-xylene, the protonation of which requires stronger superacid $HF–SbF_5$. This effect is suggested to be due to the π,π repulsion inherent in octafluoro[2.2]

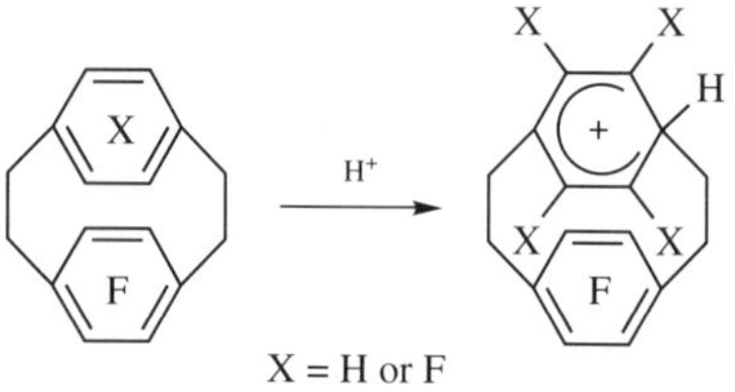

X = H or F

26 (X = H)

Scheme 16

para-cyclophane or/and by the stabilizing transannular interaction of protonated and nonprotonated rings in the benzenium cation **26**[29] (Scheme 16).

Arenium ions with fluorine atoms both at sp^3-hybridized carbon and in the unsaturated moiety, particularly perfluorinated arenium cations **27–29**, were obtained by dissolving respective polyfluorinated dihydroaromatic precursors in neat SbF_5 and its solutions in SO_2 or SO_2ClF[30] (Schemes 17–19). The perfluorinated benzenium **27** (Scheme 17) and naphthalenium **28** (Scheme 18) ions were obtained as well the first long-lived arenium ions with electronegative substituting heteroatom bonded at the saturated carbon. In all cases of arenium cations containing substituents other than fluorine, the distribution of isomers (for naphthalenium cations this concerns isomers differing by substituent location in the same ring) equilibrated through the fluoride anion addition-elimination mechanism dominated by the trend to have two fluorines at sp^3-hybridyzed carbon and fluorine at the *para* position (see below).

Polyfluorinated hydroxyarenium ions with the ring —CFX— unit (X = F, Cl, Br) were prepared by dissolving appropriate cyclohexadienones in superacids.[31] This

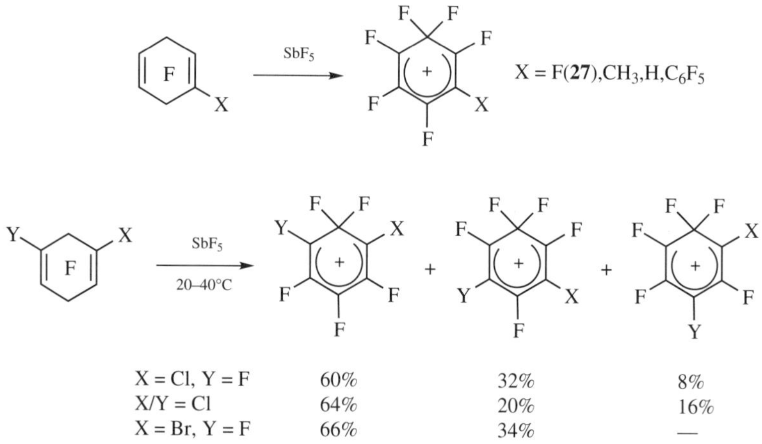

SbF_5

X = F(**27**),CH_3,H,C_6F_5

SbF_5, 20–40°C

X = Cl, Y = F	60%	32%	8%
X/Y = Cl	64%	20%	16%
X = Br, Y = F	66%	34%	—

Scheme 17

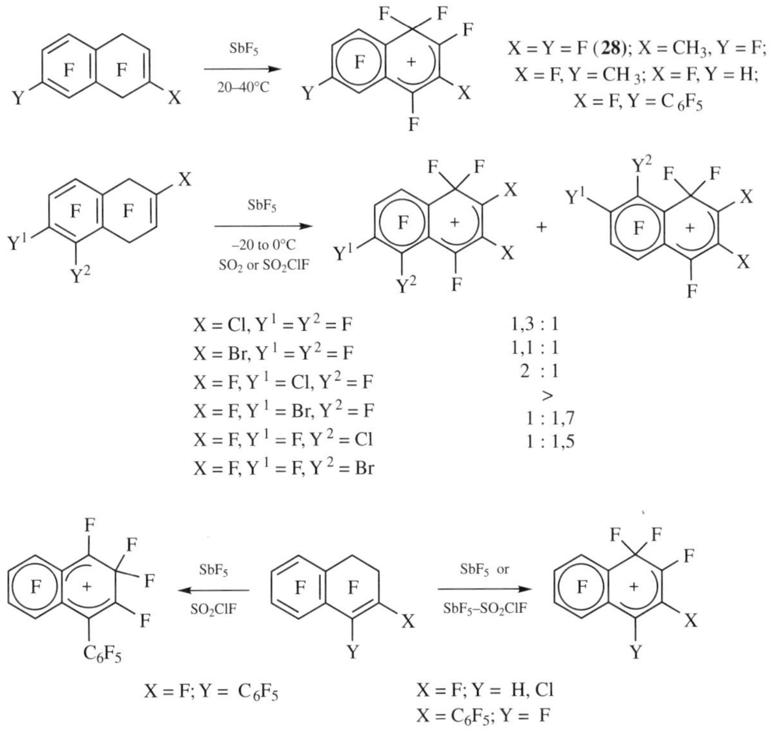

SbF_5, 20–40°C

X = Y = F (**28**); X = CH_3, Y = F;
X = F, Y = CH_3; X = F, Y = H;
X = F, Y = C_6F_5

SbF_5, −20 to 0°C, SO_2 or SO_2ClF

X = Cl, $Y^1 = Y^2$ = F	1,3 : 1
X = Br, $Y^1 = Y^2$ = F	1,1 : 1
X = F, Y^1 = Cl, Y^2 = F	2 : 1
X = F, Y^1 = Br, Y^2 = F	>
X = F, Y^1 = F, Y^2 = Cl	1 : 1,7
X = F, Y^1 = F, Y^2 = Br	1 : 1,5

SbF_5, SO_2ClF; SbF_5 or SbF_5–SO_2ClF

X = F; Y = C_6F_5

X = F; Y = H, Cl
X = C_6F_5; Y = F

Scheme 18

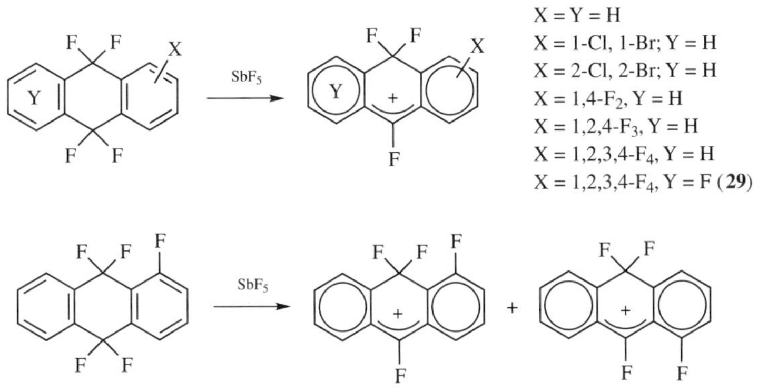

SbF_5

X = Y = H
X = 1-Cl, 1-Br; Y = H
X = 2-Cl, 2-Br; Y = H
X = 1,4-F_2, Y = H
X = 1,2,4-F_3, Y = H
X = 1,2,3,4-F_4, Y = H
X = 1,2,3,4-F_4, Y = F (**29**)

SbF_5

Scheme 19

provided an evidence for the ability of heavier halogens to be held at the position *ipso* to fluorine (Schemes 20 and 21).

Following these studies, a series of long-lived polyfluorinated arenium cations with chlorine added *ipso* to fluorine were prepared by direct chlorination of the variety of polyfluoroarenes with Cl_2 and SbF_5 in SO_2ClF solution at −80 to −60°C[32] (Scheme 22).

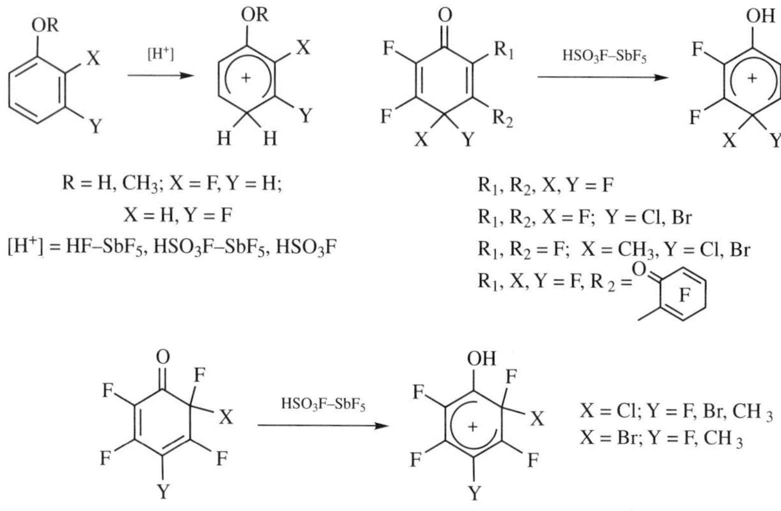

$[H^+]$; $HSO_3F–SbF_5$

R = H, CH_3; X = F, Y = H;
X = H, Y = F
$[H^+]$ = HF–SbF_5, HSO_3F–SbF_5, HSO_3F

R_1, R_2, X, Y = F
R_1, R_2, X = F; Y = Cl, Br
R_1, R_2 = F; X = CH_3, Y = Cl, Br
R_1, X, Y = F, R_2 =

$HSO_3F–SbF_5$

X = Cl; Y = F, Br, CH_3
X = Br; Y = F, CH_3

Scheme 20

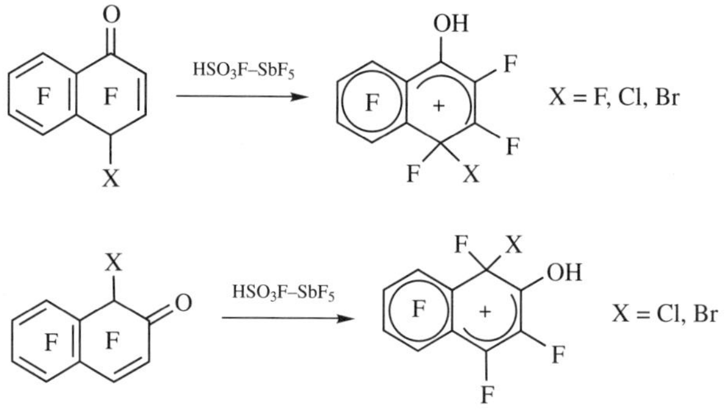

$HSO_3F–SbF_5$

X = F, Cl, Br

$HSO_3F–SbF_5$

X = Cl, Br

Scheme 21

In most cases regioselective chlorination was in line with the data on the relative stability of isomeric 1,1-difluorobenzenenium and 1,1-difluoronaphthalenium ions, which are formed from polyfluorinated cyclohexadienes or dihydronaphthalenes when dissolved in SbF_5 (see Schemes 17 and 18). The only exception is chlorination of chloro- and bromopentafluorobenzene resulting in the predominant formation of 3-X-1-chloropentafluorobenzenenium ions in a mixture of isomers (X = C1, Br). Since there is a reason to consider such an orientation, at least in the case of polyfluorinated 1-chloro-1-fluorobenzenenium ions, to reflect the equilibrium ratio of isomeric ions (by virtue of easy chlorine migration; see below), this exception is in line with the inversion of relative 2- and 3-X-substituted (X = C1, Br) isomer stability on going from polyfluorinated 1,1-difluorobenzenium ions (Scheme 17) to their 1-chloro-1-fluoro analogs (Scheme 22). This change can be due to an

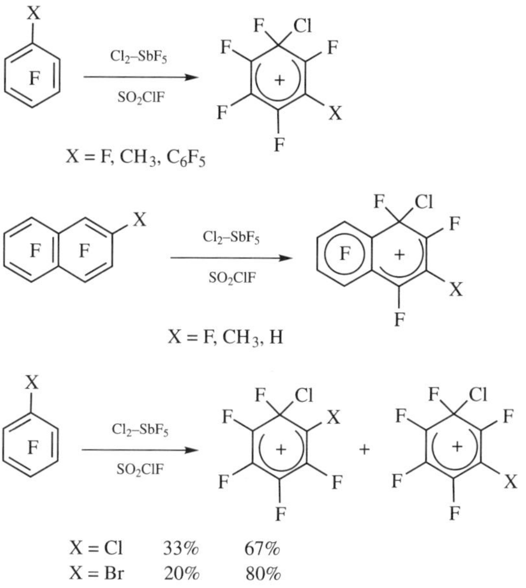
Cl2–SbF5
SO2ClF
X = F, CH3, C6F5
X = F, CH3, H
X = Cl 33% 67%
X = Br 20% 80%

Scheme 22

increase in steric hindrance in the 2-X isomer in case of geminal substitution going from fluorine to chlorine and/or to the alteration of electronic density distribution in the unsaturated part of an ion concomitant with the change of electronic effect of a substituent positioned at the sp^3-hybridized carbon.[4]

Pentafluorobenzene and 1-H-heptafluoronaphthalene are chlorinated under the aforementioned conditions at the unsubstituted position.[32c] However, whereas only the formation of chloropentafluorobenzene was observed in the former case, with a naphthalene precursor the formation of the 1-chloro-2,3,4,5,6,7,8-heptafluoronaphthalenium cation **30** has been recorded, which represents a rare case of a genuine intermediate of hydrogen substitution by an electronegative heteroatom. Other documented events of the same kind deals with the generation of arenium ions stabilized by *ortho* and *para* located hydroxy or dialkylamino groups.[33] The stability of the cation **30** with respect to proton loss is apparently of kinetic origin and caused by steric hindrance for chlorine to enter into the plane of the naphthalene framework. On quenching the solution with methanol 1-chloroheptafluoronaphthalene was formed as a main product[32c] (Scheme 23).

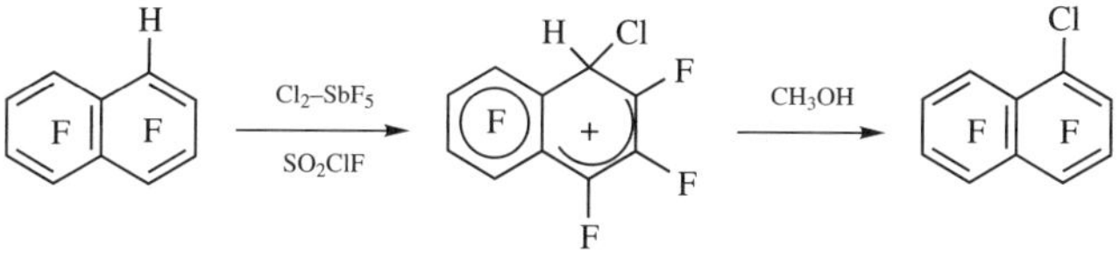
Cl2–SbF5
SO2ClF
CH3OH

30

Scheme 23

The formation of polyfluorinated arenium ions formally corresponding to electrophilic *ipso* fluorination of polyfluoroarene was observed on the action of the XeF_2–SbF_5 complex on hexafluorobenzene[34a,b], decafluorobiphenyl, and polyfluoromethylbenzenes[34b] (Scheme 24).

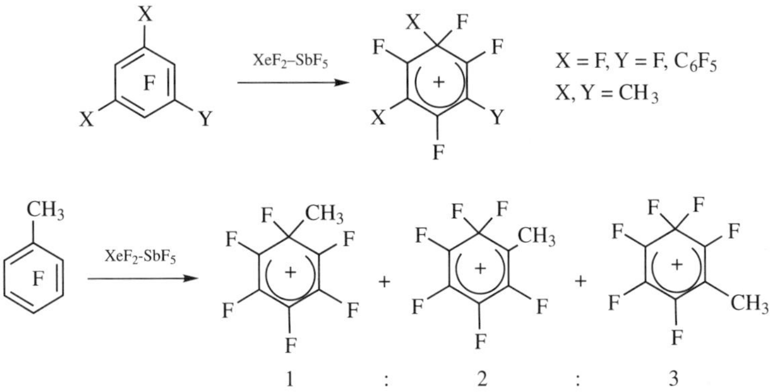

X
F
X Y
XeF_2–SbF_5
X F
F F
+
X Y
F
X = F, Y = F, C_6F_5
X, Y = CH_3

CH_3
F
XeF_2-SbF_5
F CH_3
F F
+
F F
F
+
F F
F CH_3
+
F F
F
+
F F
F F
+
F CH_3
F

1 : 2 : 3

Scheme 24

On the action of methyl fluoride and antimony pentafluoride, polyfluoromethylbenzenes are methylated at the positions *ipso* to the methyl group or fluorine to give polyfluorinated 1-methylarenium ions[35] (Scheme 25).

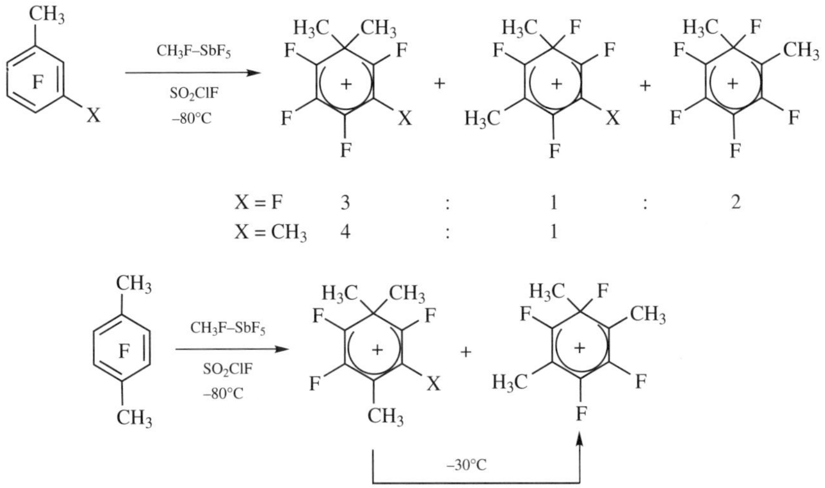

CH_3
F
X
CH_3F–SbF_5
SO_2ClF
–80°C
H_3C CH_3
F F
+
F X
F
+
H_3C F
F F
+
H_3C X
F
+
H_3C F
F CH_3
+
F F
F

X = F 3 : 1 : 2
X = CH_3 4 : 1

CH_3
F
CH_3
CH_3F–SbF_5
SO_2ClF
–80°C
H_3C CH_3
F F
+
F X
CH_3
+
H_3C F
F CH_3
+
H_3C F
F
–30°C

Scheme 25

Olah, Prakash, and coworkers have realized an intramolecular version of *ipso*-to-alkyl group alkylation of polyfluoroarene to obtain tetrafluorobenzonortricyclyl cation **31**[36] (Scheme 26).

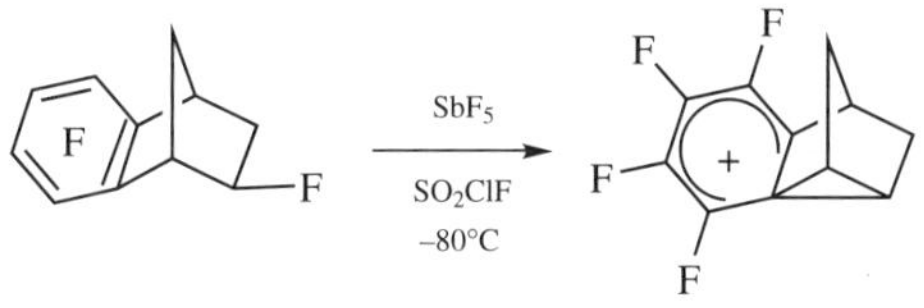

31

Scheme 26

The quoted data on the generation of polyfluorinated arenium ions offer a reasonable basis for the suggested mechanism of such reactions as polyfluoroarene *ipso* nitration,[4b,37] isomerization of polyfluorinated dihydroarenes,[38] and fluorination of polyfluoroarenes with antimony pentafluoride,[39] involving related intermediates.

The further contraction of the conjugated system in polyfluorinated carbocations has been carried out concurrently by groups of (1) Chambers, who reported the formation of 1-methoxy- **32**, 1-phenyl- **33**, and 1-*para*-methoxyphenyltetrafluoro*allyl* **34** cations through the ionization of appropriately substituted pentafluoropropenes (Scheme 27) as well as cyclobutenyl cations **35** with $R^1 = R^2 =$ Me or OMe,

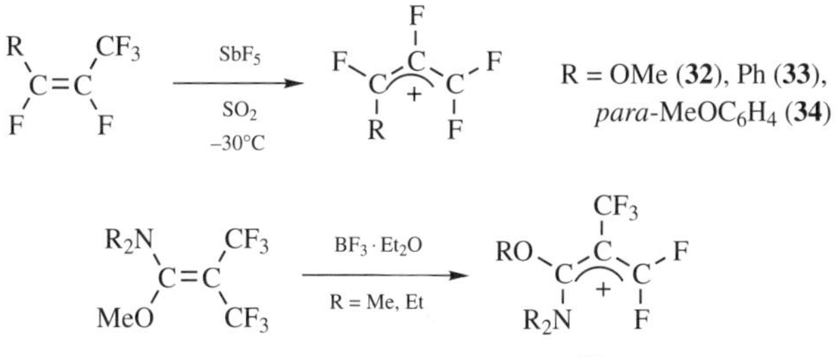

R = OMe (**32**), Ph (**33**), *para*-$MeOC_6H_4$ (**34**)

36

Scheme 27

respectively, through the ionization of polyfluorocyclobutenes (Scheme 28) with SbF_5,[40] and (2) Knunyants, who showed that the interaction of aminoacetals bis (trifluoromethyl)ketenes with BF_3 results in the formation of 1-methoxy-1-dialkylaminopentafluoroallyl cations (**36**)[41] (Scheme 27).

Olah,[15] Smart,[42] Chambers,[43] and their coworkers significantly expanded the diversity of polyfluorinated cyclobutenyl cations (Scheme 28).

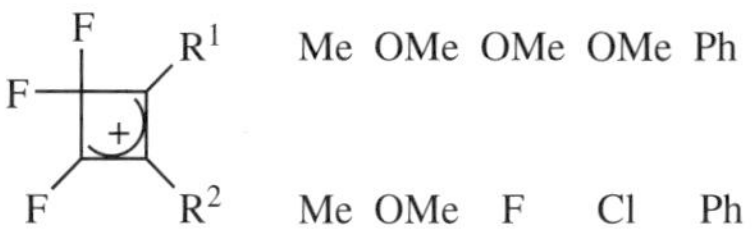

R^1	Me	OMe	OMe	OMe	Ph
R^2	Me	OMe	F	Cl	Ph

35

Scheme 28

In pursuance of these results, Knunyants' students have prepared perfluorinated allyl and pentadienyl cations as well as a wide diversity of substituted polyfluorinated acyclic and cyclic allyl cations through the ionization of relevant polyfluoroalkenes[5] (Scheme 29).

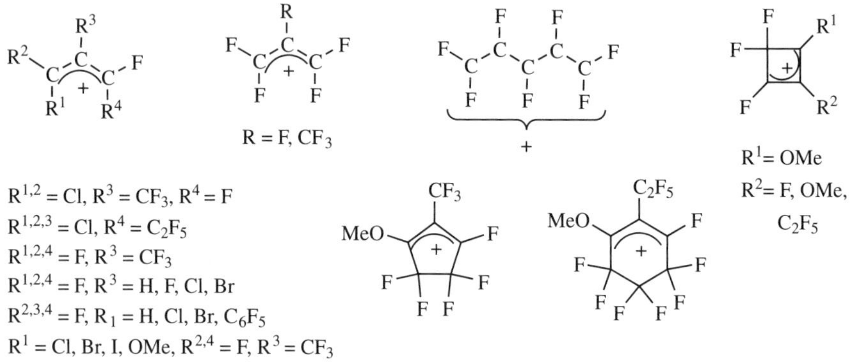

R = F, CF_3

$R^{1,2}$ = Cl, R^3 = CF_3, R^4 = F
$R^{1,2,3}$ = Cl, R^4 = C_2F_5
$R^{1,2,4}$ = F, R^3 = CF_3
$R^{1,2,4}$ = F, R^3 = H, F, Cl, Br
$R^{2,3,4}$ = F, R_1 = H, Cl, Br, C_6F_5
R^1 = Cl, Br, I, OMe, $R^{2,4}$ = F, R^3 = CF_3

R^1= OMe
R^2= F, OMe, C_2F_5

Scheme 29

7.3 STRUCTURAL CHARACTERISTICS OF POLYFLUORINATED CARBOCATIONS BASED ON NMR SPECTROSCOPY DATA

A remarkable feature of polyfluorinated carbocations consists in the opportunity of using ^{19}F NMR spectroscopy to study their structure, including chemical shift values (δ) and fluorine–fluorine spin coupling constants (J). Of special efficacy is a mutually complementing use of ^{19}F, ^{13}C, and ^{1}H NMR spectroscopies. A large body of theoretical and experimental data thus far accumulated for fluorinated carbocations unequivocally show that for fluorines at positions providing an opportunity for cationic charge delocalization by a substituent via p–π conjugation (resonance positions, π-backdonation effect), the fluorine chemical shifts (δ_F) and fluorine–fluorine spin couplings (J_{FF}), as well as fluorine couplings with carbons (J_{CF}) and hydrogens (J_{HF}) also located in resonance positions, are intimately ralated to the degree of charge delocalization, far exceeding the respective values for similarly located fluorines in neutral compounds. To minimize contributions not related to the electron demand coming from a carbocation framework, the changes of chemical shifts ($\Delta\delta_F$) on going to carbocation from the properly chosen neutral precursor are considered. In the same manner with short-distance contributions not related to substituent influence on π-charge distribution, the fluorine atom used as a probe should be located as far away from the site of substituent variation as possible.

All the NMR characteristics associated with fluorines located at resonance positions vary in a wide range from the values that are close to those in neutral compounds to unprecedentedly large ones (20–350 ppm for $\Delta\delta_F$, 20–150 Hz for J_{FF}, and 5–440 Hz for J_{CF}), thus reflecting the variation of fluorine backdonation with the electron demand coming from a carbocation framework depending on the degree of delocalization of carbocation π system, the position occupied by fluorine, and the nature and location of other substituents. This is easily seen even from only matching the structures of cations and their NMR parameters depicted by the Scheme 30. These parameters change according to the regular qualitative notion

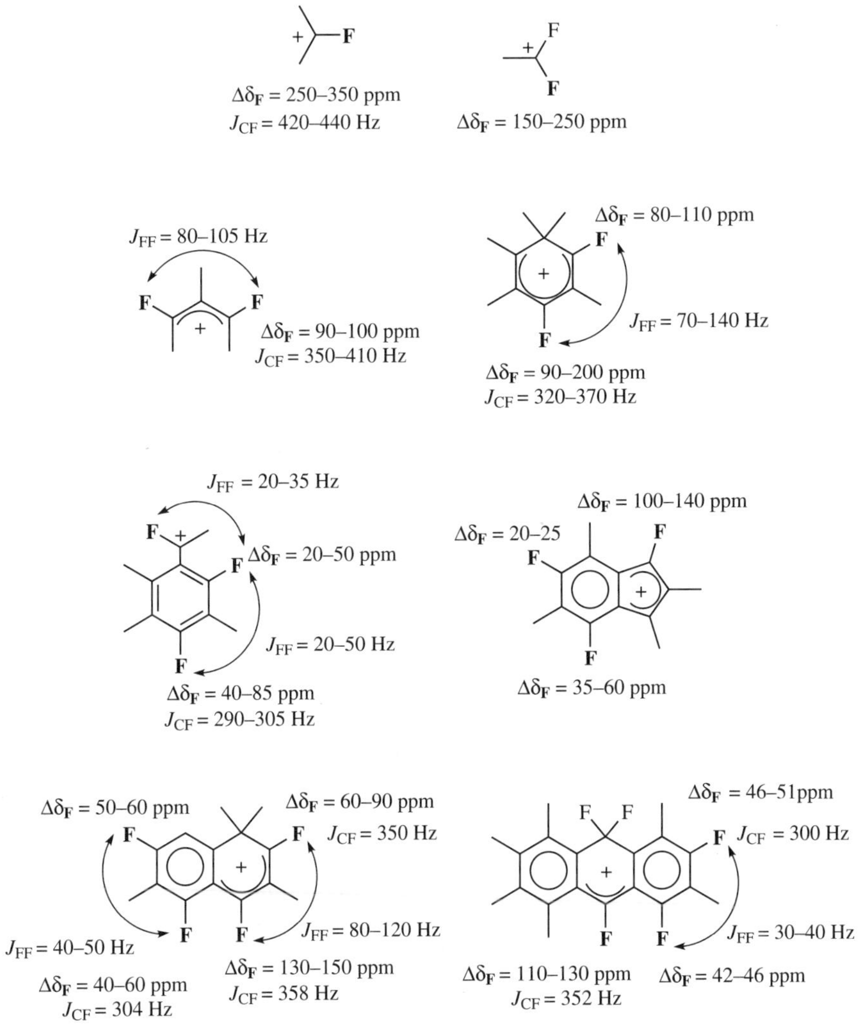

Scheme 30

of the variation of positive charge delocalization influenced by a particular cation framework and the nature of substituents. (The ^{19}F NMR data are taken from the papers referred to above in connection with cation generation; the J_{CF} values for allyl cations are taken from Ref. 5; for arenium cations, from Refs. 44 and 45; and for benzyl cations, from Ref. 45.)

Olah and coworkers gave an elegant demonstration of the dependence of fluorine backdonation on the electron demand by the NMR study of norbornyl cations.[21a] Nonclassicality of the 2-norbornyl cation is decreased by fluorine substitution at the carbocationic center. In turn, the fluorine chemical shift suggests the residual nonclassicality inherent in 2-fluoro-2-norbornyl cation to diminish fluorine backdonation as compared with diverse cyclic α-fluorosubstituted carbocations[7] (Scheme 31).

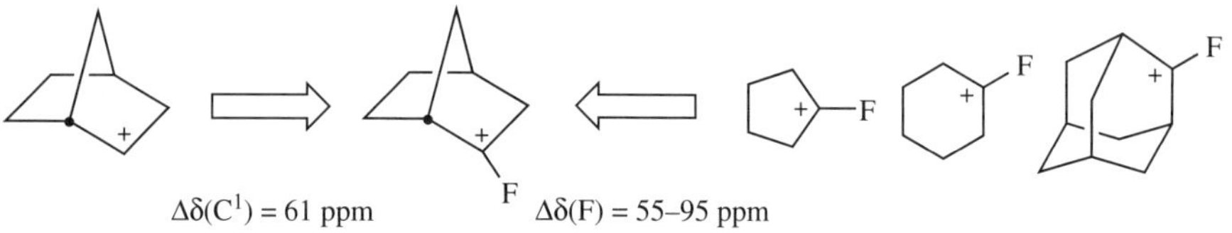

$\Delta\delta(C^1) = 61$ ppm   F $\Delta\delta(F) = 55–95$ ppm

Scheme 31

This relationship is expressed by linear correlations relating fluorine NMR characteristics and substituent σ parameters. The linear correlations hold through the wide series of polyfluorinated aryl and diarylmethyl cations between $\Delta\delta_F$ and J_{FF} for α-, *p*-, and *o*-fluorines, effectively accommodating the positive charge via π backdonation[25] (see Table 7.1), which is reflected by the sequence **A** > **B** > **C** of the respective resonance structure contributions (Scheme 32).[9,19,24,25]

Further, there is a common linear relationship (see the last equation in Table 7.1) between *p*-fluorine chemical shifts and *p*-F-*o*-F spin coupling for a wide range of C_6F_5R-type compounds, including cations $C_6F_5CFX^+$ ($X = H, F$), $(C_6F_5)(4\text{-}X\text{-}C_6F_4)CF^+$ ($X = H, F, Cl, Br, CH_3, CF_3$), $(C_6F_5)_2(CF_3)C^+$, $(C_6F_5)_3C^+$, and 61 neutral compounds.[25] The similar ratio was independently established for neutral C_6F_5R compounds whereof δ_F and $J_{F_oF_p}$ values, in turn, show linear correlations with the substituent R σ_R^o and σ_I parameters. This reveals these NMR ^{19}F characteristics to be directly related to substituent electronic effects.[46] Moreover, the variation of $J_{F_oF_p}$ values was shown to be dominated by the orbital contribution.[47] The qualitative interpretation of these relationships implies that the degree of contribution of the resonance structures **A**, **B**, and **C** in case of arylmethyl cations, and **D** and **E** in case of neutral compounds, determines both shielding of the respective fluorines and the amount of their direct resonance interaction. The latter makes, apparently, a principal contribution to the J_{FF} values. Despite these empirical formulations are far away from a rigorous theoretical description of the mechanism of

Table 7.1 Linear correlations between fluorine NMR parameters of polyfluorinated arylmethyl cations

Equation	r^a	s^a	Number of Points
$J_{F^2F^4} = -0.52\phi_{F^4}{}^b + 80.1$	0.985	1.6	11
$J_{F^2F^4} = -0.34\phi_{F^{2,6}} - 0.33\phi_{F^4} + 98.8$	0.992	1.1	11
$J_{F^\alpha F^4} = -0.60\phi_{F^4} + 70.6$	0.978	2.1	8
$J_{F^\alpha F^4} = -0.52\phi_{F^4} + 62.3$	0.993	0.6	6^c
$J_{F^\alpha F^4} = -0.94\phi_{F^\alpha} - 23.3$	0.979	1.1	6^c
$J_{F^\alpha F^4} = -0.37\phi_{F^4} - 0.30\phi_{F^\alpha} + 37.0$	0.995	0.3	6^c
$J_{F^2F^4} = -0.48\phi_{F^4} + 76.3$	0.997	0.9	72^d

[a] r = correlation coefficient; s = mean-square deviation.
[b] $\delta_F = 163.0 - \phi_F$.
[c] $J_{F^\alpha F^4}$ and ϕ_F values are included only for polyfluorinated diphenylmethyl cations.
[d] $J_{F^2F^4}$ and ϕ_F values for 61 neutral C_6F_5X type compounds[46] are included.
Source: Ref. 25.

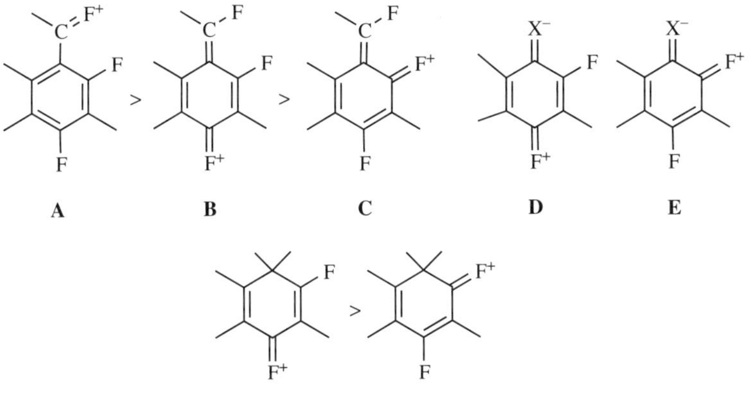

Scheme 32

spin information transfer, it seems to be useful for experimental chemists to systematically study these relationships. In case of arenium cations, the forgoing concerns the resonance structures such as **F** and **G**.

Mamatyuk et al. have found in their study of ^{13}C NMR spectra of the series of fluorinated arenium ions, including 4-fluorobenzenium **37**, 4-fluoro-1,2,3,5,6-pentamethylbenzenium **38**, 2,4,6-trifluorobenzenium **39** (Scheme 33), and perfluorinated benzenium **27**, 1-naphthalenium **28** and 9-anthracenium **29** ions, a correlation between the charge distribution and the values of δ_C and J_{CF} as well as their differences from those of parent arenes ($\Delta\delta_C$ and ΔJ_{CF}, respectively) corresponding to the pattern inferred from ^{19}F NMR spectra. For the resonance positions these values substantially exceed the respective ones for the nonresonance positions. Quantitatively, this is reflected by the linear correlation $\Delta\delta_C = 0.32\Delta\delta_F$ (± 3.5 ppm, $r = 0.978$). This may indicate an approximate consistency of a positive charge partition inside a ring C—F fragment at a rather wide variation of a degree of cationic charge delocalization.[44]

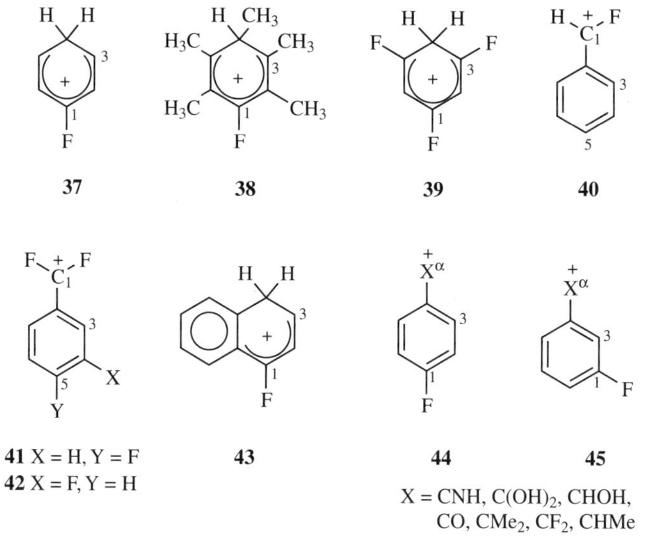

41 X = H, Y = F
42 X = F, Y = H

44 **45**
X = CNH, $C(OH)_2$, CHOH, CO, CMe_2, CF_2, CHMe

Scheme 33

A close relationship for the mode of π-electronic charge distribution, similar to that outlined above for fluorine–fluorine spin coupling, has been shown for the J_{CF} values characterizing the interaction between fluorines and ring carbons involved in the resonance charge delocalization in fluorinated benzyl and arenium cations.[45] The J_{C_1F}, J_{C_3F}, and $J_{C\alpha F}$ values in the benzyl cations **5,40–42** and **44** (Scheme 33) are considerably larger than in the compounds 4-XC_6H_4F (X = NH_2, H, NO_2, CHO, COMe, COOH, COOMe) and exhibit dependence on the degree of charge delocalization onto respective fluorine atoms. Such a relationship is also distinctly revealed by comparison of the couplings with those for *meta*-fluorinated cations **45** containing the same substituents at the α carbon. For the latters the J_{C1F} range is 4 times and the J_{C3F}, J_{C5F} and $J_{C\alpha F}$ ranges are 7–10 times smaller than for *para*-fluorinated cations and practically of the same magnitudes as in the compounds 3- and 4-XC_6H_4F. A further illustration of the correlation is provided by the variation of J_{C1F} values in the series encompassing arenium cations **37–39,43** and benzyl cations **5,40–42**, which is in agreement with the relevant structural

Table 7.2 Linear correlations between fluorine and carbon NMR parameters of fluorinated arylmethyl and arenium cations

Equation	r^a	SD
$J_{C1F} = -0.848\phi_F + 342.9^b$	−0.997	1.9
$J_{C1F} = 1.68\delta_C + 31.2^b$	0.996	2.1
$J_{C1F} = -0.892\phi_F + 347.6^c$	−0.985	0.8
$J_{C3F} = -0.151\phi_F + 25.7$	−0.994	4.3

[a] r = correlation coefficient; SD = standard deviation.

[b] For *para*-fluorinated benzyl cations **41**,**44** and compounds 4-XC_6H_4F ($X = NH_2$, H, NO_2).

[c] For *para*-fluorinated benzyl cations **41**,**44**, *meta*-fluorinated benzyl cations **45**, compounds 3- and 4-XC_6H_4F ($X = NH_2$, H, NO_2), arenium cations **37–39**, **43** and benzylic fluorines of cations **5**,**40–42**.

Source: Ref. 45.

variations. This is shown by the linear relationships between the J_{CF} values and chemical shifts of coupling carbon and fluorine atoms (Table 7.2).

Overall, the data presented above provide evidence for the direct relation between fluorine NMR characteristics and the degree of fluorine π-resonance interaction with carbocationic framework, as dependent on the electron demand on fluorine.

The effect of multiple fluorination on the charge distribution and the ratios of substituent electronic effects in carbocations have been investigated. At the carbocationic centers of the ethyl cation[8,9a,48] and pentafluorophenylmethyl cation,[23,24] there is a decrease in the backdonation of other fluorine atom located in a resonance position (Scheme 34). More recently, this NMR finding was confirmed by ab initio calculations of the trifluoromethyl cation.[49]

The complete substitution of hydrogen by fluorine allows one to reveal the effect of fluorine substitution into nonresonance positions. As exemplified by the complete fluorination of the phenyl ring attached to a carbocationic center, fluorine substitution at nonresonance positions is unfavorable for the charge dispersion onto a benzene ring and overwhelms the effect of fluorines at resonance positions, thus forcing positive charge to move to the remote part of a molecule. For example, this enhances the electron demand for fluorine attached to the benzylic carbon as it can be inferred from the comparison of the $\Delta\delta_{F\alpha}$ values for the α,α-difluorobenzyl cation **5** and α-fluorodiphenylmethyl cation **3**[6a] on one hand, and the perfluorinated benzyl **18** and diphenylmethyl **19** cations[24,25] on the other hand. Such modification of one benzene ring of 9,9,10-trifluoro-9-anthracenium cation; that is, the shift from **46** to **47** enforces transfer of the part of positive charge onto the carbocationic site located in the central ring and probably into the other benzene

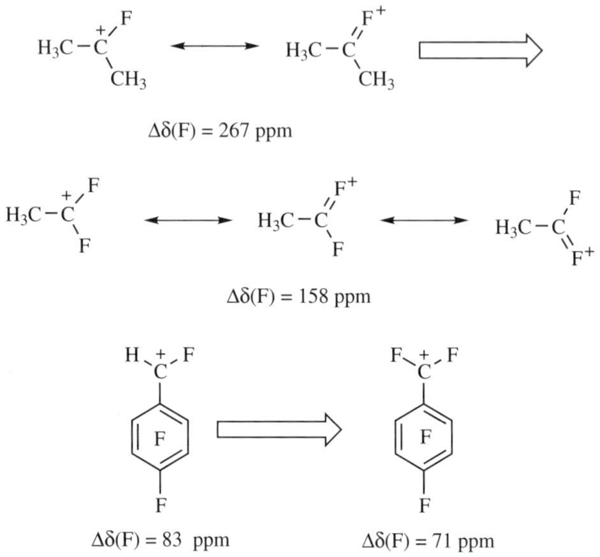

$\Delta\delta(F) = 267$ ppm

$\Delta\delta(F) = 158$ ppm

$\Delta\delta(F) = 83$ ppm $\Delta\delta(F) = 71$ ppm

Scheme 34

ring, whereas on complete fluorination of this cation (the change from **47** to **29**) there is no alternative other than to increase the backdonation of fluorine atoms at all benzene ring resonance positions[30e,f] (Scheme 35).

A similar finding is that in 2-phenyl-2-norbornyl cation, the complete phenyl ring fluorination increases the participation of the C^1–C^6 bond in the charge delocalization or, in another words, the extent of nonclassicality in much same degree with one or two *meta*-located CF_3 groups or *para*-located $NHMe_2^+$ group.[21] Overall, the sum of the effect of complete fluorine substitution in resonance and nonresonance positions is unfavorable for the charge dispersion into benzene ring.

Identical to this is the effect of multiple fluorine substitution in benzenium cations, which, moreover, increases the difference between the $\Delta\delta_{Fo}$ and $\Delta\delta_{Fp}$ values, suggesting the progressive positive charge shift from the *ortho* to the *para* positions with fluorine accumulation in the unsaturated moiety of benzenium cation. This corresponds to a gradual increase in the contribution of structure **F** to the detriment of structure **G** (Scheme 32). A similar change occurs with increase of the electron-withdrawing character of a substituent attached to the *ipso* carbon (Scheme 36).[2,4,28,30–32,34,50–52]

In the series of $4\text{-}XC_6H_4CF_2^+$ cations the change of fluorine shielding at the benzylic carbon is consistent with the backdonation of *para*-substituent increasing in the order H < Br < Cl < F,[9] which is confirmed by the ratio $\Delta\delta_F = -75.6 - 9.3\sigma_R^+$ [24b] and by the linear coorelation of $\delta_{F\alpha}$ values versus σ^+ substituent parameters including, besides halogen substituents, $X = CH_3$, CF_3 and $3,5\text{-}(CF_3)_2$.[53] However, after complete fluorination, which, as expected, is accompanied by the increase of electron demand from the framework, the *para*-halogens show shielding

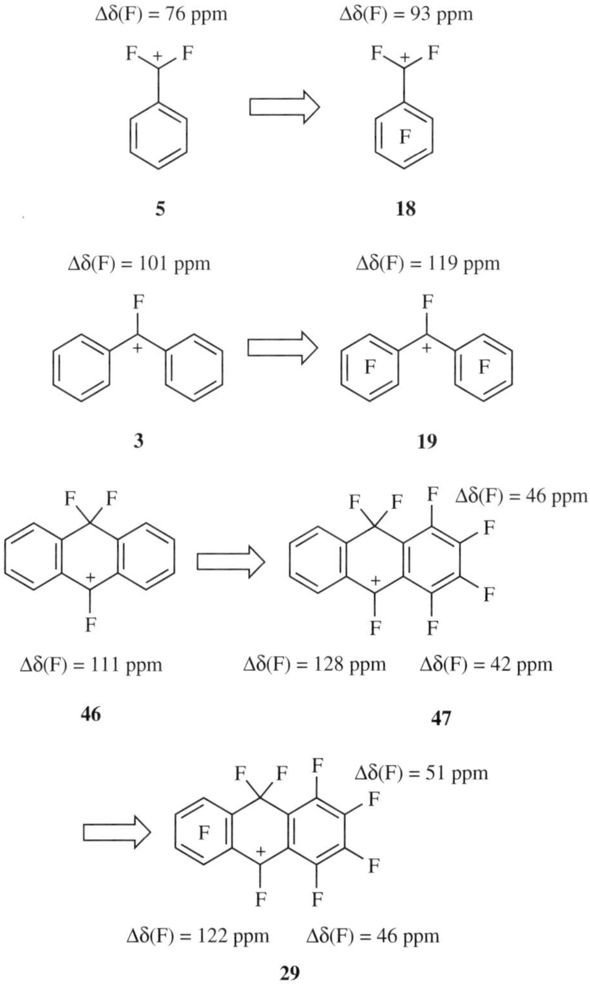

Δδ(F) = 76 ppm
Δδ(F) = 93 ppm

5 **18**

Δδ(F) = 101 ppm
Δδ(F) = 119 ppm

3 **19**

Δδ(F) = 46 ppm
Δδ(F) = 111 ppm
Δδ(F) = 128 ppm
Δδ(F) = 42 ppm

46 **47**

Δδ(F) = 51 ppm
Δδ(F) = 122 ppm
Δδ(F) = 46 ppm

29

Scheme 35

effects corresponding to the inverse sequence of effects of backdonation: H < F < Cl < Br < CH_3[24b] (Scheme 37).

The same sequence was found for halogens located at resonance positions of polyfluorinated arenium ions (Scheme 38; fluorine used as a NMR probe is in bold type). Because they are located in nonresonance positions, substituents show the correlation of fluorine shielding effects, which is in line with the commonly accepted order of their inductive effects (Scheme 39).[30,31,51,54]

The regularities found suggest an opposite order of the halogen substituent backdonation effects in carbocations depending on the framework electron demand, and thus multiple fluorine substitution can be one of the factors causing the advantage of "heavy" halogens over fluorine. Another reason for the change of the electron

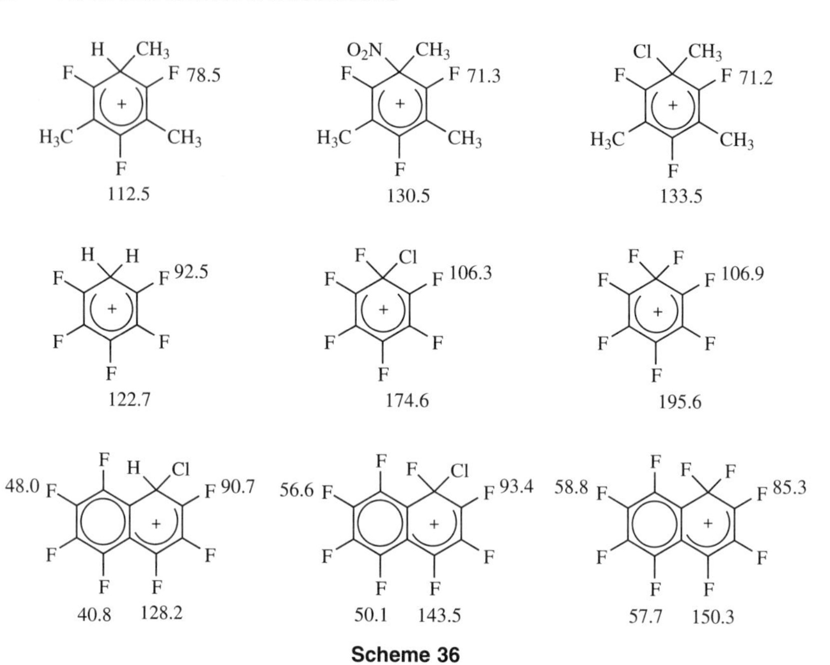

Scheme 36

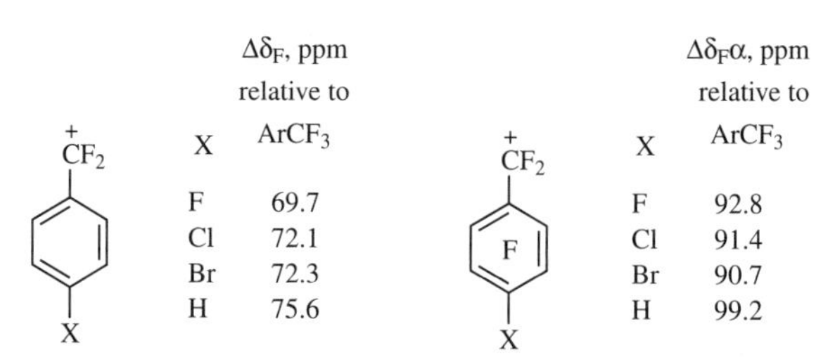

Scheme 37

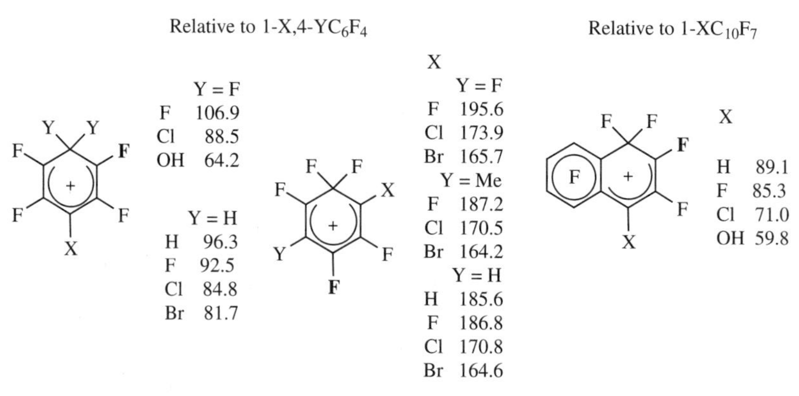

Scheme 38

Δδ(F), ppm
relative to 1-X,4-YC_6F_4

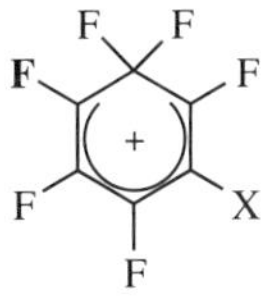

X	
F	106.9
Cl	93.7
Br	89.6
H	83.1
Me	80.8

Scheme 39

demand is the variation of the location of substituents. For example, Smart and Reddy have shown that contribution of the bicyclobutanoid resonance structure **H** and, respectively, the X substituent backdonation, depicted by the resonance structure **I** (Scheme 40), to increase in the order Cl < F < OCH_3.[41] In contrast, in resonance positions of acyclic allyl cations halogens exhibit the inverse backdonation order.[55]

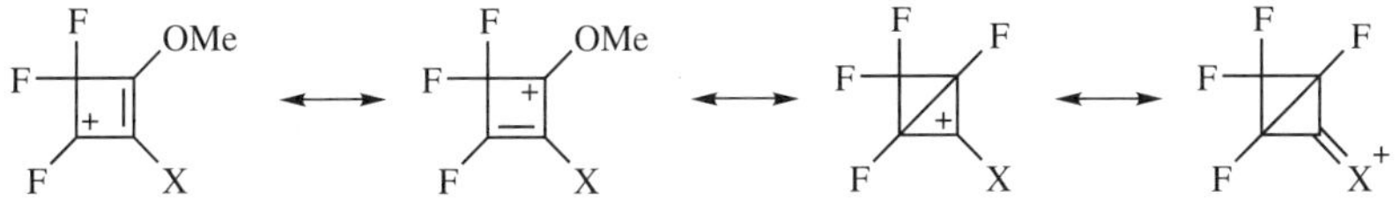

Scheme 40

As discussed, the J_{FF} values for coupling of fluorines located at resonance positions correlate directly to the degree of charge delocalization. Especially large J_{FF} values are observed when they combine contributions of interaction through conjugation and directly through space in case of spatial proximity of interacting fluorines. Thus, spin coupling of two α-fluorines in benzyl cations with different *ortho* substituents shows the value $J_{FF} = 245$–255 Hz.[9b,24] The similar value in a series of allyl cations changes in the range of 160–245 Hz, whereas for polyfluoroalkenes it is in the range of 10–60 Hz, which corresponds to an increase of 150–190 Hz on going from the neutral precursors to cations.[5,55] The contribution from the through-space interaction manifests itself also in sharp difference of J_{FF} values for spatially proximal and remote α- and *o*-fluorines in the benzyl cations **48**,[9b] **49**,[23,24] and **50**;[24] thus, this contribution can be assessed as 80–120 Hz. In this

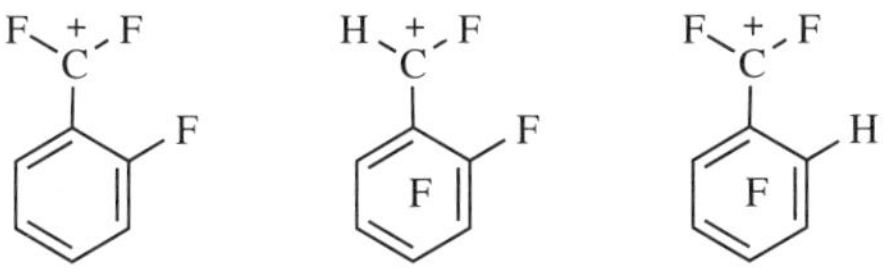

48 **49** **50**

respect, the values $J_{FF} = 130$–150 Hz are indicative for the interaction of *peri*-located fluorines in polyfluorinated 1-naphthalenium and 9-anthracenium ions,[30a–f] which exceed by about 70 Hz the analogous values in polyfluorinated naphthalenes[56] and anthracenes.[57] The increase in this value exhibited by cations over those in neutral compounds dominated by the direct through-space interaction is suggested to be due mostly to the strong resonance interaction inherent in a cationic π-electronic system. Obviously, the same holds for allyl cations X-FCCFCF_2^+; for the interaction of spacially proximal fluorines located at positions 1 and 3 of the perfluoroallyl cation, the J_{FF} value of 109 Hz was reported, and overall for the ions of this type it ranges from 100–144 Hz.[5,55]

The combined consideration of a plethora of NMR characteristics also affords a favourable opportunity for insight into stereochemichal features of polyfluorinated cations. The nonequivalence of the *ortho* proton chemical shifts indicates the rotation around the C_{arom}–C_{α} bond in the α-fluoro-α-methylbenzyl cation **2** to be restricted at $-70°C$,[9a] which is generally inherent in benzyl cations.[58] Indicative in this connection is a substituent influence on the conformational characteristics of polyfluorinated arylmethyl cations, based on consideration of F–F coupling constants of the α- and *ortho*-fluorines ($J_{F\alpha Fo}$). Consequently, in a series of 4-X-substituted hexafluorobenzyl cations,[24b] the $J_{F\alpha Fo}$ constants decrease with electron-releasing character of the X substituent. This corresponds to the reduced contributions of the structures **A** and **C** (Scheme 32), thus reflecting the decrease in α- and *ortho*-fluorine backdonation. However, the fact that this change is rather weak suggests the opposite influence of increase in the degree of cation coplanarity and, accordingly, of the approach of coupling fluorines to each other, which likely compensates for most of the effect of reduced backdonation.

Unlike polyfluorinated benzyl cations, which do not undergo free rotation around the C_{arom}–C_{α} (at least up to $+40°C$),[24b] the majority of polyfluorinated dirylmethyl cations, owing to the greater charge delocalization and, accordingly, the smaller order of the C_{arom}–C_{α} bond, undergo free rotation about this bond even at $-20°C$.[25] However, in the case of cations (4-X-C_6F_4)(4′-$CF_3C_6F_4$)CF^+ the electron-withdrawing substituent effect sharply reduces the charge delocalization onto the CF_3-substituted ring with concomitant enhancement of such delocalization onto the other ring and renders these cations to a greater degree benzylic rather than diarylmethyl in nature. As a result, the temperature of averaging the NMR signals of *ortho*-fluorines located in the ring bearing substituent X increases to $+40°C$ for X = F and to $+80°C$ for X = CH_3, indicating the relative height of energy barriers for ring rotation. As a whole, because of the substituent influence in the cations (4-X-C_6F_4)(4′-YC_6F_4)CF^+ on the characteristics of their ^{19}F NMR spectra, the delocalization of the positive charge by *para* substituents decreases in the order $CH_3 > Br > Cl > F > H > CF_3$, which is opposite halogen sequence, similar to the case of polyfluorinated benzyl cations as compared to the one observed for 4-X-$C_6H_4CF_2^+$ cations. At the same time, the influence of the nature of the substituent on values δ_F and J_{FF} in the series of diarylmethyl cations (4-X-C_6F_4)(4′-YC_6F_4)CF^+ indicates that rotationally averaged conformations along the bonds of two rings with the α carbon, are mutually dependent and sensitive to a combination of the *para* substituents. The increase in the degree of electron-donating character of a substituent X in the abovementioned sequence in the cations with X = Y brings the rotationally averaged conformation more close to coplanarity with the plane of the $C1C1'CF^+$ skeleton for both benzene rings, whereas in cations with $X \neq Y$ only for the X-substituted ring, at the expense of the Y-substituted one. Judging from ^{19}F NMR characteristics, the influence of two moieties of the dication **22** on each other is essentially weakened, apparently, owing to their mutual noncoplanarity inherent to polyfluorinated diphenyls (see, e.g., Ref. 59).

Similar regularities have been revealed for polyfluorinated allyl cations.[5a] Under the usual conditions of recording NMR spectra there is no free rotation about C—C bonds for the X-$CFCFCF_2^+$ cations with X = H and Hlg, so the cations with $X \neq F$ on generating through the ionization of respective fluoroalkenes (Scheme 27) reproduce the geometry of their precursors. However, similar to the case found for polyfluorinated benzyl cations (discussed above), the pentafluorophenyl group substitution for fluorine at the carbocationic centre sharply reduces the rotation barrier. Consequently, with $X = C_6F_5$, the cation was obtained as the *trans*-isomer, irrespective of which a geometric isomer of the starting alkene was utilized as a precursor.

On the basis of the detailed analysis of the Raman spectra of the crystal salts $C_3F_3^+Sb_2F_{11}^-$ and $C_3F_3^+B_2F_7^-$,[60] it was concluded that the perfluorocyclopropenyl cation **6** is a regular triangle. The calculational results of normal vibration frequencies of the fluorinated cyclopropenyl cations $C_3F_nH_{3-n}^+$ show the C—C bond force constant to increase with the fluorine substitution for hydrogen at carbon forming this bond and to decrease with the similar substitution at carbon, which is not included into this bond. Owing to the mutual compensation of these effects the C—C bond force constants in cations $C_3F_3^+$ and $C_3H_3^+$ are essentially the same.

7.4 INFLUENCE OF THE STRUCTURE OF POLYFLUORINATED CARBOCATIONS ON THEIR RELATIVE STABILITY

The variation in a carbocation's electronic structure as revealed by NMR spectroscopy should be matched with its relative stability in order to reveal the physical nature of carbocation stabilization. As experimental evidence on relative stability of fluorinated carbocations has grown, preference is evident for fluorine to be located at the resonance positions of the carbocation compared with the nonresonance ones. The most pronounced illustration of this trend was in the formation of 2-fluoro-2-butyl cation on attempted generation of fluoro-*tert*-butylcation[61] (Scheme 41).

$$H_3C-\underset{OH}{\overset{CH_3}{C}}-CH_2F \xrightarrow[SO_2,\ -80^\circ C]{FSO_3H-SbF_5} H_3C-H_2C-\overset{+}{C}(F)CH_3$$

Scheme 41

As for the effect of α-fluorination, from gas-phase and quantum-chemical studies one could infer that fluorine stabilizes methyl cation but the effect of multiple fluorination is not regular. For example, there is a disparity of correlations characterizing the variation of the stability of fluorinated methyl cations with fluorine substitution as deduced from the experimental gas-phase carbocation affinity for the hydride ion ($CHF_2^+ > CH_2F^+ > CF_3^+ > CH_3^+$),[62] on one side, and both experimental[62] and ab initio calculated[49,63] gas-phase affinity for the fluoride ion ($CH_2F^+ > CHF_2^+ > CF_3^+ > CH_3^+$), on the other side, which probably is due to the position of equilibrium being substantially determined by the fluorination influencing the energy of a precursor rather than a carbocation. By contrast, as one may conclude from the amounts of Lewis acid necessary for the complete transformation of precursors into cations in a solution, in the case of cyclopropenyl cations, progressive fluorine substitution was found to regularly destabilize the carbocation[60] (Scheme 42).

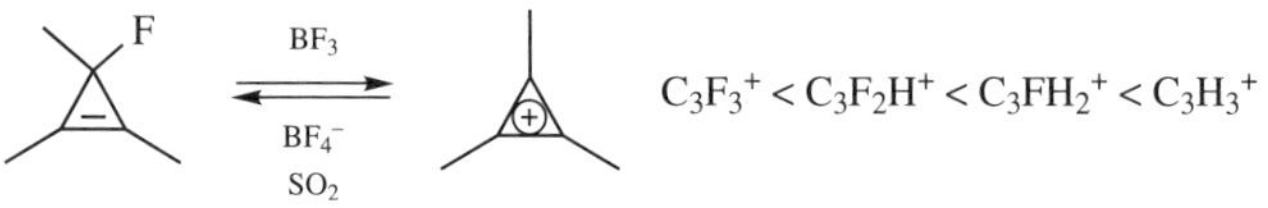

BF_3 ⇌ BF_4^-, SO_2

$C_3F_3^+ < C_3F_2H^+ < C_3FH_2^+ < C_3H_3^+$

Scheme 42

The destabilizing effect of multiple fluorine substitution exhibited by this sequence, however, cannot evidently be generalized to the aromaticity of cations with cyclic π-system. As the authors[26] believe, total fluorination results in acquiring aromaticity by indenyl cations. This conclusion was based on the finding that perfluorinated cations of this type are more stable than their isomers with no cyclic π system in a five-membered ring (Scheme 43).

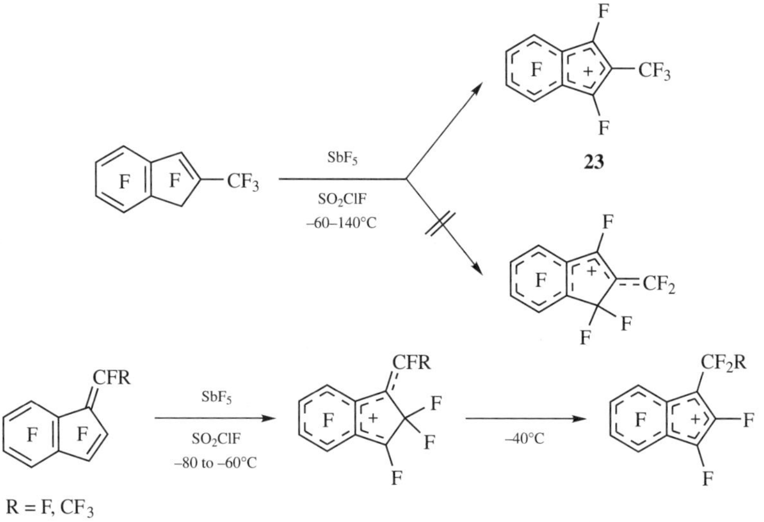

SbF_5, SO_2ClF, –60–140°C

23

SbF_5, SO_2ClF, –80 to –60°C; –40°C

R = F, CF_3

Scheme 43

Table 7.3 pK_{R+} values of triphenylcarbinols

X	Y	Z	pK_{R^+}	Ref.
H	H	H	−6.6	64
F	H	H	−10.5	65
F	F	H	−13.7	65
F	F	F	−17.5	66

In general, the effect of fluorine substitution at a single or different carbocationic centers must be presumed to be of nonuniform nature.

In contrast, the fluorine accumulation simultaneously at the resonance and non-resonance positions basically results in general destabilization of carbocations. In the series of triphenylmethyl cations this is illustrated by the variation of the pK_R^+ values of three to four negative units with each sequential substitution of the phenyl group by the pentafluorophenyl one (Scheme 44 and Table 7.3).

$\pm H^+, \mp H_2O$

Scheme 44

In the series of fluorinated benzenium cations the effects of both multiple fluorination and mutual fluorine locations are elucidated by the fact that in $HSO_3F–SbF_5$ solution practically complete protonation takes place for fluorobenzene, 1,3- and 1,4-difluorobenzenes, 1,2,4- and 1,3,5-trifluorobenzenes, 1,2,3,5-tetrafluorobenzene, whereas for protonation of 1,2-di-, 1,2,3,4-tetra- and pentafluorobenzene the even stronger acid, $HF–SbF_5–SO_2ClF$ is required. Hexafluorobenzene is not protonated even in this medium.[28a,50] These observations are in line with the results of quantum-chemical calculations.[67] Detailed consideration of these data, combined with isomer distributions in the equilibrium mixtures of isomeric arenium ions derived from ionization of polyfluorinated dihydroarenes in SbF_5-containing media,[30] strictly exhibit the location of fluorine in resonance, first the *para* position, to be preferable, whereas *meta* and, probably, mutually neighboring locations of fluorine substituents[28b] as well as fluorine accumulation on the whole are strongly destabilizing.

Using the competitive method of generation, it was demonstrated for the 9,9,10-trifluoro-9-anthracenium cations that fluorine substitution into the position 2 is somewhat stabilizing whereas destabilizing occurred simultaneously at positions 1 and 4 that are *para*, *meta*, and *ortho* positions to the principal carbocationic center, respectively. Further destabilization occurs on sequential total fluorination of one and both benzene rings.[4,68] All these effects are compliant with the influence of progressive fluorine substitution on positive charge distribution in the 9,9,10-trifluoro-9-anthracenium cations as revealed by the variation of fluorine shieldings.[30e]

In concert with these data, in the automerization of 9,10-dimethyl-9-arylphenantrenium cations via aryl 1,2 shift, so as the benzenium like transition state **51** is believed, the pentafluorophenyl group migrates 10^3 times slower as compared to the phenyl one.[69] Analogous conclusions can be drawn from the kinetic studies on the isomerization of polyfluorinated benzenenium ions occurring by hydrogen migration[50] (Scheme 51).

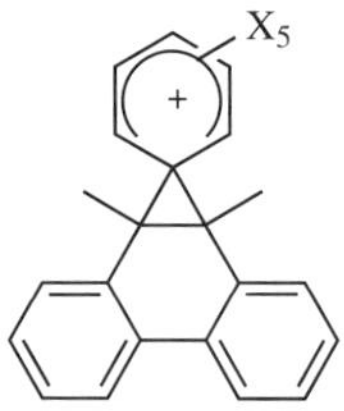

X = H, F

51

A similar study reveals the difference in the stability of isomeric benzenenium ions differing in the methyl substituent location to be much less significant.[50] As a consequence, the combination of fluorine and methyl substituent effects favors the isomer to form, in which the fluorine occupies the resonance position and methyl group—the nonresonance one,[28b,c] even though the methyl group located at the resonance position is a stronger stabilizing substituent than fluorine.[30g,70] These trends were the basis for predicting the generation of 2-halogen-5-R-pentafluorobenzenenium cations (R = CH_3, H) from polyfluorinated cyclohexadienes in SbF_5-containing media[30g] (Scheme 45).

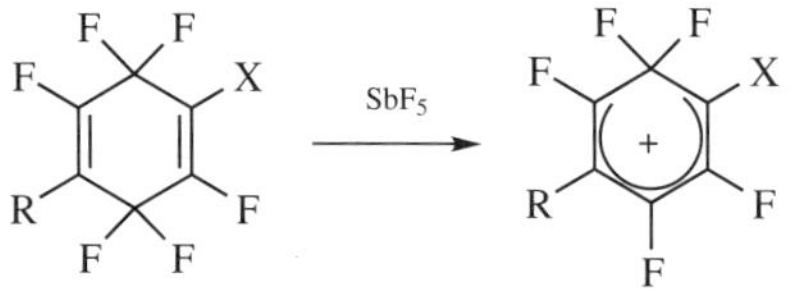

SbF_5

R = Me, X = Cl, Br, Me
R = H, X = Cl, Br, H, Me

Scheme 45

The isomer ratios reveal the relative stabilizing effects of certain substituent combinations; however, in order to determine the relative effects of individual substituents equilibrium measurements should be done. With this objective, for polyfluorinated benzyl and benzenium cations, the equilibria were studied based on the competitive interaction of two precursors with SbF_5 used in an amount not sufficient for their complete conversion into respective cations (competitive generation).[30g,70] Chlorine and bromine are better stabilizers at the *ortho* position and weaker destabilizers at the *meta* position of polyfluorinated benzenenium ions

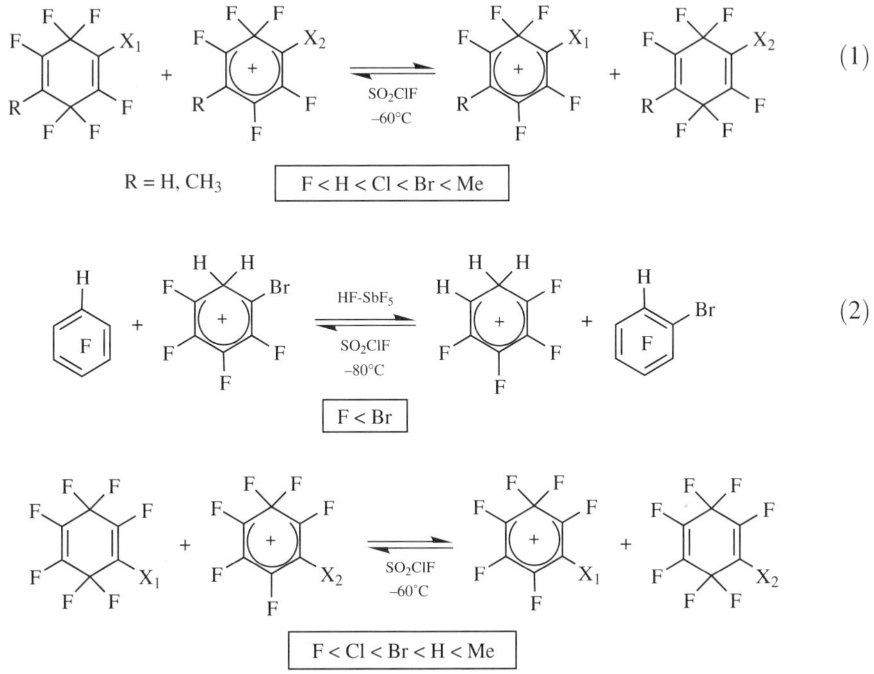

(1) SO_2ClF, −60°C

R = H, CH_3 F < H < Cl < Br < Me

(2) HF-SbF_5, SO_2ClF, −80°C

F < Br

SO_2ClF, −60°C

F < Cl < Br < H < Me

Scheme 46

(Scheme 46) and in both cases were in compliance with the correlations of electronic effects of halogen substituent as revealed by the NMR spectroscopy.

Analogous regularities were observed for polyfluorinated allyl and metallyl cations.[5,55] The prevalence of "heavy" halogen-stabilizing effects over the fluorine effect was also established for polyfluorinated cyclopropenyl cations[13] (Scheme 47).

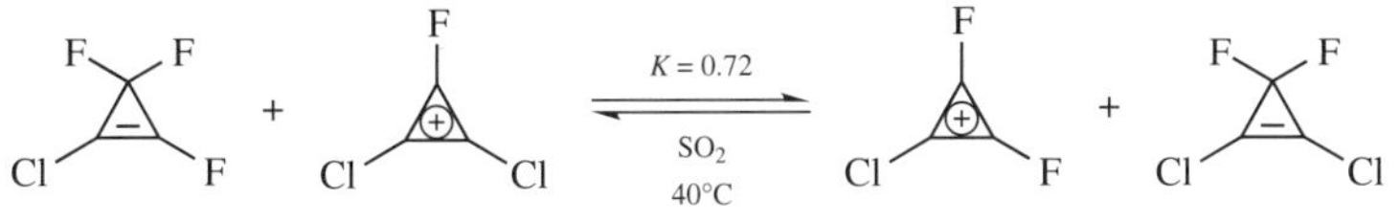

$K = 0.72$, SO_2, 40°C

Scheme 47

The stabilizing effects of halogen substituents at the *para* position of α,α-difluorobenzyl cations also turned out to qualitatively correspond to the correlation of their effects on the α-fluorine shielding in this series[70] (Scheme 37).

Unlike this, no relationship was found between the orders of electronic effects, as revealed by NMR spectroscopy, and stabilizing effects for halogen substituents at the *para* position of polyfluorinated benzyl and benzenium cations. Whereas moving from α,α-difluorobenzyl cations to their hexafluorosubstituted analogs, the aforementioned inversion of the order of *para*-halogen effects on α-fluorine shielding takes place, the correlation of their stabilizing effects, although changing in the same direction, reaches only an approximate leveling off. Moreover, in the *para* position fluorine shows greater stabilizing influence than chlorine and bromine, which is contrary to the correlation of their shielding effects[70] (Scheme 48).

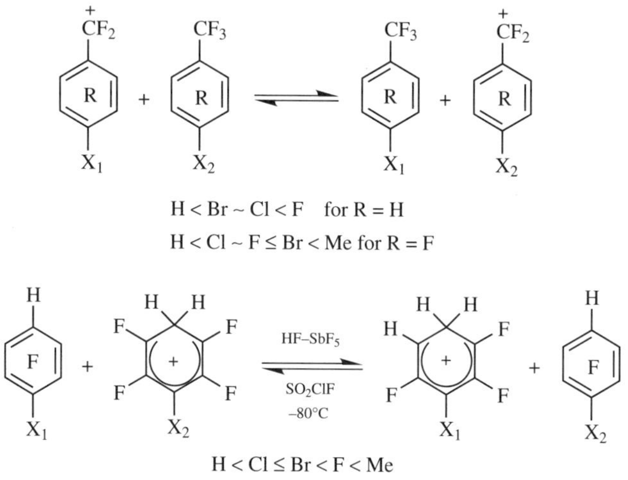

$H < Br \sim Cl < F$ for $R = H$

$H < Cl \sim F \leq Br < Me$ for $R = F$

$HF\text{–}SbF_5$, SO_2ClF, $-80°C$

$H < Cl \leq Br < F < Me$

Scheme 48

This observation seems to indicate that two counteracting factors contribute into the substituent's stabilizing effect. One contribution is believed to be backdonation, the variation of which along a series of *para*-substituted α,α-difluorobenzyl cations, as one may judge from the α-fluorine shielding, reverses on complete replacement of hydrogen by fluorine. The second component is probably in favor of fluorine substituent and defies inversion of the halogen substituent order on total fluorination. It was also supposed to be the local solvation of *para* substituent bearing an appreciable portion of positive charge.[51] For *ortho* substituents, this interaction is apparently less significant both because of a smaller fraction of the positive charge in this position and for steric reasons.

Due to such variability, the halogen-stabilizing effects should be considered to be very sensitive to the structure of carbocation, particularly to the charge delocalization mode and the position occupied by a halogen substituent. For example, as seen from the equilibrium ratios of isomeric 1,1-difluorobenzenenium and -naphthalenium ions (Schemes 17 and 18), chlorine or bromine in cyclohexadienyl moiety preferably occupies the resonance *ortho* position rather than the nonresonance *meta* position, whereas the opposite tendency is characteristic for halogens on the benzene ring of polyfluorinated 1-naphthalenium ions, in the resonance positions of which the electron demand should be weaker than with the cyclohexadienyl moiety (Scheme 18).

One can also anticipate the electron demand experienced by the substituents at the resonance positions of arenium ions to be affected by the nature of *ipso* substituents increasing with the electron-withdrawing effect of this substituent.

Accordingly, the larger relative stability of *ortho*-bromo- in comparison with *ortho*-fluorosubstituted polyfluorinated benzenenium cations (Scheme 46) is characterized by the competitive generation through the reaction (1) by the equilibrium constants $K > 300$, whereas at competitive protonation of 1-bromo-2,3,4,5-tetrafluorobenzene and pentafluorobenzene [reaction (2)], the similar difference is characterized by the value of only ~1,5.[30g]

A further case of changing electron demand is the buildup of electron deficiency in the ring in the course of precursor ionization so that the variation of a ratio of substituent-stabilizing effects in moving from a transition state to a carbocation. In compliance with this expectation, disparity has been observed between the kinetically controlled and equilibrium ratios of isomeric chloro- and bromohexafluorobenzenenium ions[71] (Scheme 49).

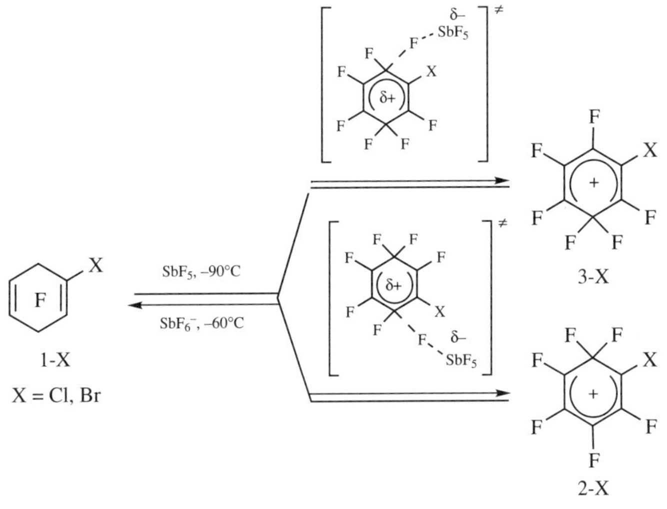

1-X (X = Cl, Br) $\xrightarrow[-90°C]{SbF_5}$ 2-X : 3-X = 1 : 2.3 for X = Cl; 1 : 3.5 for X = Br $\xrightarrow{-60°C}$ 2-X : 3-X = 2 : 1 for X = Cl; 3 : 1 for X = Br

Scheme 49

The experimental trends of the dependencies of the orders of halogen electronic and stabilizing effects on the electron demand placed on a substituent are in accordance with gas-phase studies exhibiting opposite orders of halogen stabilizing effects at the *para* position of benzyl cations (Cl < F < H[72]) and in the halomethyl cations (H < F < Cl < Br < I[73]) also on going from the alpha to the *para* position of arylmethyl cations.[74] It is instructive in the light of the abovementioned notion of the importance of solvation as a factor contributing in *para*-substituent-stabilizing effect, that in the gas phase *para* halogens are destabilizing substituents in the benzyl cations whereas in solution they were found to be stabilizing substituents in the α,α-difluorobenzyl cations.

Initially, this pronounced dependence of the order of halogen-stabilizing effects on the ring electron demand was suggested to be due to the respective dependence of the participation of heavy halogen *d*-atomic orbitals.[73] However, ab initio calculations[49,75] revealed the reversal of heavy halogen inductive effects on strong electron demand inherent in methyl cation so that in halomethyl cations heavy halogens are both stronger π and σ backdonors than fluorine (Scheme 50). The reversed π-backdonation order of *para*-halogen substituents is believed to be reflected by the α-fluorine shielding in the polyfluorinated difluorobenzyl cations because of a long distance between a varied substituent and α-fluorine probe.

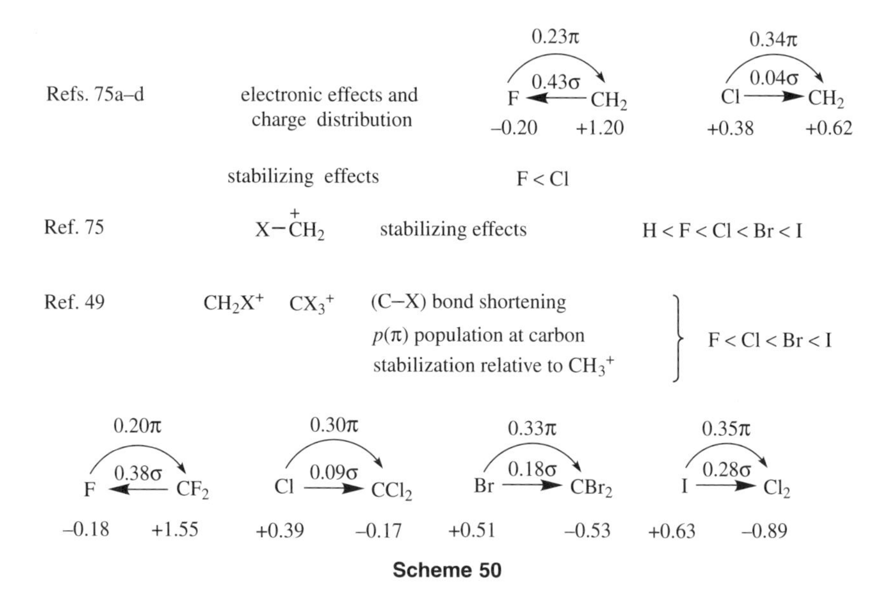

Scheme 50

Since the effective electronegativity of the carbocationic center should change in a wide range with the delocalization degree of carbocationic charge and the position occupied by a substituent, the wide variation of the correlation of electronic and stabilizing effects of the contiguous heteroatomic substituents can be predicted. The present understanding of these issues is, however, limited.

The regularities of the relative stability of polyfluorinated carbocations are important for planning the synthesis of fluoroorganic compounds through reactions involving intermediacy of polyfluorinated carbocations. For example, the regioselectivity of the electrophilic addition of nitration reagents to polyfluorarenes[4b,37] and the selective hydrolysis of substituted 9,9,10,10-tetrafluoro-9,10-dihydroanthracenes[30i] agree with the pattern of the relative stability of polyfluorinated arenium cations.

7.5 CHEMICAL PROPERTIES AND SYNTHETIC APPLICATIONS OF POLYFLUORINATED CARBOCATIONS

7.5.1 Isomeric Transformations

Historically, the chemistry of long-lived polyfluorinated carbocations began with the mutual transformations of isomers, which are often concomitant with their generation and provide much information on the relative carbocation stability. Several types of isomeric transformation have been revealed for the polyfluorinated carbocations, and most of them were observed for polyfluorinated arenium cations. The first observed example was isomerization proceeding through the *ipso-substituent migration*, in particular, such as mutual transformations of the cations formed by protonation of partially fluorinated benzenes via the 1,2 hydrogen shift from one unsubstituted ring position into the other. The kinetics of this process favors fluorine location at the *para* position; the *ipso*-to-fluorine location was not involved in this migration[50] (Scheme 51).

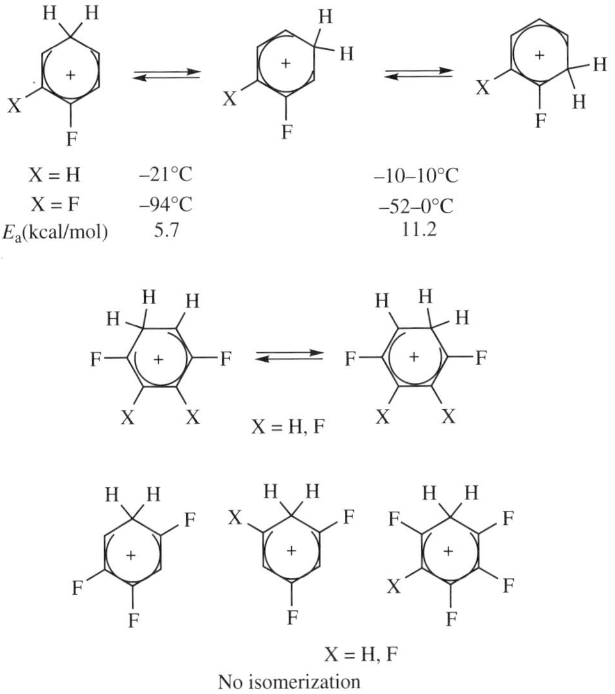

X = H	–21°C	–10–10°C
X = F	–94°C	–52–0°C
E_a(kcal/mol)	5.7	11.2

X = H, F

X = H, F

No isomerization

Scheme 51

In contrast to this scenario, the *ipso*-to-methyl group position can be involved[76] (Scheme 52).

However, the involvement of the positions occupied by fluorine was observed for the 1,2 chlorine migration[77] (Scheme 53) equilibrating the isomer distribution derived through polyfluoroarene *ipso* chlorination (Scheme 22).

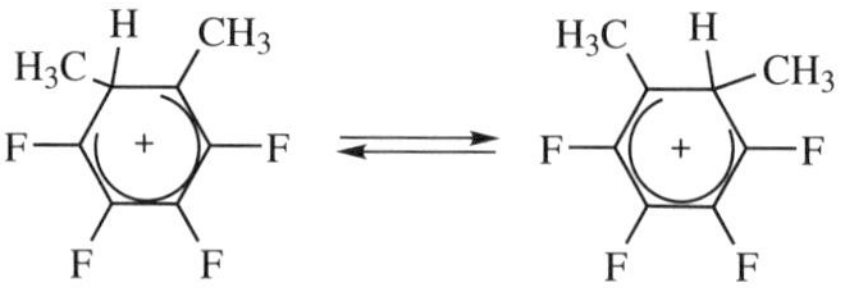

$\Delta F^{\neq}$ (25 °C) = 10.5 kcal/mol

Scheme 52

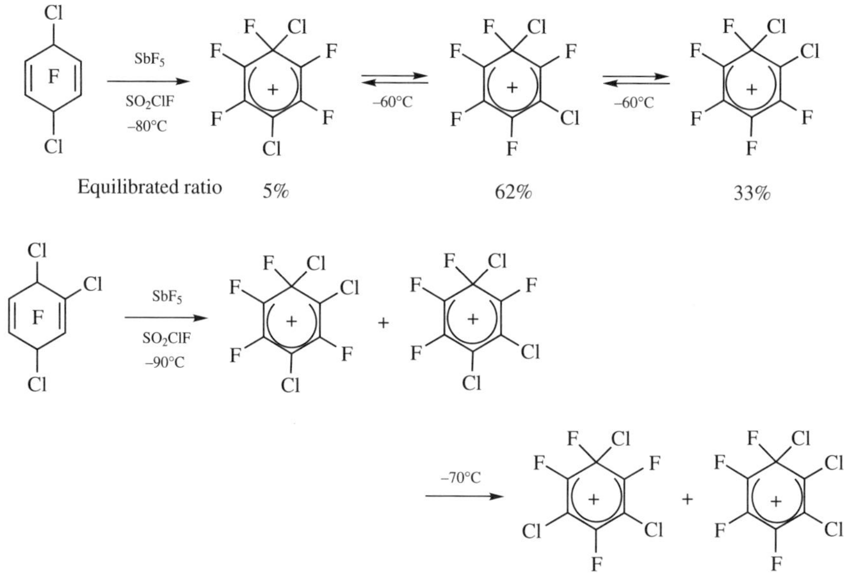

Equilibrated ratio 5% 62% 33%

Scheme 53

Interestingly, two ions formed in the latter case, unlike other polyfluorinated benzenium ions containing the —CC1F— unit, do not transform into isomers with two fluorine atoms at sp^3-hybridized carbon below −30°C.[77] This observation probably indicates the increase in stability of polyfluorinated arenium cations with the replacement of fluorine at the *ortho* and *meta* positions by chlorine, which is in line with the correlation of their stabilizing effects on the location in these positions as revealed by the competitive generation experiments (Scheme 46).

For the ion **52**, obtained by the chlorination of a superacid solution of 3,4,5,6-tetrafluoro-*ortho*-xylene,[78] the degenerate rearrangement due to the 1,2 chlorine shift from one position bearing the methyl group into the other was found to proceed much easier than the chlorine migration into the adjacent position occupied by fluorine. The chlorine moving from one *ipso*-to-fluorine position into the other was not observed under experimental conditions (Scheme 54). This apparently reveals the significant height of isomerization barrier for such a process.

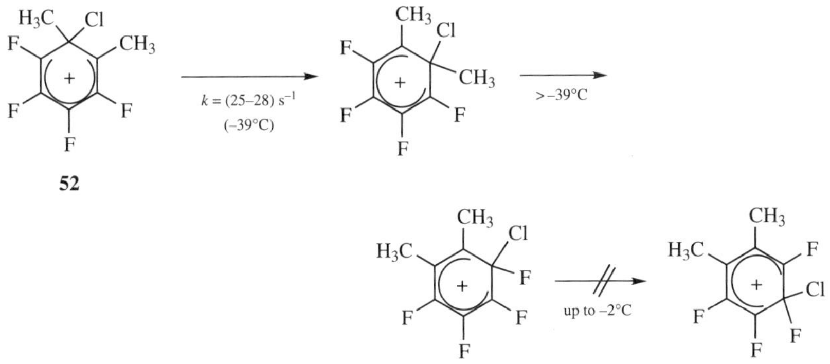

$k = (25\text{–}28)\ s^{-1}$
(−39°C)
>−39°C
up to −2°C

52

Scheme 54

Analogously, the isomerization of polyfluorinated arenium ions can be a result of the methyl group migration[35] (Scheme 25).

In contrast, the transformation of 1-bromo-2-hydroxy-5-methyl-1,3,4,6-tetrafluorobenzenium ion **53** into 1-bromo-1-methyl-4-hydroxy isomer **54** proceeds most likely through intermolecular bromine transfer, since the addition of heptafluoro-2-naphthol to the reaction mixture gave 1-bromo-2-hydroxy-heptafluoro-1-naphthalenium ion and 2,3,5,6-tetrafluoro-*para*-cresol.[31c] On attempted generation of the respective 1-hydroxy-2-naphthaleneium ion by dissolving 1-keto-2-bromoheptafluoro-1,2-dihydronaphthalene in $HSO_3F–SbF_5–SO_2ClF$ at −70 to −40°C, only the product of its isomerization (1-bromo-4-hydroxyheptafluoro-1-naphthalenium ion **55**) was obtained, undoubtedly due to a greater stability of the 1-naphthaleneium ion in comparison to the 2-naphthalenium one and the ease of the bromine migration[31c] (Scheme 55).

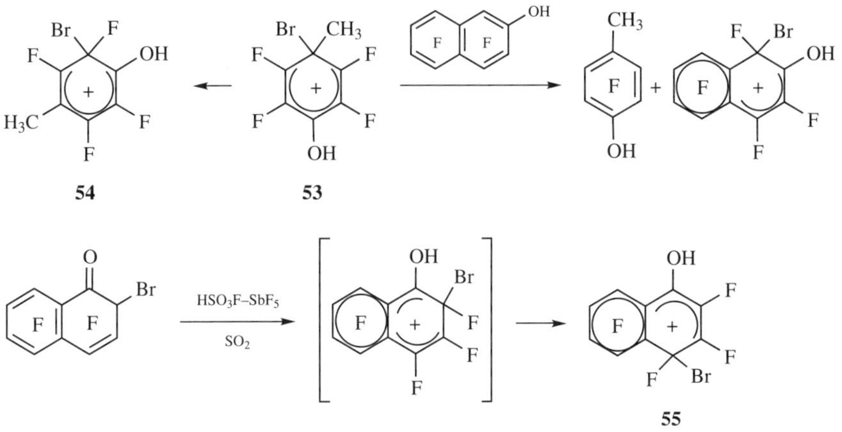

54 **53**

$HSO_3F–SbF_5$
SO_2

55

Scheme 55

Unlike the scenario above, 1-hydroxy-2-chloroheptafluoro-2-naphthalenium ion **56** was obtained through the protonation of 1-keto-2-chloroheptafluoro-1,2-dihydronaphthalene, which also isomerizes to the more stable 4-hydroxy-1-naphthalenium ion **57**, but through the migration of fluorine rather than chlorine,[31c] whereas the last process can be anticipated based on generally known relative migratory aptitudes of these two halogens. Most likely the fluorine migration proceeds through the specific mechanism of isomerization either of the ion or its precursor proceeding with the participation of HF through a cyclic transition state such as **58**[79] (Scheme 56).

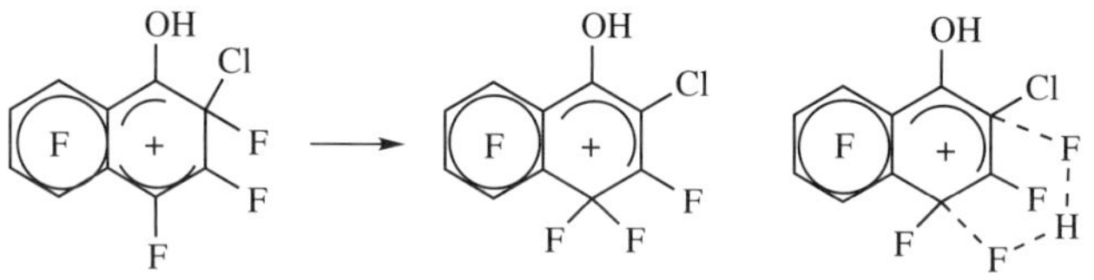

56 **57** **58**

Scheme 56

With regard to polyfluorinated arenium ions, the relative migratory aptitude of various groups is obviously important in terms of view of the interpretation of the regioselectivity observed in the reactions of polyfluoroarenes with electrophiles. In the polyfluoroarene series the most extensively investigated are the reactions with nitrating agents, in the course of which both unsubstituted and *ipso* positions, including those occupied by fluorine, can be involved.[4,36] In this connection of special interest is the question of the ability of nitro group to migrate on the face of a polyfluorinated skeleton. In fact, the 1-nitro-1,3,5-trimethyl-2,4,6-trifluorobenzenenium cation **59**—the only polyfluorinated arenium cation prepared by the direct nitration of polyfluoroarene—does not show the propensity to rearrange,[80] whereas its completely methylated analog—the 1-nitrohexamethylbenzenenium ion **60**—automerizes easily through a mechanism of intramolecular migration of the nitro group.[80,81] The lack of the rearrangement in the former case apparently should be attributed to the essentially lower stability of the arenium cation formed by the nitronium cation addition to the *ipso*-to-fluorine position that causes the high activation barrier of nitro group shift from the position *ipso* to methyl group to the adjacent one occupied by fluorine (Scheme 57).

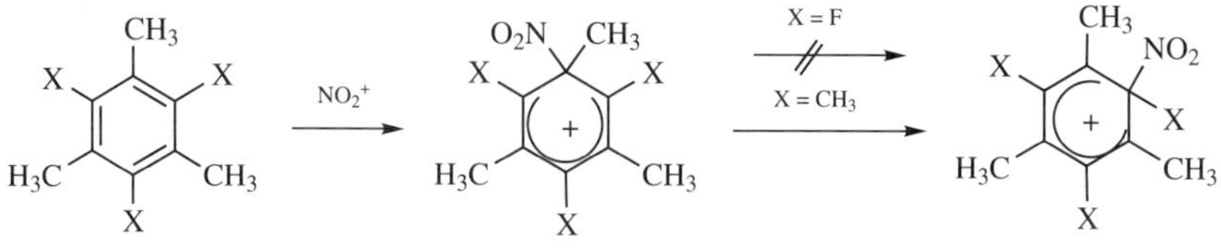

59 X = F
60 X = CH_3

Scheme 57

In this connection the results of the attempted generation of an arenium ion with the ring $-CFNO_2-$unit through the protonation of an appropriate cyclohexadienone are instructive. With HSO_3F the degree of protonation was insignificant. The addition of SbF_5 to augment acidity led to denitration with the formation of pentafluorophenol; 2 equiv of SbF_5 were required (Scheme 58). Moreover, unlike the successful preparation of polyfluorinated arenium cations through the direct *ipso*-to-fluorine chlorination and methylation (Schemes 22 and 25), numerous attempts to achieve an analogous result through the nitration of polyfluoroarenes failed.[4,82] These data are interesting in the context of the activity of nitronium cation as electrophile under a variety of acidic conditions.

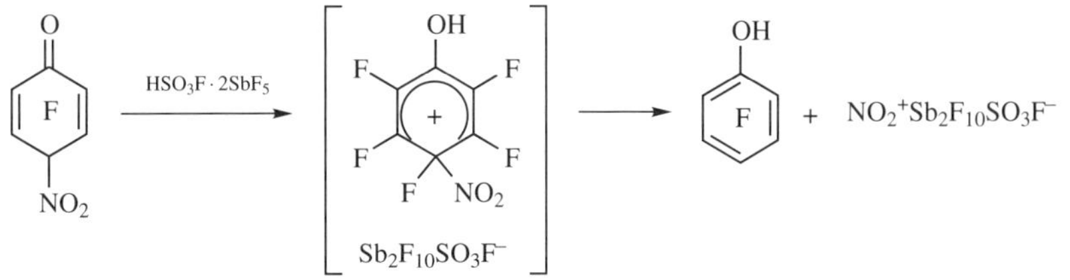

$HSO_3F \cdot 2SbF_5$; $Sb_2F_{10}SO_3F^-$; $NO_2^+Sb_2F_{10}SO_3F^-$

Scheme 58

The isomeric distributions of arenium ions formed through the ionization of polyfluorinated dihydroarenes with SbF_5 (Schemes 17–19) are completely or partially equilibrated by the sequence of the cation quenching with a fluoride anion followed by reionizing the intermediately formed polyfluorodihydroarene. This *fluoride anion addition–elimination* mechanism is corroborated, for example, by the sequence of the mutual transformations of isomeric arenium ions derived from 3-chloroheptafluorocyclohexa-1,4-diene in the SbF_5–SO_2ClF solution on increasing temperature: **61** → **62** → **63** + **64**,[32c] which reveals stronger electrophilicity of the *para* position that of the *ortho* one (Scheme 59).

Further studies showed the mutual transformations of isomeric polyfluorinated naphthalenium ions bearing a substituent in different rings. For this isomerization to occur, a fluoride anion should be added to the resonance positions of a benzene ring, which are less electrophilic than those in a cyclohexadienyl moiety, and for this reason even harsher conditions are necessary than for an isomerization changing the substituent location within the same ring[30c,d;54] (Scheme 60).

The transformations presented above expand the preceding pattern of thermodynamic relationships between isomeric polyfluorinated naphthalenium ions. Thus, 6-X-octafluoro-1-naphthalenium ions with $X = CH_3$ or H completely isomerize to the appropriate 3-X-octafluoro-1-naphthalenium ions,[32c,54] and with $X = C_6F_5$ the equilibrium ratio such as 3-X : 6-X = 2 : 1 is established.[32d] With X = Cl or Br, the substituent location is also preferable in the cyclohexadienyl moiety rather than in the benzene ring, which is most clearly illustrated by a practically complete conversion of 6,7-dichloroheptafluoro-1-naphthalenium ion **65** into the 2,3-dichloro isomer **66**[54] (Scheme 61).

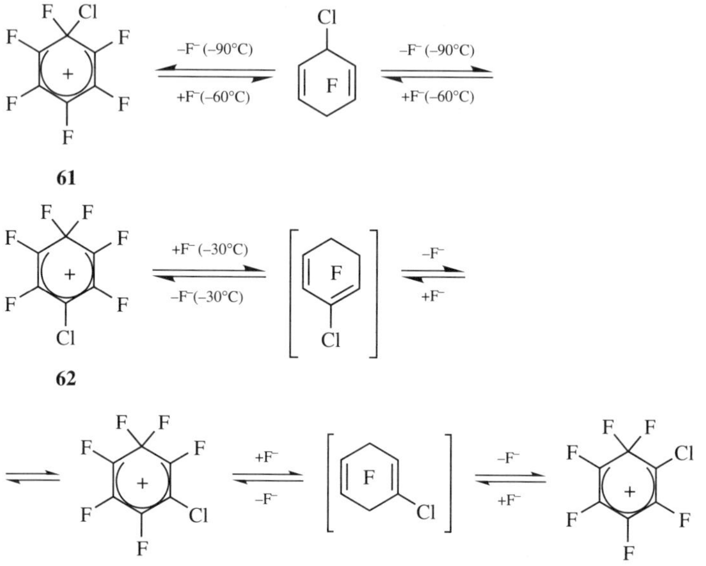

Scheme 59

In all these cases the relocation of substituent X from the benzene ring to the cyclohexadienyl one is accompanied by the reverse movement of fluorine; the latter is the most destabilizing among the studied substituents when located in the *ortho* or *meta* position of a cyclohexadienyl moiety. Therefore, it seems reasonable to consider the thermodynamic relationships manifesting themselves in these cases as an indication that the influence of a substituent on the relative stability of a naphthalenium cation is more essential when a substituent is located in the cyclohexadienyl moiety rather than in the benzene ring. Contrary to this, when a substituent other than fluorine occupies one of the alpha positions of the naphthalene framework, the reverse transfer of a substituent from the cyclohexadienyl ring to the benzene ring has been observed. This was suggested to occur as a result of the preference for fluorine to occupy the *para* position on a cyclohexadienyl ring and minimizing the steric repulsion of *peri* substituents[54,83] (Scheme 62).

Such typical carbocation transformations as isomerizations with the change of a carbon skeleton were found for various polyfluorinated compounds under conditions that leave no doubt that skeletal rearrangements of the intermediately formed polyfluorinated carbocations do occur. The first examples of this kind are the contractions of a six-membered ring to a five-membered one on heating 2-bromoheptafluoronaphthalene[84] and perfluoro-2-X-tetralines (X = Cl, Br)[85] with SbF_5. There is good reason to assume that in the former case the reaction includes fluorination of a ring-bearing bromine to give perfluoro-2-bromotetralin **67**, which is ionized with SbF_5 to perfluorotetrahydronaphthalenyl cation **68**. The latter isomerizes to the

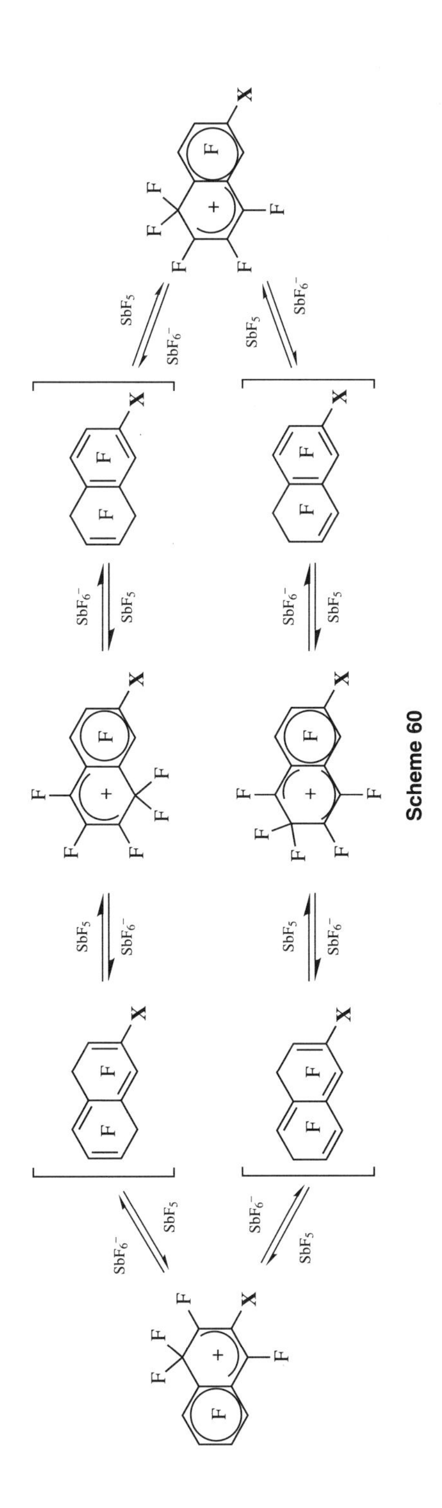

Scheme 60

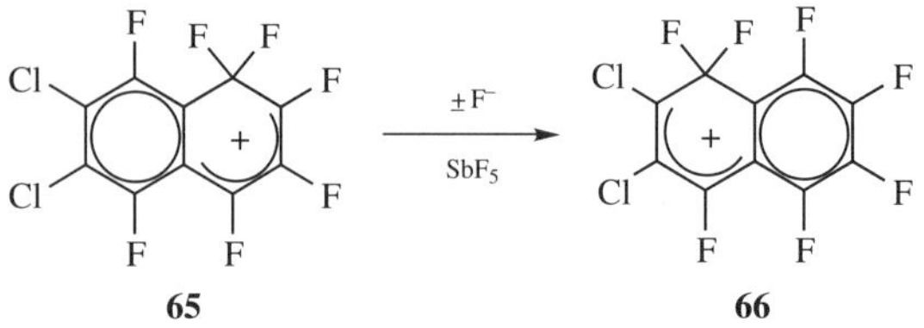

Scheme 61

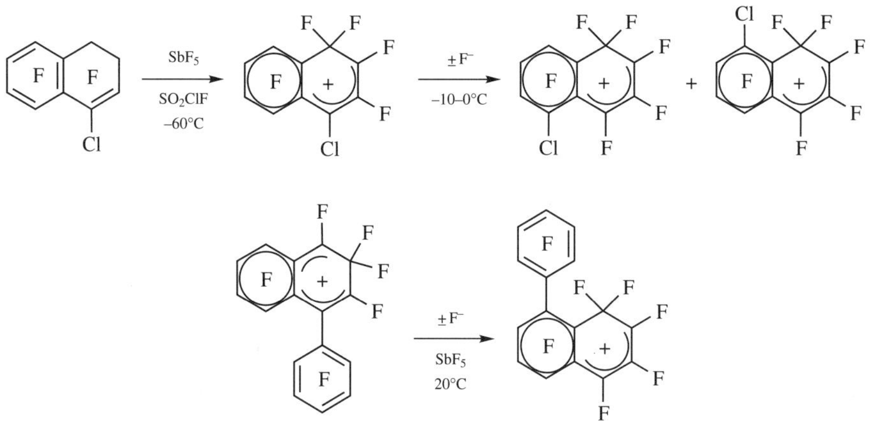

Scheme 62

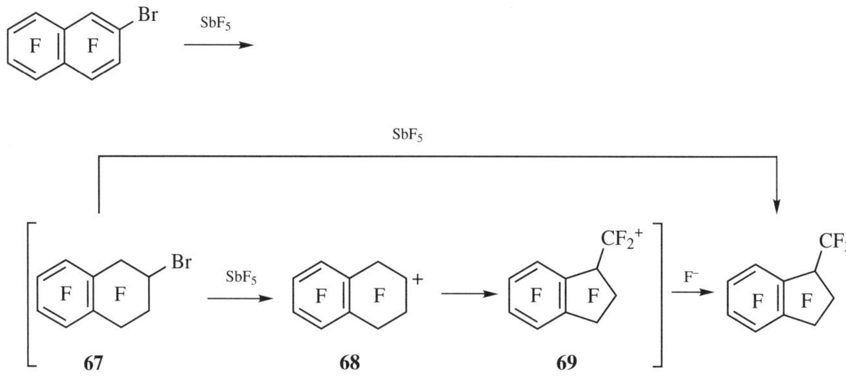

Scheme 63

cation **69**; the reaction is completed by quenching of this cation by fluoride anion (Scheme 63).

The assumption of eliminating a "heavy" halogen anion from **67** is justified by the formation of the α,α-difluorobenzyl cation **5** from α-chloro-α,α-difluorotoluene[6a] and the perfluorobenzenium cation **27** from 3-bromoheptafluoro-1,4-cyclohexadiene when ionized with SbF_5[84] (Scheme 64).

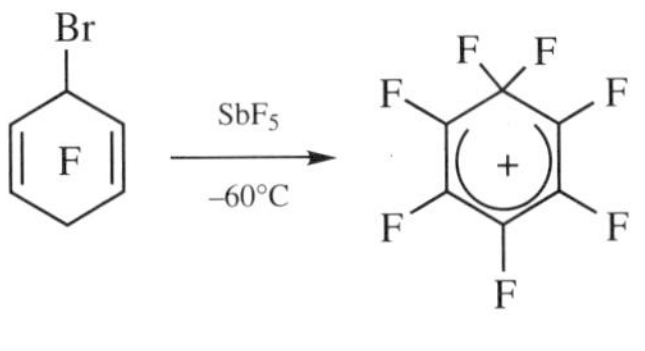

Br F SbF_5 −60°C F F F F F F + 27

Scheme 64

Subsequently similar transformations were realized for a wide range of polyfluorinated benzocycloalkanes.[5] In these not only the contractions but also the expansions of a fluorocarbon cycle were found[86] (Scheme 65).

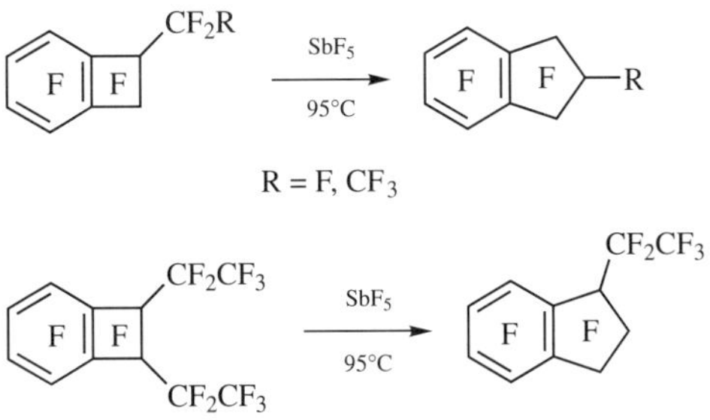

CF_2R F F SbF_5 95°C F F R

R = F, CF_3

CF_2CF_3 F F CF_2CF_3 SbF_5 95°C F F CF_2CF_3

Scheme 65

7.5.2 Polyfluorinated Carbocation as Electrophiles

The use of long-lived carbocations as synthons can provide the basis for a fruitful methodology in organic synthesis, which, however, is still in its infancy. The hydrolysis of SbF_5 solutions of polyfluorinated α,α-difluorobenzyl hexafluoroantimonates was demonstrated as a method for preparing polyfluorinated benzene carboxylic acids from the precursors of these cations, namely, polyfluorinated benzotrifluorides[87] (Scheme 66).

Hydrolysis and alcoholysis of cyclic polyfluorinated cations with fluorine at the carbocationic center allows one to obtain polyfluorinated compounds with oxygen-containing functions (Scheme 67).

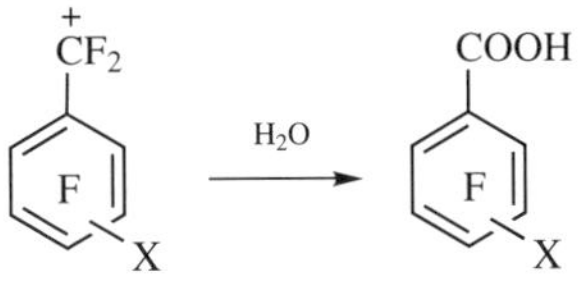

$\overset{+}{C}F_2$ F X H_2O COOH F X

Scheme 66

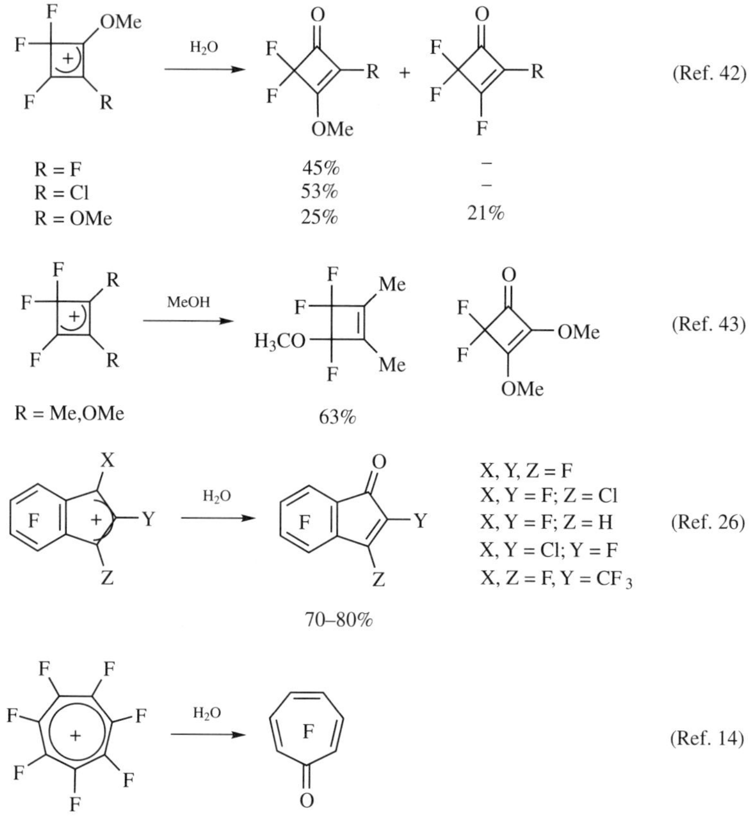

(Ref. 42)

R = F	45%	–
R = Cl	53%	–
R = OMe	25%	21%

R = Me,OMe — 63% (Ref. 43)

X, Y, Z = F
X, Y = F; Z = Cl
X, Y = F; Z = H
X, Y = Cl; Y = F
X, Z = F, Y = CF_3

70–80% (Ref. 26)

(Ref. 14)

Scheme 67

Numerous cyclohexadienones were prepared by the hydrolysis of arenium ions.[4,30,36,88] The latter reaction in combination with the preparation of arenium cations through electrophilic *ipso* reactions of polyfluoroarene ions and reduction of polyfluorinated cyclohexadienones affords the general approach to polyfluorinated hydroxyarenes that are inaccessible by the traditional means of nucleophilic substitution of fluorine[37,88] (Scheme 68). The treatment of arenium ions with hydrogen chloride results in polyfluorinated 3-chlorocyclohexa-1,4-dienes[77] (Scheme 69).

Of special significance with respect to application in the synthetic utility of polyfluorinated carbocations as reagents is alkylation with the formation of a C–C bond. A pioneering example of this application is the reaction of chlorodifluorocyclopropenyl cation **70** with benzene followed by hydrolysis[13] to obtain diphenylcyclopropenone **71** (Scheme 70).

Diarylmethyl cations were generated by alkylating polyfluoroarenes with α,α-difluorobenzyl cations in SbF_5 (Scheme 11) and applied to prepare a series of polyfluorinated benzophenones, whereas the treatment with the excess of HF to obtain their immediate precursors—polyfluorinated diarylmethanes[87] (Scheme 71). Analogously, perfluorofluorenyl cation **24** prepared from biphenyl **72** through

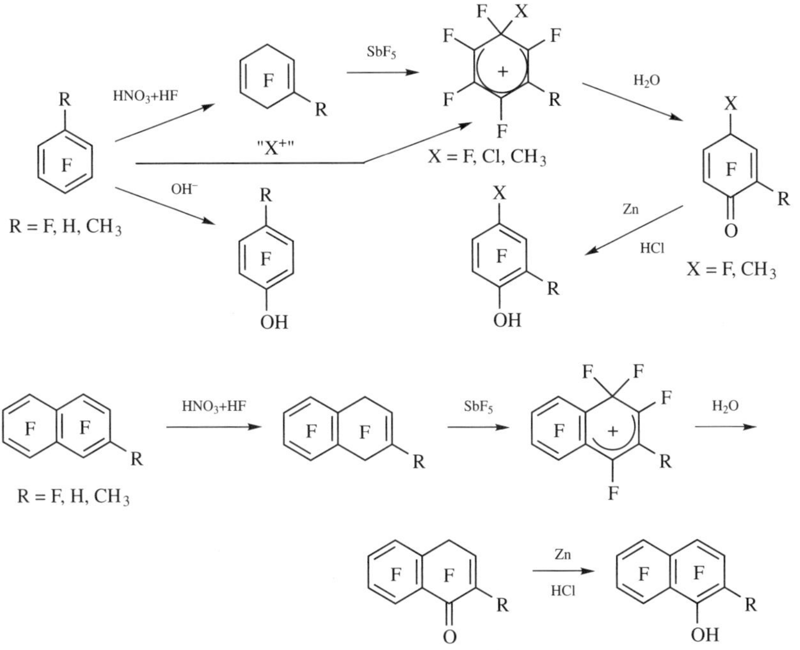

Scheme 68

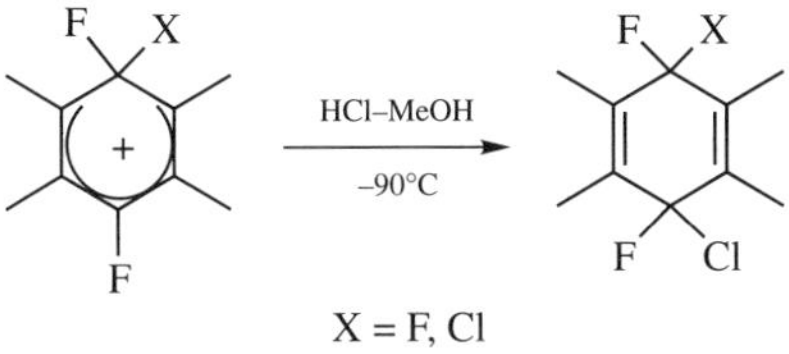

Scheme 69

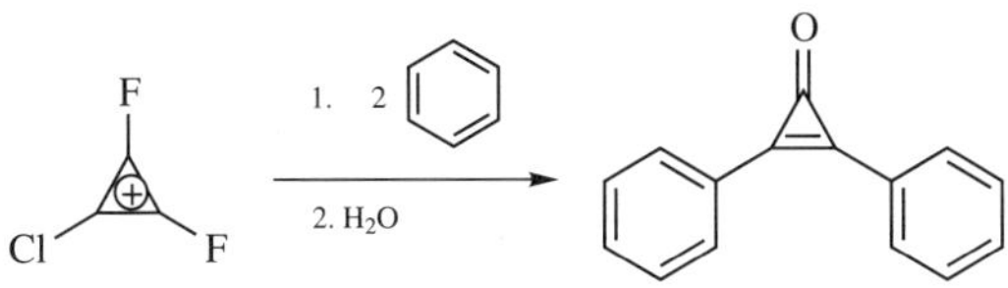

Scheme 70

intramolecular alkylation was hydrolyzed and quenched with HF to yield perfluorofluorenone and perfluorofluorene, respectively[27] (Scheme 72).

More recently the α,α-difluorobenzyl cation was shown to couple with its precursor, benzotrifluoride, to give *meta*-trifluoromethyl-α-fluorodiphenylmethyl cation[89] (Scheme 73).

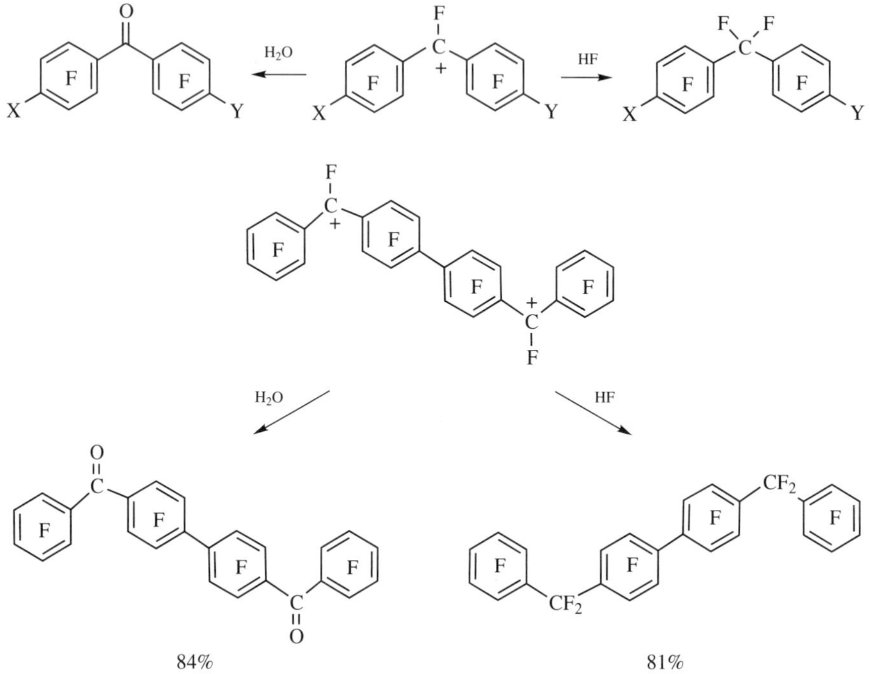

H$_2$O HF

84% 81%

Scheme 71

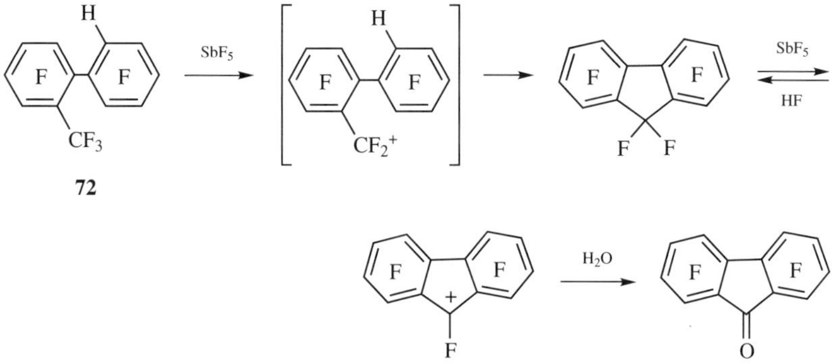

72

SbF$_5$ SbF$_5$ HF H$_2$O

24

Scheme 72

SbF$_5$ SO$_2$ClF

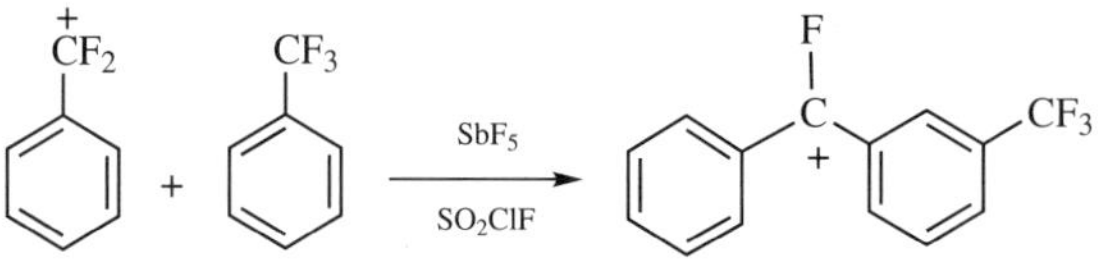

Scheme 73

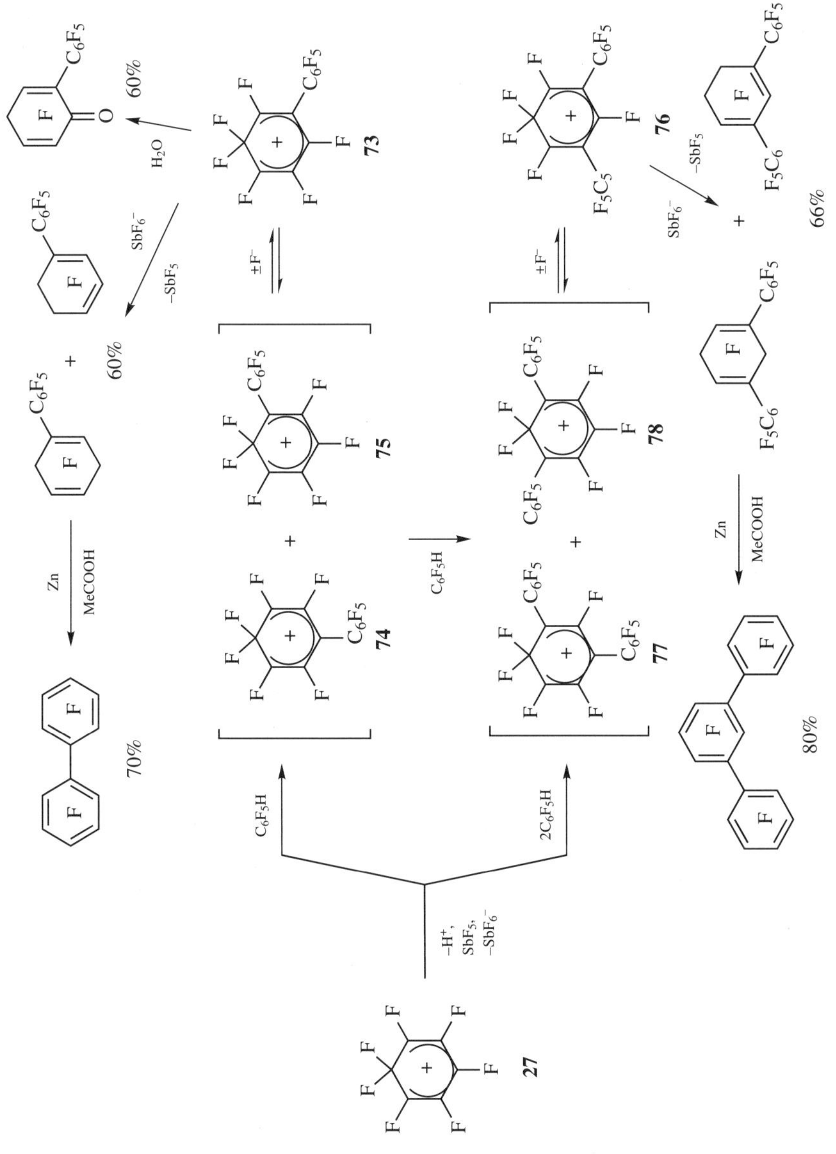

Scheme 74

An analogous application of polyfluorinated arenium cations was demonstrated to be a general approach to the *meta*-polyfluoroaryl-substituted arenium cations and further to polynuclear polyfluo- dihydroarenes and arenes[30d,83,90,91] (Scheme 74). For example, the interaction of pentafluorobenzene with perfluorobenzenium ion **27** gave perfluoro-3-phenyl benzenium cation **73**, which must have been derived from the first formed 4- and 2-isomers, **74** and **75** by the isomer equilibration through the sequence of fluoride anion addition–elimination steps.[30d] The interaction of the cation **27** with 2 equiv of pentafluorobenzene in SbF_5 gives perfluoro-3,5-diphenyl benzenium cation **76**, again through the intermediacy of polyfluorobenzenium cations **77** and **78**.[91]

Surprisingly, the reactions of the cation **27** with 3 equiv of 4-XC_6F_4H (X = F, Br) lead directly to completely aromatized products—polyfluorinated 1,3,5-triphenylbenzenes **79** and **80**.[90,91] Taking into account the tendency of pentafluorophenyl group to occupy the *meta* position in polyfluorinated benzenium ions (Scheme 17) and the stability of the cations with —CF_2— units, this was considered to be a consequence of specific properties of the transient cation **81** with a polyfluoroaryl moiety *ipso* to fluorine (Scheme 75).

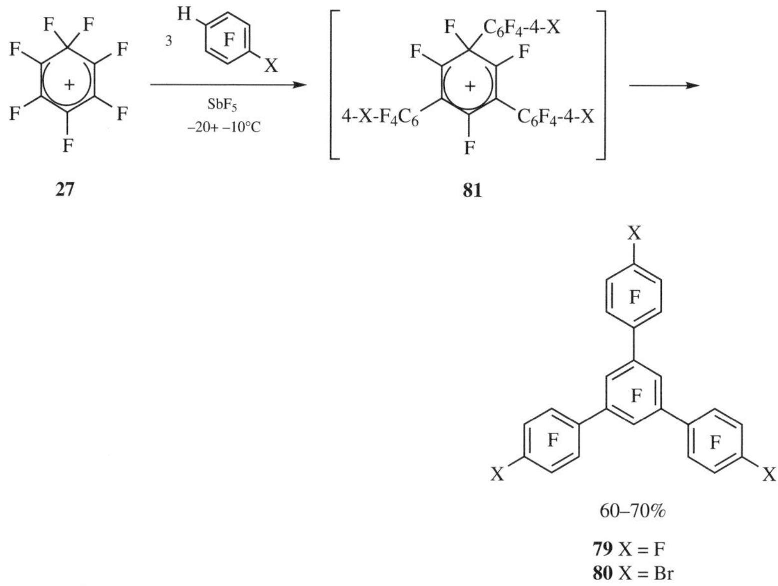

Scheme 75

Probably cations of the same type are responsible as intermediates for the formation of perfluorotriphenylene in the reaction of the cation **27** with 2,2′-di-H-octafluorobiphenyl and of perfluoro-1,5-diphenylnaphthalene in the reaction of the naphthalenium cation **28** with pentafluorobenzene[83,90] (Scheme 76). This assumption is in accord with the results of interactions of octafluoronaphthalene with CH_3F–SbF_5, suggested to proceed through intermediacy of cations **82** and **83**[92] (Scheme 77) and of octafluoronaphthalene radical cation with polyfluorobenzenes[83,90] in neat SbF_5.

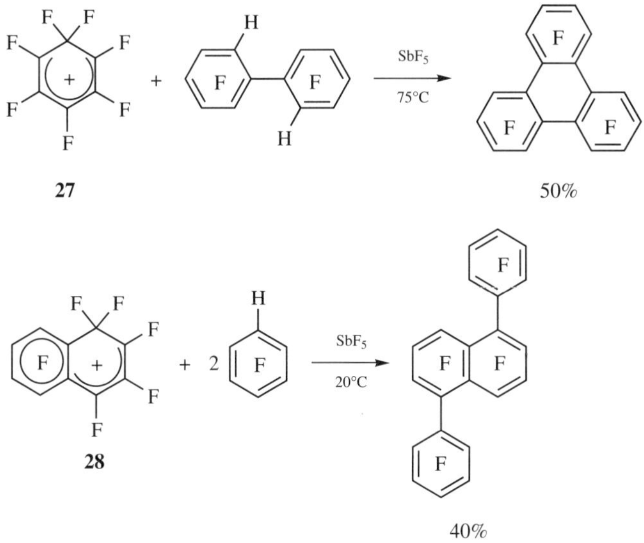

27

SbF_5 75°C

50%

28

SbF_5 20°C

40%

Scheme 76

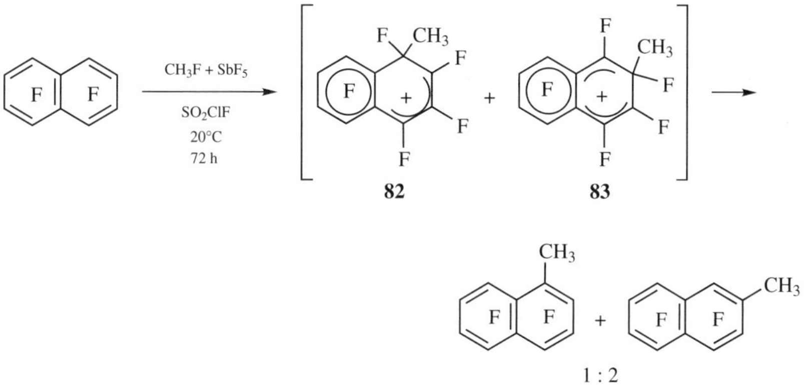

$CH_3F + SbF_5$

SO_2ClF 20°C 72 h

82

83

1 : 2

Scheme 77

There is reason to anticipate polyfluorinated arenium cations of being capable of alkylating the polyfluorinated conjugated diene system. Thus, the observed disproportionation of heptafluorobenzenium hexafluoroantimonate[93] seems likely to proceed through its addition to perfluoro-1,3-cyclohexadiene, which exists in equilibrium with the starting cation, resulting in the cyclohexenyl cation **84**. The further sequence of fluoride anion addition–elimination steps leads to the arenium cation **85** with perfluorocyclohexenyl moiety *ipso* to fluorine, which decays to hexafluorobenzene and cyclohexenyl cation **86**. The latter cation captures fluoride ion to complete the reaction (Scheme 78).

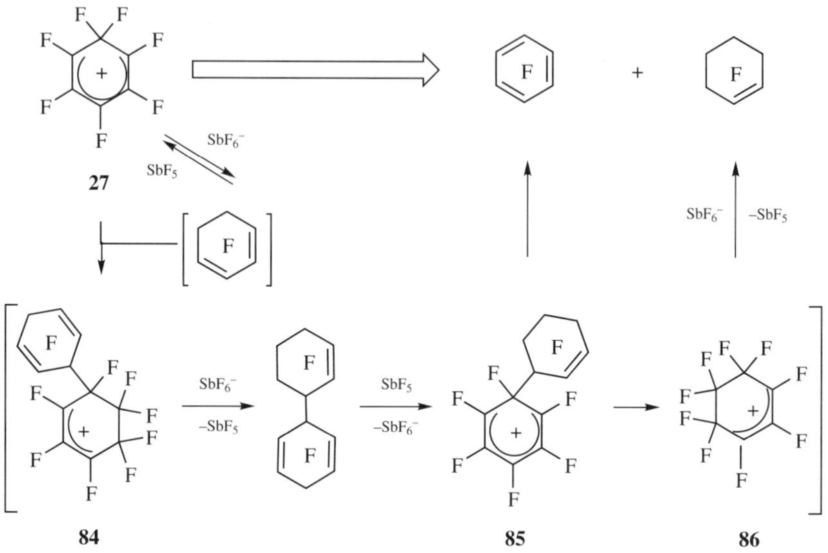

Scheme 78

Mechanistically, the Friedel–Crafts chemistry of long-lived polyfluorinated arylmethyl and arenium cations in superacidic media should be considered on the grounds of the concept of superelectrophilic activation developed by Olah[94] and applied to account for the seemingly unusual reactivity of trihalomethyl cations.[95] From this point of view, polyfluorinated carbocations seem to be challenging systems to study since they generally provide a series of nucleophilic sites for the coordination with an acid. Some of these sites are more basic than others, thus affording considerable concentration of the superelectrophile, but coordination of less basic ones with an acid may provide stronger electrophilicity that can be more significant for subsequent reaction with the nucleophile. Thus, one could anticipate a wide variation of reactivity of these cations with the reaction conditions.

By now the wide range of polyfluoroalkylations and isomerizations of polyfluorinated unsaturated compounds catalyzed by strong Lewis acids, first of all by SbF_5, and involving, in all probability, the intermediate formation of polyfluorinated alkyl cations, has been documented. The numerous examples of such reactions are given in reviews.[5]

In summary, intriguing progress has been made in the area of polyfluorinated carbocations, but numerous crucial issues relevant to the electronic structure of polyfluorinated carbocations still remain unsolved. There is no doubt that the application of X-ray methodology to the salts of longlived polyfluorinated cations will offer outstanding new possibilities to move in this direction. The scope of the application of long-lived polyfluorinated carbocations as synthons is just starting but looks very promising for the development of the synthesis of new polyfluorinated systems and their functionalized derivatives.

ACKNOWLEDGMENT

The generous support of Professors G. A. Olah and G. K. S. Prakash in preparation of this chapter is greatly appreciated.

REFERENCES

1. G. A. Olah, *Angew. Chem. Int. Ed. Engl.* **34**, 1393 (1995); G. A. Olah, in *Stable Carbocation Chemistry*, G. K. S. Prakash and P. v. R. Schleyer, eds., Wiley, New York, 1997, p. 1.
2. G. A. Olah and Y. K. Mo, *Adv. Fluorine Chem.* **7**, 69 (1973).
3. V. D. Shteingarts, *Izv. Sib. Otd. AN SSSR, Ser. Khim. Nauk* **3**(7), 53 (1980).
4. (a) V. D. Shteingarts, *Usp. Khim.* **50**, 1407 (1981); (b) V. D. Shteingarts, L. S. Kobrina, I. I. Bilkis, and V. F. Starichenko, *Khimiya Polyftorarenov: Mekhanism Reactsii, Intermediaty*, Nauka, Novosibirsk, 1991.
5. (a) V. I. Bakhmutov and M. B. Galakhov, *Usp. Khim.* **57**, 1467 (1988); (b) A. D. Allen and T. T. Tidwell, in *Advances in Carbocation Chemistry*, 1989, Vol. 1, p. 1; (c) G. G. Belen'kii, *J. Fluorine Chem.* **77**, 107 (1996); (d) C. G. Crespan and V. A. Petrov, *Chem. Rev.* **96**, 3269 (1996); (e) V. A. Petrov and F. Davidson, *J. Fluorine Chem.* **95**, 5 (1999).
6. (a) G. A. Olah, C. A. Cupas, and M. B. Comisarov, *J. Am. Chem. Soc.* **88**, 362 (1966); (b) G. A. Olah, R. D. Chambers, and M. B. Comisarov, *J. Am. Chem. Soc.* **89**, 1268 (1967).
7. G. A. Olah, G. Liang, and Y. K. Mo, *J. Org. Chem.* **39**, 2394 (1974).
8. G. A. Olah and Y. K. Mo, *J. Org. Chem.* **37**, 1028 (1972).
9. (a) G. A. Olah and M. B. Comisarov, *J. Am. Chem. Soc.* **91**, 2955 (1969); (b) G. A. Olah and Y. K. Mo, *J. Org. Chem.* **38**, 2686 (1973).
10. G. A. Olah, L. Heiliger, and G. K. S. Prakash, *J. Am. Chem. Soc.* **111**, 8020 (1989); G. A. Olah, G. Rasul, L. Heiliger, and G. K. S. Prakash, *J. Am. Chem. Soc.* **118**, 3580 (1996).
11. F. T. Prochaska and L. Andrews, *J. Am. Chem. Soc.* **100**, 2102 (1978).
12. P. B. Sargeant and C. G. Krespan, *J. Am. Chem. Soc.* **91**, 415 (1969).
13. D. C. F. Law, S. W. Tobey, and R. West, *J. Org. Chem.* **38**, 768 (1973).
14. W. P. Dailey and D. M. Lemal, *J. Am. Chem. Soc.* **106**, 1169 (1984).
15. G. A. Olah and J. S. Staral, *J. Am. Chem. Soc.* **98**, 6290 (1976).
16. G. Paprott, S. Lehmann, and K. Seppelt, *Chem. Ber.* **121**, 727 (1988).
17. R. D. Chambers, M. Salisbury, G. Apsey, and G. Moggi, *J. Chem. Soc. Chem. Commun.* 680 (1988).
18. G. A. Olah and C. U. Pittman, *J. Am. Chem. Soc.* **88**, 3310 (1966).
19. G. A. Olah and M. B. Comisarov, *J. Am. Chem. Soc.* **89**, 1027 (1967).
20. G. A. Olah and Y. K. Mo, *J. Am. Chem. Soc.* **89**, 6827 (1973).
21. (a) G. A. Olah, G. K. S. Prakash, and G. Liang, *J. Am. Chem. Soc.* **99**, 5683 (1977); (b) G. A. Olah, G. K. S. Prakash, and T. N. Rawdah, *J. Am. Chem. Soc.* **102**, 6127 (1980).
22. V. A. Barkhash, *Neklassicheskie Karbokationy*, Nauka, Novosibirsk, 1984; V. A. Barkhash, *Top. Curr. Chem.* (Springer-Verlag, Berlin) **116/117**, 1 (1984).
23. G. A. Olah and Y. K. Mo, *J. Org. Chem.* **38**, 2682 (1973).

24. (a) Yu. V. Pozdnyakovich and V. D. Shteingarts, *Zh. Org. Khim.* **6**, 1753 (1970); (b) Yu. V. Pozdnyakovich and V. D. Shteingarts, *J. Fluorine Chem.* **4**, 283 (1974).
25. Yu. V. Pozdnyakovich and V. D. Shteingarts, *J. Fluorine Chem.* **4**, 297 (1974).
26. V. M. Karpov, V. E. Platonov and G. G. Yakobson, *Izv. AN SSSR, Ser. Khim.* 2647 (1976); V. M. Karpov, V. E. Platonoy, and G. G. Yakobson, *Tetrahedron* **34**, 3215 (1978).
27. Yu. V. Pozdnyakovich and V. D. Shteingarts, *Zh. Org. Khim.* **14**, 603 (1978).
28. (a) G. A. Olah and T. E. Kiovsky, *J. Am. Chem. Soc.* **89**, 5692 (1967); (b) G. A. Olah and T. E. Kiovsky, *J. Am. Chem. Soc.* **90**, 2583 (1968); (c) D. M. Brouwer, Rec. Trav. Chim. **87**, 342 (1968).
29. K. Laali, J. Fluorine Chem. **43**, 415 (1989).
30. (a) V. D. Shteingarts, Yu. V. Pozdnyakovich, and G. G. Yakobson, *Chem. Commun.* 1264 (1969); (b) V. D. Shteingarts and Yu. V. Pozdnyakovich, *Zh. Org. Khim.* **7**, 734 (1971); (c) T. V. Chuikova, A. A. Shtark, and V. D. Shteingarts, *Zh. Org. Khim.* **10**, 1712 (1974); (d) Yu. V. Pozdnyakovich, T. V. Chuikova, and V. D. Shteingarts, *Zh. Org. Khim.* **11**, 1689 (1975); (e) B. G. Oksenenko and V. D. Shteingarts, *Zh. Org. Khim.* **10**, 1190 (1974); (f) B. G. Oksenenko, V. I. Mamatyuk, and V. D. Shteingarts, *Zh. Org. Khim.* **12**, 1322 (1976); (g) A. A. Shtark and V. D. Shteingarts, *Zh. Org. Khim.* **13**, 1662 (1977); (h) A. A. Shtark and V. D. Shteingarts, *Zh. Org. Khim.* **12**, 1449 (1976); (i) B. G. Oksenenko, M. M. Shakirov, and V. D. Shteingarts, *Zh. Org. Khim.* **12**, 1313 (1976).
31. (a) N. G. Kostina and V. D. Shteingarts, *Zh. Org. Khim.* **10**, 1705 (1974); (b) N. E. Akhmetova and V. D. Shteingarts, *Zh. Org. Khim.* **11**, 1226 (1975); (c) N. E. Akhmetova and V. D. Shteingarts, *Zh. Org. Khim.* **13**, 1269 (1977).
32. (a) V. D. Shteingarts, *Zh. Org. Khim.* **11**, 461 (1975); (b) V. D. Shteingarts and P. N. Dobronravov, *Zh. Org. Khim.* **12**, 2005 (1976); (c) P. N. Dobronravov and V. D. Shteingarts, *Zh. Org. Khim.* **13**, 1679 (1977).
33. P. Menzel and F. Effenberger, *Angew. Chem. Int. Ed. Engl.* **11**, 922 (1972); L. P. Kamshii and V. A. Koptyug, *Izv. AN SSSR, Ser. Khim.* 236 (1974).
34. (a) V. V. Bardin, G. G. Furin, and G. G. Yakobson, *Zh. Org. Khim.* **17**, 999 (1981); (b) P. N. Dobronravov and V. D. Shteingarts, *Zh. Org. Khim.* **17**, 1556 (1981).
35. P. N. Dobronravov and V. D. Shteingarts, *Zh. Org. Khim.* **19**, 995 (1983).
36. G. A. Olah, P. V. Reddy, and G. K. S. Prakash, in *The Chemistry of the Cyclopropyl Group*, 1995, Vol. 2, p. 813.
37. V. D. Shteingarts, in *Synthetic Fluorine Chemistry*, G. A. Olah, R. D. Chambers, and G. K. S. Prakash, eds., Wiley, New York, 1992, p. 259.
38. T. V. Chuikova and V. D. Shteingarts, *Zh. Org. Khim.* **9**, 1733 (1973).
39. Yu. V. Pozdnyakovich and V. D. Shteingarts, *Zh. Org. Khim.* **14**, 2237 (1978).
40. R. D. Chambers, R. S. Matthews, and A. Parkin, *Chem. Commun.* 509 (1973).
41. I. L. Knunyants, Yo. G. Abdulganiev, E. M. Rokhlin, P. O. Okulevich, and N. I. Karpushina, Tetrahedron **29**, 595 (1973).
42. B. F. Smart and G. S. Reddy, *J. Am. Chem. Soc.* **98**, 5593 (1976).
43. R. D. Chambers, A. Parkin, and R. S. Matthews, *J. Chem. Soc. Perkin Trans. I* 2107 (1976).
44. V. I. Mamatyuk, Yu. V. Pozdnyakovich, B. G. Oksenenko, V. I. Buraev, E. V. Malykhin, and V. D. Shteingarts, *Izv. AN SSSR, Ser. Khim.* 1626 (1975).
45. R. J. Spear, D. A. Forsyth, and G. A. Olah, *J. Am. Chem. Soc.* **98**, 2493 (1976).

46. M. G. Hogben and W. A. G. Graham, *J. Am. Chem. Soc.* **91**, 283 (1969).
47. I. Brown and D. W. Davies, *J. Chem. Soc. Chem. Commun.* 939 (1972).
48. G. A. Olah, Y. K. Mo, and Y. Halpern, *J. Am. Chem. Soc.* **94**, 3351 (1972).
49. G. Frenking, S. Fau, C. M. Marchand, and H. Grützmacher, *J. Am. Chem. Soc.* **119**, 6648 (1997).
50. G. A. Olah and Y. K. Mo, *J. Org. Chem.* **38**, 3212 (1973).
51. A. A. Shtark, Yu. V. Pozdnyakovich, and V. D. Shteingarts, *Zh. Org. Khim.* **13**, 1671 (1977).
52. A. N. Detsina and V. A. Koptyg, *Zh. Org. Khim.* **8**, 2158 (1972).
53. G. K. S. Prakash, L. Heiliger, and G. A. Olah, *J. Fluorine Chem.* **49**, 33 (1990).
54. P. N. Dobronravov, T. V. Chuikova, Yu. V. Pozdnyakovich, and V. D. Shteingarts, *Zh. Org. Khim.* **16**, 796 (1980).
55. M. V. Galakhov, V. A. Petrov, G. G. Belen'kii, L. S. German, E. I. Fedin, V. F. Snegirev, and V. I. Bakhmutov, *Izv. AN SSSR, Ser. Khim.* 1072 (1986); S. D. Chepik, M. V. Galakhov, G. G. Belen'kii, V. A. Petrov, L. S. German, and V. I. Bakhmutov, *Izv. AN SSSR, Ser. Khim.* 2761 (1986)
56. R. Chambers, M. Hole, W. Musgrave, R. A. Storey, and M. B. Iddon, *J. Chem. Soc. C* 2331 (1966); D. Price, H. Suschitzky, and J. I. Hollies, *J. Chem. Soc. C* 1967 (1969); R. A. Fletton, R. D. Lapper, and L. F. Thomas, *Chem. Commun.* 1049 (1969); L. S. Kobrina, V. D. Shteingarts, and L. N. Shchegoleva, *Izv. Sib. Otd. AN SSSR, Ser. Khim. Nauk* **1**(14), 68 (1974).
57. B. G. Oksenenko, V. D. Shteingarts, and G. G. Yakobson, *Zh. Org. Khim.* **7**, 745 (1971).
58. G. A. Olah, M. B. Comisarov, E. Namanworth, and B. Ramsey, *J. Am. Chem. Soc.* **89**, 5254 (1967); G. A. Olah, M. B. Comisarov, and C. J. Kim, *J. Am. Chem. Soc.* **91**, 1458 (1969).
59. N. N. Neronova, *Zh. Struct. Khim.* 9 (1968); N. Boden, J. W. Emsley, J. Feeney, and L. H. Sutcliff, *Mol. Phys.* **8**, 467 (1964).
60. N. C. Craig, G. F. Fleming, and J. Pranata, *J. Am. Chem. Soc.* **107**, 7324 (1985).
61. G. A. Olah and J. M. Bollinger, *J. Am. Chem. Soc.* **90**, 947 (1968).
62. R. J. Blint, T. B. McMahon, and J. L. Beauchamp, *J. Am. Chem. Soc.* **96**, 1269 (1974).
63. G. A. Olah, G. Rasul, L. Heiliger, and G. K. S. Prakash, *J. Am. Chem. Soc.* **118**, 3580 (1996).
64. N. C. Deno, J. J. Jaruzelski, and A. Schriesheim, *J. Am. Chem. Soc.* **77**, 3044 (1955).
65. T. N. Gerasimova, A. G. Badashkeeva, E. G. Lubenets, V. A. Barkhash, and N. N. Vorozhtsov, Jr., *Zh. Org. Khim.* **5**, 2199 (1969).
66. R. Filler, C.-S. Wang, M. A. McKinney, and F. N. Miller, *J. Am. Chem. Soc.* **89**, 1026 (1967).
67. D. T. Clark and D. M. Lilley, *J. Chem. Soc. Chem. Commun.* 603 (1970); L. D. Kispert, C. U. Pittman, D. Le Allison, T. B. Patierson, C. W. Gilbert, C. Mains, and I. Prather, *J. Am. Chem. Soc.* **94**, 5979 (1972); W. L. Hehre and P. C. Hiberty, *J. Am. Chem. Soc.* **96**, 7163. (1974).
68. B. G. Oksenenko, *The Study of Properties of Fluorinated Anthracenes, Dihydroanthracenes, and Anthracenium Ions*, Thesis, Novosibirsk Institute of Organic Chemistry, Siberian Division of the USSR Academy of Sciences, Novosibirsk, 1974.
69. V. G. Shubin, D. V. Korchagina, B.G. Derendyaev, G. I. Borodkin, and V. A. Koptyug, *Zh. Org. Khim.* **9**, 1041 (1973).

70. A. A. Shtark, Yu. V. Pozdnyakovich, and V. D. Shteingarts, *Zh. Org. Khim.* **13**, 1671 (1977).
71. P. N. Dobronravov, T. V. Chuikova, Yu. V. Pozdnyakovich, and V. D. Shteingarts, *Zh. Org. Khim.* **16**, 796 (1980).
72. A. G. Harrison, P. Kebarle, and F. P. Lossing, *J. Am. Chem. Soc.* **83**, 777 (1961).
73. R. W. Taft, R. H. Martin, and F. W. Lampe, *J. Am. Chem. Soc.* **87**, 2492 (1965).
74. H. Volz and W. D. Mayer, *Lieb. Ann.* 835 (1975).
75. (a) F. Bernardi, I. G. Csizmadi, and N. D. Epiotis, *Tetrahedron* **31**, 3085 (1975); (b) F. Bernardi, A. Mangini, N. D. Epiotis, J. R. Larson, and S. Shaik, *J. Am. Chem. Soc.* **99**, 7465 (1977); (c) F. Bernardi, I. A. Bottoni, and N. D. Epiotis, *J. Am. Chem. Soc.* **100**, 7205 (1978); (d) F. Bernardi, I. A. Bottoni, and A. Venturini, *J. Am. Chem. Soc.* **108**, 5395 (1986); (e) J. Kapp, C. Schade, A. D. El-Nahasa, and P. v. R. Schleyer, *Angew. Chem. Int. Ed. Engl.* **35**, 2236 (1996).
76. V. F. Loktev and V. G. Shubin, *Izv. Sib. Otd. AN SSSR, Ser. Khim. Nauk*, **1**(2), 108 (1981).
77. P. N. Dobronravov and V. D. Shteingarts, *Zh. Org. Khim.* **13**, 461 (1977).
78. V. F. Loktev, D. V. Korchagina, and V. G. Shubin, *Izv. Sib. Otd. AN SSSR, Ser. Khim. Nauk*, **1**(2), 53 (1981).
79. A. S. Kende and P. MacGergor, *Chem. Ind.* 460 (1962).
80. G. A. Olah, H. C. Lin, and Y. K. Mo, *J. Am. Chem. Soc.* **94**, 3367 (1972).
81. A. N. Detsina and V. A. Koptyg, *Zh. Org. Khim.* **8**, 2215 (1972).
82. A. A. Shtark and V. D. Shteingarts, *Zh. Org. Khim.* **22**, 831 (1986).
83. B. A. Selivanov, Yu. V. Pozdnyakovich, T. V. Chuikova, O. I. Osina, and V. D. Shteingarts, *Zh. Org. Khim.* **16**, 1910 (1980).
84. Yu. V. Pozdnyakovich, V. V. Bardin, A. A. Shtark, and V. D. Shteingarts, *Zh. Org. Khim.* **15**, 656 (1979).
85. V. V. Bardin, G. G. Furin, and G. G. Yakobson, *J. Fluorine Chem.* **14**, 455 (1979).
86. V. M. Karpov, T. V. Mezhenkova, V. E. Platonov, and G. G. Yakobson, *J. Fluorine Chem.* **28**, 115 (1985); V. M. Karpov, T. V. Mezhenkova, V. E. Platonov, and G. G. Yakobson, *Izv. AN SSSR, Ser. Khim.* 1918 (1987); V. M. Karpov, T. V. Mezhenkova, V. E. Platonov, and G. G. Yakobson, *Izv. AN SSSR, Ser. Khim.* 645 (1990); V. M. Karpov, T. V. Mezhenkova, V. E. Platonov, and G. G. Yakobson, *Izv. AN SSSR, Ser. Khim.* 1114 (1990).
87. Yu. V. Pozdnyakovich and V. D. Shteingarts, *J. Fluorine Chem.* **4**, 317 (1974).
88. L. S. Kobrina and V. D. Shteingarts, *J. Fluorine Chem.* **41**, 111 (1988).
89. K. O. Christe, X. Zhang, R. Bau, J. Hegge, G. A. Olah, G. K. S. Prakash, and J. A. Sheehy, *J. Am. Chem. Soc.* **122**, 481 (2000).
90. Yu. V. Pozdnyakovich, T. V. Chuikova, V. V. Bardin, and V. D. Shteingarts, *Zh. Org. Khim.* **12**, 690 (1975).
91. Yu. V. Pozdnyakovich and V. D. Shteingarts, *Zh. Org. Khim.* **13**, 1911 (1976).
92. P. N. Dobronravov and V. D. Shteingarts, *Zh. Org. Khim.* **17**, 2245 (1981).
93. Yu. V. Pozdnyakovich and V. D. Shteingarts, *Zh. Org. Khim.* **7**, 2002 (1971).
94. G. A. Olah, *Angew. Chem. Int. Ed. Engl.* **32**, 767 (1993).
95. M. Vol'pin, I. Akhrem, and A. Orlinkov, *New. J. Chem.* **13**, 771 (1989); G. A. Olah, G. Rasul, A. K. Yudin, G. K. S. Prakash, A. L. Chistyakov, I. V. Stankevich, I. S. Akhrem, N. P. Gambaryan, M. E. Vol'pin, *J. Am. Chem. Soc.* **118**, 1446 (1996).

8

CARBOCATIONS, FAST REARRANGEMENT REACTIONS, AND THE ISOTOPIC PERTURBATION METHOD

Martin Saunders

Department of Chemistry
Yale University
New Haven, Connecticut

Olga Kronja

Faculty of Pharmacy and Biochemistry
University of Zagreb
Zagreb, Croatia

Carbocation Chemistry, Edited by George A. Olah and G. K. Surya Prakash
ISBN 0-471-28490-4

8.1 INTRODUCTION

Carbocations were originally proposed by Meerwein[1] in order to account for a number of molecular rearrangements that occurred in reactions of terpenes and other classes of compounds. The term *carbocation* is used to describe cations containing only carbon and hydrogen. Trivalent (classical) and hypervalent (nonclassical) ions are both included in this description.

The high reactivity of these species with nucleophiles is the main reason why they have not been studied with the classical methods of organic chemistry (isolation, crystallization, degradation, melting or boiling point determination). They were therefore studied through solvolytic reactions where they were presumed to be intermediates. Reaction rate and product studies, kinetic isotope effect determinations, trapping experiments, and other assays were employed. The application of solution kinetic studies by Ingold, Winstein, and many others established our principal ideas regarding the structures, relative energies, and reactivity of these carbocations. Additional energetic and reaction rate information was obtained through mass spectroscopic studies in the gas phase. Mass spectroscopic methods have the disadvantage that they do not yield structural information directly.[2]

About 50 years ago, it became possible to produce stable solutions of simple tertiary and some secondary carbocations due to the work of Olah and a group at Shell in Amsterdam. NMR spectroscopy of these solutions became the main tool used in studying these species.

The first published NMR spectrum of a carbocation, presented in Figure 8.1, was the proton spectrum of the heptamethylbenzenonium ion (**1**) published in 1958. Ion **1**, which was generated by exhaustive Friedel–Crafts methylation of toluene, is stable in aqueous acid.[3] The line broadening that appears in the spectrum at higher temperatures indicates rapid rearrangement. It was concluded that a 1,2-methide shift was occurring rapidly.

The first spectrum of a non-classical carbocation was that of norbornadienyl (**2**).[4] The idea that in some carbocations the neighboring group participation can occur originates from the fact that their solvolysis rates are remarkably enhanced in comparison with the rates of the corresponding compound lacking the neighboring group. Such a case is the acetolysis of norbornadienyl tosylate calculated to be 10^{14} faster than that of the corresponding saturated norbornyl analog.[5] In order

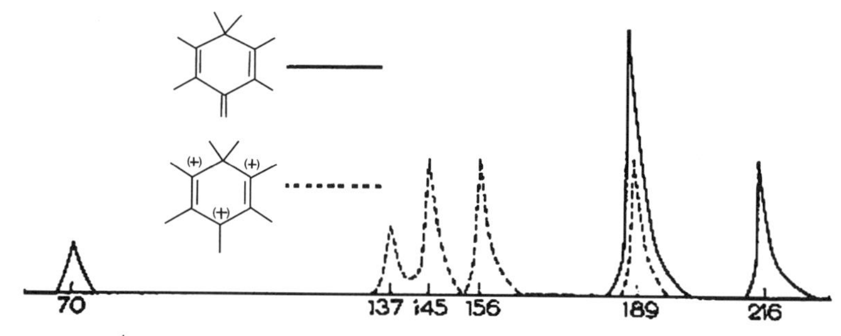

Figure 8.1 ^{1}H NMR spectrum of the heptamethylbenzenonium ion and of its precursor.

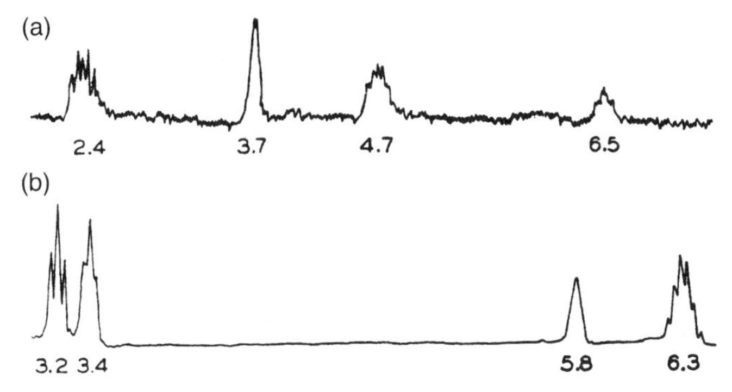

Figure 8.2 ^{1}H NMR spectrum of norbornadienyl tetrafluoroborate (a) and chloride (b) in sulfur dioxide.

to rationalize these observations, the concept of the nonclassical carbocation intermediate was introduced. However, conclusions regarding the exact structure of the nonclassical ion could not be based on the solvolysis results, since several candidates for the structure were often consistent with solvolytic behavior. Such a rapid solvolysis rate suggested that the carbocation intermediate should be stable. It was found possible to prepare the norbornadienyl cation by treatment of the corresponding chloride with $AgBF_4$ in SO_2 solution (Scheme 1). This procedure yielded the

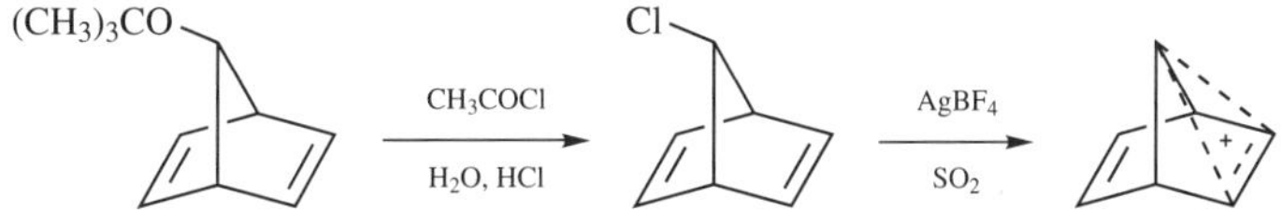

$(CH_3)_3CO$ —[CH_3COCl / H_2O, HCl]→ Cl —[$AgBF_4$ / SO_2]→ **2**

Scheme 1

cation as the tetrafluorborate salt, stable at temperatures below −50°C. The spectra of the parent chloride (below) and the tetrafluoroborate salt are presented at Figure 8.2. In the spectrum of the chloride, two olefin areas with slightly different frequencies can be observed that correspond to the double bonds *syn* and *anti* to the chlorine. The olefin peaks in the carbocation are split much more, suggesting that they are magnetically quite different; that is, two olefinic protons are considerably shifted upfield in comparison with the olefinic protons of the parent chloride. The structure is constituent with structure **2**, in which participation of one double bond occurs on giving an unsymmetric nonclassical carbocation.

8.2 REARRANGEMENTS VIA PROTONATED CYCLOPROPANE INTERMEDIATES

Many rearrangements involved in carbocations can be rationalized if protonated cyclopropanes are considered as reaction intermediates.[6] The common Wagner–Meerwein 1,2-methide shift must go through the geometry of a corner-protonated cyclopropane.

The simplest carbocation prepared in stable solution is the isopropyl cation **3** by Olah et al.[7] Saunders and Hagen[8] found that line broadening caused by exchange of the methine hydrogen with the six methyl protons was seen when solutions of this ion were observed at temperatures near 0°C. The barrier for that process is 16.4 kcal/mol. That observation is consistent with the mechanism I (Scheme 2). First, the methyl hydrogen is shifted to the central carbon to yield the primary propyl cation as an intermediate. If the hydrogen, which was originally the methine hydrogen, moves to the methyl carbon, exchange occurs. The experiments carried out with ^{13}C labeling revealed that the carbons interchange with a similar rate to the hydrogens. Those two processes were directly compared starting with dilabeled isopropyl cation (2-propyl cation-2-D-2-^{13}C), and it was found that the rate of the hydrogen scrambling and the carbons scrambling is almost the same, indicating similar barriers.[9]

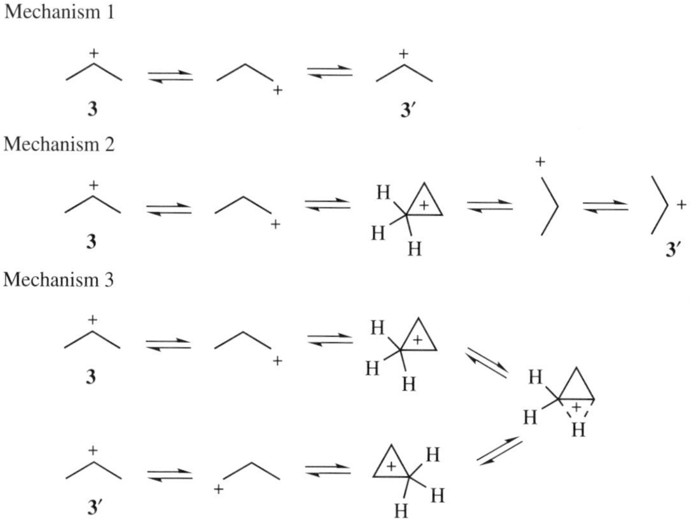

Mechanism 1

3 3′

Mechanism 2

3 3′

Mechanism 3

3 3′

Scheme 2

Carbon scrambling can be accounted for if the protonated cyclopropane intermediate, obtained by closure of the three membered ring from the primary ion, is presumed. The intermediate structure is a corner-protonated cyclopropane. Carbon scrambling occurs if the ring opens by breaking the C2–C3 bond, or by migration of the hydrogen to other corner via the edge-protonated cyclopropane intermediate (mechanism 3, Scheme 2). All these conclusions were supported by quantum-mechanical calculations by Pople and Schleyer.[10] They found that the corner-protonated cyclopropane is the most stable intermediate. The route from the primary ion to the protonated cyclopropane is a downhill process, while going to the edge-protonated ion "cost" only 1 kcal/mol.

The next carbocation to consider in sequence is the 2-butyl cation **4**. This ion was first prepared in stable solution at Yale with SbF_5 in SO_2ClF.[11] The proton and carbon spectra show that the ion undergoes a rapid degenerate 1,2-hydride shift.

That process is so fast that the three inside hydrogens give a time-averaged single peak as do the six methyl protons. The inside and the outside carbons, respectively, are averaged yielding a carbon spectrum of only two peaks. The free energy of activation for this hydride shift must be less than 2.4 kcal/mol since the process could not be slowed down out even at very low temperatures to cause line broadening. At temperatures above −100 °C, a second process has been seen that interchanges the set of three inside hydrogens with the set of six methyl hydrogens. The barrier of that process is 7.5 kcal/mol. A mechanism that can account for these results is presented in Scheme 3. The first step is closure to the protonated cyclopropane, followed by a corner-to-corner proton shift and then reopening to *sec*-butyl cation. The overall effect is scrambling of the protons and the carbons.

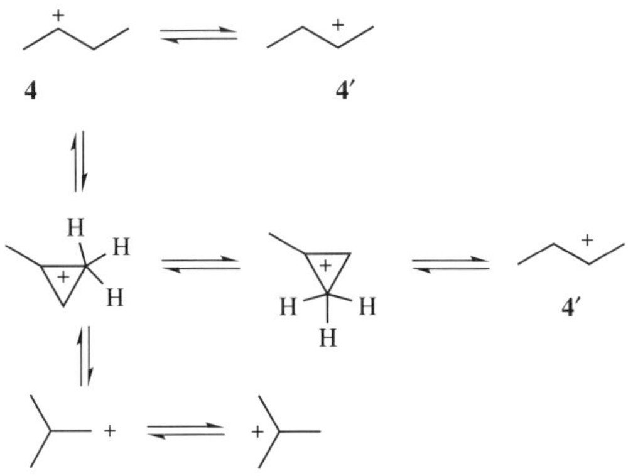

4 **4′**

H H H H H H **4′**

Scheme 3

When the cation is heated to −40°C, it very rapidly goes to *tert*-butyl, presumably via isobutyl cation. This process is irreversible and the barrier for the return to sec-butyl cation is about 30 kcal/mol.

tert-Amyl cation (**5**) (2-methyl-2-butyl cation) is an extremely stable ion.[12] When its solution in SbF_5 is heated to well over 100 °C and then cooled down again, the spectrum is unchanged. Warming a cold solution to about −40°C caused broadening of the two methyl peaks. Eventually these peaks overlap and coalesce. At the same time, there was no visible change of the methylene peak in the spectrum. By further heating of the cation solution to 80 °C, a second process was observed that interchanges all hydrogens. Mechanisms consistent with these observations are presented in Scheme 4. The first step is 1,2-hydride shift uphill to the secondary ion, followed by 1,2-methide shift through the corner-protonated cyclopropane. Reopening the cyclopropane ring permits return to *t*-amyl cation. This overall process interchanges only the methyl protons, and is in accord with the lower barrier process (15.4 kcal/mol). The second process, which interchanges all protons, can be accounted for if a corner-to-corner proton shift is considered, followed by opening into the unbranched secondary 2-pentyl cation. If the unbranched ion closes again and returns to *t*-amyl, it interchanges methylene with the methyl protons. The barrier of this process was found to be 18.2 kcal/mol. If the slower process is followed

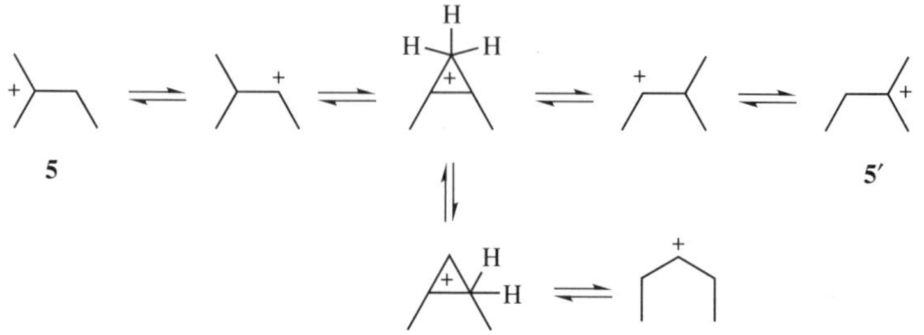

5 5′

Scheme 4

carefully, it could be discovered that this process does not scramble the carbons from the inside to the outside positions. We were curious to see whether a third process occurs and observed the spectrum of the ^{13}C-labeled *t*-amyl cation. We found that interchange of the inside and outside carbons occurs with essentially the same barrier as for the slower process.[13]

Many attempts have been made to prepare stable solutions of the cyclohexyl cation, but all yielded only methylcyclopentyl cation **6**. Ion **6** is an extremely stable ion,[14] and its behavior is analogous to that of the *t*-amyl ion. The hydrogens interchange via a lower barrier process, while all hydrogens are mixed via a barrier of 18.2 kcal/mol, which is the same as the barrier for the slower process in the *t*-amyl cation. Essentially, those two processes are similar (Scheme 5). After the uphill

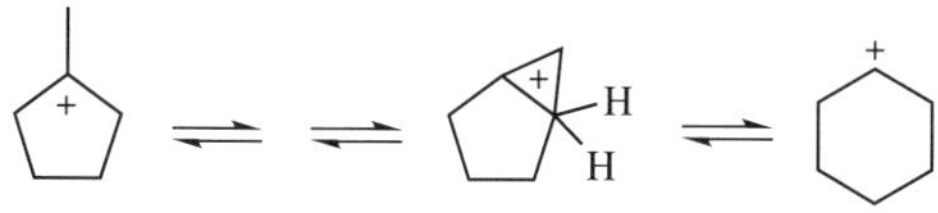

6

Scheme 5

hydrogen migration to the secondary cation and formation of the cyclopropane derivative, corner-to corner migration of the hydrogen occurs to an ion that can than open to the secondary cyclohexyl cation. When this process is reversed to return to methylcyclopentyl, the carbons and hydrogens are scrambled.

These mechanistic steps account for a general phenomenon that puzzled workers in carbocation chemistry for a long time. How does interconversion of branched and unbranched structures occur? A number of mechanisms have been suggested. We believe that these conversions most probably go through protonated cyclopropane intermediates as shown. Catalysts that generate carbocation intermediates can produce change in the branching of hydrocarbons.[15]

8.3 1,3-HYDRIDE SHIFTS AND SHIFTS TO MORE DISTANT CARBONS

The 2,4-dimethyl-2-pentyl cation **7** is a readily prepared tertiary carbocation. At low temperatures the spectrum is completely consistent with its structure. However, when the solution of the cation was warmed, broadening of the methyl peaks occurred, indicating that a process that interchanges the methyl groups takes place.[16] The barrier was found to be 7.5 kcal/mol. One way of accounting for this is to propose a 1,3-hydride shift interchanging the functionality of the two different kinds of methyls (Scheme 6, mechanism 1). But this is not the only way.

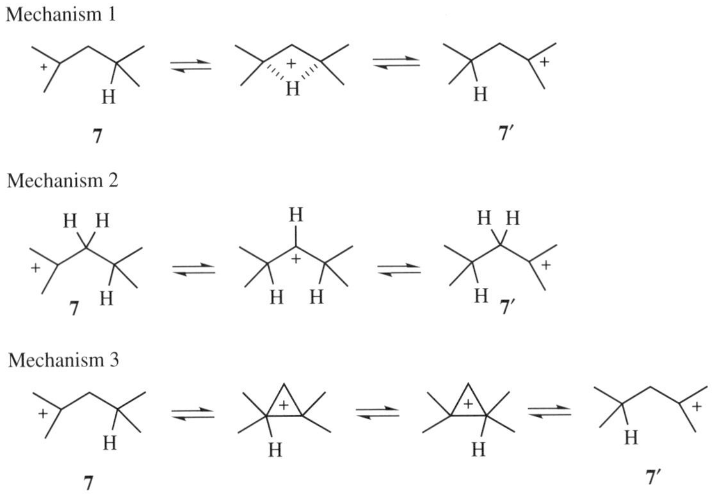

Mechanism 1

7 ⇌ ⇌ 7′

Mechanism 2

7 ⇌ ⇌ 7′

Mechanism 3

7 ⇌ ⇌ ⇌ 7′

Scheme 6

Another possibility is a series of 1,2 shifts (mechanism 2). The first 1,2 shift would be uphill, yielding a secondary carbocation intermediate. A further 1,2 shift would give a rearranged tertiary cation. Two observations clearly rule out this second mechanism: (1) a mechanism that goes through the secondary cation should have the barrier similar to that obtained for the amyl cation, which is ~15 kcal/mol—the barrier of this process is only 8.5 kcal/mol; and (2) the methylene does not show any broadening under conditions where the methyl peaks are broadened. The reaction may also proceed via mechanism 3, in which formation of the protonated cyclopropane intermediate is followed by 1,2-proton shift. The methylene is not involved so it is consistent with the observation that it does not go through the secondary ion and does not scramble the methylene protons. These two mechanisms yield the same result, but we estimate that inclusion of a protonated cyclopropane would require more energy than that found experimentally. Ab initio calculations also support the mechanism with a direct 1,3-hydride shift.

More distant hydride shifts were also studied. 1,4[17] and 1,5-hydride shifts[18] were found to occur in compounds that are homologous to **7**. These rearrangements are also too fast to go through secondary cation intermediates. If those shifts went via mechanism analogous to 3, the intermediate structure would be protonated cyclobutane and protonated cyclopentane intermediates, respectively, which are predicted to be very high in energy. Therefore these rearrangements probably proceed via mechanisms analogous to mechanism 1 in Scheme 6.

8.4 ALLYL CATIONS

Solvolysis studies of the three isomeric chlorodimethylcyclopropanes revealed that ring opening occurs and that all three give derivatives of *cis* and *trans* 2-pentene (Scheme 7). Since the cyclopropyl cation is very high in energy, and having in mind that ring opening occurred in solvolysis, it is reasonable to conclude that the solvolysis is concerted with ring opening to stable allyl cations. There are three isomers of the allyl cation: *cis–cis*, *cis–trans*, and *trans–trans*. These can react with the solvent and give the products isolated. When the all-*cis* dimethylchlorocyclopropane was ionized, only the *cis–cis* allyl cation was formed. On warming, this *cis,cis* ion is converted first to *cis,trans* and than to *trans,trans*. (Fig. 8.3).[19] The barrier for these processes are 17.5 and 24.0 kcal/mol, respectively, which represent rotation barriers around the partial double bond in the allyl cation.

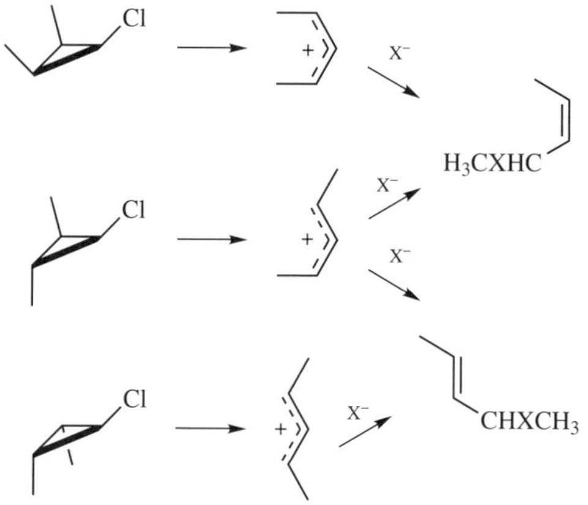

Cl
X⁻
H_3CXHC
X⁻
Cl
X⁻
Cl
X⁻
$CHXCH_3$

Scheme 7

Another extremely stable allyl cation is the cyclopentenyl cation (**8**).[20] When a solution of this ion is heated above 100 °C and cooled down again, the original spectrum is obtained. The spectrum taken at room temperature is presented in Figure 8.4. At higher temperatures, the hydrogen peaks broaden (observed at 85–112 °C) and the barrier for the hydrogen exchange of is found to be 19 kcal/mol. This could occur if the hydride shift happens to give the nonconjugated ion. Another hydride shift can return to a conjugated structure with the atoms scrambled. Even though theoretical calculations indicated much higher barrier, a new, independent experiment produced exactly the same barrier for the hydrogen migration.[21]

8

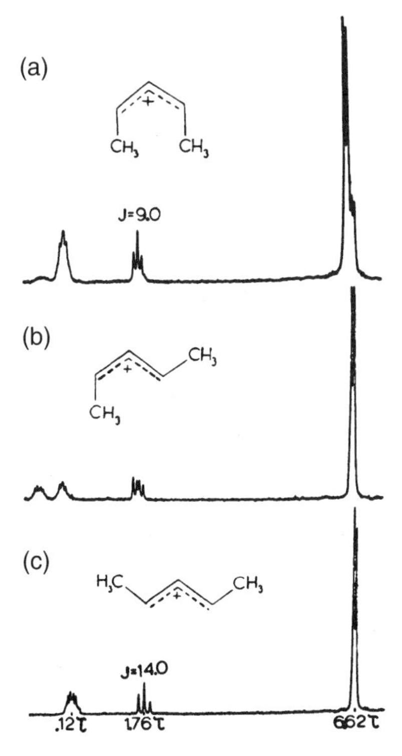

Figure 8.3 ^1H NMR spectra of cations generated from 2,3-dimethylcyclopropyl chloride: (a) initially formed *cis,cis*-pentenyl cation; (b) *cis,trans*-pententl cation; (c) the ultimate most stable *trans,trans*-pentenyl cation.

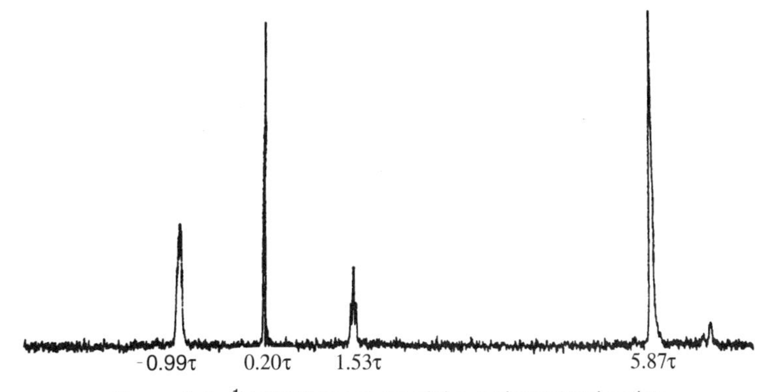

Figure 8.4 ^1H NMR spectrum of the cyclopentenyl cation.

8.5 ISOTOPIC PERTURBATION IN CARBOCATIONS

CH_2D CH_2D

9-d_1 $\rightleftharpoons$ **9′**-d_1

The 2,3-dimethylbutyl cation **9** is a very stable cation that undergoes a rapid degenerate 1,2-hydride shift. On the basis of line broadening of the peak due to the central carbons at lower temperature, a barrier of 3.5 kcal/mol was obtained. Therefore, a single averaged peak for the methyl protons and a peak for the methyl carbons, respectively, were found in the spectra. When one deuterium was introduced, the unexpected spectrum, presented in Figure 8.5a, was obtained.[22] The peak in the center corresponds to some starting material in which the methine group is deuterated, so the deuterium goes back and forth in a degenerate equilibrium. Beside that singlet, there are two doublets, due to the average coupling between the methine hydrogen and the methyl hydrogens. In order to show that this is the case, the methine hydrogen was irradiated causing collapse of the doublets (Fig. 8.5b). The splitting of the signals is due to nonequivalent interchanging structures. In one structure, the deuterium is next to the positive charge, and in the other the deuterium is away from the positive charge. Hyperconjugation should lower the vibration frequencies of hydrogens on the methyls next to the positive charge. A deuterium would therefore prefer to be on a methyl away from the charge. Through introduction of the isotope, the degeneracy of the process is disturbed and the equilibrium constant is changed from unity. The two methyls on one side are therefore at an average shift different from the two methyls on the other side. The magnitude of the splitting depends on the equilibrium constant, which is a function of the temperature; specifically, the splitting decreases with higher temperatures. This was the first report on isotopic perturbation of equilibrium.

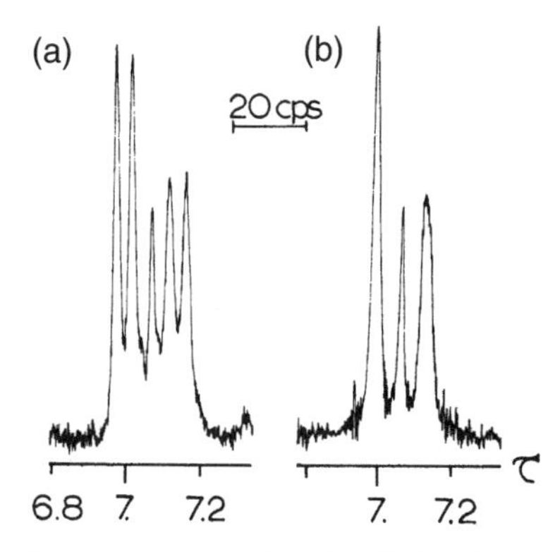

Figure 8.5 (a) Spectrum of monodeuterated 2,3-dimethylbutyl cation and (b) the decoupled ^1H NMR spectrum of the monodeuterated 2,3-dimethylbutyl cation in SbF_5/SO_2ClF at −56 °C.

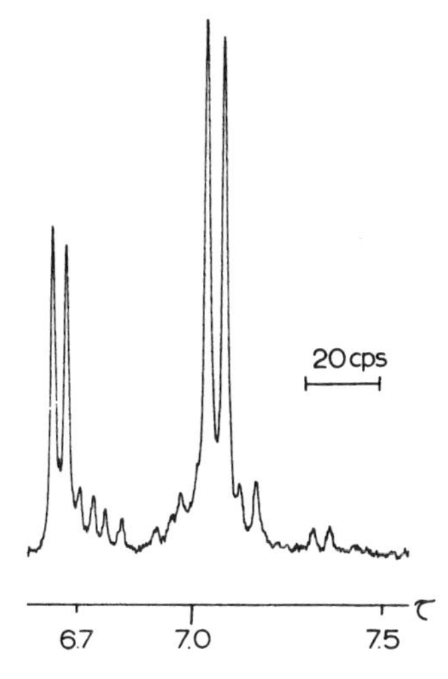

Figure 8.6 ^{1}H NMR spectrum of hexadeuterated 2,3-dimethylbutyl cation in SbF_5/SO_2ClF at −60°C.

The isotopomer **9** with two deuteromethyl groups introduced on one side was also prepared (**9**-d_6).[23] The spectrum is presented in Figure 8.6. The methyl groups interchange rapidly. Again, there are two doublets in the spectrum due to the coupling the methyls with the inside hydrogen, but the splitting is considerably bigger. The larger peak due to the compound with one deuterated methyl on each side, and the smaller is due the compound with both deuteriums at one side.

CD_3 H + CD_3 ⇌ + CD_3 H CD_3

9-d_6 **9′**-d_6

The next step was examination of the mixture of isotopomers by means of carbon NMR.[24] A mixture of deuterated ions **9**-d_n (deuteriums restricted to no more than two methyl groups) was prepared from partially deuterated starting material. Among 15 different isotopomers (Scheme 8), three give a single line for the averaged methine carbon because of symmetric deuteration. Others show splitting of the methine signals, and the magnitude of that splitting depends on the excess of deuteriums on one side over the other side. In order to identify the separate peaks, quantum-chemical calculations were carried out.

The energy minimum structure obtained at the HF 6-31G(d) level of theory (Fig. 8.7) has C_1 symmetry with strong indications of hyperconjugation of hydrogens and carbons. The Gaussian output was used as input to program QUIVER,[25] which calculates the equilibrium isotope effect from the reduced partition

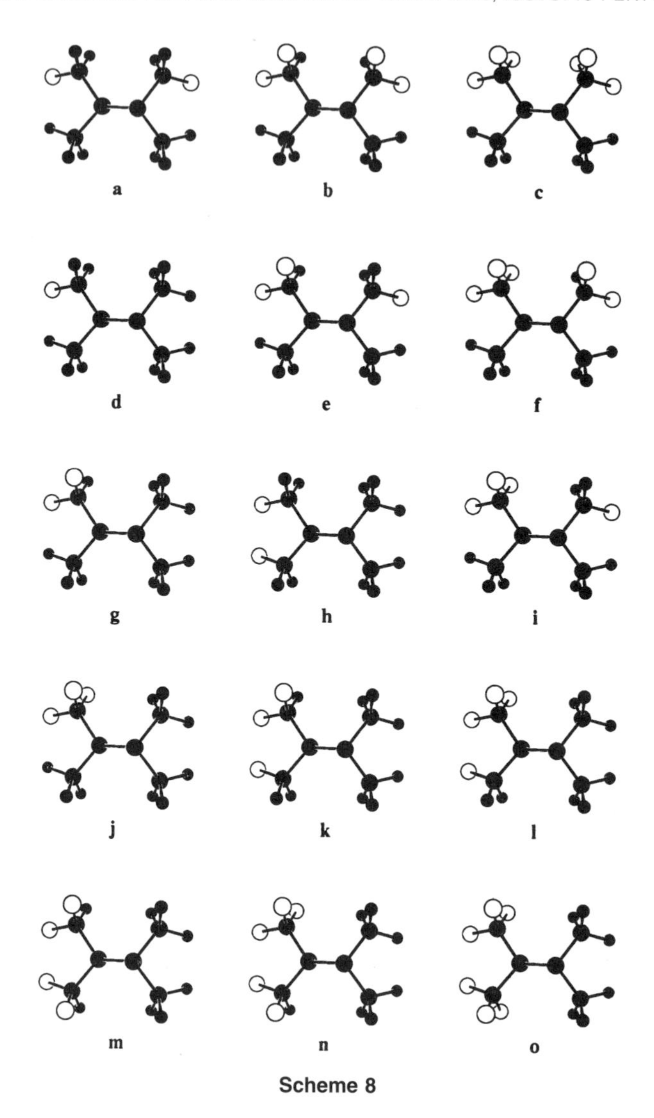

Scheme 8

functions. The calculated data are presented in Figure 8.8a; the experimental spectrum is presented in Figure 8.8b. It is obvious that these spectra match quite well. Once the assignment of the spectrum was done, the experimental equilibrium isotope effect, which is the ratio between the two interchanging structures ([**9**]/[**9′**]), were calculated according to the following equation.

$$K = \frac{\Delta + \delta}{\Delta - \delta}$$

where δ is the observed chemical shift difference splitting between the two interchanging carbons and Δ represents the chemical shift difference in the frozen

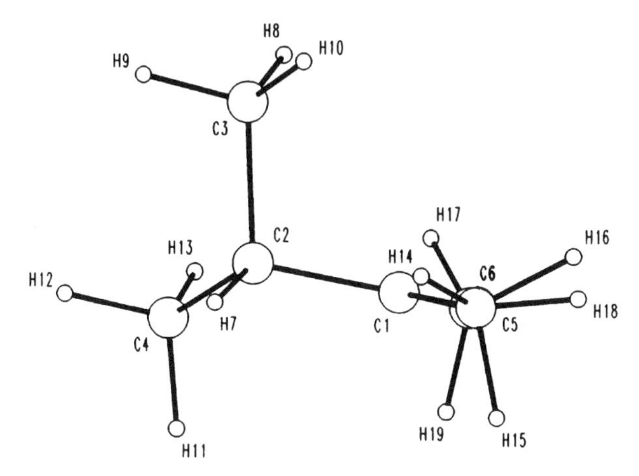

Figure 8.7 The energy minimum structure of 2,3-dimethylbutyl cation at HF/6-31G level of theory.

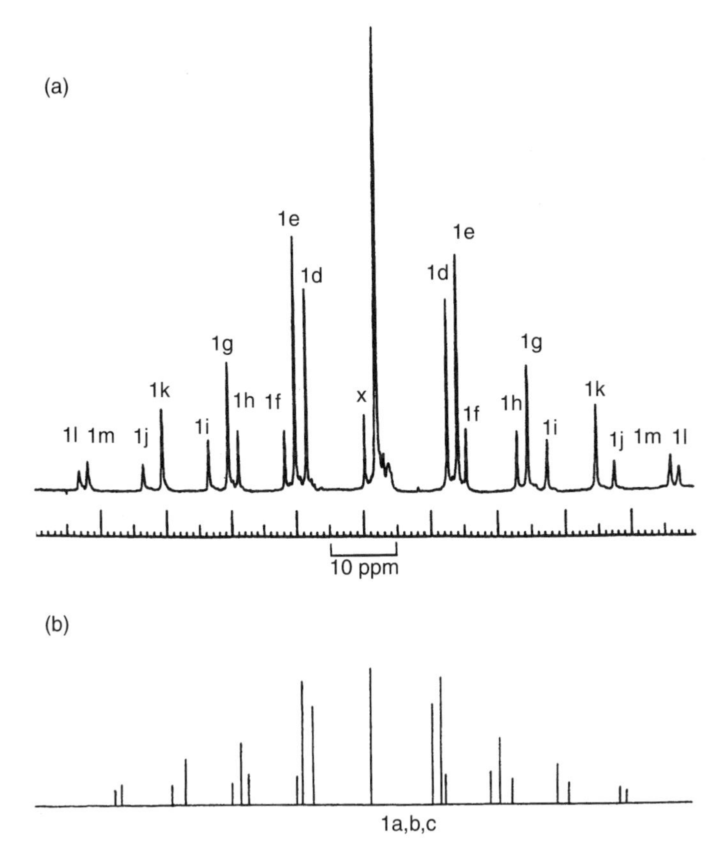

Figure 8.8 (a) Proton decoupled ^{13}C NMR spectra centered at $\delta \sim 200$ ppm of the average cation/methine resonances of the mixtrue of isotopomers of 2,3-dimethylvutyl cation; (b) spectra predicted by QUIVER.

equilibrium of 9 and 9′. The energy difference between the isotopomers was calculated from the values of K.

Attention has also focused on the ^{13}C isotope effect, specifically, to determine the direction of the primary equilibrium if the hydrogen interchanges between ^{12}C and ^{13}C.[26] The difficulty is that only the ^{13}C can be seen, while ^{12}C is invisible. There must be a reference signal to see if the ^{13}C peak is shifted. For the reference, the double labeled cation has been made with a label at C2 and C3 (Scheme 9).

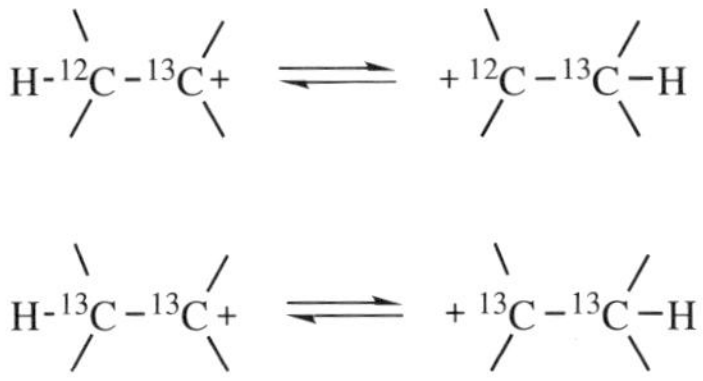

$$\text{H}-{}^{12}\text{C}-{}^{13}\text{C}^{+} \rightleftharpoons {}^{+}\,{}^{12}\text{C}-{}^{13}\text{C}-\text{H}$$

$$\text{H}-{}^{13}\text{C}-{}^{13}\text{C}^{+} \rightleftharpoons {}^{+}\,{}^{13}\text{C}-{}^{13}\text{C}-\text{H}$$

Scheme 9

In that case the hydride shift is a degenerate process and the equilibrium constant is unity. The spectrum of the mixture of the monolabeled and dilabeled ions is presented in Figure 8.9. The inside peak of the monolabeled cation is shifted downfield from peak of the symmetrically double-labeled isotopomer, indicating that ^{13}C prefers to be at the cationic center.

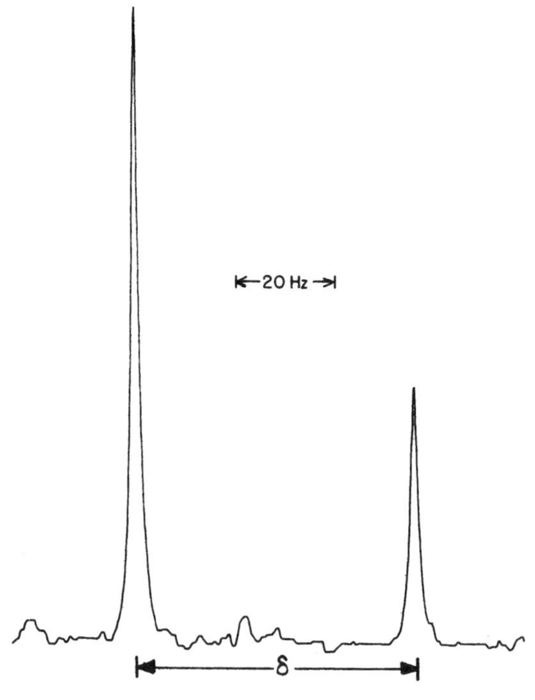

Figure 8.9 ^{13}C NMR spectrum of the central carbons of the 1 : 1 mixture presented in Scheme 9.

In isotopic perturbation, one starts with a degenerate equilibrium and alters it by introducing an isotope to change the equilibrium constant. One can observe the effect in the carbon NMR spectrum and check on it by predicting the splitting from ab initio calculations followed by using the program QUIVER.

The isotopic perturbation method is also used for distinguishing rapidly equilibrating molecules from symmetric molecules. As shown above, in species where rapid degenerate equilibrium occur, deuterium substitution results in characteristically large splitting of the interchanging carbons in ^{13}C NMR. However, if the system posseses true symmetry and not time-averaged molecular symmetry, the splitting is much smaller and is due to the intrinsic isotope effect since the deuterium substitution usually shifts a carbon slightly.

The bicyclo[2.1.1]hexyl cation **10** is a case where there might be a degenerate alkide shift. Another possibility is that the structure in between these two is the minimum-energy structure (Scheme 10). The question here is do we have a single

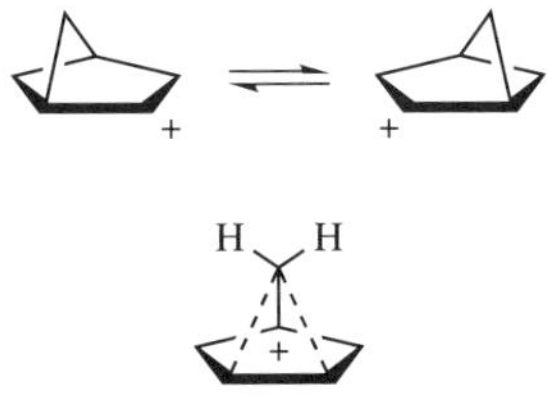

10

Scheme 10

minimum or double minimum energy surface. The spectrum obtained with the methine deuterated compound is presented in Figure 8.10. The deuterium introduced resulted in splitting in the carbon spectrum of 1.2 ppm. This is a far smaller value than predicted for the rapidly equilibrating system but is consistent with the usual upfield shift due to the intrinsic effect of the heavier isotope. Therefore, the conclusion is that there is a single energy minimum with a symmetric bridged, nonclassical, structure.[27]

The most famous case where this question arises is that of the norbornyl cation. The problem of deciding whether the norbornyl cations rapidly equilibrating or a nonclassical structure is the most intensively discussed case. Possible structures are presented in Scheme 10; above are the two classical structures and below is the nonclassical symmetrically bridged structure. The difficulty of applying the isotopic perturbation method comes from an additional rapid 6,2-hydride shift, which proceeds over a barrier of 5.9 kcal/mol and results in some line broadening in the spectrum. Thus, in the protio compound the downfield carbon line (due to former C2 and C6) was ~2 ppm wide at −150 °C (Fig. 8.11). When one deuterium was introduced to C2 or when two deuteriums were introduced to C3, no additional splitting or broadening was recorded. Thus the isotopic splitting must be less than 2 ppm. This value is much smaller than the splitting in case of equilibrating classical ions, strongly suggesting the symmetric nonclassical ion as an minimum-energy structure **11** (Scheme 11).[28]

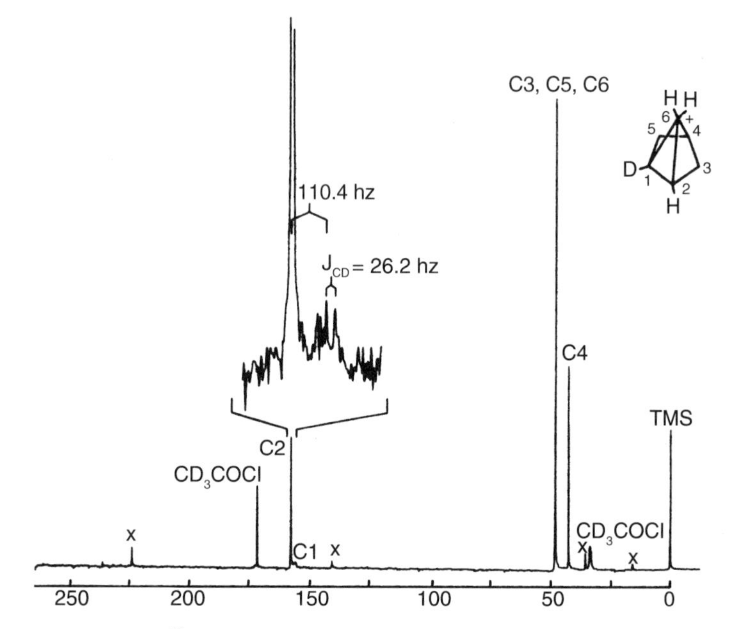

Figure 8.10 ^{13}C NMR spectrum of bicylo[2.1.1]hexyl cation at −115 °C.

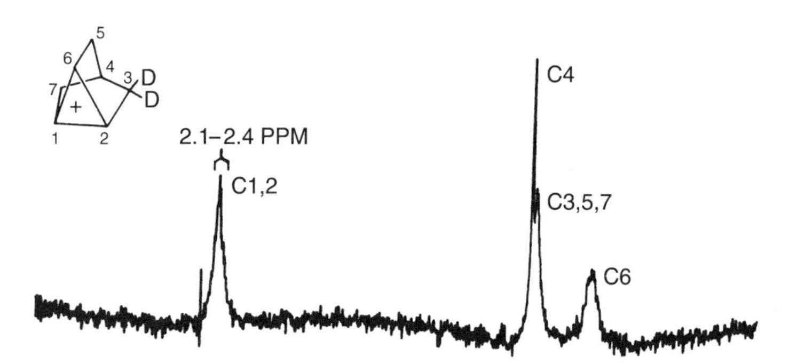

Figure 8.11 ^{13}C NMR spectrum of the norbornyl cation at about −150 °C.

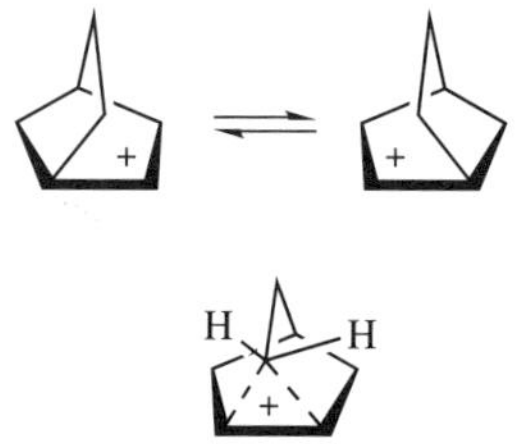

11

Scheme 11

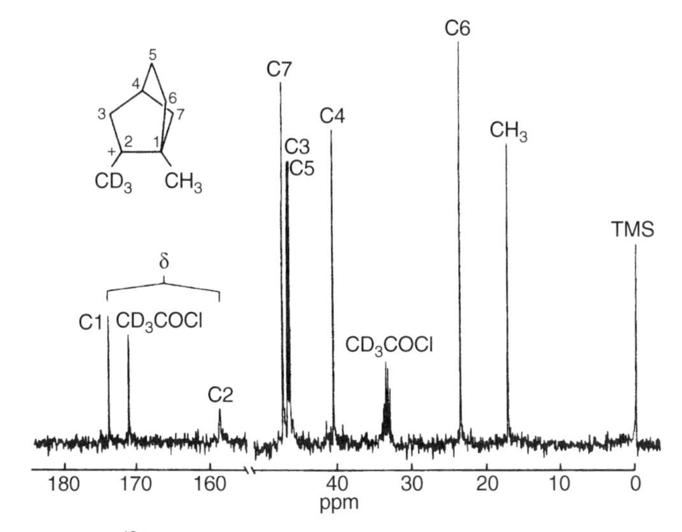

Figure 8.12 ^{13}C NMR spectrum of the dimethyl norbornyl cation at −71 °C.

An interesting case is the structure of the 1,2-dimethylnorbornyl cation **12**.[29] The cation was deuterated on the methylene group and caused an intermediate value of splitting of 24 ppm (Fig. 8.12). This splitting indicates rapidly equilibrating structures. The relatively small value for the splitting was assigned to partial bridging in the ion. This was found later in X-ray structures of similar ions published by Laube.[30] It is worth comparing the magnitude of the splitting observed in the three closely related ions, presented in Scheme 12.[31] Into all three ions the two deuteriums are introduced adjacent to the carbonation center. The splitting of 104 ppm,

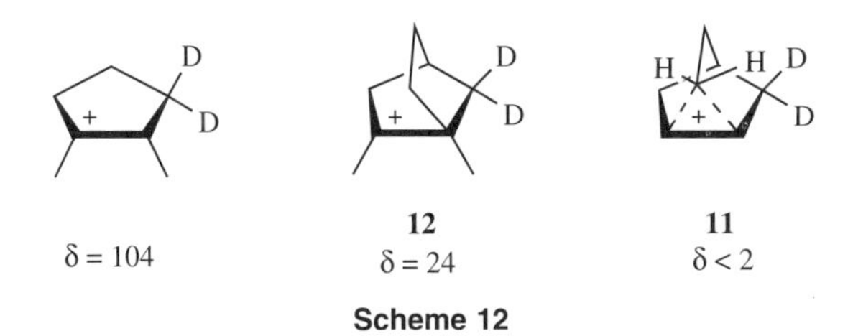

Scheme 12

obtained in the 2,3-dimethylpentyl cation, is the value typical for rapidly equilibrating classical cations. The splitting in the 2,3-dimethylnorbornyl cation was reduced by a factor of ~4, and the splitting in the norbornyl cation was negligible. Pronounced bridging produced a reduction of the isotope effect. The symmetrically bridged structure shows no splitting caused by the isotope.

From the norbornenyl and the nortricyclyl precursor the same cation is obtained in superacid medium. Deuterium was introduced on C3. The deuterated cation showed a splitting of only 0.7 ppm. It was concluded that the cation consists mostly of the nonclassical bridged nortricyclyl cation, and that less than 1% of equilibrating norbornenyl ion might be present, all presented in Scheme 13.[32] Ab initio calculations further support the idea that both cation are minimum-energy structure, and that the nortricyclyl is more stable by 5.7 kcal/mol at MP2/6-31G(d)//6-31G(d) level of theory.

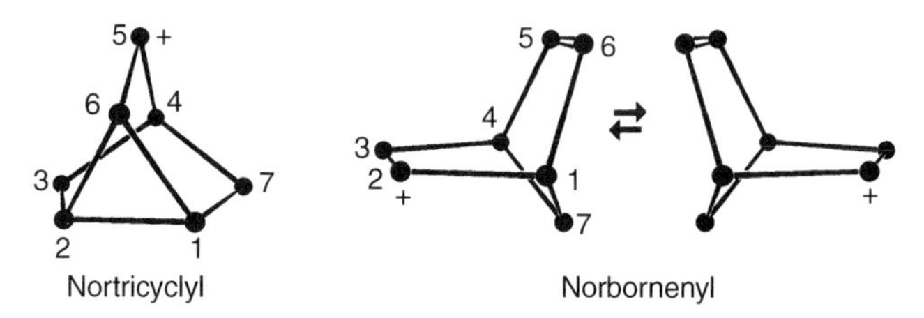

Scheme 13

The isotopic perturbation method was also used to determine the structure of the cyclpropylcarbinyl/bicyclobutonium case. The proton spectrum of $C_4H_7^+$ shows two sets of three protons and one additional proton, while the ^{13}C NMR spectrum consists of a single peak only. These data can be attributed to a single threefold symmetry structure or to the degenerate equilibrium among three ions.[33] An equilibrium between the cyclopropylcarbinyl **13** and bicyclobutonium cation **14** was considered as the most probable:

13 **14**

The spectrum obtained of the mixture of cations monodeuterated at *endo* and *exo* positions of the methylene indicated that *endo* and *exo* isomers produced equilibrium isotope effects of opposite signs (Fig. 8.13).[34] Even though the experimental results suggested that the bicyclobutonium cation is the main component in the mixture, this was not known with certainty, since ab initio calculations showed that the two ions have approximately the same energy.[35] The final conclusion was reached by calculation of the theoretical isotope effect with QUIVER.[25] Calculations showed that bicyclobutonium cation should have different isotope effects for *exo*- and *endo*-deuterated isomers, while the cyclopropylcarbinyl would have about the same isotope effect. Also the magnitude of the effect for the bicyclobutonium cation was also found to quantitatively agree with the experimental value, indicating that the bicyclobutonium cation is the major component of the mixture.

Similar behavior was observed with the methylcyclobutyl cation. The ^{13}C NMR spectrum at $-70°C$ showed complete averaging of the three methylene groups due a fast scrambling. This rearrangement can be frozen out at -154 °C. The puckered methylcyclobutyl cation **15** was suggested to be a dominant structure.[36] Siehl

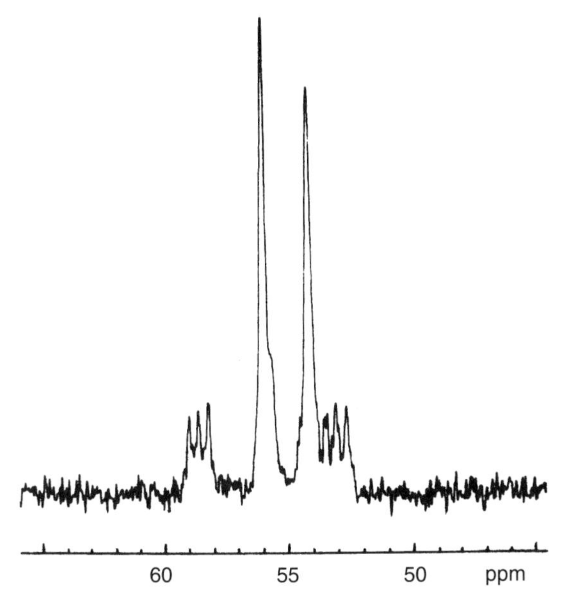

Figure 8.13 ^{13}C NMR spectrum of the monodeuterated cyclobutyl/cyclopropylcatinyl cation.

generated the mono and dideuterated cation and showed that the isotopes strongly influence the equilibrium, indicating rapid interchange of the structures.[37] On the other hand, the spectrum taken with the frozen-out structure (at −153 °C) showed only and intrinsic isotope effect, which is a further evidence that a static structure interchanges at higher temperature. On the basis of these results, the nonplanar 1-methylbicyclobutonium structure **15** was proposed with cyclopropylcarbinyl **16** as a minor species in equilibrium.

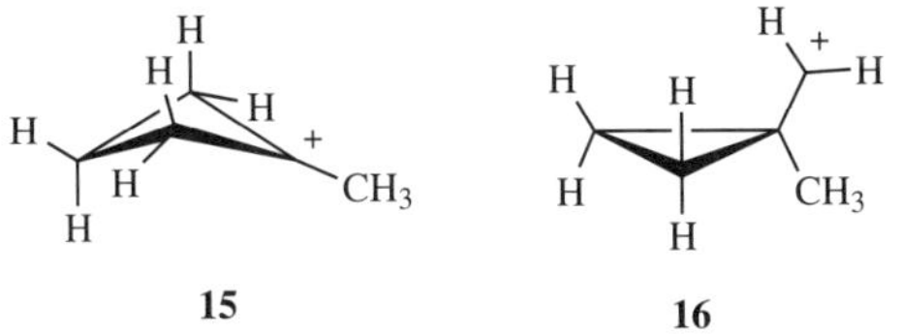

Nonplanarity of the methylcylobutyl was shown using a trideuterated isotopomer of the cation.[38] The deuteriums were introduced onto each methylene carbon in such way that two were on one side and the third was at the other side of the ring (**15**-d_3). This arrangement perturbs the degenerate inversion of the puckered ring and produces an equilibrium isotope effect. The spectrum obtained at −80 °C is presented on Fig. 8.14. It can be seen that two peaks for the methylene protons in the ratio 2 : 1 are shifted downfield with respect to the methylene protons of the unlabeled ion; this result is interpreted as a strong indication of the nonplanar geometry of the ion, since the planar would not have given an isotope effect at all.

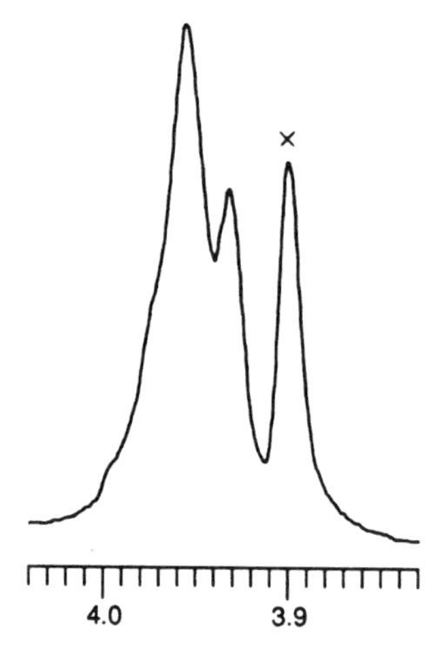

Figure 8.14 ^{13}C NMR spectrum of the trideuterated methylcyclobutyl cation at $-80°C$.

A very challenging case concerns the structure of the nonamethylcyclopentyl cation **17**:

CD_3 α β γ CH_3 +

17-d_3

This is a tertiary cation, which, because of the possible methide shifts, might show an averaged peak for five carbons and an averaged peak for the nine methyl groups. Mayr did obtain this expected spectrum at $-80°C$. However, at $-135°C$ the methyl peak splits into two with the ratio 5 : 4, while the ring carbons exhibit a single sharp peak, which does not show broadening with lower temperatures, indicating a barrier of less than 2 kcal/mol. He explained this by proposing that five methyls remained fixed to the ring carbons, while the other four chased each other around the ring.[39] Introduction of one deuteromethyl group breaks the degeneracy of the circumambulatory migration of the methyl and the peak area of the five split, as shown on Figure 8.15. The pattern of the splitting of the ring carbons (2 : 2 : 5 : 1) indicates that the minimum structure is not symmetrically bridged.[40]

Quantum-chemical calculations [B3LYP/6-31G(d) level] revealed that a minimum-energy structure resembles to a trivalent classical ion, which is stabilized by hyperconjugation, and partial bridging of two quasiaxial methyl groups (Fig. 8.16) occurs. The calculated chemical shifts and the equilibrium isotope effect obtained with QUIVER showed reasonable agreement with the experimental results. This bridging is related to the unique dynamic behavior of the ion.

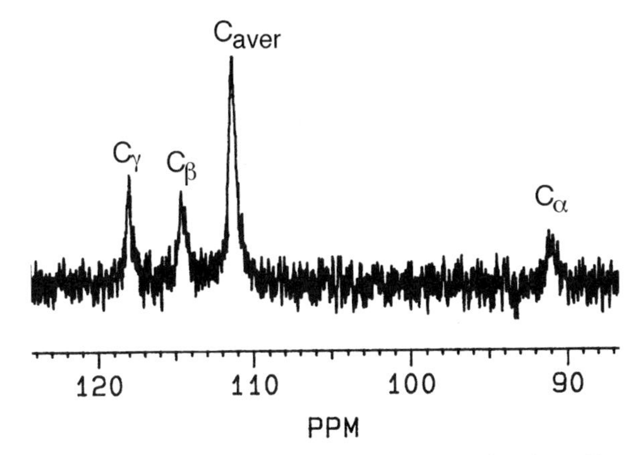

Figure 8.15 ^{13}C NMR spectrum of the nonamethylcfyclopentyl cation with one trideuterated methyl group taken at −133°C.

The isotopic perturbation method was used to determine the mechanism of the biomimetic five-membered ring expansion in isopropylcyclopentyl/cyclopentyl-2-propyl cation, **18A/18B** (Scheme 14). At low temperatures this cation undergoes a fast nondegenerate hydride shift over a barrier of ~5 kcal/mol in which the more stable structure has the positive charge located in the ring (**18a**). At about −100°C the ring expansion process starts to yield the 2,3-dimethylcyclohexyl cation.[41] There are two possible pathways for the reaction to proceed, as illustrated in Scheme 14.

Kinetic isotope effects were calculated from the ring expansion rates of the hexadeuterated isotopomer **16**-d_6 and the tetradeuterated isotopomer **16**-d_4. Again, the experimental results were compared with the theoretical results obtained with QUIVER and showed excellent agreement (experimental $k_H/k_D = 2.40$ vs. theoretical $k_H/k_D = 2.24$ for **16**-d_6; experimental $k_H/k_D = 0.18$ vs. theoretical $k_H/k_D = 0.24$ for **16**-d_4) with the process in which first the methyl migration takes place in the more stable isomer, followed by the ring expansion process. The other

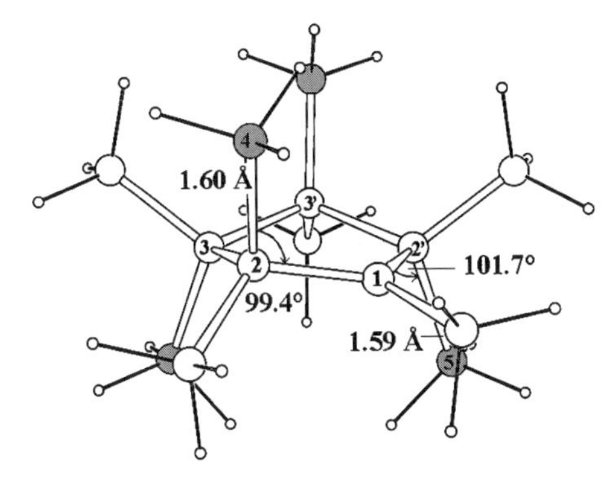

Figure 8.16 The optimized geometry on the B3LYP/6-31G(d) level of the nonamethylcyclopentyl cation (the migrating methyl groups are shaded).

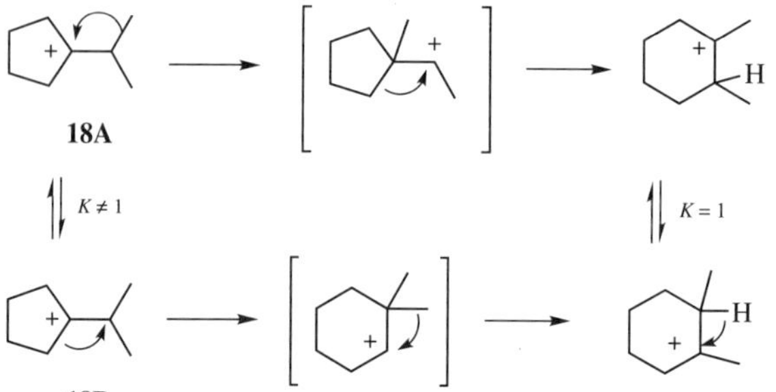

18A

$K \neq 1$

$K = 1$

18B

Scheme 14

possible way in which the first the ring expansion step could occur was ruled out as a major reaction pathway on the grounds of a theoretical isotope effect predicted with QUIVER ($k_H/k_D = 0.59$ for **16**-d_6 and $k_H/k_D = 1.18$ for **16**-d_4).[42]

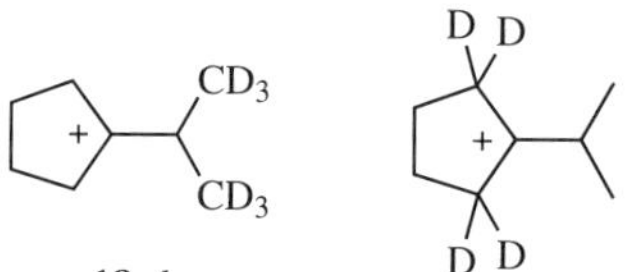

18-d_6

18-d_6

REFERENCES

1. H. Meerwein and K. van Emster, *Chem. Ber.* **55**, 2500 (1922).
2. See, for example (a) C. J. Collins and N. S. Bowman, eds., *Isotope Effects in Chemical Reactions*, ACS Monograph 167, Van Nostrand Reinhold, New York, 1970; (b) D. Bethell, in M. Jones, Jr. and R. A. Moss, eds., *Reactive Intermediates*, Vol. 1, Wiley, New York, 1988, p. 117; (c) C. K. Ingold, *Structure and Mechanism in Organic Chemistry*, Cornell Univ. Press, Ithaca, NY, 1953; (d) S. Winstein and D. Trifan, *J. Am. Chem. Soc.* **74**, 1147 (1952).
3. W. von E. Doering, M. Saunders, H. G. Boynton, H. W. Earhart, E. F. Wadley, and G. Laber, *Tetrahedron* **4**, 178 (1958).
4. P. R. Story and M. Saunders, *J. Am. Chem. Soc.*, **82**, 6199 (1960).
5. S. Winstein and C. Ordroneau, *J. Am. Chem. Soc.* **82**, 2084 (1960).
6. M. Saunders, P. Vogel, E. L. Hagen, and J. C. Rosenfeld, *Acc. Chem. Res.* **6**, 53 (1973).
7. G. A. Olah, E. B. Baker, J. C. Evans, W. S. Tolgyesi, J. S. McIntyre, and I. J. Basteion, *J. Am. Chem. Soc.* **86**, 1360 (1964).
8. M. Saunders and E. L. Hagen, *J. Am. Chem. Soc.* **90**, 6881 (1968).
9. M. Saunders, A. P. Hewett, and O. Kronja, *Croatica Chem. Acta* **65**, 673 (1992).
10. W. Koch, P. von R. Schleyer, P. Buzek, and B. Lin, *Croatica Chem. Acta* **65**, 655 (1992).

11. M. Saunders, E. L. Hagen, and J. C. Rosenfeld, *J. Am. Chem. Soc.* **91**, 6882 (1969).
12. (a) M. Saunders and E. L. Hagen, *J. Am. Chem. Soc.* **90**, 2436 (1968); (b) M. Saunders and J. C. Rosenfeld, *J. Am. Chem. Soc.* **91**, 7756 (1969).
13. Unpublished results
14. M. Saunders and J. C. Rosenfeld, *J. Am. Chem. Soc.* **92**, 2548 (1970).
15. M. Saunders, P. V. Vogel, L. Hagen, and J. Rosenfeld, *Acc. Chem. Res.* **6**, 53 (1973).
16. M. Saunders and J. Stofko, *J. Am. Chem. Soc.* **95**, 252 (1973).
17. V. Vrček, I. Vinković Vrček, and H.-U. Siehl, *J. Phys. Chem. A* **106**, 1604 (2002).
18. (a) F. Sun and T. S. Sorensen, *J. Am. Chem. Soc.* **115**, 77 (1993); (b) H.-U. Siehl, *Adv. Phys. Org. Chem.* **23**, 63 (1987).
19. P. von R. Schleyer, T. M. Su, M. Saunders, and J. C. Rosenfeld, *J. Am. Chem. Soc.* **91**, 5174 (1969).
20. M. Saunders and R. Berger, *J. Am. Chem. Soc.* **94**, 4094 (1972).
21. T. Sorensen, personal communication
22. M. Saunders, M. H. Jaffe, and P. Vogel, *J. Am. Chem. Soc.* **93**, 2558 (1971).
23. M. Saunders and P. Vogel, *J. Am. Chem. Soc.* **93**, 2561 (1971).
24. (a) M. Saunders, G. W. Cline, and M. Wolfsberg, *Z. Naturforsch.*, **44a**, 480 (1989); (b) M. Saunders and G. W. Cline, *J. Am. Chem. Soc.* **112**, 3955 (1990).
25. M. Saunders, K. E. Laidig, and M. Wolfsberg, *J. Am. Chem. Soc.* **111**, 8989 (1989).
26. M. Saunders, M. R. Kates, and G. E. Walker, *J. Am. Chem. Soc.* **103**, 4623 (1981).
27. G. Seybold, P. Vogel, M. Saunders, and K. B. Wiberg, *J. Am. Chem. Soc.* **95**, 2045 (1973).
28. M. Saunders and R. Kates, *J. Am. Chem. Soc.* **102**, 6867 (1980).
29. M. Saunders, L. A. Telkowski, and M. R. Kates, *J. Am. Chem. Soc.* **99**, 8070 (1977).
30. T. Laube, *J. Am. Chem. Soc.* **111**, 9224–9232 (1998).
31. M. Saunders, L. A. Telkowski, and M. R. Kates, *J. Am. Chem. Soc.* **99**, 8070 (1977).
32. M. Saunders, R. M. Jarret, and P. Pramanik, *J. Am. Chem. Soc.* **109**, 3735 (1987).
33. (a) G. A Olah, D. P. Kelly, C. J. Leuell, and R. D. Porter, *J. Am. Chem. Soc.* **93**, 2544 (1970); (b) J. S. Starat, G. D. Roberts, G. K. S Prakash, D. J. Donovan, and G. A. Olah, *J. Am. Chem. Soc.* **100**, 8016 (1978).
34. M. Saunders and H.-U. Siehl, *J. Am. Chem. Soc.* **102**, 6868 (1980).
35. (a) P. von R. Schleyer, K. E. Laidig, K. B. Wiberg, M. Saunders, and M. Schindler, *J. Am. Chem. Soc.* **110**, 300 (1988); (b) W. Koch, B, Liu, and D. J. DeFrees, *J. Am. Chem. Soc.* **110**, 7325 (1988).
36. T. S. Sorensen and R. P Kirchen, *J. Am. Chem. Soc.* **99**, 6687 (1977).
37. H.-U. Siehl, *J. Am. Chem. Soc.* **107**, 3390 (1985).
38. M. Saunders and N. Krause, *J. Am. Chem. Soc.* **110**, 8050 (1988).
39. H. Mayr and R. Koschinsky, *J. Am. Chem. Soc.* **111**, 2305 (1989).
40. O. Kronja, T.-P. Kühli, H. Mayr, and M. Saunders, *J. Am. Chem. Soc.* **122**, 8067 (2000).
41. (a) V. Vrček, H.-U. Siehl, and O. Kronja, *J. Phys. Org. Chem.* **13**, 616 (2000); (b) H.-U. Siehl, V. Vrček, and O. Kronja, *J. Chem. Soc. Perkin Trans. 2* 106 (2002).
42. V. Vrček, M. Saunders, and O. Kronja, *J. Org. Chem.* **68**, 1859 (2003).

9

STABLE ION CHEMISTRY OF POLYCYCLIC AROMATIC HYDROCARBONS(PAHs); MODELING ELECTROPHILES FROM CARCINOGENS

Kenneth K. Laali
Department of Chemistry
Kent State University
Kent, Ohio

Carbocation Chemistry, Edited by George A. Olah and G. K. Surya Prakash
ISBN 0-471-28490-4

9.1 PROLOGUE

In January 1974 I went to UMIST as a postgraduate student to work with Robert N. Haszeldine (then head of the chemistry department). I interviewed with three of his "faculty lieutenants" who offered me projects in high-pressure chemistry, ESR work on fluorinated nitroxides, and electrophilic aromatic substitution utilizing strong acids in particular exploiting Haszeldine's CF_3SO_3H "triflic acid." I chose to work on strong-acid chemistry (Ph.D. thesis title *Carbocation Chemistry Catalyzed by Trifluoromethanesulphonic Acid*; co-advised by Brian L. Booth). The obvious connection between my thesis work and Olah's groundbreaking studies in electrophilic aromatic substitution in the 1970s, his pioneering investigations on persistent carbocations and onium ions in superacid media, and a massive number of papers coming out of Olah's group at Case Western opened my eyes as a graduate student to an exciting and dynamic area of chemistry. I religiously read and studied a large number of Olah's papers while working on my Ph.D. (1974–1977). A postdoctoral stint with the late Victor Gold at King's College got me more intimately involved with superacids and NMR (quantitative acidity measurements by DNMR). George was a visiting Professor at King's College, and I was fortunate to meet him there on two occasions to discuss chemistry and to seek his advice during small group meetings. I became seriously interested in moving to the United States to work with him, a career plan that took several years to materialize! While he moved to USC, I went to Strasbourg to work on hydrocarbon activation with Jean Sommer and then to Amsterdam to work with Hans Cerfontain on aromatic sulfonation and protonation, and finally to ETH—Zurich to study dediazoniation reversibility and aryl cations with H. Zollinger. I finally joined George's group at USC in mid-1982. The three productive years there, until my independent career began, provided me with a more diverse and broad-based experience, which in retrospect was more rewarding and career setting than if I had gone there earlier with less prior research accomplishments.

On this wonderful occasion of the centennial of stable carbocations chemistry, I dedicate this chapter with great respect to Professor George Olah for the inspiration, ever-lasting influence that he has had on shaping my research career, and for long-lasting friendship and continuous encouragement and support.

It is indeed a pleasure and honor for me to participate in "100 Years of Carbocations." Special thanks also to Professor Prakash, a long-time colleague, collaborator, and friend for the invitation to contribute a chapter.

9.2 INTRODUCTION

9.2.1 Bridging Carbocation Chemistry with Biology

Active research on chemical carcinogenesis during the 1970s and 1980s led to major advances in understanding the metabolic pathways and the events that ultimately lead to mutagenesis and carcinogenesis. Since polyarenes are most widely distributed class of human carcinogens in the environment, much of these efforts focused on understanding the mechanism of metabolic activation of polycyclic aromatic hydrocarbons (PAHs). An interdisciplinary field to which toxicologists, synthetic, mechanistic, and computational chemists greatly contributed.[1–3] Development of the "bay-region theory" and subsequent progress in developing structure–bioreactivity relationships, stereoselective synthesis of model diol-epoxides, and their DNA intercalation and binding studies, coupled to extensive model kinetic studies, product analysis, computational work, and X-ray analysis, created a wealth of data.[4–9] These studies established that in a number of classes of PAHs, carbocations are the key intermediates that can bind to DNA:

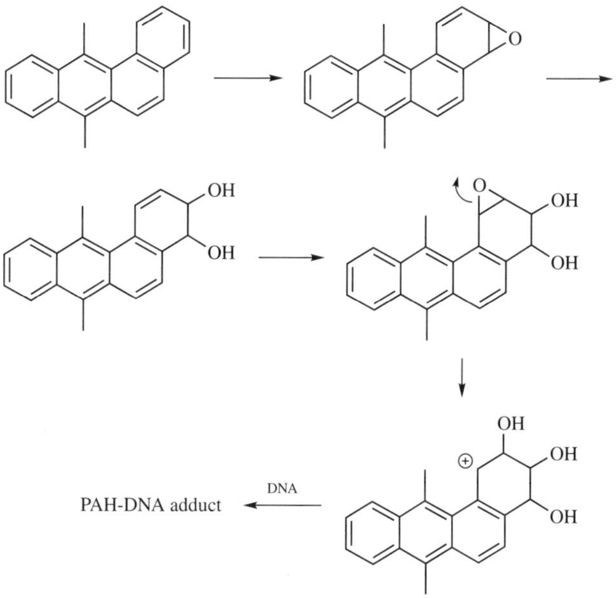

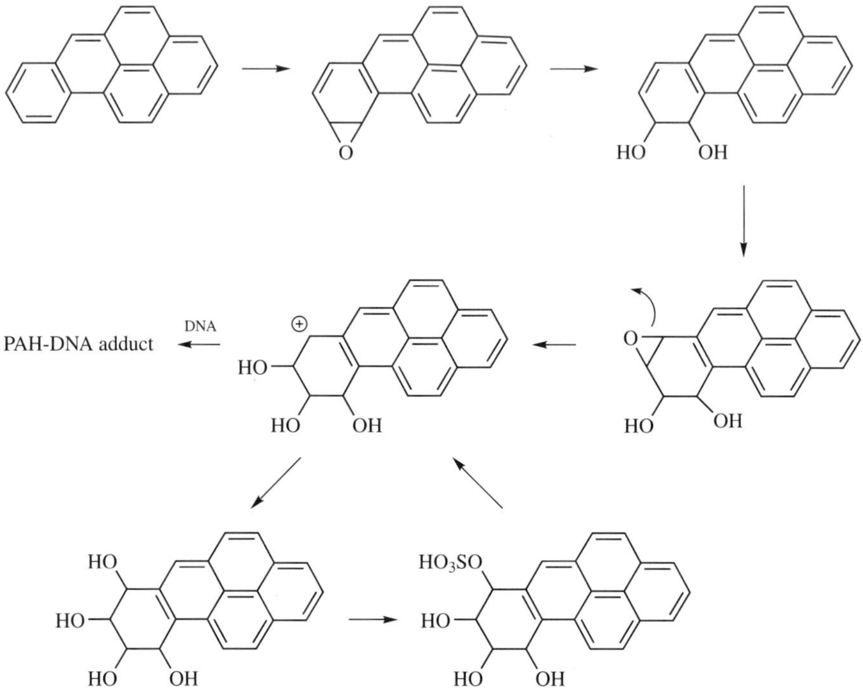

For larger PAHs with lower ionization potentials, direct biological oxidation to the radical cation could become competitive and the resulting $PAH^{+\cdot}$ may undergo nucleophilic attack by nucleotides to give the PAH-DNA adduct:[10]

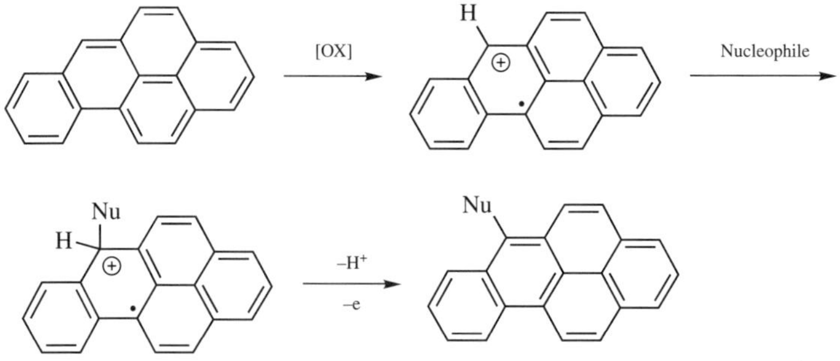

Formation of benzylic carbocations from methylated PAHs via sidechain oxidation to form benzylic alcohol followed by esterification and solvolysis was also established:[3e,5e,10]

bioalkylation → CH_3 → [OX] → CH_2OH

esterification → CH_2OSO_3H → $\overset{\oplus}{C}H_2$

Despite a great deal of progress, no stable ion studies to model biological carbocations existed. The opportunity presented itself as a natural progression of our studies on hindered PAH arenium ions in the early–mid-1990s, and work aimed at bridging PAH–carbocation chemistry and biology took off.

9.3 RATIONALE

In stable ion studies, protonation is used as a mimic reaction for generation of electrophiles from the PAHs, whereas α-PAH-substituted carbocations serve as models for carbocations formed by epoxide ring opening. Certain PAHs are direct acting mutagens, but many require metabolic activation via oxidative pathways before they can bind to DNA. The H^+ concentration mapping in nucleic acids has identified the most acidic sites in the major and minor grooves of DNA.[12] These sites correlate with the observed covalent binding sites for benzo[*a*]pyrene-epoxide to guanine and fit the model that epoxide protonation gives a benzylic carbocation that binds to DNA.[12a] Given the high local acidity in DNA,[12] equilibrium protonation as a means to generate in vivo carbocations is probably not far fetched, especially for larger, more reactive (highly basic) PAHs. This process could be relevant in PAHs that are direct-acting mutagens.

9.4 PROTOCOLS

PAH–arenium ions are typically generated in low concentrations in FSO_3H/SO_2ClF and dications in $FSO_3H.SbF_5$ (4 : 1)/SO_2ClF. The use of higher acidity, more oxidizing, $FSO_3H.SbF_5$ (1 : 1)/SO_2ClF often leads to competing oxidation and/or polymerization.

NMR studies determine the site(s) of attack and the $\Delta\delta^{13}C$ values produce a charge delocalization path for the resulting carbocations. These studies allow substituent effects on carbocation stability/charge delocalization to be probed. These projects require specific assignments of the ring carbons, including the ring junction carbons (using HMQC and HMBC experiments). Assignments are assisted via NOED experiments that also provide conformational information in relevant cases. NMR studies are complemented with theoretical work. Given the size of the systems, AM1 minimizations and single point calculations are first performed. Relative arenium ion energies are usually computed for all possible carbocations for comparison with experiment (this method has proved to work quite well). More recently, computational studies have been upgraded to density functional methods (1) to calculate natural charges from which changes in charges are mapped out for comparison with the NMR-based conclusions, (2) GIAO-NMR to predict the chemical shifts for comparison with the experimental results, and (3) NICS calculations to probe "relative aromaticity" in different rings. The long-term goal in this area is to build up a database of structure–reactivity and substituent effects for comparison with the biological data. In selected case, parallel DNA binding studies (e.g., in MCF-7 human mammary carcinoma cells) are performed starting with the covalent precursors of model benzylic carbocations (typically alcohols and chlorides) for comparative purposes. The aim of these projects is to find out whether a carbocation-based structure/activity database could be used to predict biological activity, and to explore whether charge delocalization patterns could identify certain "unifying modes" that may be associated with increased potency.

9.5 SCOPE OF THE REVIEW

The discussion focuses on the progress that has been made in the area of PAH–arenium ions since the review article in 1996,[13] which emphasized earlier fundamental studies on protonation as well as oxidation (both radical cations and stable dications) of polycyclic arenes.

9.5.1 Phenanthrenium Ions

Detailed ^{13}C NMR studies and charge delocalization mapping in representative methyl and dimethylphenanthrenium mono- and dications have been conducted.[14] Directive effect of the methyl groups are seen in representative examples below:

CH_3 2 1 3 4 10 9 5 6 8 7

FSO_3H

SO_2ClF

(>95%)

CH_3 + H H

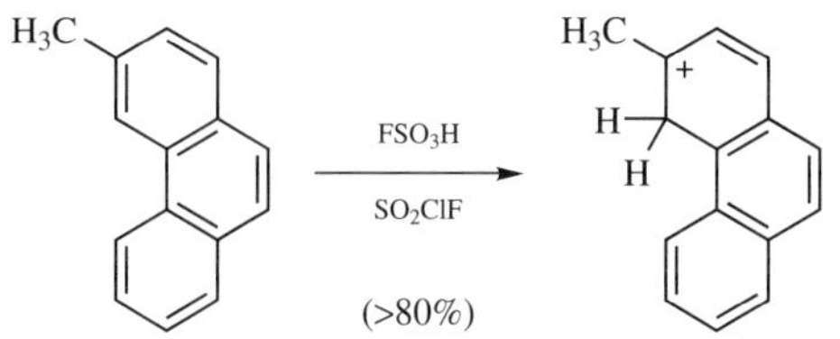

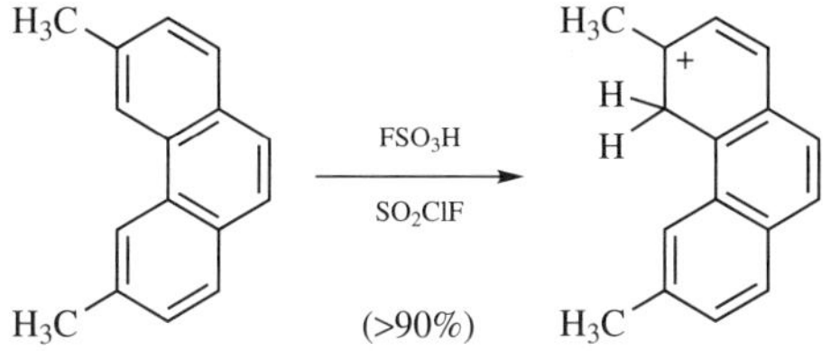

H_3C → [FSO$_3$H / SO$_2$ClF] → H_3C, + , H, H

(>80%)

H_3C, H_3C → [FSO$_3$H / SO$_2$ClF] → H_3C, +, H, H, H_3C

(>90%)

With the FSO$_3$H.SbF$_5$ (1 : 1)/SO$_2$ClF system with 2-methyl- and 3-methylphenanthrenes mixtures of dication and monocation are formed, whereas with 3,6-dimethylphenanthrene a symmetrical dication is produced:

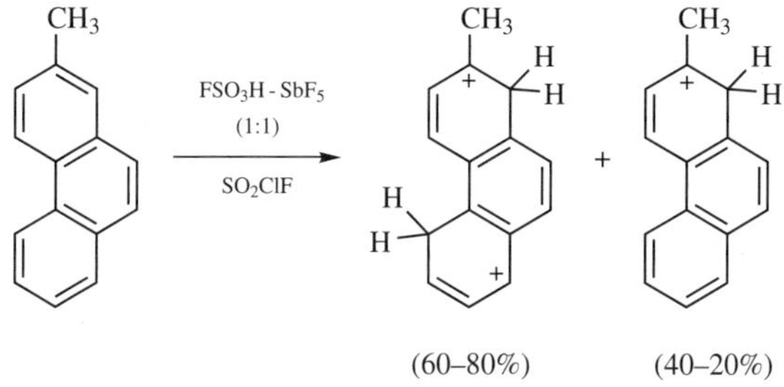

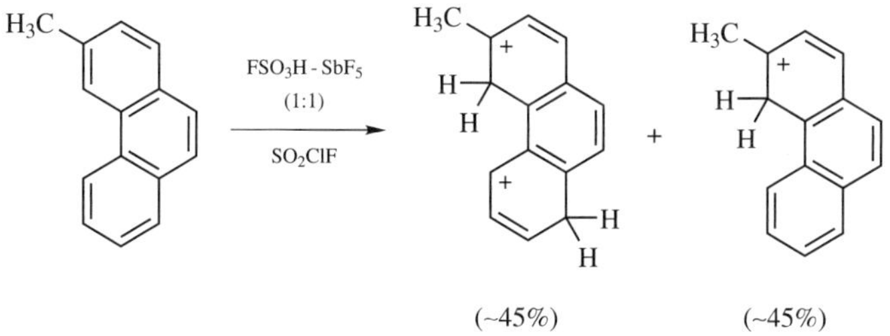

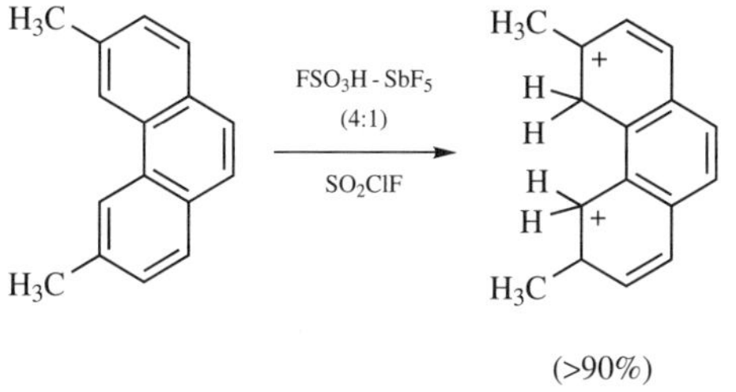

CH_3 → [FSO$_3$H - SbF$_5$ (1:1) / SO$_2$ClF] → CH_3, +, H, H, H, H, + + CH_3, +, H, H

(60–80%) (40–20%)

H_3C → [FSO$_3$H - SbF$_5$ (1:1) / SO$_2$ClF] → H_3C, +, H, H, +, H, H + H_3C, +, H, H

(~45%) (~45%)

H_3C, H_3C → [FSO$_3$H - SbF$_5$ (4:1) / SO$_2$ClF] → H_3C, +, H, H, H, H, +, H_3C

(>90%)

The charge delocalization path in the monocations is limited, but in the dications there is extensive charge delocalization, which in some cases involves adjacent ring positions:

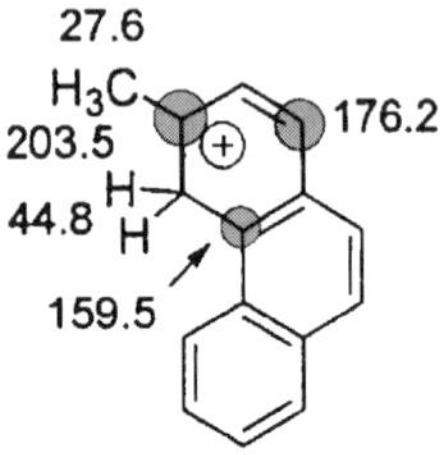

^{13}C NMR data for 3-methlphenanthrenium ion (size of the dark circles is roughly proportional to magnitude of $\Delta\delta^{13}C$ values)

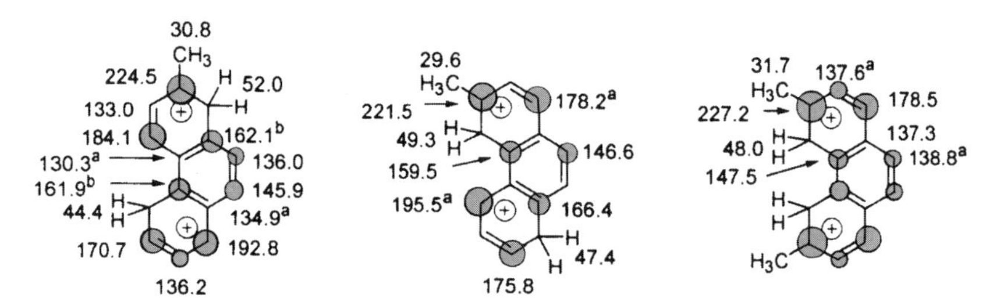

^{13}C NMR data for methylphenanthrenium dications (size of the dark circles is roughly proportional to magnitude of $\Delta\delta^{13}C$ values; a and b denote interchangeable assignments)

9.5.2 Regioisomeric α-Phenanthrenyl-Substituted Carbocations and Carboxonium Ions

A series of regioisomeric α-phenanthrenyl-substituted carbocations and carboxonium ions have been generated to serve as simplified models of various phenanthrene–epoxide ring opening.[15] The analogy between these model cations and those from diol epoxide metabolites is illustrated below:

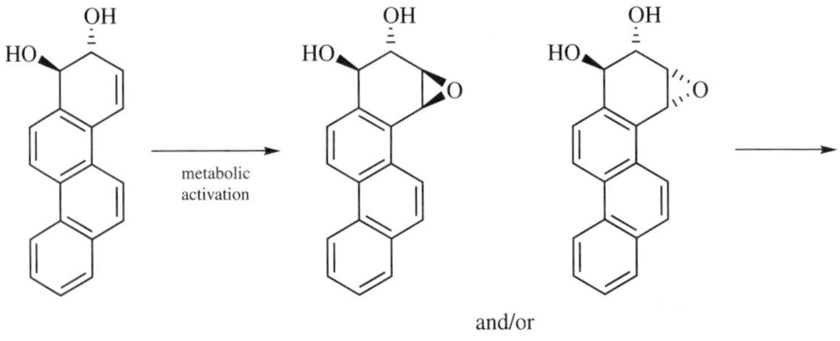

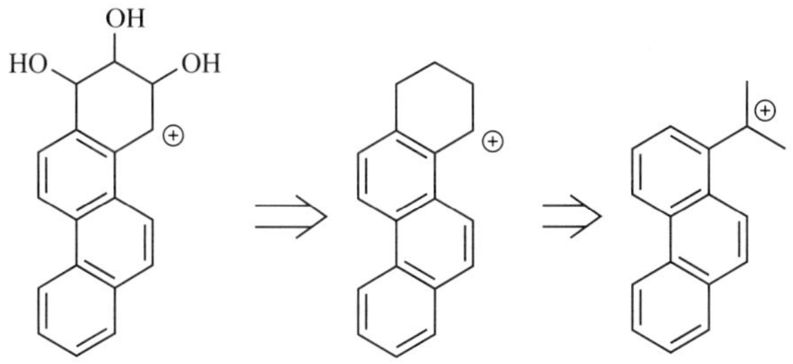

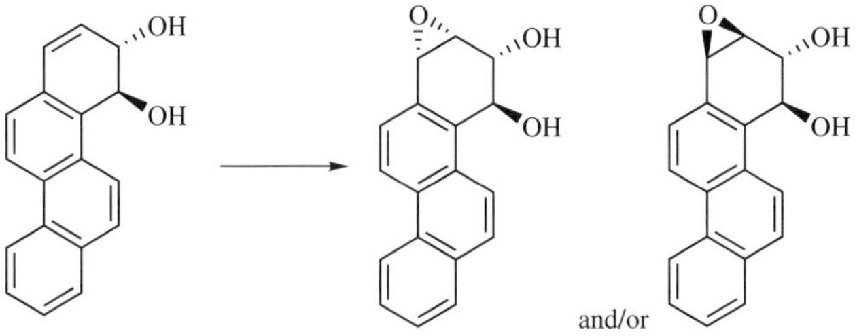

OH HO OH ⊕ ⇒ ⊕ ⇒ ⊕

OH OH → O OH OH O OH OH and/or

⊕ ⇒ ⇐ O

OH OH O ⇒ ⊕ O OH OH ⇒ ⊕

OH HO O ⇒ ⊕

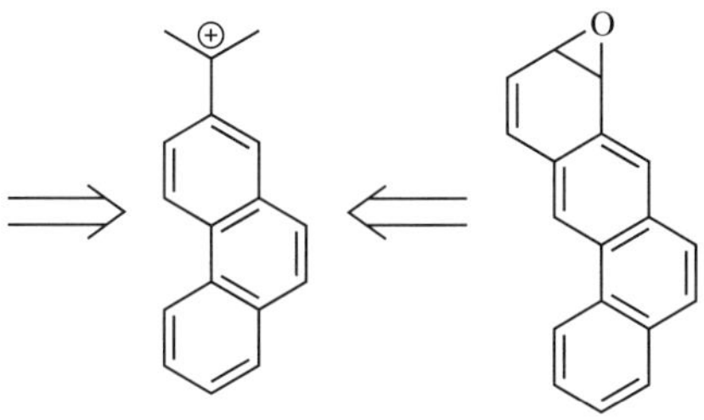

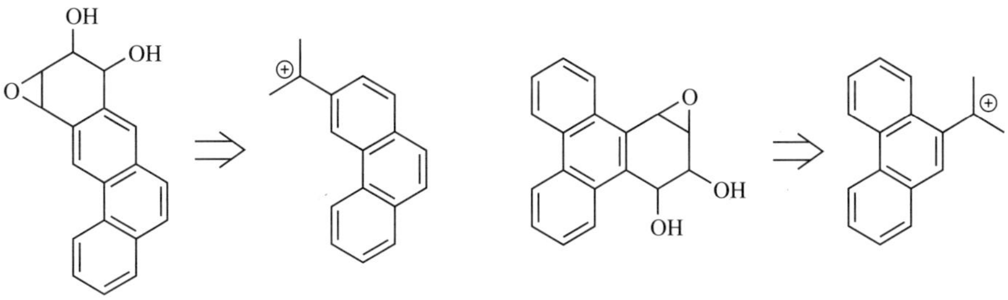

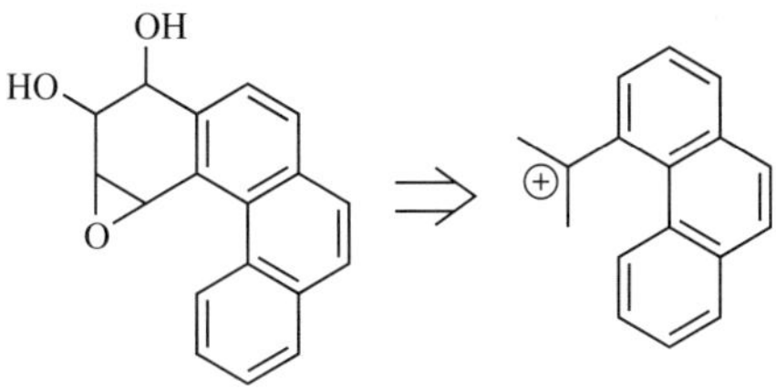

Charge delocalization mapping shows that positive charge is most effectively delocalized from the *meso* position and least effectively from C2. The carboxonium ions exhibit minimal charge delocalization into the phenanthrene moiety:

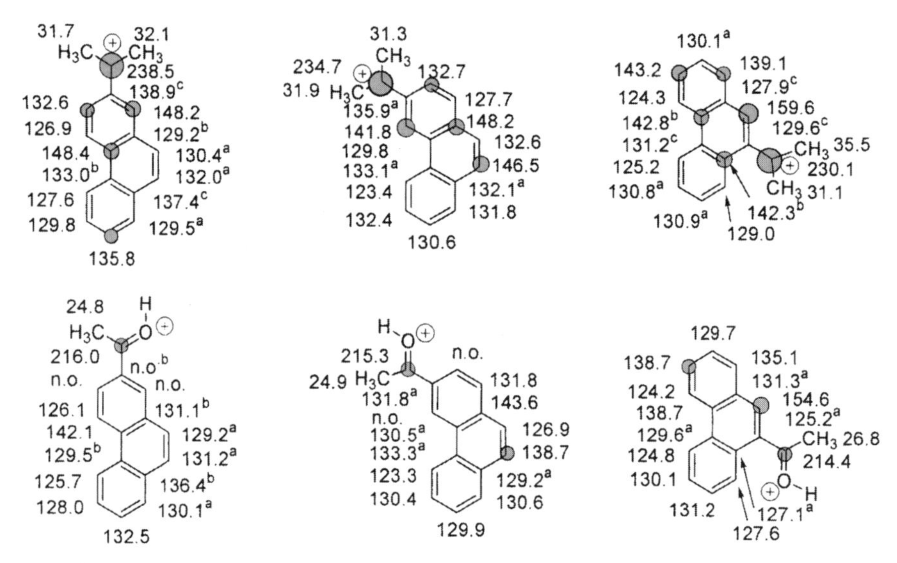

^{13}C NMR data for α-phenanthrenyl-substituted carbocations and carboxonium ions (size of the dark circles is roughly proportional to magnitude of $\Delta\delta^{13}$C values; a and b denote interchangeable assignments; n.o. = not observed)

9.5.3 Cyclopenta[*a*]phenanthrenium Cations

Because of their structural similarity with natural steroids, cyclopenta[*a*]phenanthrene $C_p[a]P$ derivatives have been the subject of numerous synthetic, mechanistic, and biological studies. The A ring in $C_p[a]P$ is metabolized to the 3,4-dihydrodiol and the *anti*- and *syn*-diol-epoxides:

R OH OH R O OH OH R O OH OH

Detailed stable ion studies have been carried out on a series of $C_p[a]P$ derivatives.[16,17] Persistent monoarenium ions are formed by protonation of the parent hydrocarbon, its methylated analogs, and the dimethyl derivatives having a double bond in the five-membered ring. The bay-region methyl derivative is *ipso*-protonated. The bay-region-methoxy derivative (a potent carcinogen) is protonated *ortho/para*. The keto derivatives give carboxonium ions:

17 16 13 15 12 14 11 8 7 9 10 6 1 5 2 4 3

$\xrightarrow[SO_2ClF]{FSO_3H}$

$\xrightarrow[SO_2ClF]{FSO_3H}$

$\xrightarrow[SO_2ClF]{FSO_3H}$

$\xrightarrow[SO_2ClF]{FSO_3H}$

$\xrightarrow[SO_2ClF]{FSO_3H}$

The ^{13}C NMR data and the derived charge delocalizaion path are shown for representative cases:

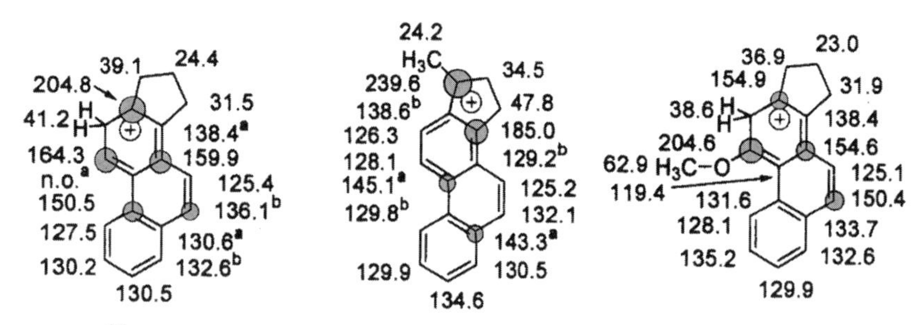

^{13}C NMR data for carbocations derived from cyclopenta[*a*]phenanthrenes (size of the dark circles is roughly proportional to magnitude of $\Delta\delta^{13}C$ values; a and b denote interchangeable assignments; n.o. = not observed)

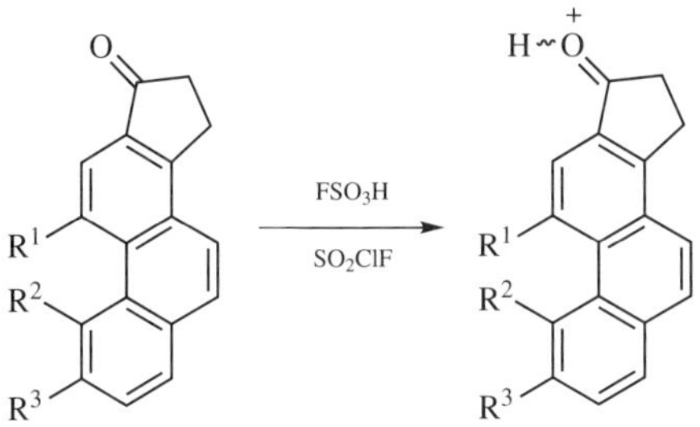

FSO_3H / SO_2ClF

$R^1 = R^2 = R^3 = H$
$R^1 = Me;\ R^2 = R^3 = H$
$R^2 = Me;\ R^1 = R^3 = H$
$R^3 = Me;\ R^1 = R^2 = H$
$R^1 = CF_3;\ R^2 = R^3 = H$

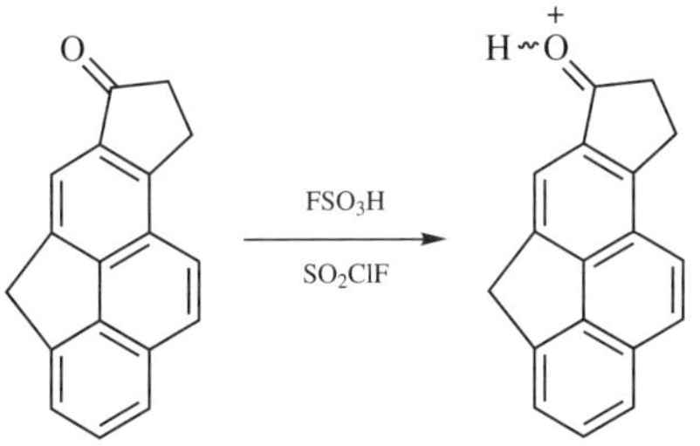

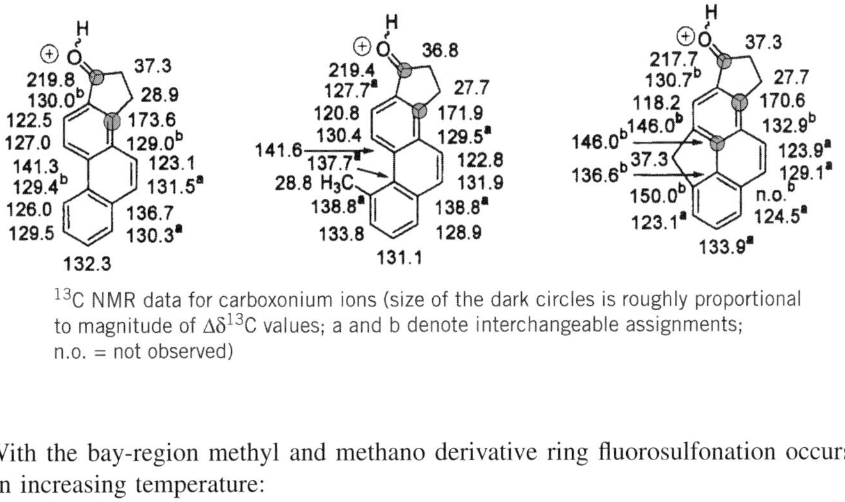

FSO_3H / SO_2ClF

^{13}C NMR data for carboxonium ions (size of the dark circles is roughly proportional to magnitude of $\Delta\delta^{13}C$ values; a and b denote interchangeable assignments; n.o. = not observed)

With the bay-region methyl and methano derivative ring fluorosulfonation occurs on increasing temperature:

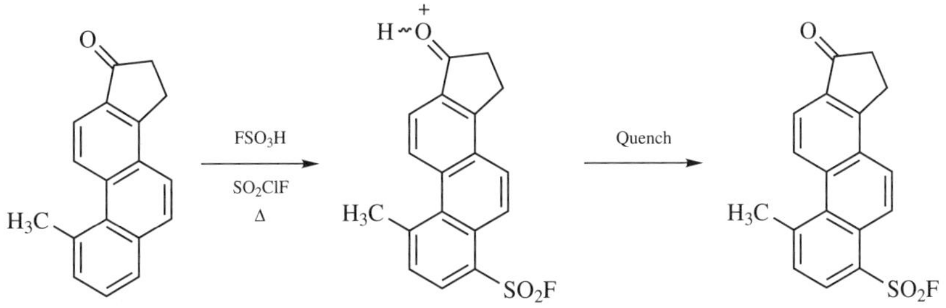

FSO_3H / SO_2ClF / Δ ; Quench

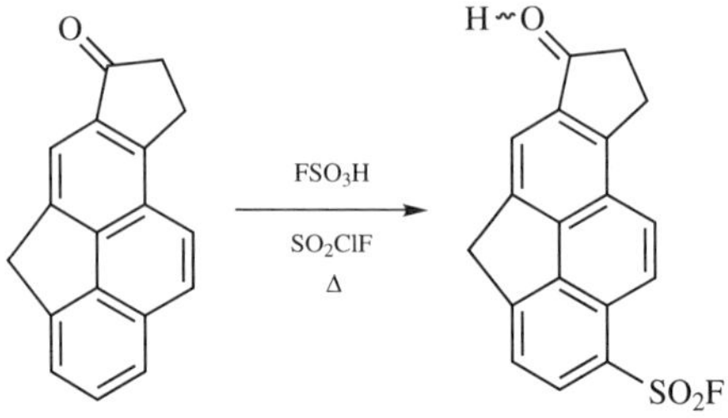

FSO$_3$H
SO$_2$ClF
Δ
H~O
SO$_2$F

The dimethylated dimer cleaves to give the tertiary carbocation, whereas the dimethoxy dimer is stable in superacid media and is diprotonated to give three regioisomeric dications:

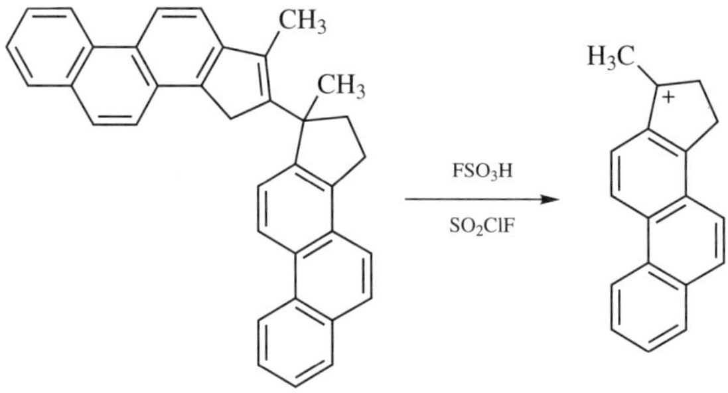

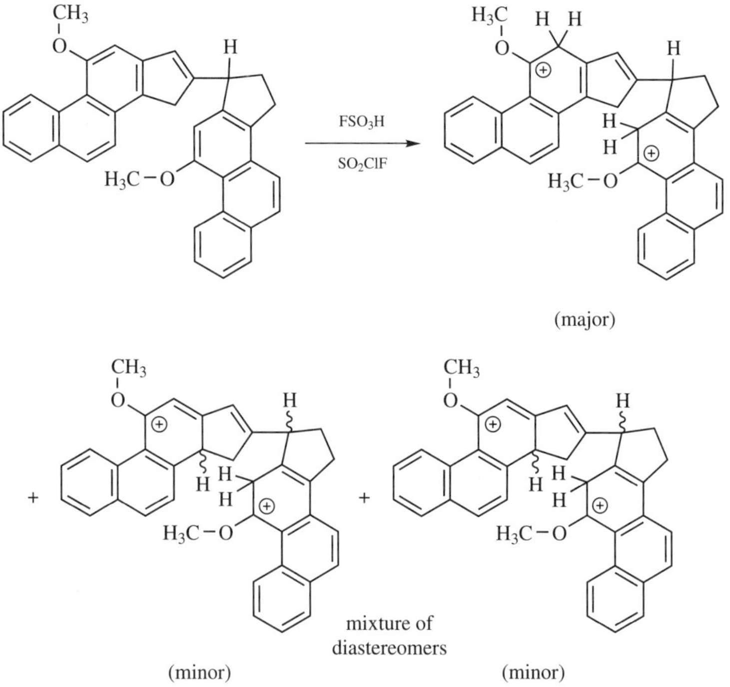

CH$_3$
CH$_3$
FSO$_3$H
SO$_2$ClF
H$_3$C

CH$_3$
O
H
H$_3$C−O
FSO$_3$H
SO$_2$ClF
H$_3$C H H
O
H
H
H
H$_3$C−O

(major)

+ + mixture of diastereomers

(minor) (minor)

The α-carbocation derived from the 15-ol is persistent and delocalized. On quenching, it undergoes in situ elimination and forms a stable PAH-dimer. On the other hand, the carbocation derived from the 17-ol derivative is unstable; it forms the dimer in situ that is then diprotonated (see reactions shown above). The 15-ol derivative undergoes elimination on reflux in $CHCl_3$. The solvolytic carbocation (in dichloroethane solvent) has been trapped with imidazole and analyzed as an imidazolium salt after methylation:

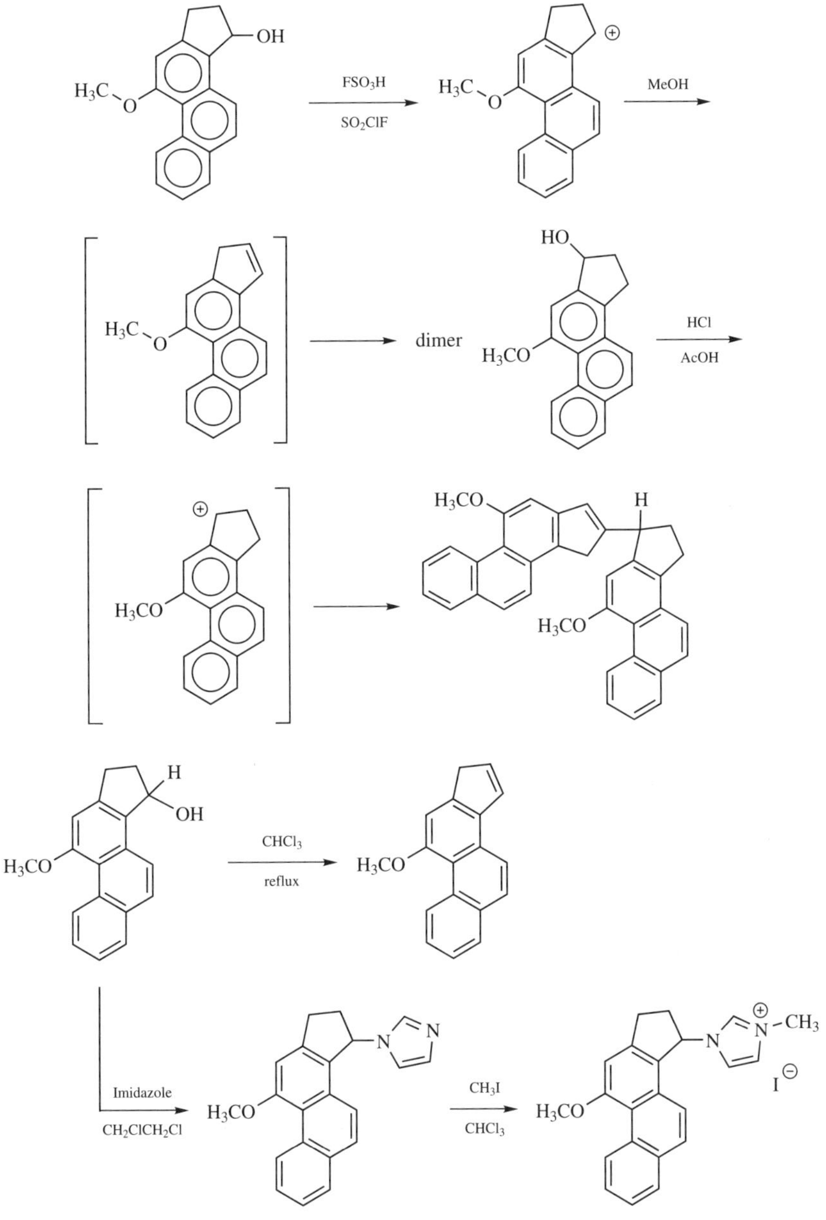

H_3C–O, OH; FSO_3H / SO_2ClF; H_3C–O, ⊕; MeOH

H_3C–O → dimer; HO, H_3CO; HCl / AcOH

⊕, H_3CO → H_3CO, H, H_3CO

H, OH, H_3CO; $CHCl_3$ / reflux; H_3CO

Imidazole / CH_2ClCH_2Cl; N, N, H_3CO; CH_3I / $CHCl_3$; N, ⊕N–CH_3, $I^{\ominus}$, H_3CO

9.5.4 Chrysenium Ions

Chrysene (*Ch*) is an important building block of carcinogenic PAHs that metabolize via the diol-epoxide route. 5-Methylchrysene is as potent as benzo[*a*]pyrene. Stable ion studies have been carried out on a series of substituted *Ch* in order to understand their structure/reactivity relationships and the charge delocalization modes in the resulting carbocations.[18,19]

Parent *Ch* and its 6-halo derivatives are protonated at C12 and the 6-acetyl derivative gives a carboxonium ion:

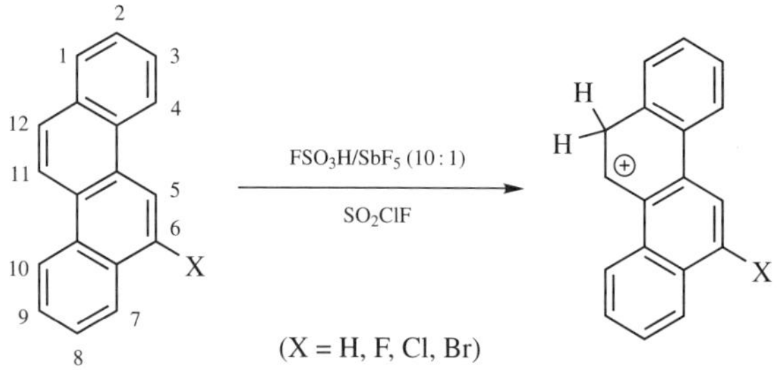
2
1
3
4
12
11
5
6
X
10
9
7
8
FSO3H/SbF5 (10 : 1)
SO2ClF
(X = H, F, Cl, Br)
H
H
X

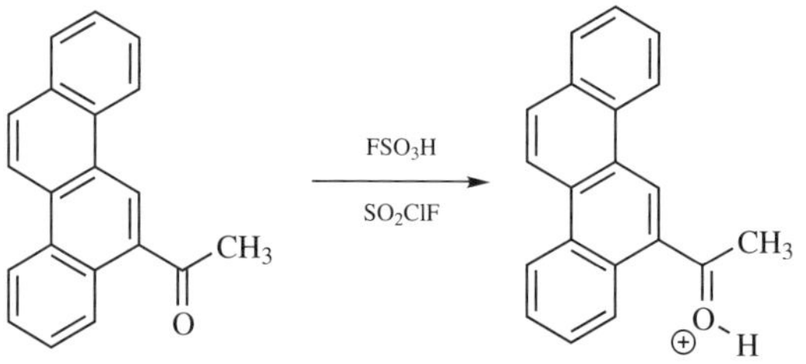
FSO3H
SO2ClF
CH3
O
CH3
O
H

The 5-methyl derivative is protonated at C6 and C12. A methoxy substituent at C2 directs the attack to C1. The resulting arenium ion exhibits significant carboxonium ion character, with the OMe group existing in two distinct conformations. Introduction of methyl at C5 or C12 (bay regions) does not change the preference for attack *ortho* to methoxy:

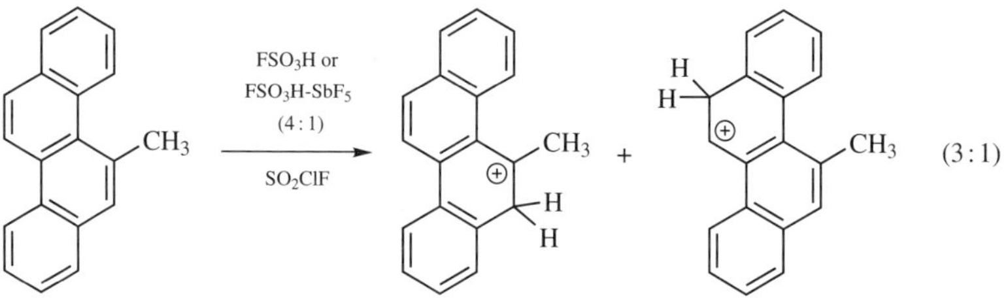
CH3
FSO3H or
FSO3H-SbF5
(4 : 1)
SO2ClF
CH3
H
H
+
H
H
CH3
(3 : 1)

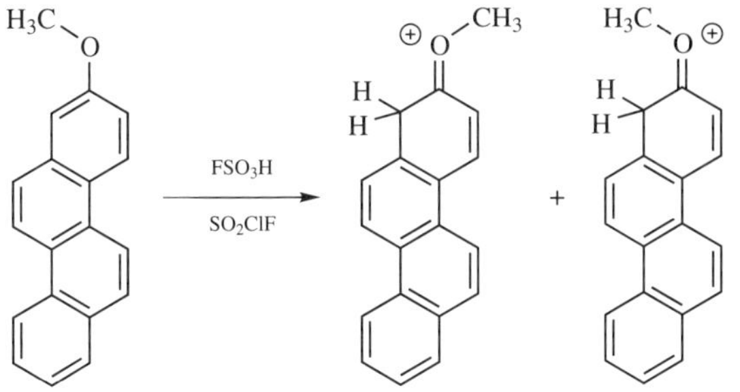

$\xrightarrow[SO_2ClF]{FSO_3H}$ (3 : 2)

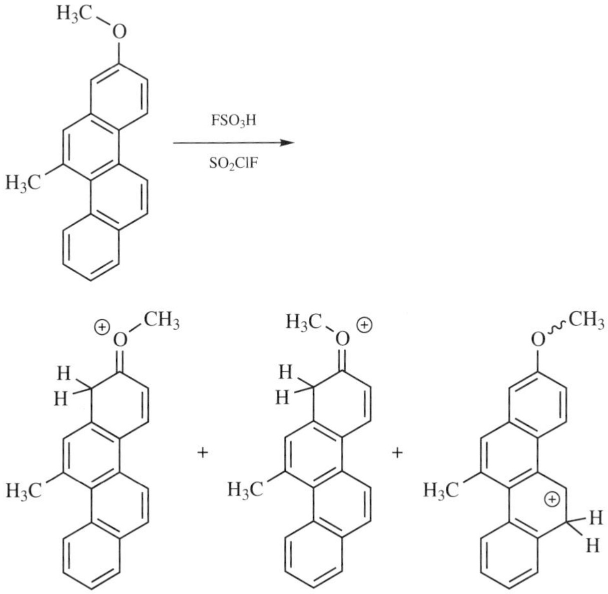

$\xrightarrow[SO_2ClF]{FSO_3H}$ (6 : 3 : 1)

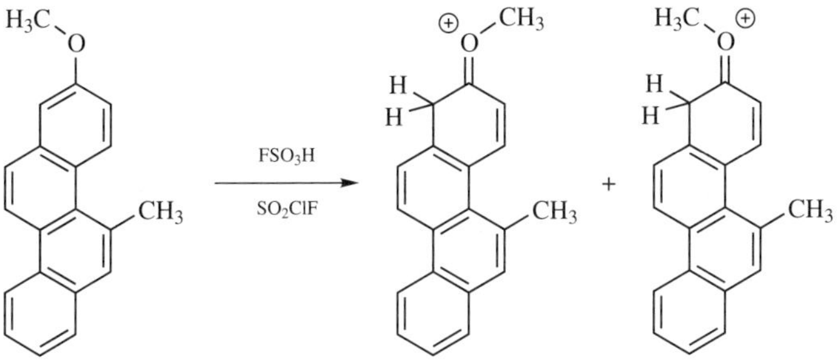

$\xrightarrow[SO_2ClF]{FSO_3H}$ (4 : 1)

Charge delocalization mode in the studied chrysenium cations are sketched below:

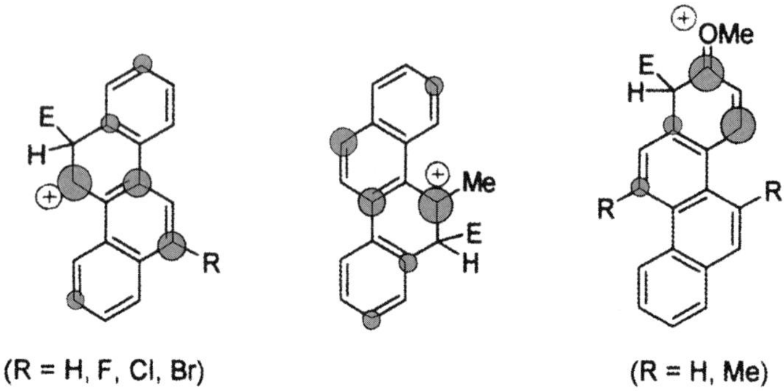

(R = H, F, Cl, Br) (R = H, Me)

^{13}C NMR-derived charge delocalization mode in the carbocations

9.5.5 Benzo[*a*]anthracenium BA Cations and Related α-Carbocations and Carboxonium Ions

Whereas parent BA is a weak carcinogen, 7,12-dimethyl-BA and 3-methylcholanthrene are potent carcinogens. Several examples of persistent carbocations derived from the BA skeleton have been generated.[20] Parent BA is protonated at C7/C12 to give two arenium ions in 3:1 ratio, which does not change with time. Introduction of methyl at C1 changes the ratio to 10:1. The 7,12-dimethyl-BA gives a mixture of two *ipso*-protonated carbocations in equal amounts at low temperature, but in time this ratio changes to 50 : 1 in favor of protonation at C12. 3-Methyl-cholanthrene is protonated at C6:

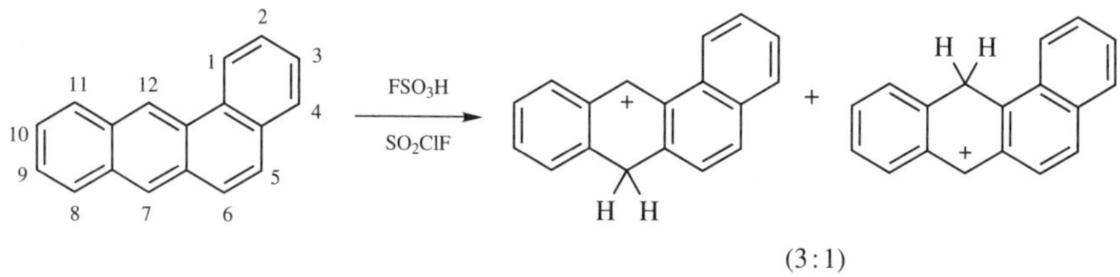

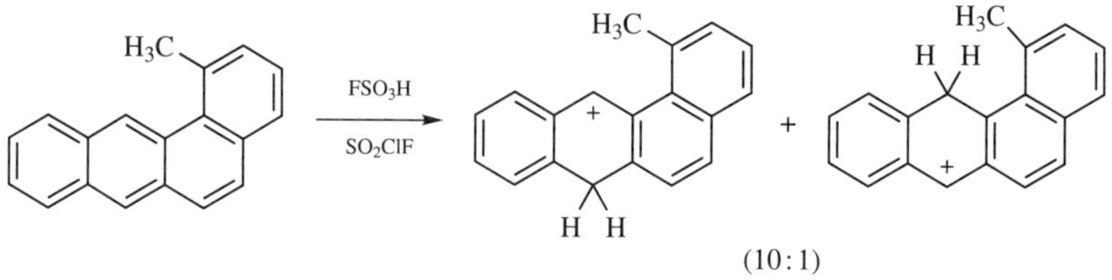

FSO_3H / SO_2ClF

(3:1)

FSO_3H / SO_2ClF

(10:1)

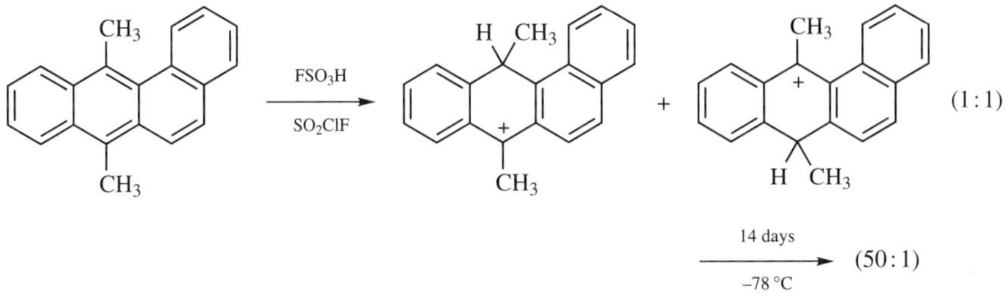

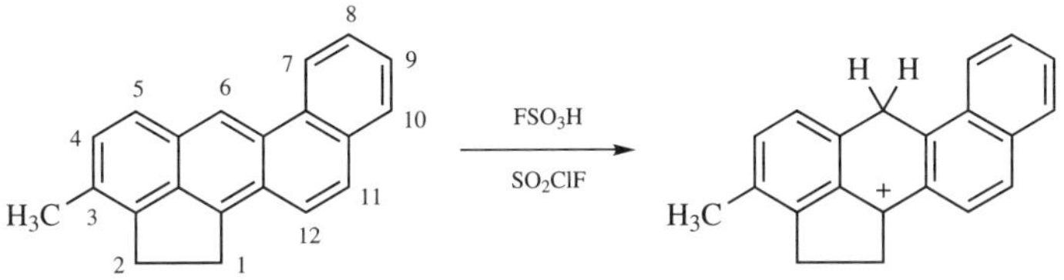

$$\xrightarrow[SO_2ClF]{FSO_3H} \quad + \quad (1:1)$$

$$\xrightarrow[-78\ °C]{14\ days} (50:1)$$

$$\xrightarrow[SO_2ClF]{FSO_3H}$$

The carbocations and carboxonium ions shown below were generated as models for benzo[*a*]anthracene and cholanthrene-epoxide ring opening; to determine the influence of the annelated ring, α-anthracene-substituted carbocations and carboxonium ions have also been studied:

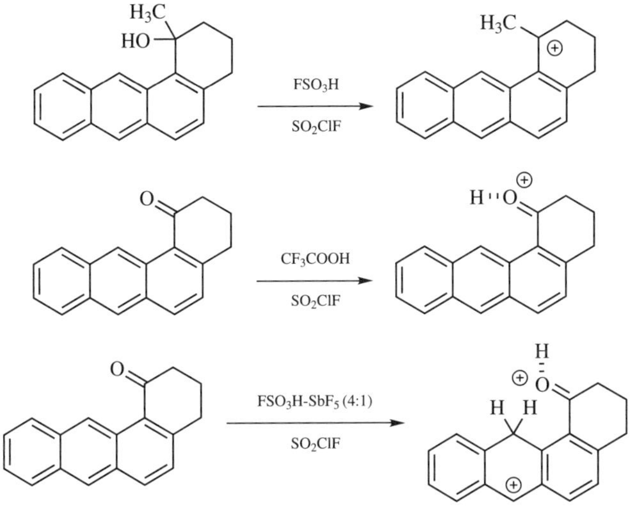

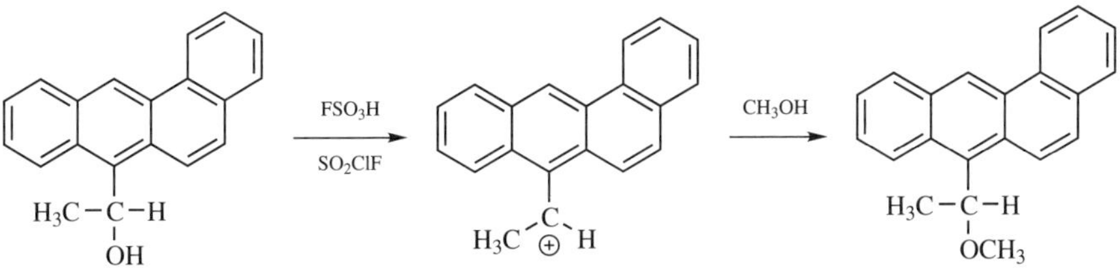

$$\xrightarrow[SO_2ClF]{FSO_3H}$$

$$\xrightarrow[SO_2ClF]{CF_3COOH}$$

$$\xrightarrow[SO_2ClF]{FSO_3H\text{-}SbF_5\,(4:1)}$$

$$\xrightarrow[SO_2ClF]{FSO_3H} \quad \xrightarrow{CH_3OH}$$

$H_3C-C(=O)$ (benz[a]anthracene) $\xrightarrow[SO_2ClF]{FSO_3H}$ $H_3C-C(=O^+\cdot\cdot H)$

$H_3C-C(OH)-H$ (anthracene) $\xrightarrow[SO_2ClF]{FSO_3H}$ H_3C-C^+-H

$H_3C-C(=O)$ (anthracene) $\xrightarrow[SO_2ClF]{FSO_3H}$ $H_3C-C(=O^+\cdot\cdot H)$

Charge delocalization mapping in the resulting carbocations and carboxonium ions establishes strong anthracenium ion character. Selected examples are show below. Stable ion studies have identified a unifying charge delocalization pattern involving C4a, C6, C7, C8, C10, C11a, and C12a carbons:

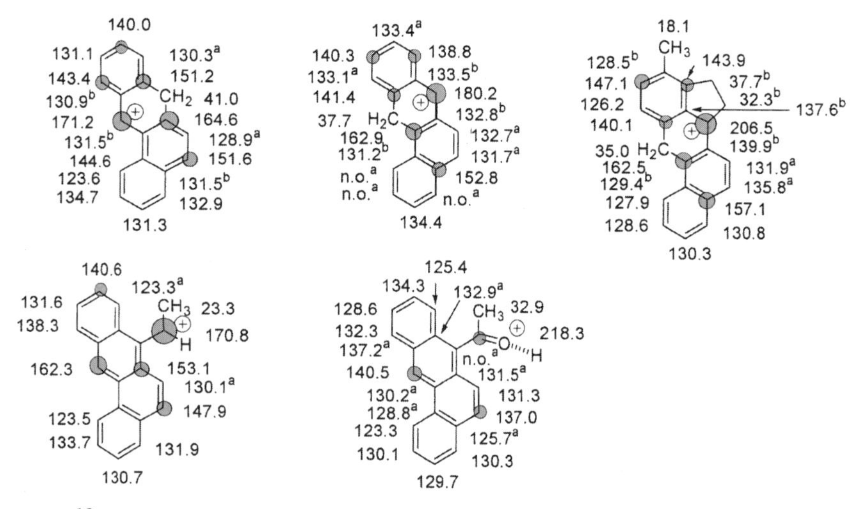

^{13}C NMR data for carbocations and a carboxonium ion from benzo[*a*]anthracene and cholanthrene derivatives (size of the dark circles is roughly proportional to magnitude of $\Delta\delta^{13}C$ values; a and b denote interchangeable assignments; n.o. = not observed)

These patterns are in overall qualitative agreement with the AM1-derived changes in charges. The AM1-derived relative arenium ion energies are also in concert with experiment.

9.5.6 Benzo[*c*]phenanthrenium and Benzo[*g*]chrysenium Cations

There is much evidence in the literature to establish a link between increased diol-epoxide reactivity and increasing structural deformation in the PAH. Structural deformaty/nonplanarity can be induced by increasing steric crowding at or in the vicinity of bay regions as in 5-methyl-*Ch*, 1-methyl-BA, and 12-methyl-*BA* (see discussion earlier). A more effective way to cause structural deformation is to create fjord regions by appropriate benzannelation as in benzo[*c*]phenanthrene *BcPh* and benzo[*g*]chrysene *BgCh*. Strained *fiord*-region diol-epoxides of *BcPh* and *BgCh* are potent DNA alkylating agents. On the basis of AM1 relative arenium ion energies, protonation at C5 is most favored.[19] Whereas attempts to directly observe parent *BcPh*-H^+ have not been successful, persistent carboxonium ions have been generated from its 3-methoxy and 3-hydroxy derivatives. In these cases attack is directed to C4. With the 3-OMe derivative, the carboxonium ion exists in two conformationally distinct forms, whereas with the 3-OH compound only an averaged structure is observed. Model nitration and benzoylation reactions on *BcPh* give substitution at C5:

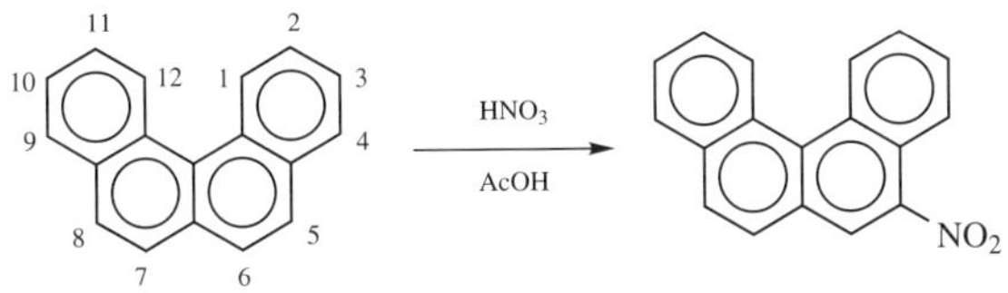

11 2 10 12 1 3 9 4 8 5 7 6

HNO_3 / AcOH → NO_2

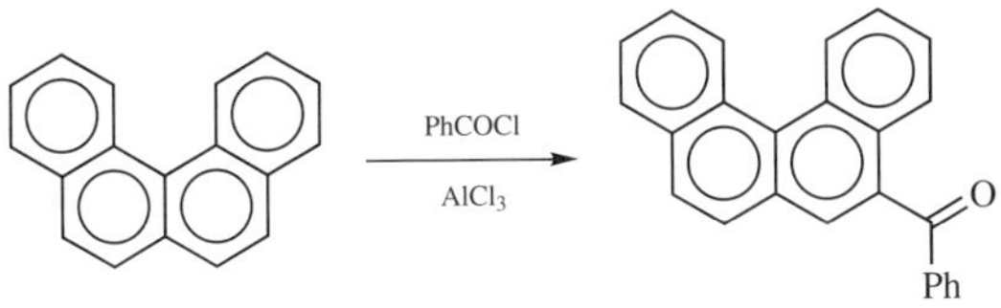

PhCOCl / $AlCl_3$ → O, Ph

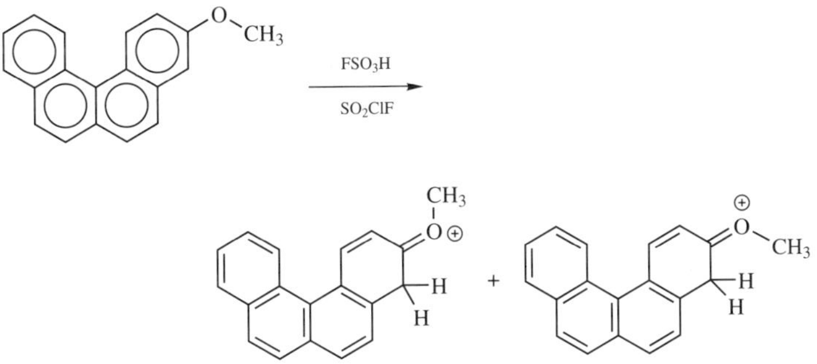

O CH_3 — FSO_3H / SO_2ClF →

CH_3 O⊕ H H + ⊕ O CH_3 H H

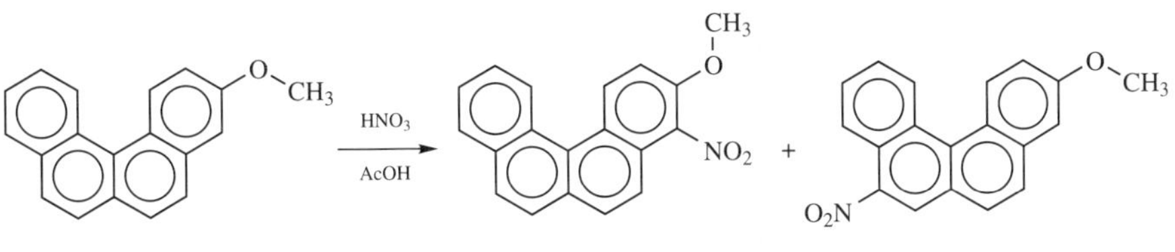

O CH_3 — HNO_3 / AcOH → CH_3 O NO_2 + O CH_3 O_2N

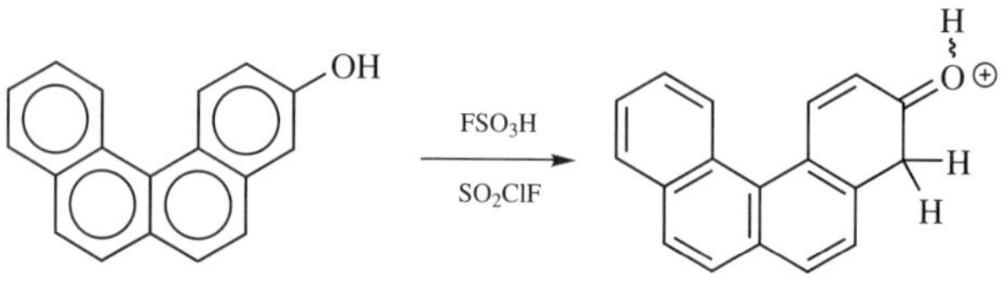

A persistent carbocation has been generated from parent *BgCh*.[19] The protonation site is C10, and this is in agreement with AM1. The 12-OMe and 12-OH derivatives of *BgCh* have also been generated; they behave analogously to *BcPh*. The 1,2,3,4-tetrahydro-1-one derivative of *BgCh* gives a carboxonium ion in the fjord:

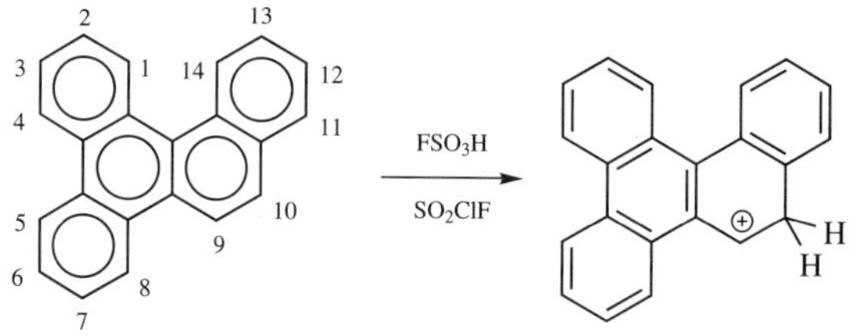

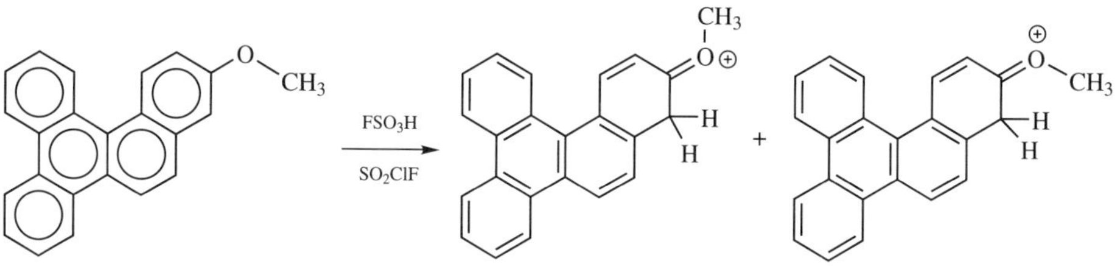

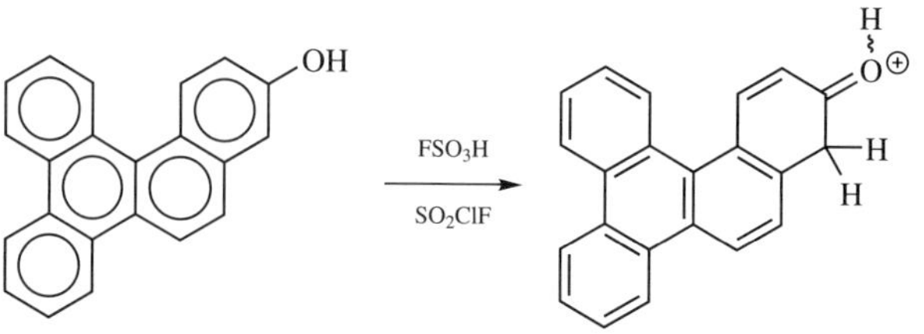

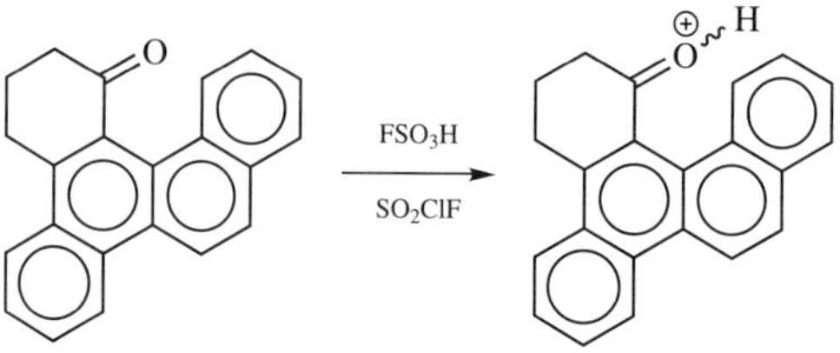

Charge delocalization in the *BcPh* carboxonium ions is limited to the substituted ring and one conjugated carbon, which identifies C1 (in the fjord) and C6 as favorable sites of subsequent nucelophilic attachment. This agrees with in vitro DNA binding studies for *BcPh*-diol-epoxides whereby covalent linkage occurs between the exocyclic amino group of the nucleoside residues and C1.

For *BgCh*-H^+, positive charge is delocalized within the *BcPh* moiety and does not involve the C ring. The magnitude of $\Delta\delta^{13}C$ values indicates that C9 (bay region) is the most likely candidate for nucleophilic attack on the carbocation, followed by C1/C14 (fjord region) and C3/C12. In the *BgCh* carboxonium ions positive charge is localized in the substituted ring and one conjugated carbon, with C14 and C9 identified as most probable sites for nucleophilic attack. These carbocation-based conclusions are in good agreement with in vitro DNA binding studies on *anti-BgCh*-dihydrodiol-13,14-epoxide, which was shown to connect to DNA via C14:

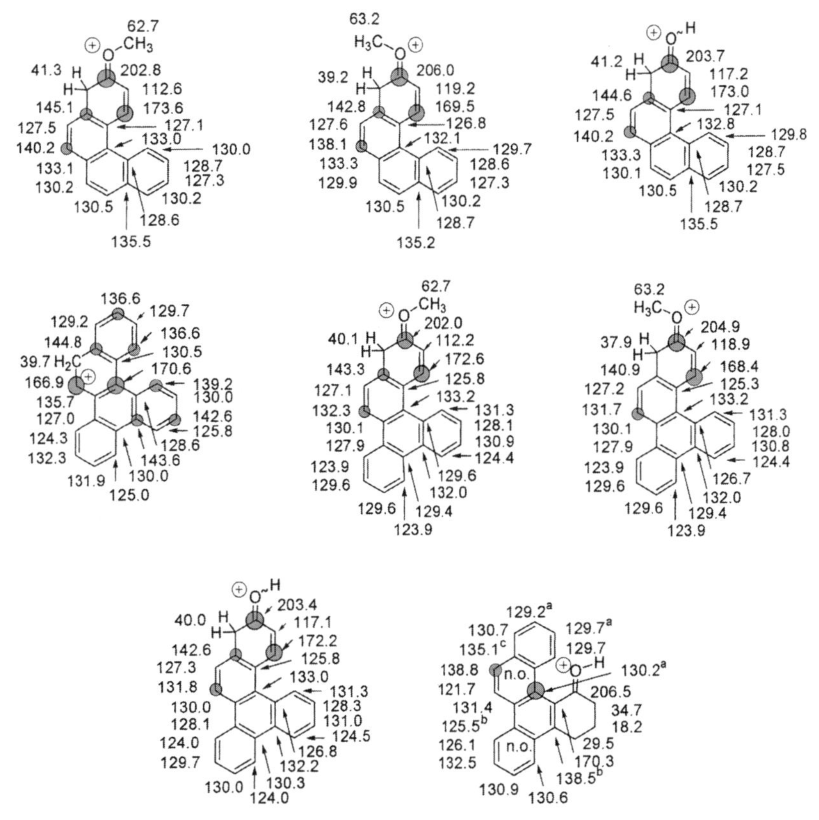

^{13}C data for carbocations and carboxonium ions from benzo[*a*]phenanthrene and benzo[*g*]chrysene derivatives (size of the dark circles is roughly proportional to magnitude of $\Delta\delta^{13}C$ values; a and b denote interchangeable assignments; n.o. = not observed)

GIAO-NMR has been used to derive charge delocalization mode based on computed $\Delta\delta^{13}C$ values for *BcPh*, *BgCh*, as well as *Ch*, as models of carcinogenic

PAH–epoxide ring opening. Relative aromaticity in various rings has been evaluated using NICS. These model studies illustrate that carbocations formed by opening of the bay-region and fjord-region PAH–epoxides are logical candidates for linkage to nucleophile via the bay and fjord regions. Computed NICS values show that rings which are remote to the carbocation and, therefore, less involved in delocalization remain most aromatic:

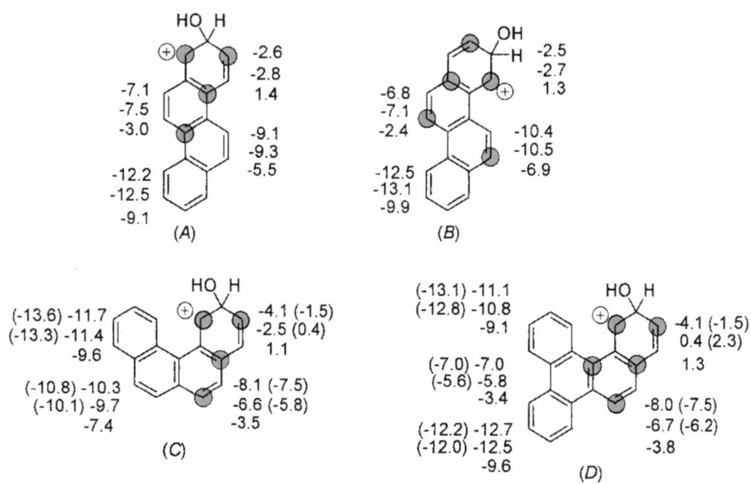

GIAO-NMR-derived charge delocalization mode and calculated NICS data at 1.0 Å and 0.5 Å above each ring centroid and at each ring centroid, respectively. Values outside parentheses are measured with OH group on the same side and inside parentheses with OH group on the opposite side

9.5.7 Dihydropyrenium(ethanophenanthrenium) Cations

Dihydropyrene is protonated *peri* to the ethano bridge.[21] On warming, it is transformed into pyrenium cation. The 2,7-di-*tert*-butyl derivative behaves similarly. Changing the superacid from FSO_3H/SO_2ClF to $FSO_3H.SbF_5$ (4 : 1)/SO_2ClF or $FSO_3H.SbF_5$ (1 : 1)/SO_2ClF produces symmetric dications:

FSO_3H / SO_2ClF ; Δt

FSO_3H–SbF_5 (1 : 1)/SO_2ClF or FSO_3H–SbF_5 (4 : 1)/SO_2ClF

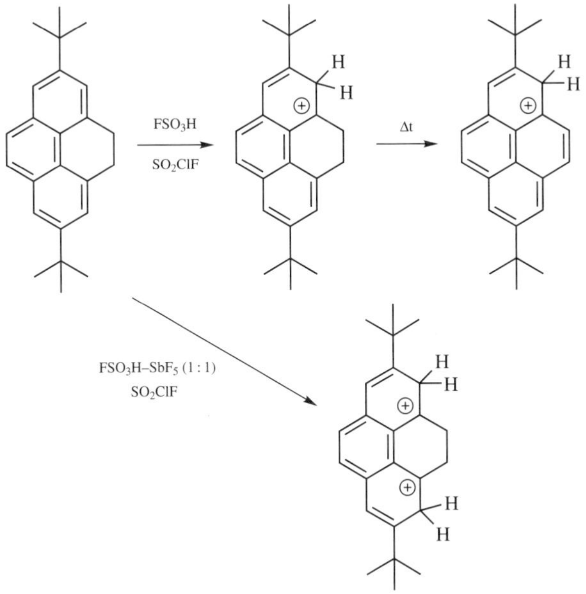

FSO_3H / SO_2ClF → Δt

$FSO_3H–SbF_5$ (1 : 1) / SO_2ClF

The charge delocalization patterns in the mono and dications are sketched below for comparison:

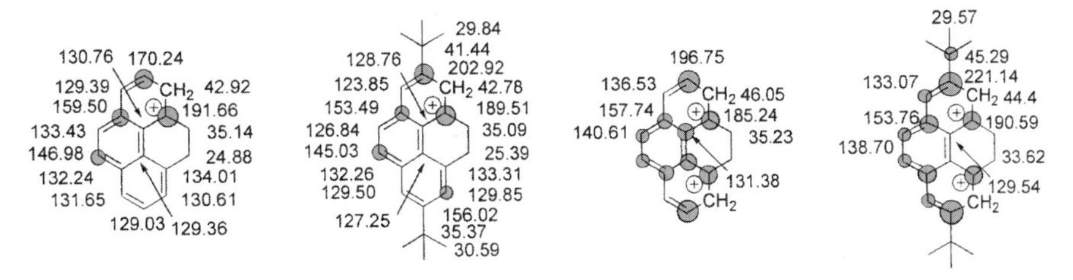

^{13}C NMR data for dihydropyrenium cations (size of the dark circles is roughly proportional to magnitude of $\Delta\delta^{13}C$ values)

9.5.8 Regioisomeric α-Pyrenyl-Substituted Carbocations

A series of regioisomeric secondary and tertiary α-pyrenyl-substituted carbenium ions have been studied:[22]

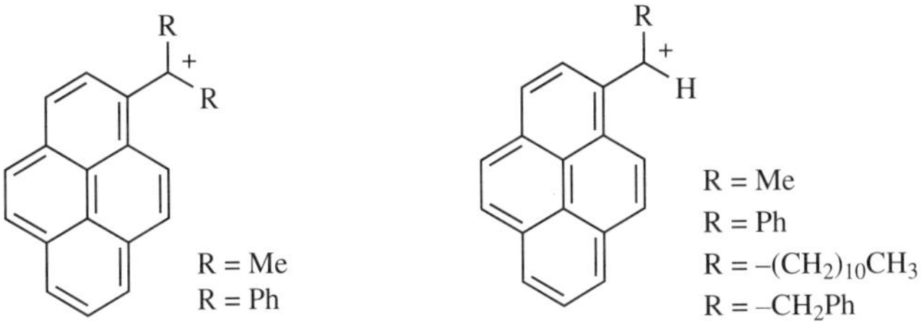

R = Me
R = Ph

R = Me
R = Ph
R = $-(CH_2)_{10}CH_3$
R = $-CH_2Ph$

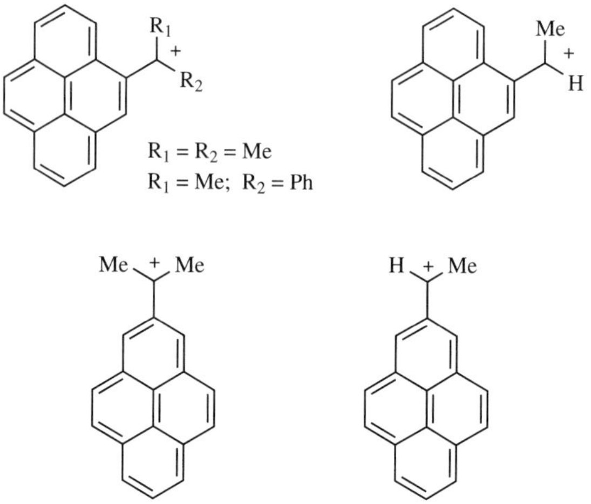

$R_1 = R_2 = Me$
$R_1 = Me;\ R_2 = Ph$

These delocalized carbocations serve as simplified models for the ring opening of diol-epoxide metabolites derived from benz[*a*]pyrene (a potent carcinogen) and benzo[*e*]pyrene (not a carcinogen). The analogy or approach is illustrated below:

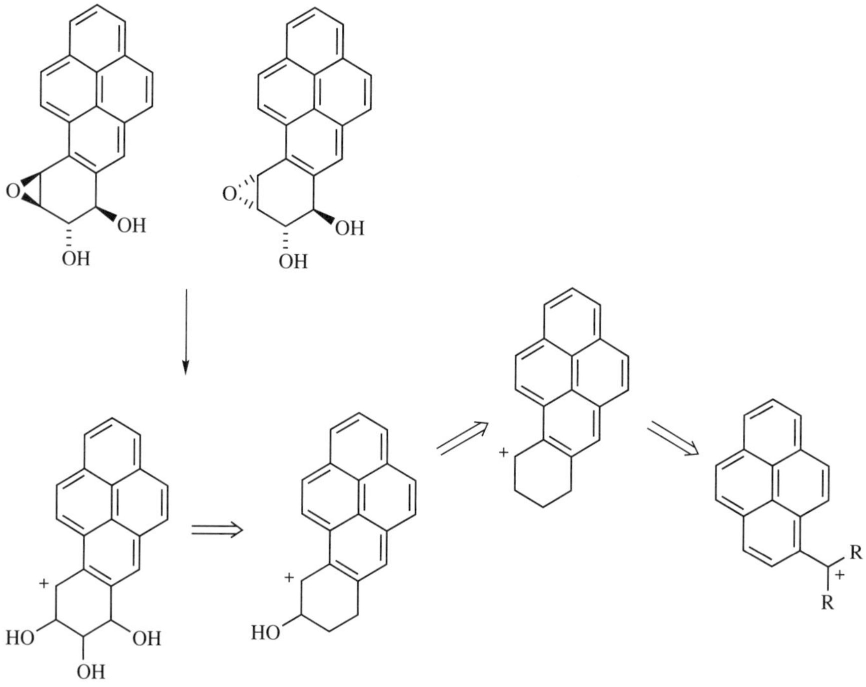

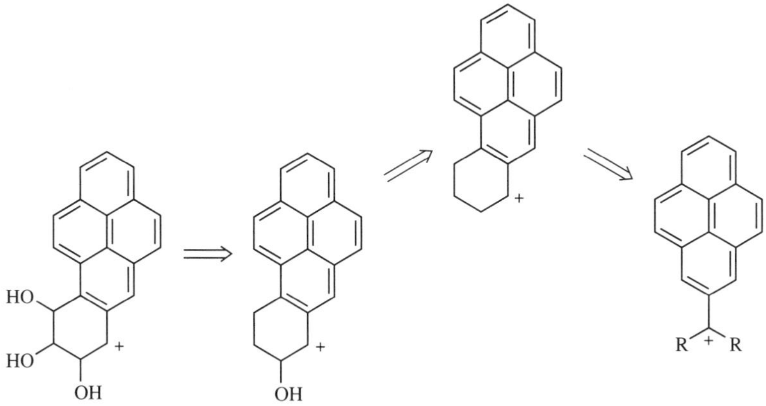

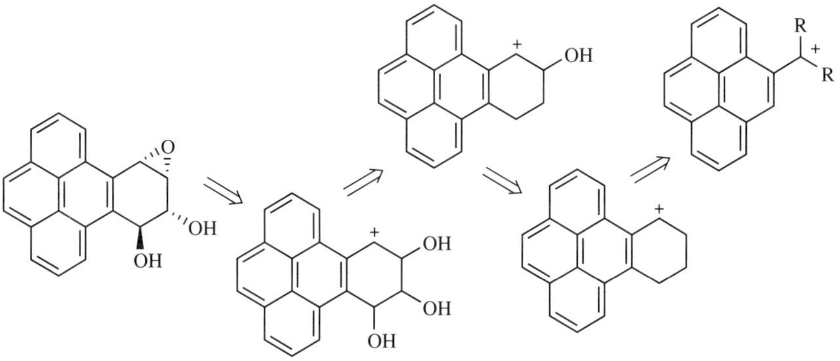

9.5.9 (1-Pyrenyl)diphenylmethyl Cation, 1,6- and 1,8-Bis(diphenylmethylenium)pyrene Dications

The (1-pyrenyl)diphenylmethyl cation is a significantly more crowded analog of trityl cation.[23] Conformational studies show that it undergoes a conventional two-ring flip (*ph,ph*). The 1,6- and 1,8-bis(diphenylmethylenium)pyrene dications undergo a one-ring flip (*py*) followed by a two-ring flip (*ph,ph*) with a higher rotational barrier. These crowded and highly delocalized mono- and dications have been generated from their carbinol precursors by reaction with TFA and are indefinitely stable at room temperature:

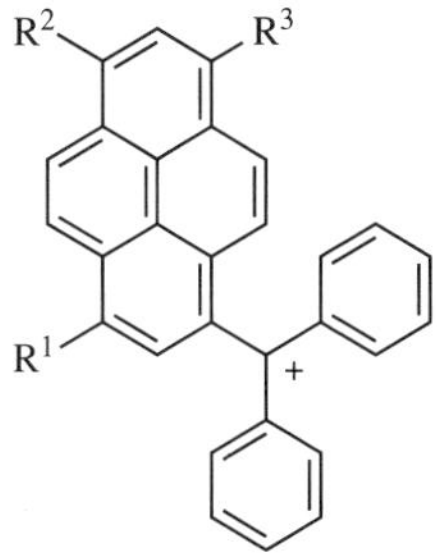

$R^1 = R^2 = R^3 = H$
$R^1 = R^3 = H$ $R^2 = {}^+C(Ph)_2$
$R^1 = R^3 = Br$ $R^2 = {}^+C(Ph)_2$
$R^1 = R^2 = H$ $R^3 = {}^+C(Ph)_2$
$R^1 = R^2 = Br$ $R^3 = {}^+C(Ph)_2$

9.5.10 Persistent $ArC^+(R)CF_3$ Carbocations

Several examples of regioisomeric α-pyrenyl- and α-phenanthrenyl-carbenium ions that are destabilized by an α-CF_3-substituent have been successfully generated.[24] In these cations, increasing electron demand leads to increased π participation and arenium ion character which are reflected in increased $\Delta\delta^{13}C$ values and in increased double-bond character, which shields the carbocation:

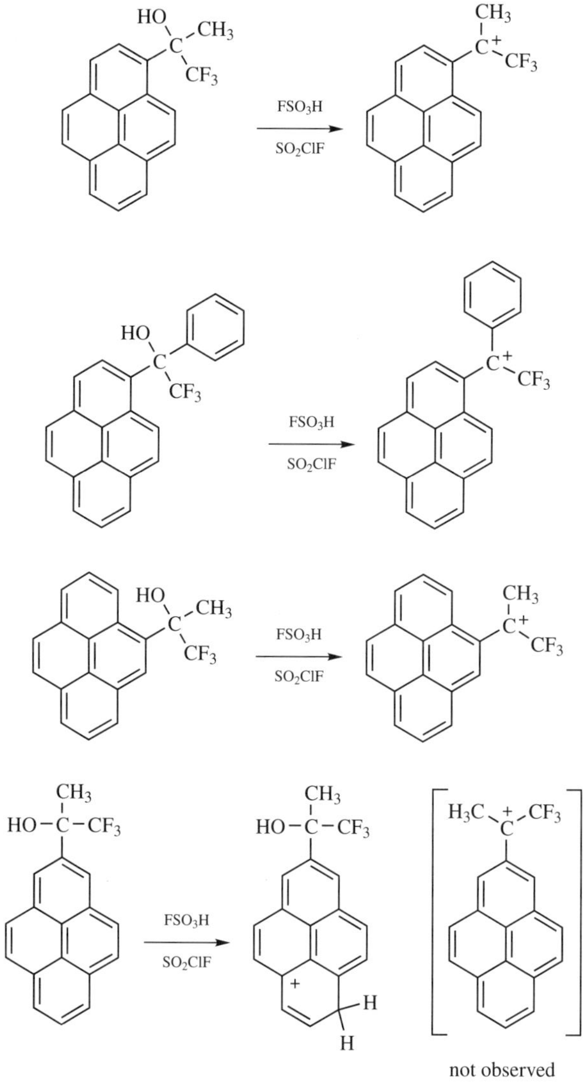

The results demonstrate that in cases where π participation is effective, the resulting destabilized carbocations are still persistent (1-pyrenyl-, 4-pyrenyl-, 9-phenanthrenyl-), whereas when π-participation is poor (2-pyrenyl-, 2-phenanthrenyl- and

3-phenanthrenyl-), either the precursor refuses to ionize and protonation moves to a different site (as in 2-pyrenyl-) or the in situ–formed carbocation immediately alkylates its precursor and dimerization/polymerization ensues.

Charge delocalization mode in the resulting carbocations are sketched below:

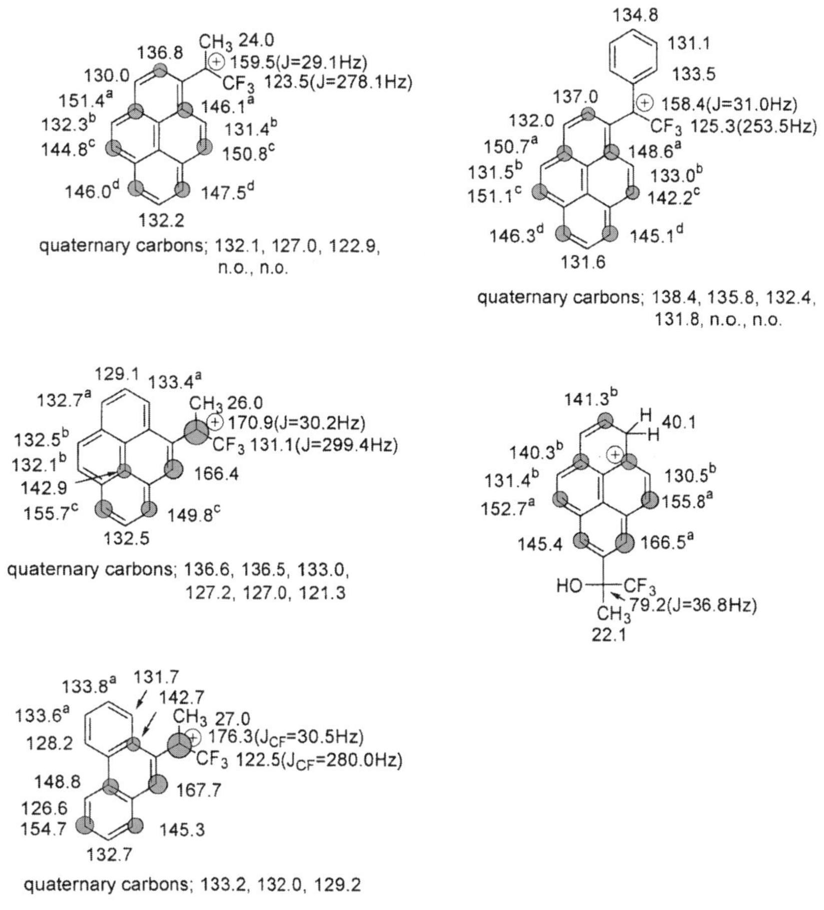

^{13}C NMR data for the CF_3-substituted cations (a,b and c denote interchangeable assignments; n.o. = not observed; C/F couples in parentheses)

9.5.11 Carboxonium-Substituted Pyrenium Ion

Regioisomeric carboxonium-substituted pyrenium ions have been generated by diprotonation of isomeric acetyl- and benzoylpyrenes in order to further test and tune the degree of π participation as a function of substitution position on a pyrene ring.[25] The configuration of the carboxonium group has been gauged via NOED experiments:

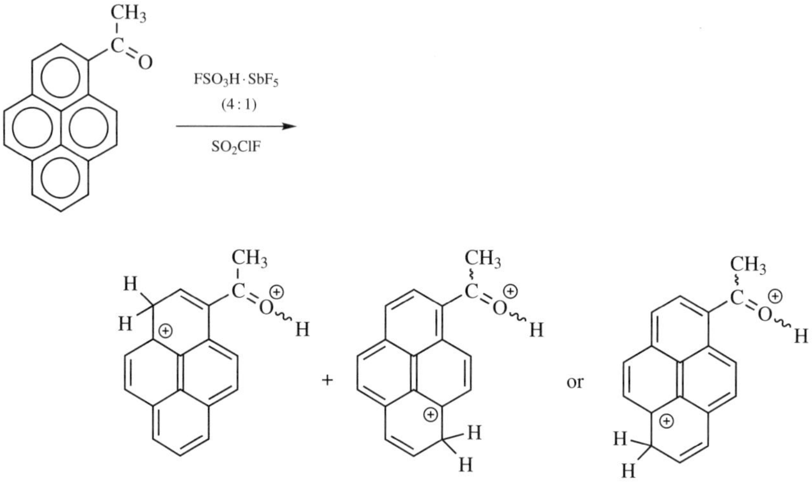

$FSO_3H \cdot SbF_5$ (4 : 1)

SO_2ClF

(93%) + or (7%)

$FSO_3H \cdot SbF_5$ (4 : 1)

SO_2ClF

−60°C

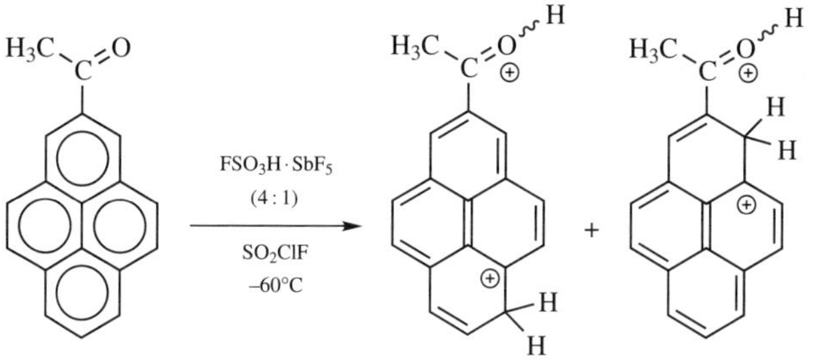

98 : 2 (−60°C)
67 : 33 (−40°C)

$FSO_3H \cdot SbF_5$ (4 : 1)

SO_2ClF

$FSO_3H \cdot SbF_5$ (4 : 1)

SO_2ClF

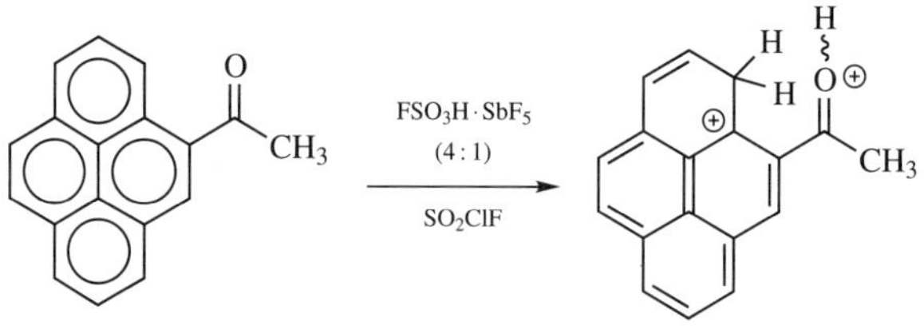

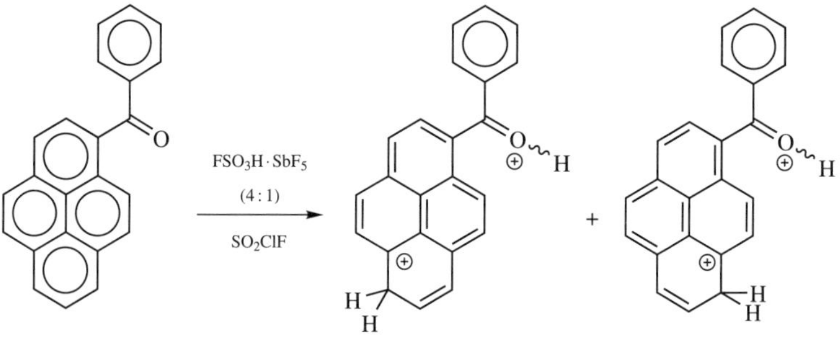

With isomeric diacetyl- and dibenzoylpyrenes, diprotontion leads to the formation of biscarboxonium dications:

$$\xrightarrow[SO_2ClF]{FSO_3H}$$

$$\xrightarrow[SO_2ClF]{FSO_3H}$$

When the carboxonium group is at an α position there is extensive charge delocalization and alternation at the pyrenium moiety. Charge delocalization modes in different regioisomeric pyrenium–carboxonium dications are sketched below for comparison; these studies indicate that the carboxonium group attached to a PAH acts as a strong electron sink, but the magnitude of its electronic response is sensitive to steric factors:

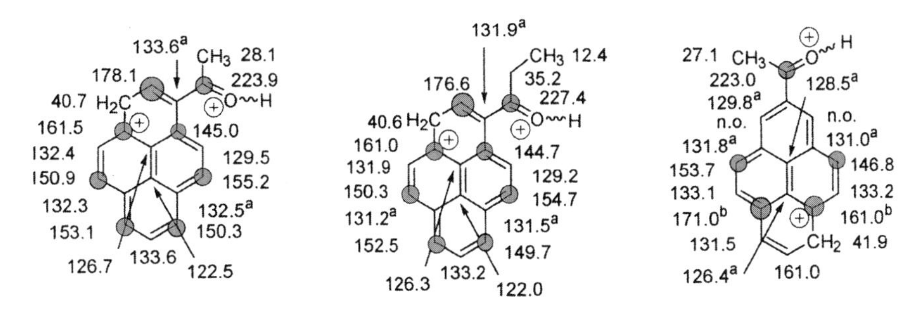

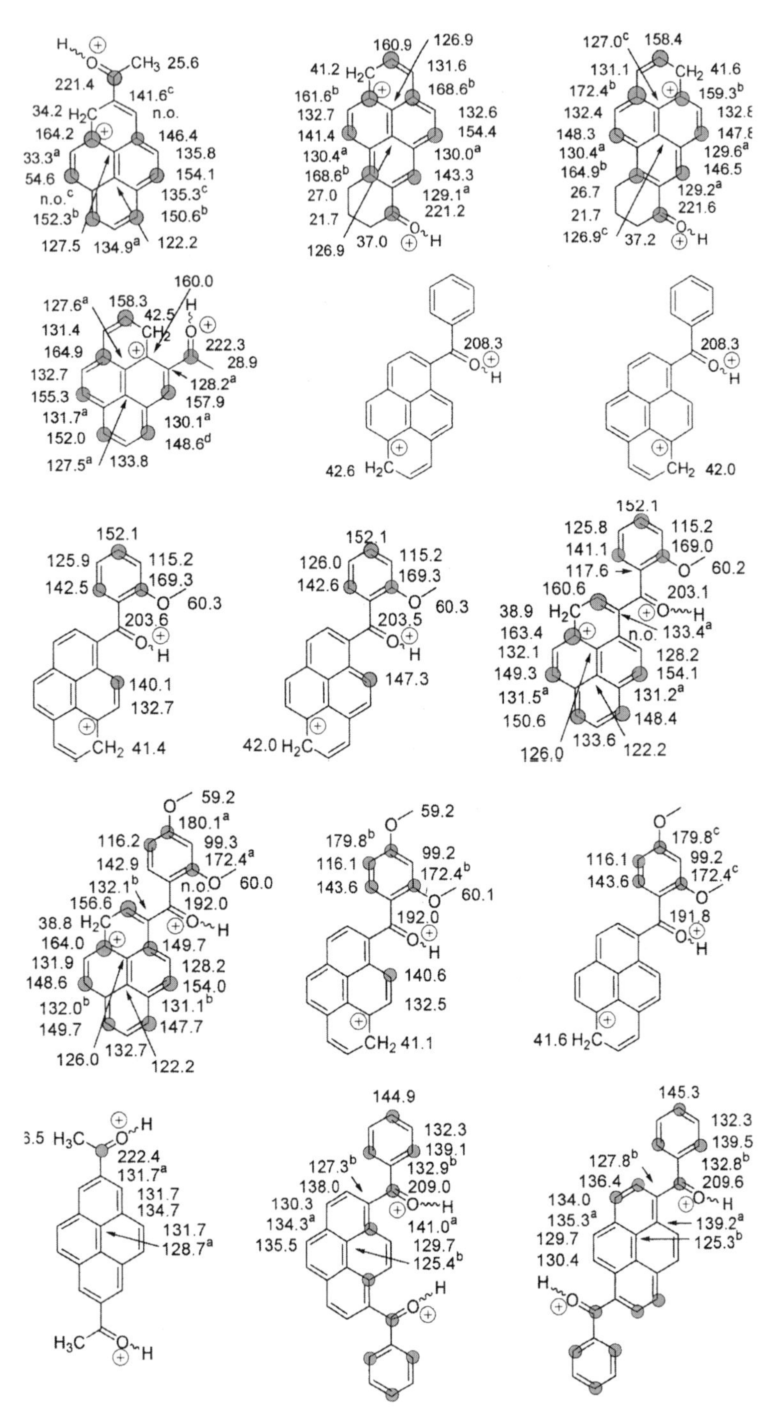

^{13}C NMR data for pyrenium-carboxonium dications and carboxonium dications (size of the dark circles is roughly proportional to magnitude of $\Delta\delta^{13}C$ values; a,b, and c denote interchangeable assignments; n.o. = not observed)

9.5.12 Dihydrocyclobuta[*e*]pyrenium Ions

Dihydrocyclobuta[*e*]pyrene is an in situ source of *o*-quinodimethane that allows the C60–pyrene adduct to be synthesized via [4 + 2]cycloaddition:[26]

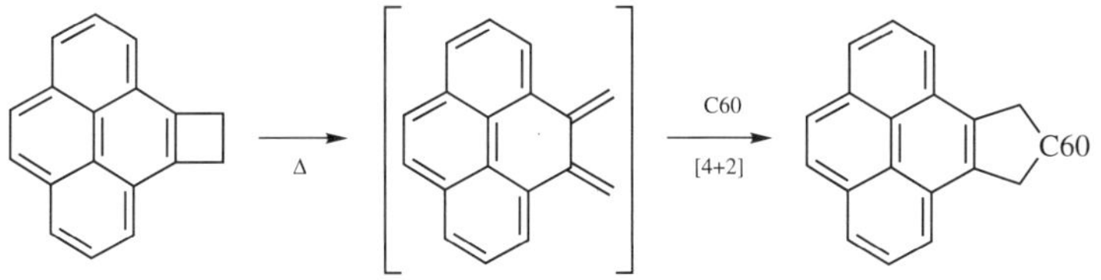

Protonation of dihydrocyclobuta[*e*]pyrene gives a 2 : 1 mixture of two arenium ions of α protonation:

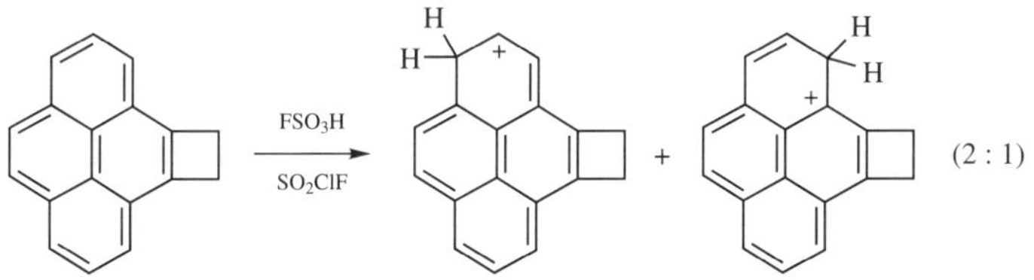

The C60–pyrene adduct is similarly protonated to give regioisomeric bound-arenium ions; however, solubility problems prevented full NMR study:

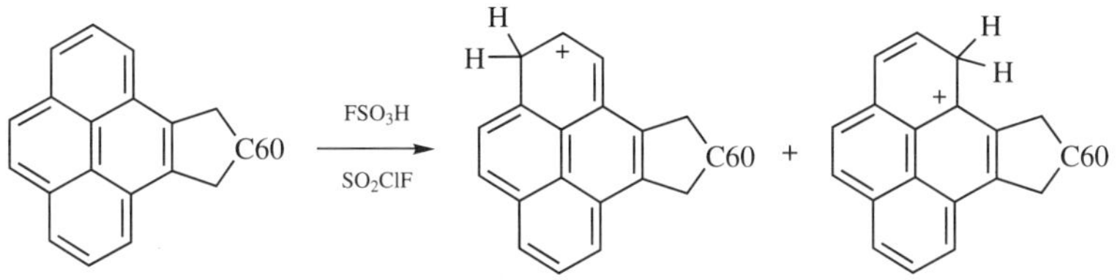

9.5.13 Methylene-Bridged PAH–Arenium Ions

Introduction of the methano bridge increases electrophilic reactivity in the PAH and confers increased tumorigenic and carcinogenic activity.

Protonation and nitration were used as mimic reactions for generation of biological electrophiles from a series of *nonalternant* methylene-bridged PAHs.[27] The site(s) of attack and charge delocalization modes in the resulting arenium ions have been probed as a function of structure and benzannelation mode. The results are sketched below:

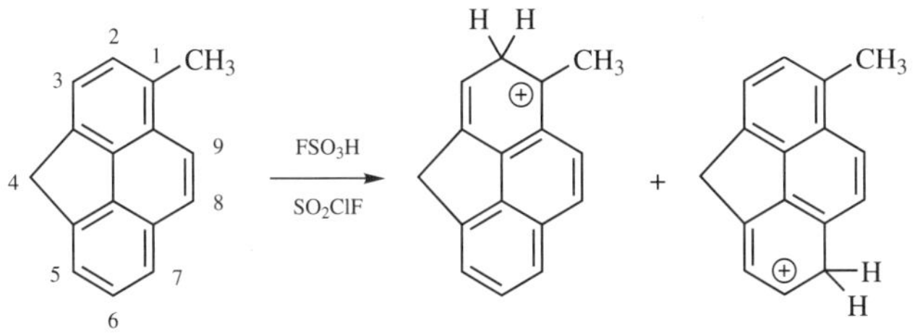

$\xrightarrow[SO_2ClF]{FSO_3H}$

$\xrightarrow[SO_2ClF]{FSO_3H}$

$\xrightarrow[SO_2ClF]{FSO_3H}$

$\xrightarrow[AcOH]{HNO_3}$ NO_2

$\xrightarrow[SO_2ClF]{FSO_3H}$

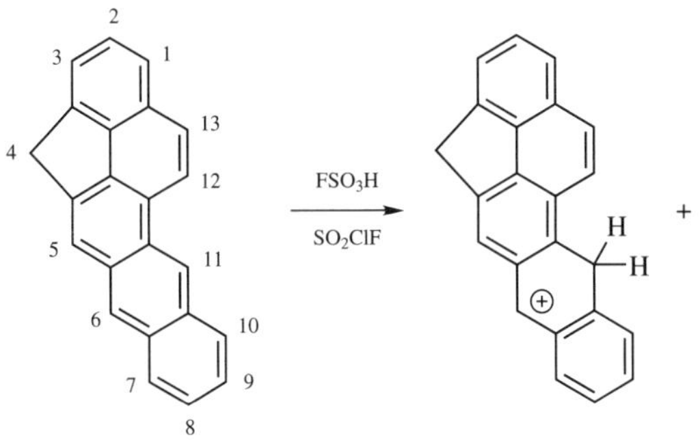

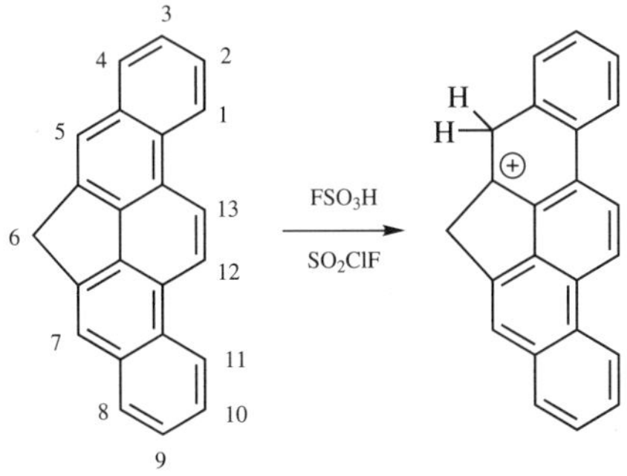

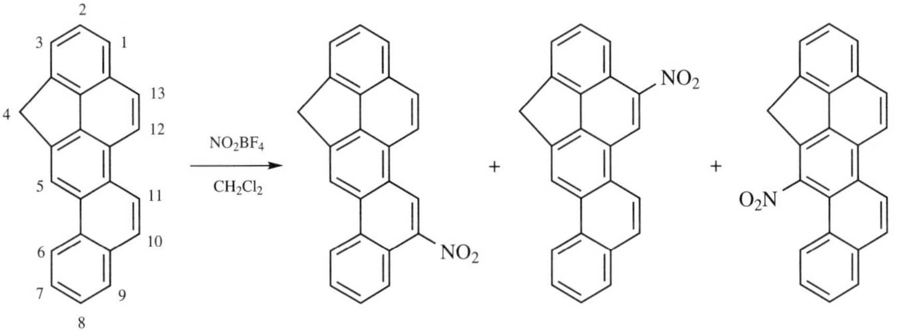

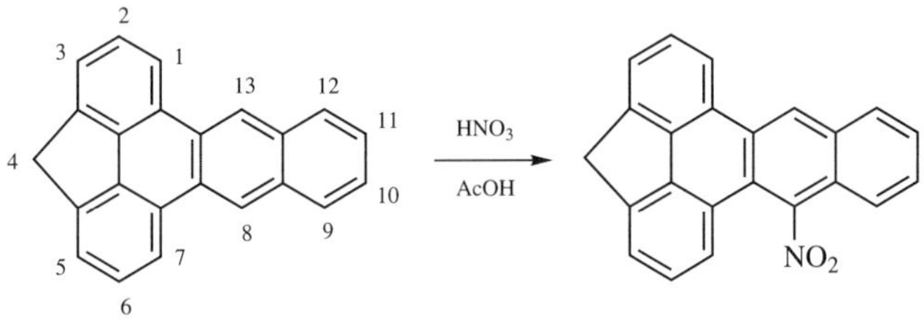

The ^{13}C NMR–derived charge delocalization modes in the carbocations are illustrated for comparison:

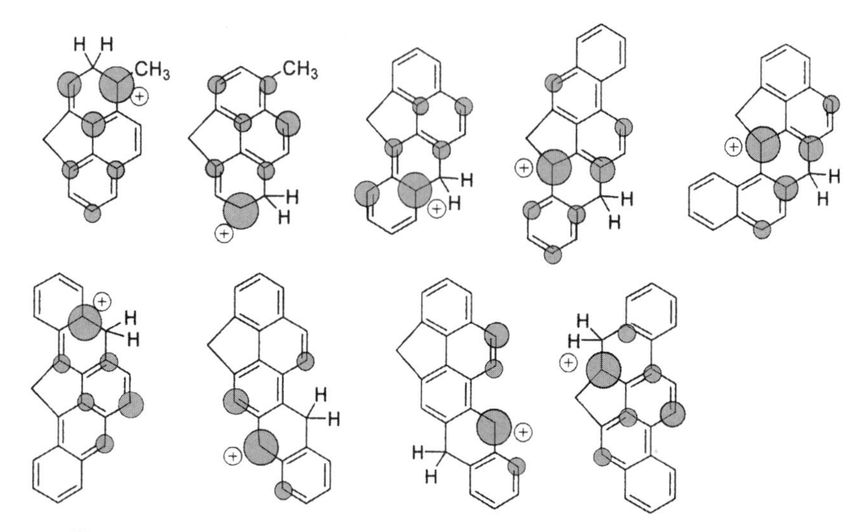

^{13}C NMR-derived charge delocalization mode in methylene-bridge arenium ions

These findings point to the importance of the methano bridge in directing electrophilic attack, which diminishes as the bridge moves from the more central "inner" positions to more peripheral "outer" positions. Using the derived charge delocalization patterns in the carbocations, preferred epoxidation sites and ring-opening modes have been predicted for comparison with biological reactivity studies.

9.5.14 Carbocations from Fluoranthene PAHs

Because parent fluoranthene is known to metabolize via the diol-epoxide pathway, its triol-carbocations attracted the attention of toxicologists. Stable ion studies[28] have shown that the sites of attack in parent fluoranthene and its more active benzannelated derivative benz[*e*]acephenanthrylene are the same. A methoxy substituent in benz[*e*]acephenanthrylene has a strong directive effect (protonation and nitration). The resulting *nonalternant* carbocations are paratropic, and most proton resonances are shielded relative to their neutral precursors. Indeno[1,2,3-*cd*]pyrene, which acts as complete carcinogen, is protonated at an α position of the pyrene moiety:

$\xrightarrow[SO_2ClF]{FSO_3H}$

5 6 4 7 3 8 2 9 1 12 10 11

FSO_3H / SO_2ClF →

⊕ H H

FSO_3H / SO_2ClF →

O CH_3

H H O ⊕ CH_3 + H H O CH_3 ⊕

NO_2BF_4 / $CHCl_3$ →

O CH_3

NO_2 O CH_3

FSO_3H / SO_2ClF →

F

⊕ H H F

9 8 10 7 10 6 11 5 1 4 2 3

FSO_3H / SO_2ClF →

H H ⊕

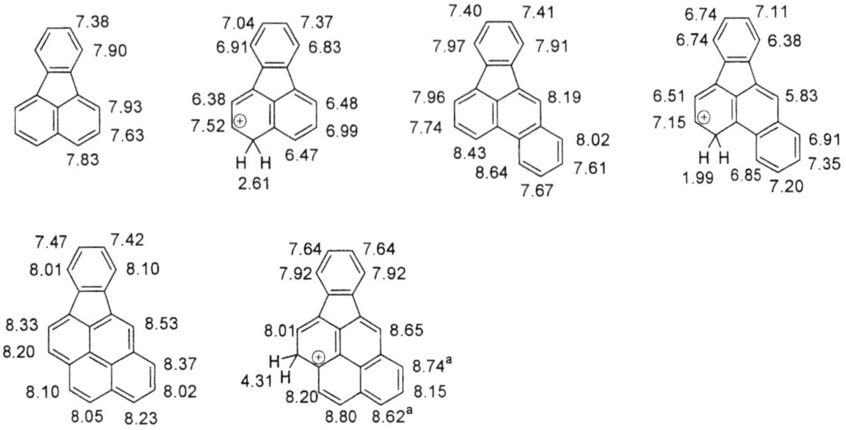

^{1}H NMR data for flouranthenium cations and their precursors (a denotes interchangeable assignments)

9.6 CONCLUDING REMARKS AND CONTINUATION PLAN

This review has demonstrated that considerable progress has been achieved in the area of stable arenium ions from PAHs. Since the early 1990s a large number of persistent carbocations from several classes of *alternant* and *nonalternant* PAHs ranging in activity from highly potent to inactive have been generated and studied. A group of α-PAH-substituted carbocations have also been studied that serve as simple models of epoxide ring opening. The sites of electrophilic attack and charge delocalization modes in the resulting carbocations and carboxonium ions have been determined. Conformational aspects and substituent effects on charge delocalization and π participation have also been probed as a function of structure. These investigations represent the groundwork for assembling carbocation-based structure–reactivity relationships.

In continuation, numerous other classes of PAHs and their substituted derivatives including biologically active analogs that possess twisted/nonplanar framework remain to be examined. These studies should be complemented with DFT calculations to determine geometry and relative carbocation stability, and to deduce charge delocalization maps based on GIAO-NMR and the NPA-derived changes in charges. These studies will also allow quantitative correlations between the computed and experimental NMR chemical shifts to be examined for various classes of delocalized PAH–carbocations.

Quenching experiments between model benzylic carbocations and biologically relevant bases should be performed in order to isolate and characterize the PAH–nucleophile adducts. In parallel, DNA-binding studies are to be carried out on suitable covalent precursors to determine the extent of binding as a function of structure. Protonation studies on PAH-epoxides and dihydrodiols (proximate carcinogens) will provide additional and complementary stable ion data. Charge delocalization maps could enable certain epoxides and epoxide ring-opening modes to be singled out relative to alternatives and certain sites to be identified as most probable sites for covalent attachment to nucleotides. These trends are to be compared with DNA binding and carcinogenicity data to ascertain their relationships and correspondence, which in due course could become a basis for "predicting" biological potency.

ACKNOWLEDGMENTS

Former postdoctoral fellows Dr. Sandro Hollenstein, Dr. Mutsuo Tanaka, and Dr. Takao Okazaki were instrumental in advancing various PAH–arenium ion projects. I especially thank my former Postdoc Dr. Takao Okazaki (Kyoto University) for his exemplary productivity and for assistance in preparation of this manuscript. Fruitful collaborations with Professors P. E. Hansen (Roskilde) and S. E. Galembeck (Sao Paulo) are also acknowledged.

Support of our work in the area of reactive intermediates of carcinogenesis of PAHs by the NCI of NIH (R15 CA 78235-01A1 and R15 CA 63595-01A1) is gratefully acknowledged.

REFERENCES

1. *Monographs*: (a) R. G. Harvey, *Polycyclic Hydrocarbons and Carcinogenesis*, ACS Symposium Series 283, ACS, Washington DC, 1985; (b) R. G. Harvey, *Polycyclic Aromatic Hydrocarbons Chemistry and Carcinogenicity*, Cambridge Monographs on Cancer Research, Cambridge, UK, 1991; (c) R. G. Harvey, *Polycyclic Aromatic Hydrocarbons*, Wiley-VCH, New York, 1997; (d) M. M. Coombs and T. S. Bhatt, *Cyclopenta[a]phenanthrenes*, Cambridge Monographs on Cancer Research, Cambridge, UK, 1987.
2. *Proceedings*: (a) M. Cooke, K. Loening, and J. Merritt, *Polynuclear Aromatic Hydrocarbons: Measurements, Means, and Metabolism;* 11th Int. Symp. Polynuclear Aromatic Hydrocarbons, Battelle Press, Columbus, OH, 1999; (b) D. M. Jerina, J. M. Sayer, D. R. Thakker, H. Yagi, W. Levin, A. W. Wood, and A. H. Conney, *Jerusalem Symp. Quantum Chem. Biochem.* **13**, 1–12 (1980); (c) P. W. Jones and P. Leber, *Polynuclear Aromtic Hydrocarbons*; 3rd Int. Symp. Chemistry and Biology—Carcinogenesis and Mutagenesis; Ann Arbor Science, Ann Arbor, MI, 1979.
3. *Reviews*: (a) R. G. Harvey and N. E. Geacintov, *Acc. Chem. Res.* **21**, 66–73 (1998); (b) R. G. Harvey, *Acc. Chem. Res.* **14**, 218–226 (1981); (c) T. C. Bruice and P. Y. Bruice, *Acc. Chem. Res.* **9**, 378–384 (1976); (d) H. Bartsch, *Mutat. Res.* **462**, 255–279 (2000); (e) Y.-J. Surh, *Chem.-Bio. Interact.* **109**, 221–235 (1998); (f) A. Dipple, *Carcinogenesis* **16**, 437–441 (1995); (g) J. P. Lowe and B. D. Silverman, *Acc. Chem. Res.* **17**, 332–338 (1984).
4. *Notable kinetic/solvolytic studies*: (a) N. T. Nashed, S. K. Balani, R. J. Loncharich, J. M. Sayer, D.Y. Shipley, R. S. Mohan, D. L. Whalen, and D. M. Jerina, *J. Am. Chem. Soc.* **113**, 3910–3919 (1991); (b) N. T. Nashed, J. M. Sayer, and D. M. Jerina, *J. Am. Chem. Soc.* **115**, 1723–1730 (1993); (c) N. T. Nashed, A. Bax, R. J. Loncharich, J. M. Sayer, and D. M. Jerina, *J. Am. Chem. Soc.* **115**, 1711–1722 (1993); (d) P. Y. Bruice, T. C. Bruice, P. M.

Dansette, H. G. Selander, H. Yagi, and D. M. Jerina, *J. Am. Chem. Soc.* **98**, 2965–2973 (1976); (e) N. T. Nashed, T. V. S. Rao, and D. M. Jerina, *J. Org. Chem.* **58**, 6344–6348 (1993); (f) J. W. Keller and C. Heildelberger, *J. Am. Chem. Soc.* **98**, 2328–2336 (1976); (g) S. C. Gupta, N. B. Islam, D. L. Whalen, H. Yagi, and D. M. Jerina, *J. Org. Chem.* **58**, 3812–3815 (1987); (h) N. B. Islam, D. L. Whalen, H. Yagi, and D. M. Jerina, *J. Am. Chem. Soc.* **109**, 2108–2111 (1987), (i) R. E. Royer, T. A. Lyle, G. G. Moy, G. H. Daub, and D. L. Vander Jagt, *J. Org. Chem.* **44**, 3202–3207(1979); (j) R. E. Royer, G. H. Daub, and D. L. Vander Jagt, *J. Org. Chem.* **44**, 3196–3201 (1979).

5. *Structure–activity relationships—representative studies*: (a) Chapter 3 in Ref. 1b; (b) Chapter 5 in Ref. 1a; (c) R. G. Harvey and F. B. Dunne, *Nature* **273**, 566–568 (1978); (d) C. B. Huggins, J. Pataki, and R. G. Harvey, *Proc. Natl. Acad. Sci.* **58**, 2253–2260 (1967).
6. *Stereoselective synthesis of metabolites—representative cases*: (a) Chapter 14 in Ref. 1b; (b) A. S. Kiselyov, H. Lee, and R. G. Harvey, *J. Org. Chem.* **60**, 6123–6128 (1995); (c) D. R. Bushman, S. J. Grossman, D. M. Jerina, and R. E. Lehr, *J. Org. Chem.* **54**, 3533–3544 (1989); (d) F.-J. Zhang and R. G. Harvey, *J. Org. Chem.* **63**, 2771–2773 (1998); (e) W. Dai, E. Abu-Shqara, and R. G. Harvey, *J. Org. Chem.* **60**, 4905–4911 (1995); (f) J. Krzeminski, J.-M. Lin, S. Amin, and S. S. Hecht, *Chem. Res. Toxicol.* **7**, 125–129 (1994); (g) W. H. Rastetter, R. B. Nachbar, Jr., S. Russo-Rodriguez, R. V. Wattley, W. G. Thilly, B. M. Andon, W. L. Jorgensen, and M, Ibrahim, *J. Org. Chem.* **47**, 4873–4878 (1982).
7. *Examples of X-ray analysis*: (a) Chapter 5 in Ref. 1a; (b) A. K. Katz, H. L. Carrell, and J. P. Glusker, *Carcinogenesis* **19**, 1641–1648 (1998); (c) M. K. Lakshman, P. L. Kole, S. Chaturedi, J. H. Saugier, H. J. C. Yeh, J. P. Glusker, H. L. Carrell, A. K. Katz, C. E. Afshar, W.-M. Dashwood, G. Kenniston, and W. L. Baird, *J. Am. Chem. Soc.* **122**, 12629–12636 (2000); (d) L. Shimoni, H. L. Carrell, M. M. J. P. Glusker, and M. M. Coombs, *J. Am. Chem. Soc.* **116**, 8162–8168 (1994).
8. *Examples of PAH-DNA adducts*: (a) H. Kroth, H. Yagi, A. Seidel, and D. M. Jerina, *J. Org. Chem.* **65**, 5558–5564 (2000); (b) A. S. Giles, A. Seidel, and D. H. Philips, *Chem. Res. Toxicol.* **10**, 1275–1284 (1997); (c) S. K. Agrawal, J. M. Sayer, H. J. C. Yeh, L. K. Pannell, B. D. Hilton, M. A. Pigott, A. Dipple, H. Yagi, and D. M. Jerina, *J. Am. Chem. Soc.* **109**, 2497–2504 (1987); (d) H. Lee, E. Luna, M. Hinz, J. J. Stezowski, A. S. Kiselyov, and R. G. Harvey, *J. Org. Chem.* **60**, 5604–5613 (1995); (e) A. S. Kiselyov, T. Steinbrecher, and R. G. Harvey, *J. Org. Chem.* **60**, 6129–6134 (1995); (f) J. Szeliga, J. E. Page, B. D. Hilton, A. S. Kiselyov, R. G. Harvey, Y. M. Dunayevskiy, P. Vouros, and A. Dipple, *Chem. Res. Toxicol.* **8**, 1014–1019 (1995).
9. *Representative computational studies*: (a) Ref. 3g; (b) G. L. Borosky, *J. Org. Chem.* **64**, 7738–7744 (1999); (c) A. F. Lehner, J. Horne, and J. W. Flescher, *Theochem* **366**, 203–217 (1996); (d) J. R. Rabinowitz and S. B. Little, *Chem. Res. Toxicol.* **5**, 286–292 (1992); (e) B. D. Silverman, *Chem.-Biol. Interact.* **53**, 313–325 (1985).
10. (a) Chapter 11 in Ref. 1a; (b) N. V. S. RamaKrishna, E. L. Cavalieri, E. G. Rogan, G. Dolnikowski, R. L. Cerny, M. L. Gross, H. Jeong, R. Jankowiak, and G. L. Small, *J. Am. Chem. Soc.* **114**, 1863–1874 (1992); (c) E. Cavalieri and R. Roth, *J. Org. Chem.* **41**, 2679–2684 (1976); (d) R. Todorovic, F. Ariese, P. Devanesan, R. Jankowiak, G. J. Small, E. Rogan, and E. Cavalieri, *Chem. Res. Toxicol.* **10**, 941–947 (1997).
11. (a) A. A. Leon, J. C. Ball, S. Foxall- Van Aken, G. H. Daub and D. L. Vander Jagt, *Chem.-Biol. Interactions* **56**, 101–111 (1985); (b) K. H. Stansbury, J. W. Flesher, and R. C. Gupta, *Chem. Res. Toxicol.* **7**, 254–259 (1994).

12. (a) G. Lamm and G. R. Pack, *Proc. Natl. Acad. Sci. USA* **87**, 9033–9036 (1990); (b) H. S. Wong and G. R. Pack, *Biophys. J.* **72**, 291–300 (1997).
13. K. K. Laali *Chem. Rev.* **96**, 1873–1906 (1996).
14. K. K. Laali, S. Hollenstein, and P. E. Hansen, *J. Chem. Soc. Perkin Trans. 2* 2207–2213 (1997).
15. K. K. Laali and S. Hollenstein, *J. Chem. Soc. Perkin Trans. 2*. 897–904 (1998).
16. K. K. Laali, S. Hollenstein, S. E. Galembeck, and M. M. Coombs, *J. Chem. Soc. Perkin Trans. 2*. 211–220 (2000).
17. K. K. Laali, T. Okazaki, and M. M. Coombs, *J. Org. Chem.* **65**, 7399–7405 (2000).
18. K. K. Laali, S. Hollenstein, R. G. Harvey, and P. E. Hansen, *J. Org. Chem.* **62**, 4023–4028 (1997).
19. K. K. Laali, T. Okazaki, S. Kumar, and S. E. Galembeck, *J. Org. Chem.* **66**, 780–788 (2001).
20. K. K. Laali and M. Tanaka, *J. Org. Chem.* **63**, 7280–7285 (1998).
21. K. K. Laali and P. E. Hansen, *Res. Chem. Intermed.* **22**, 737–751 (1996).
22. K. K. Laali and P. E. Hansen, *J. Org. Chem.* **62**, 5804–5810 (1997).
23. P. E. Hansen, J. Spanget-Larsen, and K. K. Laali, *J. Org. Chem.* **63**, 1827–1835 (1998).
24. K. K. Laali, M. Tanaka, S. Hollenstein, and M. Cheng, *J. Org. Chem.* **62**, 7752–7757 (1997).
25. K. K. Laali, T. Okazaki, and P. E. Hansen, *J. Org. Chem.* **65**, 3816–3828 (2000).
26. K. K. Laali, S. Hollenstein, S. E. Galembeck, Y. Nakamura, and J. Nishimura, *J. Chem. Soc. Perkin Trans. 2* 2129–2132 (1999).
27. K. K. Laali, T. Okazaki, and R. G. Harvey, *J. Org. Chem.* **66**, 3977–3983 (2001).
28. K. K. Laali, T. Okazaki, and S. E. Galembeck, 621–629 (2002).

10

CHROMIUM TRICARBONYL–COORDINATED CARBOCATIONS

Bruce N. Hietbrink

Department of Chemistry
California State University, Northridge
Northridge, California

Dean J. Tantillo

Department of Chemistry
University of California, Davis
Davis, California

Kendall N. Houk and Craig A. Merlic

Department of Chemistry and Biochemistry
University of California, Los Angeles
Los Angeles, California

Carbocation Chemistry, Edited by George A. Olah and G. K. Surya Prakash
ISBN 0-471-28490-4

10.1 INTRODUCTION

Formation of chromium tricarbonyl complexes of aromatic compounds profoundly alters their reactivity.[1] This has been the basis for a great deal of study, and the novel properties of complexed arenes have been exploited by many synthetic chemists.[2] Although the metal fragment is often regarded as an electron-withdrawing substituent because of its ability to inductively stabilize anionic intermediates and to activate the arene ring toward nucleophilic aromatic substitution, it also, perhaps surprisingly, stabilizes cationic species. This pair of seemingly opposite characteristics has prompted the description of chromium tricarbonyl complexed arenes as "hermaphroditic."[3]

The origins and limitations of the stabilization of carbocations by chromium tricarbonyl were uncovered through theoretical studies of its interactions with a series of stabilized cations: benzyl cation **1**, phenonium cation **2**, and two different benzonorbornenyl cations, **3** and **4**. Chromium complexes of each of these cations (**5**–**10**) (Fig. 10.1) were examined theoretically, and these studies found that, depending on the geometric constraints of each carbocation framework, stabilization by the metal fragment can either compete with or diminish other modes of cation stabilization.[4–8] This chapter reviews these results and summarizes the understanding gained from these theoretical studies.

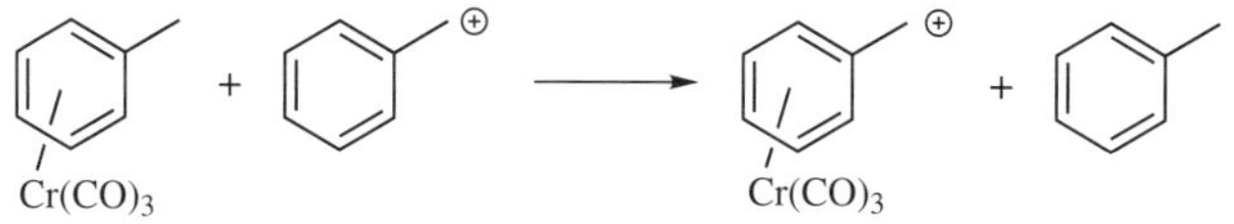

$Cr(CO)_3$ + **1** ⟶ $Cr(CO)_3$ **5** + (10.1)

$\Delta E = -12.0$ kcal/mol

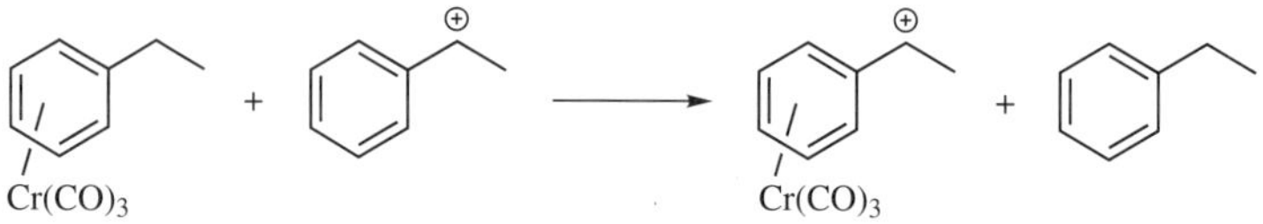

$Cr(CO)_3$ + ⟶ $Cr(CO)_3$ **11** + (10.2)

$\Delta E = -5.1$ kcal/mol

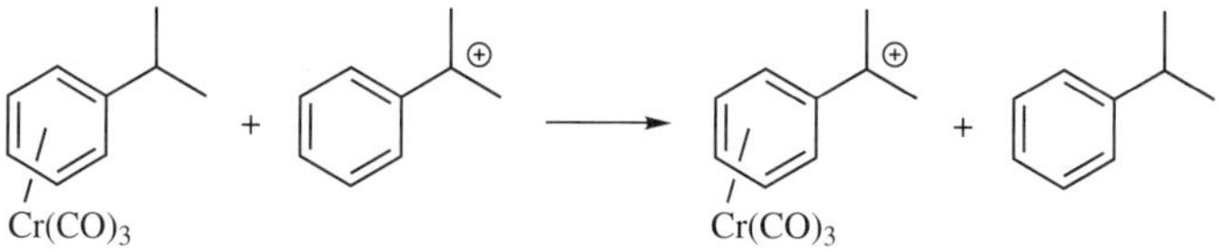

$Cr(CO)_3$ + ⟶ $Cr(CO)_3$ **12** + (10.3)

$\Delta E = -1.7$ kcal/mol

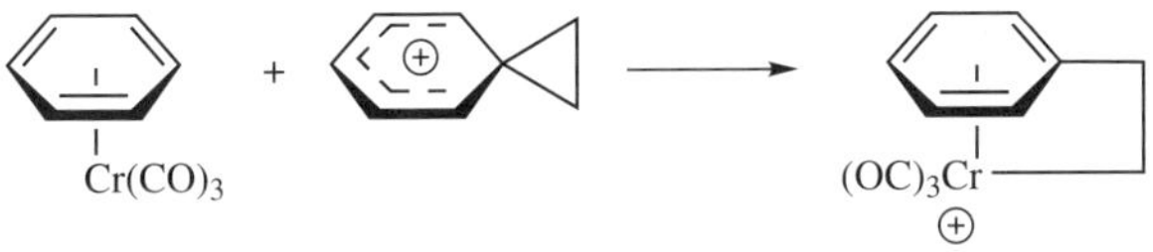

$Cr(CO)_3$ + **2** → $(OC)_3Cr$ ⊕ **6** + (10.4)

$\Delta E = -6.6$ kcal/mol

$Cr(CO)_3$ + **2** → $Cr(CO)_3$ **13** + (10.5)

$\Delta E = +5.0$ kcal/mol

$(OC)_3Cr$ + **3** → $(OC)_3Cr$ **7** + (10.6)

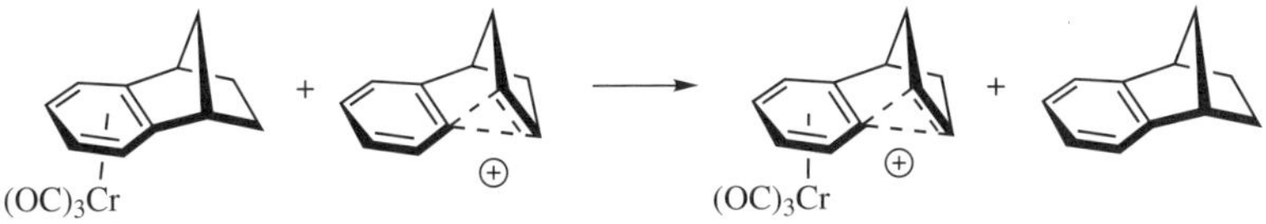

$\Delta E = +4.2$ kcal/mol

$(OC)_3Cr$ + **3** → $(OC)_3Cr$ ⊕ **8** + (10.7)

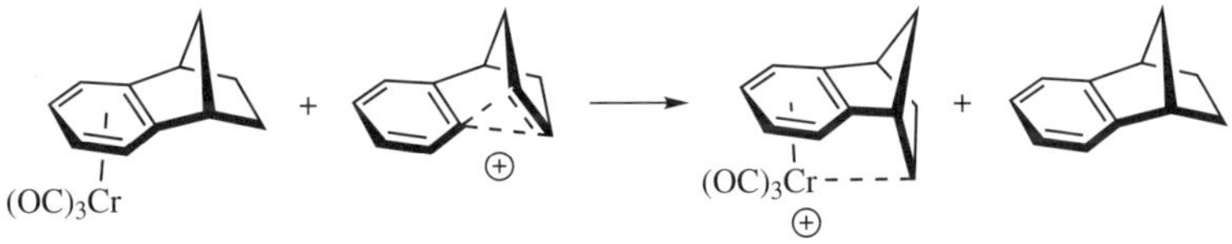

$\Delta E = +0.1$ kcal/mol

$(OC)_3Cr$ + **4** → $(OC)_3Cr$ **9** + (10.8)

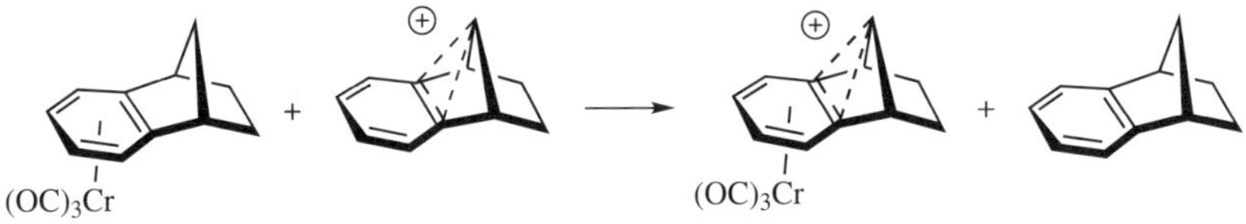

$\Delta E = +7.5$ kcal/mol

$(OC)_3Cr$ + 4 ⟶ $(OC)_3Cr$ (10) +

$$\Delta E = -3.7 \text{ kcal/mol} \tag{10.9}$$

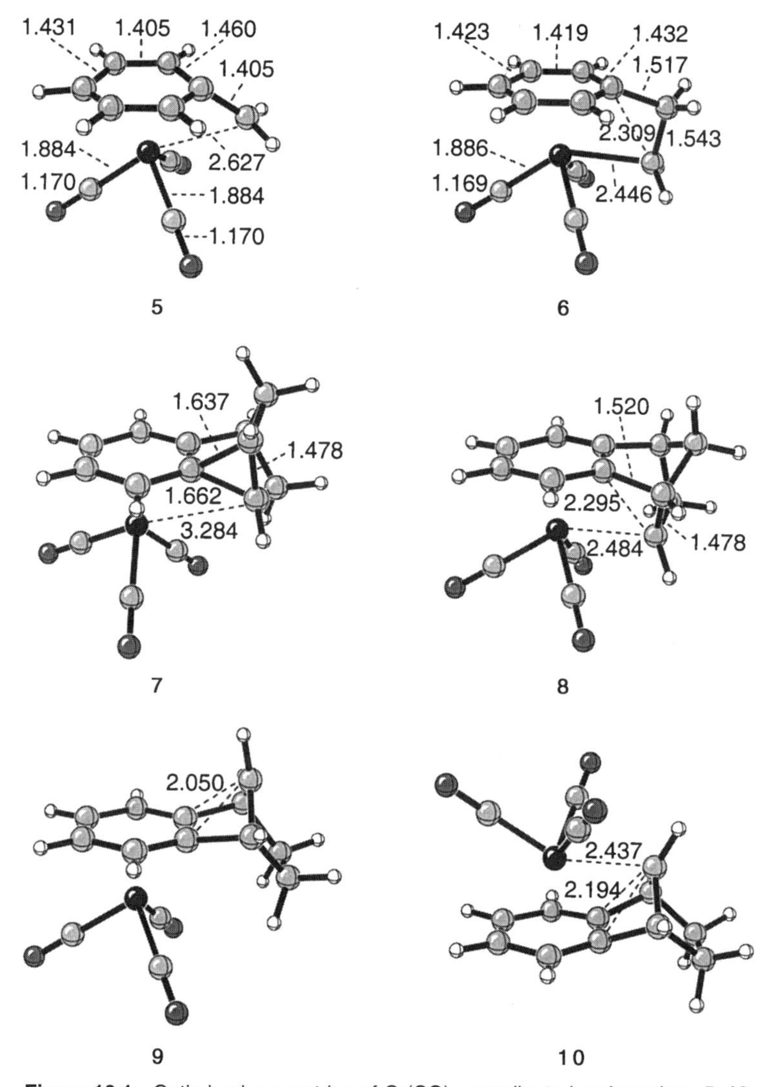

Figure 10.1 Optimized geometries of $Cr(CO)_3$-coordinated carbocations **5–10**.

10.2 BENZYLIC CATIONS

It has been known since the early 1960s that chromium tricarbonyl complexation of arenes stabilizes carbocations at the benzylic position. Holmes and coworkers found that chromium complexation increases the rate of solvolysis of benzyl chloride by a factor of 10^5.[9] Facile ionization of leaving groups at this position leads to cationic intermediates that can be trapped by a wide variety of nucleophiles.[10] In these reactions, the chromium tricarbonyl moiety is also an element of stereocontrol, as nucleophiles preferentially approach from the face of the cation opposite the metal fragment.[11] Coupled with leaving groups departing *anti* to the metal and a high barrier to rotation for the complexed benzylic cation, this chemistry provides for stereospecific substitution with net retention of stereochemistry.[12]

The mechanism of chromium stabilization of the benzylic cation in complex **5** is manifested in its structure. Two independent theoretical studies found that there is a gross distortion in the geometry of the organic fragment (Fig. 10.1).[4,5] As expected, there is a significant amount of double-bond character between the arene ring and the benzylic position of **5**, although this bond is slightly longer than that found in the uncomplexed benzyl cation **1** (1.405 Å for the chromium complex **5** vs. 1.383 Å for parent cation **1** calculated at the same level of theory). There is a calculated barrier to rotation of 45.4 kcal/mol about this bond.[5] This is consistent with NMR studies of chromium-complexed cations that show a lack of free rotation,[13] and also with trapping studies that show how such cations maintain their stereochemical integrity.[14] The most striking aspect of the structure of **5** is that the benzylic position is bent down toward the metal center by over 35° from the plane of the arene ring. This π complex is best described as an exocyclic double bond and a pentadienyl cation, resulting in η^7 coordination to the metal. Buildup of positive charge on the metal can be seen by examining the carbonyl bond lengths.[4,5] An increase in Cr–CO bond lengths and a decrease in C–O bond lengths indicates a decrease in backbonding from the metal to the carbonyls, relative to neutral chromium arene complexes. Effects of decreased back bonding are also seen in IR and NMR spectra.[15]

Chromium complexation also has a strong effect on the stabilities of benzylic cations. Homodesmotic[16] equation (10.1) has a ΔE of -12.0 kcal/mol, indicating significant stabilization of **5** due to the metal fragment.[4,6] This is consistent with experimental studies where measurement of pK_{R^+} values of benzylic species indicate that chromium-bound cations should be over 7.5 kcal/mol more stable than their uncomplexed counterparts.[17] In addition, in a competition experiment in which an in situ–generated cyclopropylcarbinyl cation could rearrange to form either a complexed or a free benzylic cation, a chromium stabilized cation was formed preferentially (a 95 : 5 ratio of products arising from the complexed and uncomplexed species, respectively, was observed).[4]

Stabilization of **5** by coordination to the metal is dependent on the degree of bending of the benzylic position toward the metal center. Calculations in which the organic fragment of the chromium complexed benzylic cation is constrained to a planar geometry show a slight destabilization relative to parent cation **1**.[4]

Additional calculations show that the stabilization imparted by the chromium tricarbonyl group is attenuated by increased substitution at the benzylic position,[6] a demonstration of the tool of increasing electron demand.[18] Equation (10.2) has a ΔE of -5.1 kcal/mol, and Eq. (10.3) has a ΔE of only -1.7 kcal/mol, indicating that stabilization due to the metal is more than cut in half in secondary cation **11**, and still further in tertiary cation **12**.[6] The geometries of **11** and **12** show less distortion from planarity than is seen in **5**, with their benzylic positions inclined toward the metal center by 21.8° and 12.7°, respectively. Experimentally it has been shown that chromium complexation enhances the rate of hydrolysis of cumyl chloride by a factor of 28,[19] in contrast to a factor of 10^5 for benzyl chloride.[9]

35.3° 21.8° 12.7°

$(OC)_3Cr$ ⊕ $(OC)_3Cr$ ⊕ $(OC)_3Cr$ ⊕

5 **11** **12**

10.3 PHENONIUM CATION

Cram proposed the intermediacy of spirocyclic phenonium ion **2** in the cationic 1,2 shift of phenyl groups.[20] Initial opposition to this proposal eventually gave way to widespread acceptance of this structure, which allows for stabilization of carbocations β to the arene ring. Bly and coworkers repeated Cram's studies on homobenzylic cation rearrangements with chromium complexed to the arenes, finding that complexation of the arene enhanced solvolyses β to the ring by a factor of 1.8.[21] The structure of the intermediate cation was unclear, though, from the available experimental evidence. It was possible to formulate the structure of this key intermediate as a complexed phenonium ion **13** (Eq. 10.5), or as chromacycle **6** (Eq. 10.4) with direct interaction between the metal and the carbocationic center.

Theoretical investigation of these possibilites showed that the preferred structure is the chromacycle **6** (Fig. 10.1).[7] All attempts to optimize a structure such as **13** led to chromacycle **6**. Indeed, a complexed phenonium ion geometry turned out to be a transition state for rearrangement in a substituted analog.[7] The geometry of chromacycle **6** showed strong interaction between the metal and the cationic carbon, with pyramidalization of this carbon and a Cr–C bond length of 2.446 Å (slightly longer than known neutral Cr–C bond lengths measured from 2.054 to 2.206 Å).[22] The benzylic carbon is also slightly bent toward the metal center to facilitate interaction with the β carbon. Bond lengths within the six-membered ring are typical of delocalized arenes.[23] A decrease in backbonding to the carbonyls similar to that seen in **5** indicates a decrease in electron density at the metal center.

Direct interaction with the chromium atom in **6** stabilizes β cations even more than electron donation from the arene π system (phenonium ion formation).[7]

Homodesmotic equation (10.4) shows that this stabilization is worth 6.6 kcal/mol. In contrast, if the structure is artificially constrained to that of chromium complexed phenonium ion **13**, calculations indicate that chromium complexation would destabilize the cation, with a ΔE of $+5.0$ kcal/mol for Eq. (10.5).

10.4 2-BENZONORBORNENYL CATION

Winstein's proposal[24] of a nonclassical structure for the 2-norbornyl cation was the start of a long debate in the chemical literature,[25] and the benzo-fused analog **3** was one of the many variants that was studied.[26] Solvolysis studies indicate that **3** is also best described as a symmetric nonclassical structure. Structure **3** contains the frameworks of both the 2-norbornyl cation and the phenonium ion **2**.

Again, the effect of chromium tricarbonyl complexation on these cations is of great interest. Solvolysis studies indicated that the geometry of complexation was important. A chromium complex of benzonorbornene bearing a mesylate at the 2 position in which the metal is *syn* to the forming cation undergoes solvolysis at a slightly greater rate (2.61 times) than the analogous uncomplexed species. In contrast, the complex with the metal *anti* to the incipient cation undergoes solvolysis at a much slower rate (0.0098 times) than the analogous uncomplexed species.[27] If the leaving group is changed to a brosylate, both the *syn* and *anti* metal-complexed species solvolyze at a slower rate than does the uncomplexed 2-brosylate, but the effect is much more pronounced in the *anti* case (solvolysis rates of 0.67 and 0.0024 for the *syn* and *anti* cases, respectively, relative to the uncomplexed brosylate).[28]

This system was examined theoretically, and two geometries were found for the cationic intermediate (Fig. 10.1).[8] One minimized structure, **7**, is best described as a chromium complex of the parent cation **3**. The symmetry of the hydrocarbon framework is broken by the presence of the metal on one face of the arene, but the difference between bond lengths in the three-membered ring is small (only 0.025 Å). A second minimized geometry, **8**, shows major deviation from the symmetric structure of **3**. In structure **8**, the cationic center does not interact with the arene ring, but instead interacts directly with the chromium center, as evidenced by the Cr–C distance of 2.484 Å and the hybridization of the carbon.

These structures differ considerably in stability as well as geometry.[8] Equation (10.6) has a ΔE of $+4.2$ kcal/mol, indicating that the formation of structure **7** is disfavored relative to the parent nonclassical species **3**. This is consistent with the decreased rate of solvolysis observed when the metal is *anti* to the forming cation. Withdrawal of electron density from the arene by the chromium moiety diminishes the ability of the arene to stabilize the carbocation. Structure **8**, on the other hand, does not impart any special stability compared to **3**, with a ΔE of 0.1 kcal/mol for Eq. (10.7). This is consistent with the trend described above for solvolysis rates since *syn* solvolyses should lead directly to **8**. Theoretical modeling of solvolysis transition states did, in fact, show a preference for solvolysis with the metal *syn* to the forming carbocation.[29] Thus, the metal's ability to assist anchimerically the ionization is dependent on the proximity of the metal to the forming carbocation. When the metal is unable to interact directly, its electron-withdrawing nature impedes ionization.

10.5 7-BENZONORBORNENYL CATION

Winstein also proposed a nonclassical structure for the 7-norbornenyl cation.[30] Benzonorbornenyl cation **4** is an analogous system to **3** in which the carbocationic center is stabilized by donation of electron density from the arene π system.[26c,31] The two arene faces in **4** are different, and therefore two different chromium complexes can be envisaged—one with the metal near the saturated two-carbon bridge of the bicyclo [2.2.1] heptane framework (**9**) and the other with the metal near the cationic center on the one-carbon bridge (**10**). Solvolysis studies indicated that a chromium tricarbonyl fragment *syn* to the one-carbon bridge enhanced the rate of ionizaton by a factor of 40 as compared to the complex with the metal *anti* to the one-carbon bridge.[32]

Complex **9**, in which chromium is unable to interact directly with the cationic center on the one-carbon bridge, closely resembles the parent cation **4** (Fig. 10.1).[8] There is a slight increase in the distance between the closest arene carbons and the cationic center in **9** versus **4** (from 1.980 to 2.050 Å), indicating that there is a decrease in donation of electron density from the arene ring due to the electron-withdrawing chromium tricarbonyl group. In complex **10**, though, the metal atom is able to interact directly with the cationic center. This can be seen in a relatively close Cr–C distance (2.437 Å) (Fig. 10.1) and in rehybridization of the cationic carbon to point its empty orbital toward the chromium.

Homodesmotic equations (10.8) and (10.9) demonstrate the energetic consequences of chromium complexation of the two faces of the arene of cation **4**.[8] Complexation *anti* to the cationic center to produce **9** is unfavorable, with $\Delta E = +7.5$ kcal/mol for Eq. (10.8). The withdrawal of electron density from the arene π system diminishes the ability of the arene to stabilize the cationic center in this case. On the other hand, complexation *syn* to the cationic center to produce **10** is preferred, with $\Delta E = -3.7$ kcal/mol for Eq. (10.9). In this case, direct donation of electron density from the metal to the cationic carbon is better than nonclassical stabilization by the arene π system. Modeling of transition states for solvolysis reactions again showed an enhancement of ionization when the chromium could directly assist the departure of the leaving group.[29]

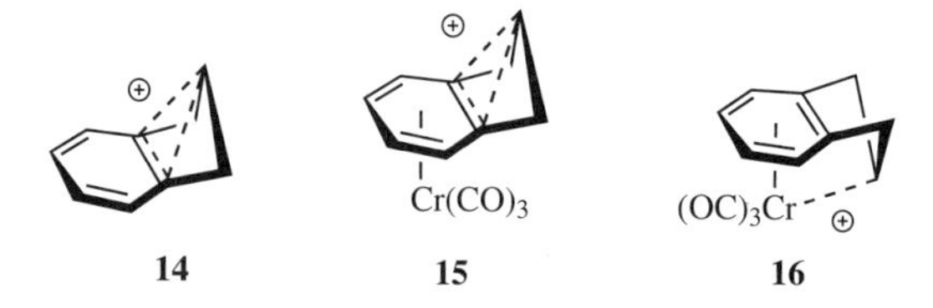

The fundamental differences between **9** and **10** are highlighted by consideration of 2-indanyl cation **14**.[8] Two possible chromium complexes of **14** could be imagined—**15**, in which the cationic center interacts with the arene π system, and **16**, in which the five-membered ring is flipped to enable the cationic center to interact directly with chromium. When these possibilities were examined theoretically, no optimized structure corresponding to **15** could be found; all optimizations led to structure **16**.[8] When structure **16** was compared to a geometry constrained to resemble **15**, the fully optimized geometry was preferred by approximately 14 kcal/mol.[6]

10.6 CONCLUSIONS

Theoretical analyses of chromium-tricarbonyl-complexed cations revealed that the metal has two opposing effects on cation stability—it is inductively electron-withdrawing but directly electron-donating. In all cases where the metal is unable to interact directly with the cationic center, whether due to the geometry of the system as in **7** and **9**, or due to artificially imposed geometric constraints as in **13** and **15**, chromium complexation destabilizes carbocations. The organometallic fragment pulls electron density from the arene and consequently diminishes its ability to stabilize a positively charged center through π donation. When the chromium atom can directly interact with the cationic center, as in **5**, **6**, **8**, **10–12**, and **16**, donation of *d*-orbital electron density and concomitant delocalization of charge onto the metal effectively competes with resonance and nonclassical stabilization. In most of these cases, stabilization due to the presence of the metal is greater than other forms of stabilization, although the effect is diminished by increasing substitution at the cationic center. In metal complex **8** the positive charge is stabilized by approximately the same amount as in nonclassical cation **3** (Eq. 10.7), presumably because the Cr–C connection in **8** adds strain to the norbornyl framework, which twists slightly to bring these atoms close together.

These studies expand our understanding of both chromium arene chemistry and stabilized carbocations. Further studies have examined the interaction of other transition metal π complexes with carbocations[29] and applied the lessons learned here to discovery of new reactions of chromium-tricarbonyl-complexed arenes.[33]

REFERENCES

1. For reviews, see (a) M. L. H. Green, Six-electron ligands, in *Organometallic Compounds*, G. E. Coates, M. L. H. Green, and K. Wade, eds., Methuen, London, 1967, Vol. 2, pp. 165–189; (b) M. J. Morris, Arene and heteroarene complexes of chromium, molybdenum and tungsten, in *Comprehensive Organometallic Chemistry II*, E. W. Abel, F. G. A. Stone, and G. Wilkinson, eds., Pergamon Press, Oxford, 1995, Vol. 5, pp. 471–549; (c) R. Davis and L. A. P. Kane-Maguire, Chromium compounds with η^2-η^8 carbon ligands, in *Comprehensive Organometallic Chemistry*, G. Wilkinson, F. G. A. Stone, and E. W. Abel, eds., Pergamon Press, Oxford, 1982, Vol. 3, pp. 953–1077; (d) M. F. Semmelhack, Transition metal arene complexes: Ring lithiation, in *Comprehensive Organometallic Chemistry II*, E. W. Abel, F. G. A. Stone, and G. Wilkinson, eds., Pergamon Press, Oxford, 1995, Vol. 12,

pp. 1017–1038; (e) A. Solladié-Cavallo, Chiral arene-chromium-carbonyl complexes in asymmetric synthesis, in *Advances in Metal-Organic Chemistry*, L. S. Liebeskind, ed., JAI Press, London, 1989, Vol. 1, pp. 99–133; (f) C. Bolm and K. Muniz, *Chem. Soc. Rev.* **28**, 51–59 (1999); (g) S. E. Gibson (née Thomas) and E. G. Reddington, *Chem. Commun.* 989–996 (2000); (h) S. G. Davies, S. J. Coote, and C. L. Goodfellow, Synthetic applications of chromium tricarbonyl stabilized benzylic carbanions, in *Advances in Metal-Organic Chemistry*, L. S. Liebeskind, ed., JAI Press Ltd., London, 1991, Vol. 2, pp. 1–57; (i) S. G. Davies and T. D. McCarthy, Transition metal arene complexes: Side-chain activation and control of stereochemistry, in *Comprehensive Organometallic Chemistry II*, E. W. Abel, F. G. A. Stone, and G. Wilkinson, eds., Pergamon Press., Oxford, 1995, Vol. 12, pp. 1039–1070; (j) M. F. Semmelhack, G. R. Clark, J. L. Garcia, J. J. Harrison, Y. Thebtaranonth, W. Y. Wulff, and A. Yamashita, *Tetrahedron* **37**, 3957–3965 (1981); (k) M. F. Semmelhack, Transition metal arene complexes: Ring nucleophilic addition, in *Comprehensive Organometallic Chemistry II*, E. W. Abel, F. G. A. Stone, and G. Wilkinson, eds., Pergamon Press, Oxford, 1995, Vol. 12, pp 979–1015; (l) A. R. Pape, K. P. Kaliappan, and E. P. Kündig, *Chem. Rev.* **100**, 2917–2940 (2000); (m) J.-F. Carpentier, F. Petit, A. Mortreux, V. Dufaud, J.-M. Basset, and J. Thivolle-Cazat,. *J. Mol. Catal.* **83**, 1–15 (1998); (n) M. Uemura, Tricarbonyl(η6-arene)chromium complexes in organic synthesis, in *Advances in Metal-Organic Chemistry*, L. S. Liebeskind, ed., JAI Press, London, 1991, Vol. 2, pp. 195–245; (o) V. N. Kalinin, *Russ. Chem. Rev.* (Engl. transl.) **56**, 682–700 (1987); *Uspekhi Khimii* **56**, 1190–1224 (1987); (p) L. S. Hegedus, Synthetic applications of transition metal arene complexes, in *Transition Metals in the Synthesis of Complex Organic Molecules*, University Science Books, Mill Valley, CA, 1994, Chapter 10.

2. For selected examples, see (a) A. Majdalani and H.-G. Schmalz, *Synlett* 1303–1305 (1997); (b) M. Uemura, K. Isobe, K. Take, and Y. Hayashi, *J. Org. Chem.* **48**, 3855–3858 (1983); (c) K. Schellhaas, H.-G. Schmalz, and J. W. Bats, *Chem. Eur. J.* **4**, 57–66 (1998); (d) S. J. Coote, S. G. Davies, D. Middlemiss, and A. Naylor, *Tetrahedron: Asymmetry* **1**, 33–56 (1990); (e) E. J. Corey and C. J. Helal, *Tetrahedron Lett.* **37**, 4837–4840 (1996); (f) M. F. Semmelhack and A. Yamashita, *J. Am. Chem. Soc.* **102**, 5924–5926 (1980).
3. G. Jaouen, S. Top, and M. J. McGlinchey, *J. Organomet. Chem.* **195**, C5-C8 (1980).
4. C. A. Merlic, J. C. Walsh, D. J. Tantillo, and K. N. Houk, *J. Am. Chem. Soc.* **121**, 3596–3606 (1999).
5. A. Pfletschinger, T. K. Dargel, J. W. Bats, H.-G. Schmalz, and W. Koch, *Chem. Eur. J.* **5**, 537–545 (1999).
6. C. A. Merlic, B. N. Hietbrink, and K. N. Houk, *J. Org. Chem.* **66**, 6738–6744 (2001).
7. C. A. Merlic, M. M. Miller, B. N. Hietbrink, and K. N. Houk, *J. Am. Chem. Soc.* **123**, 4904–4918 (2001).
8. (a) D. J. Tantillo, B. N. Hietbrink, C. A. Merlic, and K. N. Houk, *J. Am. Chem. Soc.* **122**, 7136–7137 (2000); (b) D. J. Tantillo, B. N. Hietbrink, C. A. Merlic, and K. N. Houk, *J. Am. Chem. Soc.* **123**, 5851 (2001).
9. J. D. Holmes, D. A. K. Jones, and R. Pettit, *J. Organomet. Chem.* **4**, 324–331 (1965).
10. For examples, see (a) S. Top and G. Jaouen, *J. Organomet. Chem.* **197**, 199–215 (1980); (b) S. J. Coote and S. G. Davies, *J. Chem. Soc., Chem. Commun.* 648–649 (1988); (c) S. E. Gibson (née Tomas) and G. A. Schmid, *Chem. Commun.* 865–866 (1997); (d) M. Uemura, T. Kobayashi, and Y. Hayashi, *Synthesis* 386–388 (1986); (e) A. Ceccon, *J. Organomet. Chem.* **29**, C19-C21 (1971).
11. M. T. Reetz and M. Sauerwald, *Tetrahedron Lett.* **24**, 2837–2840 (1983).

12. For example, see E. J. Corey and C. J. Helal, *Tetrahedron Lett.* **37**, 4837–4840 (1996).
13. V. Galamb and G. Pályi, *J. Chem. Soc. Chem. Commun.* 487–488 (1982).
14. (a) S. Top and G. Jaouen, *J. Chem. Soc. Chem. Commun.* 1110–1112 (1980); (b) S. E. Gibson (née Tomas), P. Ham, G. R. Jefferson, and M. H. Smith, *J. Chem. Soc. Perkin Trans. 1* 2161–2162 (1997).
15. (a) D. Seyferth, J. S. Merola, and C. S. Eschbach, *J. Am. Chem. Soc.* **100**, 4124–4131 (1978); (b) M. Acampora, A. Ceccon, M. Dal Farra, G. Giacometti, and G. Rigatti, *J. Chem. Soc. Chem. Commun.* 871–872 (1975).
16. P. George, M. Trachtman, C. W. Bock, and A. M. Brett, *Theor. Chim. Acta* **38**, 121–129 (1975).
17. W. S. Trahanovsky and D. K. Wells, *J. Am. Chem. Soc.* **91**, 5870–5871 (1969).
18. (a) P. G. Gassman and A. F. Fentiman, Jr., *J. Am. Chem. Soc.* **91**, 1545–1546 (1969); (b) H. G. Richey, Jr., J. D. Nichols, P. G. Gassman, A. F. Fentiman, Jr., S. Winstein, M. Brookhart, and R. K. Lustgarten, *J. Am. Chem. Soc.* **92**, 3783–3784 (1970); (c) G. K. S. Prakash and P. S. Iyer, *Rev. Chem. Intermed.* **9**, 65–116 (1988); (d) D. K. Wells and W. S. Trahanovsky, *J. Am. Chem. Soc.* **91**, 5871–5872 (1969).
19. S. P. Gubin, V. S. Khandkarova, and A. Z. Kreindlin, *J. Organomet. Chem.* **64**, 229–238 (1974).
20. D. J. Cram, *J. Am. Chem. Soc.* **71**, 3863–3870 (1949).
21. R. S. Bly, R. C. Strickland, R. T. Swindell, and R. L. Veazey, *J. Am. Chem. Soc.* **92**, 3722–3729 (1970).
22. (a) M. D. Fryzuk, D. B. Leznoff, and S. J. Rettig, *Organometallics* **16**, 5116–5119 (1997); (b) R. A. Heintz, S. Leelasubcharoen, L. M. Liable-Sands, A. L. Rheingold, and K. H. Theopold, *Organometallics* **17**, 5477–5485 (1998); (c) M. D. Fryzuk, D. B. Leznoff, S. J. Rettig, and V. G. Young, Jr., *J. Chem. Soc. Dalton Trans.* 147–154 (1999).
23. P. LeMaguenes, S. V. Lindeman, and J. K. Kochi, *Organometallics* **20**, 115–125 (2001).
24. S. Winstein and D. S. Trifan, *J. Am. Chem. Soc.* **71**, 2953 (1949).
25. G. A. Olah, *J. Org. Chem.* **66**, 5943–5957 (2001).
26. (a) P. D. Bartlett and W. P. Giddings, *J. Am. Chem. Soc.* **82**, 1240–1246 (1960); (b) H. Tanida, T. Tsuji, and S. Teratake, *J. Org. Chem.* **32**, 4121–4122 (1967); (c) J. W. Wilt and P. J. Chenier, *J. Org. Chem.* **35**, 1571–1576 (1970); (d) W. P. Giddings and J. Dirlam, *J. Am. Chem. Soc.* **85**, 3900–3901 (1963).
27. R. S. Bly and R. C. Strickland, *J. Am. Chem. Soc.* **92**, 7459–7461 (1970).
28. D. K. Wells and W. S. Trahanovsky, *J. Am. Chem. Soc.* **92**, 7461–7463 (1970).
29. B. N. Hietbrink, *Density Functional Theory Studies of the Interaction of Reactive Intermediates with Arenechromium Tricarbonyl Complexes and Other Transition Metal Pi Complexes and Electrocyclic Reactions of 1-Substituted Cis-1,2,4,6-Heptatetraenes*, Ph.D. thesis, Univ. California, Los Angeles, CA, 2000.
30. S. Winstein, M. Shatavsky, C. Norton, and R. B. Woodward, *J. Am. Chem. Soc.* **77**, 4183–4184 (1955).
31. H. Tanida, Y. Hata, S. Ikegami, and H. Ishitobi, *J. Am. Chem. Soc.* **89**, 2928–2932 (1967).
32. R. S. Bly and T. L. Maier, *J. Org. Chem.* **43**, 614–621 (1978).
33. C. A. Merlic and M. M. Miller, *Organometallics* **20**, 373–375 (2001).

11

CARBOCATIONS IN GOLD CHEMISTRY

Hubert Schmidbaur and Keith A. Porter
Department of Chemistry
Technical University Munich
Germany

Carbocation Chemistry, Edited by George A. Olah and G. K. Surya Prakash
ISBN 0-471-28490-4

For almost a century, carbocations have remained elusive species that have attracted the attention of generations of scientists. Once it had been recognized that these carbon-based cations were probably playing an essential role in many key reactions of organic chemistry, they became not only an intellectual but also an experimental challenge. It required innovative concepts and refined preparative and analytical techniques to finally build up a profound knowledge of this fascinating area.* With this historical and contemporary background, it was surprising to find that gold-substituted carbocations proved to be readily available and even exceedingly stable, open to investigation by physical methods under standard conditions. The present account traces this more recent development in a clearly special area of carbocation chemistry.

11.1 A BRIEF INTRODUCTION TO AUROPHILICITY

The rapid increase of structural data for gold compounds in the 1970s and later revealed an intriguing structural phenomenon that appeared to be specific for gold.[1–3] It is indeed almost impossible to ignore the evidence that there is a strong tendency for virtually all low-coordinate gold(I) complexes of the standard type L–Au–X (where L is a neutral donor ligand; X, an anionic ligand) to form aggregates in the solid state through Au–Au contacts. In modern terminology, there is an extensive supramolecular chemistry of gold based on a new variation of chemical bonding between metal atoms with closed-shell configurations. The term "aurophilicity" introduced in 1988 to draw the attention of the chemical community to this phenomenon[3–5] was quickly accepted and is now widely used as more and more examples of the effect become known.[6] Since interactions between other closed-shell metal atoms—although generally weaker than with gold—have meanwhile been traced in many other areas of coordination chemistry, "aurophilicity" was generalized first into "numismatophilicity" (for the coinage metals) and then into "metallophilicity"[7] and adopted as a means to refer to nonstandard bonding between "closed-shell" metal centers mainly with d^{10} or $d^{10}s^2$ configuration.[8,9] There is also an extensive literature on the theoretical treatment of the phenomenon that finally lead to the conclusion that "aurophilicity" can and must be taken as "real." The quantum-chemical calculations lead to satisfactory results only if relativistic effects are considered; otherwise the interactions can even be found repulsive. With correlation effects taken into account, the aurophilic bond energies can reach as much as 10 kcal per Au–Au contact,[8,9,12,13] in good agreement with experimental data.[14,15] This energy is significant for the conformation, configuration, and aggregation of many gold(I) complexes, and best compared to that of hydrogen bonding,[6,16,17] one of the most common principles of binding in supramolecular chemistry.[18]

*Two of the most recent groundbreaking studies in this area are G. A. Olah, K. K. Laali, Q. Wang, and G. K. S. Prakash, *Onium Ions*, Wiley-Interscience, New York, 1998, and G. Rasul, G. K. S. Prakash, and G. A. Olah, *J. Am. Chem. Soc.* **119**, 12984 (1997).

11.2 AURATION OF THE METHANE CARBON ATOM

The prelude above is appropriate and even essential for an essay on carbocations in gold chemistry, because some of the earliest and most convincing examples for aurophilicity were obtained during attempts to prepare species with polyaurated carbon.

While tetramercuriomethanes $C(HgX)_4$ (**A**) have been known since the early years of the twentieth century,[19] related tetragoldmethanes $C(AuL)_4$, namely, complexes of the simple pentaatomic molecule CAu_4—formed when all hydrogen atoms of methane are substituted by gold atoms (**B**)—had never been prepared. The first synthetic attempts used C–H-acidic methyl (CH_3X)[20,21] and methylene compounds (CH_2X_2)[22] as substrates and it was found that the H/AuL substitution proceeds smoothly if sufficiently powerful aurating agents are employed, mainly [LAu]F and $[LAu]BF_4$ (prepared in situ), or the oxonium salts $[(LAu)_3O]BF_4$. Three auration steps are easily accomplished as shown, for instance, in Eq. (11.1), where a silylated phosphonium cation (with a C–H-acidic methylgroup) is converted into the triply aurated species.[20] It was noted in the structure of this cation that the Au–C–Au angles were all smaller than the tetrahedral angle as if the gold atoms were drawn together. Similar observations were made with diphosphonium[22] or sulfonium salts,[24] malodinitrile,[23] and other substrates with acidic CH_2 groups.[25] In an early case with methyloxazoline, the auration gave a product where the methyl carbon atom on auration became finally pentacoordinated by four gold atoms and the oxazoline group in an octanuclear dimer [Eq. (11.2)]. This result has shown clearly that carbon atoms can become hypercoordinate, that is, reach coordination numbers larger than four, as the number of gold atoms increases:[21]

$$[H_2C(PMe_3)(SiMe_3)]^+ \xrightarrow[2.\ 3\ LAuCl/F]{1.\ base} [(LAu)_3C(PMe_3)]^+ \qquad L = PPh_3 \tag{11.1}$$

$$2\ \text{(oxazoline)}\!-\!CH_3 \xrightarrow[2.\ LAuCl/F]{1.\ base} 2\ \text{(oxazoline)}\!-\!C(AuL)_3 \xrightarrow{2\ [Au]^+} [\{(LAu)_4C\text{-(oxazoline)}\}_2Au_2]^{2+} \qquad L = PPh_3 \tag{11.2}$$

In order to allow the synthesis of a true homoleptic tetragoldmethane compound, tetrakis(dialkylboryl)methanes were chosen as substrates.[4] In the reaction with [LAu]X agents generated in situ from readily available chlorides and CsF, all boryl groups are removed from the central carbon atom and replaced by LAu functions. However, the resulting intermediate of the expected type $(LAu)_4C$ was found to be an efficient acceptor for one or even two more $[LAu]^+$ units to afford the penta- and hexacoordinated cations. Thus, with an excess of aurating agent, the hexanuclear dication is the sole product in high yield [Eq. (11.3)]. With stoichiometric amounts of the components and/or with variations in the borylmethane precursor,[27] the salts with the pentaaurated cations are the main products [Eq. (11.4)]:

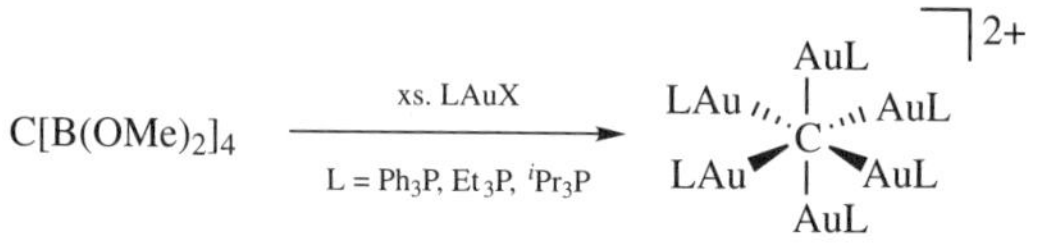

$$C[B(OMe)_2]_4 \xrightarrow[L = Ph_3P,\ Et_3P,\ {}^iPr_3P]{xs.\ LAuX} [C(AuL)_6]^{2+} \tag{11.3}$$

$$CH_2[B(OMe)_2]_2 \xrightarrow[L = Ph_3P]{xs.\ LAuX} [C(AuL)_5]^{+} \tag{11.4}$$

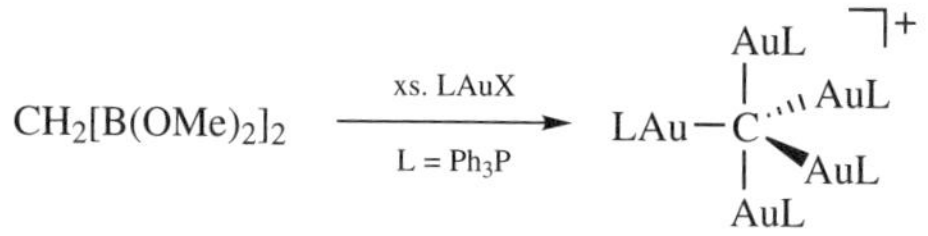

The 1,1,1-triborylated ethane was also aurated employing this synthetic strategy and found to give finally the tetranuclear cation with a pentacoordinate carbocation $[MeC(AuL)_4]^+$, which bears a methyl group in addition to four gold atoms:[28]

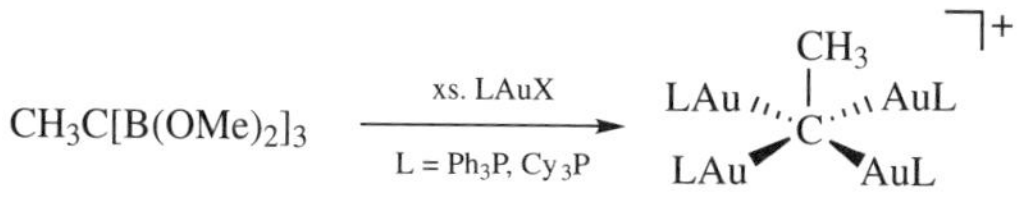

$$CH_3C[B(OMe)_2]_3 \xrightarrow[L = Ph_3P,\ Cy_3P]{xs.\ LAuX} [CH_3C(AuL)_4]^{+} \tag{11.5}$$

A few years after the first reports of these unusual hypercoordinate carbocations, a second, more convenient pathway was found using the more readily available silyldiazomethanes R_3SiCHN_2 as the substrates for auration.[29] With an excess of [LAu]X agent the hexanuclear dication is also formed in this process, but in addition a novel pentacoordinated, tetranuclear cation can be isolated as the tetrafluoroborate salt:

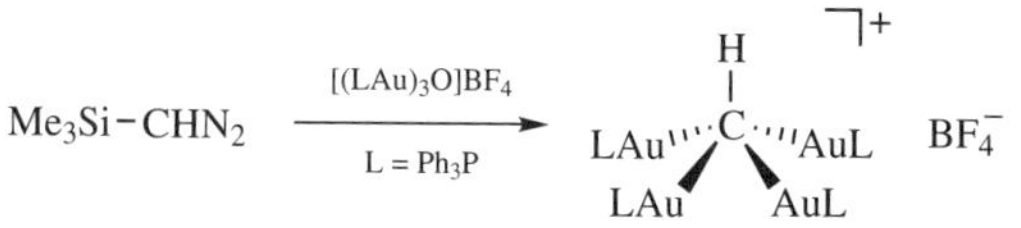

$$Me_3Si{-}CHN_2 \xrightarrow[L = Ph_3P]{[(LAu)_3O]BF_4} [HC(AuL)_4]^{+}\ BF_4^- \tag{11.6}$$

The carbocation is surrounded by four gold atoms and one hydrogen atom, suggesting—very surprisingly—that tetragoldmethane molecules are strong Brønsted acids! In separate tests it was shown that the $[HC(AuL)_4]^+$ cation is in fact not deprotonated even by strong bases like tertiary amines. Obviously it is this high basic or acceptor character of $(LAu)_4C$ molecules that induces the penta- or hexa-auration of the carbon atom in the reactions with the tetraborylmethanes (shown above).

In order to stop the auration of methane at the four-coordinate level, very bulky phosphines L were introduced at the gold atoms.[30] Of a series of products obtained in this work, the complex with four tricyclohexylphosphine ligands could be isolated and identified as a tetrahedral tetragoldmethane complex [Eq. (11.7)]. In this case the four phosphines protect the CAu_4 core from any further electrophilic attack by another bulky $[LAu]^+$ cation:

$$C[B(OMe)_2]_4 \xrightarrow[L = Cy_3P]{LAuX} C(AuL)_4 \qquad (11.7)$$

With most small or medium-sized phosphines, including triphenylphosphine, the auration of tetraborylmethanes proceeds to the hexanuclear, dicationic state.[4] Six compounds of the series were obtained in pure form and identified by analytical, spectroscopic, and structural techniques.[4,31–34] The list of salts with the pentanuclear cation is shorter, and only the triphenylphosphine complex has been completely characterized.[27] It is an isoelectronic analog of the corresponding dication with a nitrogen atom as the clustering nucleus.[35,36]

Pentacoordinate carbocations are also generated in the auration of silylated methanes.[37] Thus $(Me_3Si)_2CH_2$ can be metallated at its methylene group to give a neutral molecule $(Me_3Si)_2C(AuL)_2$ with tetrahedrally tetracoordinate carbon, but this molecule is found to accept yet another $[LAu]^+$ cation to afford $[(Me_3Si)_2C(AuL)_3]^+$ as the tetrafluoroborate salt with a pentacoordinate carbocation:

$$(Me_3Si)_2CHAuL \xrightarrow[L = Ph_3P]{[(LAu)_3O]X} (Me_3Si)_2(AuL)_2 + [(Me_3Si)_2C(AuL)_3]^+ \qquad (11.8)$$

Sulfur ylides and their corresponding sulfonium salts have also been exhaustively aurated to produce not only trinuclear sulf(ox)onium salts as observed with the corresponding phosphonium salts (above), but even tetranuclear dications with pentacoordinate carbon atoms:

$$[Me_2SO{-}CH_3]^+ \xrightarrow[L = PPh_3]{LAuX} [Me_2SO{-}C(AuL)_4]^{2+} \qquad (11.9)$$

It has been reported[23a,b] that CH_3CN and $CH_2(CN)_2$ can be aurated to give products of the stoichiometry $[(LAu)_2CHCN]^+$ and $[(LAu)_3C(CN)_2]^+$ (as the BF_4^- salts), but the structures of these salts are still unknown. Pentacoordination at carbocations is thus known to date with the substitution patterns $[C(AuL)_5]^+$, $[RC(AuL)_4]^+$, and $[R_2C(AuL)_3]^+$, but not yet for $[R_3C(AuL)_2]^+$. It appears that at least two aurophilic contacts [as, e.g., in $[(Me_3Si)_2C(AuPPh_3)_3]^+$, Eq. (11.8)] are needed in order provide enough energy for the changes in connectivity of the central atom. It should be noted that L is in all cases a tertiary phosphine. No other auxiliary ligands have been probed. The counterions are almost exclusively "innocent" tetrafluoroborate anions, with $[MeOBF_3]^-$, $[B_3F_4O_3]^-$, and $[Me_3SiF_2]^-$ as three of the few exceptions.

11.3 STRUCTURAL DATA OF POLYAURATED CARBOCATIONS

The structures of the polyaurated carbocations have been investigated in great detail by single-crystal X-ray diffraction. The present discussion focuses mainly on the connectivity of the central carbon atoms.

11.3.1 Reference Compounds

Simple methylgold(I) complexes of tertiary phosphines of the type $H_3C–Au–PR_3$ (**C**, R = alkyl or aryl) are known to have standard tetrahedral structures regarding the methyl carbon atom.[38,39] Bond distances and angles as well as vibrational characteristics[40] of these monoaurated methanes indicate conventional Au–C sigma bonding in good agreement with the concept of sp^3 hybridization. A benchmark value for an Au–CH_3 moiety is the Au–C distance of 2.12 Å in crystals of $[(Ph_3P)Au–CH_3]$:[41]

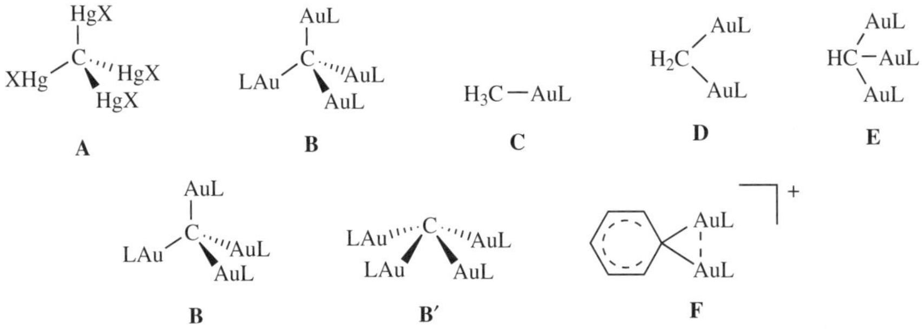

$C(HgX)_4$ **A** $C(AuL)_4$ **B** $H_3C—AuL$ **C** $H_2C(AuL)_2$ **D** $HC(AuL)_3$ **E**

$C(AuL)_4$ **B** $C(AuL)_4$ **B′** $[C_6H_5C(AuL)_2]^+$ **F**

Surprisingly, there is no report on any digoldmethane compound of the type LAu–CH_2–AuL (**D**) with a methylene bridge between two gold(I) centers.[39] Preparative work oriented toward these target molecules gave more highly aurated methane species [Eq. (11.4)], clearly a direct consequence of the aurophilicity phenomenon.[27] Experimental data are available, however, on substituted compounds derived from malodinitrile,[23] acetylacetone,[42] or other C–H-acidic precursors. It is a common structural feature of these dinuclear compounds that their Au–C–Au angle is consistently smaller than the tetrahedral standard. For example, this angle is only 84.5° in $[(Ph_3P)Au]_2C[CO(Me)]_2$ derived from acetylacetone, but the Au–C distances are largely unchanged at 2.13 Å.[42] Even in the diaurated methylene disulfone $[(Ph_3P)Au]_2C[SO_2(Me)]_2$ with more bulkyl groups the Au–C–Au angle is only 95.7°, while the Au–C distances are again at 2.10 Å (average).[43]

Triply aurated compounds with one C–H bond of methane retained, that is, of the type $HC(AuL)_3$ (**E**), are also not known, almost certainly owing to their high affinity for any excess of an aurating agent in the reaction mixtures (above). The structure analysis of substituted species shows similar anomalies as the diaurated counterparts; the Au–C–Au angles at the tetracoordinate carbon center of

$\{Me_3P–C[Au(PPh_3)]_3\}^+$ [Eq. (11.1)] are small at 101° with Au–C distances of 2.06 Å (both average values).[20]

No structural data are available on crystalline tetraauromethane complexes.[30]

11.3.2 Homoleptic Penta- and Hexanuclear Carbocations

The pentakis[(triphenylphosphine)gold]methanium(+1) cation [Eq. (11.4)] was studied in crystals of the tetrafluoroborate salt containing three CH_2Cl_2 solvent molecules. The cation has crystallographically imposed threefold symmetry (in space group $R\overline{3}$). All five Au–C bonds are equal within the limits of standard deviations [average 2.085 Å] and all Au–C–Au angles are close to 90° (ax/eq) and 120° (eq/eq), respectively. The core unit of the cation thus has almost perfect trigonal bipyramidal geometry and the carbon atom is truely pentacoordinated.[27]

The hexakis[(triphenylphosphine)gold]methanium(2+) dication has been investigated in both the $\{[BF_4]^-\}_2(CH_2Cl_2)_3$ and $\{[MeOBF_3]^-\}_2$ salts [Eq. (11.3)]. In the former the central carbon atom resides on a center of inversion,[34] while in the latter the dication has no crystallographically imposed symmetry.[4] All Au–C bond lengths are in the narrow range of 2.09–2.15 Å (average 2.12 Å), and all angles are very close to 90°, indicating virtually perfect octahedral geometry.[4,34]

Centrosymmetric dication structures were also found[33] for the tri(isopropyl)phosphine complex $\{[(^iPr)_3PAu]_6C\}^{2+}$ $\{(B_3F_4O_3]^-\}_2$ $(CH_2Cl_2)_3$ and for the corresponding compound with the ditertiary phosphine $L_2 = 1,2\text{-}C_6H_4(CH_2CH_2PPh_2)_2$, namely, $[(L_2)_3Au_6C]^{2+}$ $\{[BF_4]^-\}_2$ (C_6H_6).[31] The Au–C bond lengths are in the range 2.09–2.22 Å, and all Au–C–Au angles are close to 90°. Thus in all four representative cases the central carbon atom is octahedrally hexacoordinated to gold atoms with Au–C distances in full agreement with data for standard Au–C sigma bonds.[38]

11.3.3 Tri- and Tetranuclear Cations with a Heteroleptic Substitution Pattern

The cation of the compound $\{(Me_3Si)_2C[Au(PPh_3)]_3\}^+$ BF_4^- [Eq. (11.8)] was found to have a distorted trigonal bipyramidal structure with the two trimethylsilyl groups in equatorial positions. The equatorial and the two axial Au–C bond lengths are almost equal (average 2.21 Å), and the two Au–C–Au angles are only 76.3° (average). The two Si–C (center) distances (average 1.89 Å) show no anomalies. The central carbon atom must therefore be considered strictly pentacoordinate.[37]

The cation of $\{HC[Au(PPh_3)]_4\}^+$ BF_4^- $(CH_2Cl_2)_4$ [Eq. (11.6)] has a square pyramidal structure with the hydrogen atom in the apical position.[29] The Au–C distances are in the range 2.11–2.13 Å, and all *cis*-Au–C–Au angles are very close to 82°. A similar cation geometry was found for $\{MeC[Au(PPh_3)]_4\}^+$ BF_4^-, where the methyl group is in the apical position.[28] The Au–C distances vary between 2.14 and 2.19 Å and the Au–C–Au angles are all very close to 83°. Taking these data together with the C1–C2 distance of 1.44(5) Å, the central carbon atom is genuinely pentacoordinated.

The structure of the octanuclear dication shown in Eq. (11.2) for the tetraaurated methyloxazoline compound (a salt with rare $[Me_3SiF_2]^-$ anions and CH_2Cl_2 and Et_2O solvent molecules) is more complex, but the two core units with pentacoordinate carbon atoms in a square pyramidal environment can easily be distinguished. Their dimensions are very similar to the analogs presented above, and it therefore appears that heteroleptic pentacoordination of the type $[RC(AuL)_4]^+$ is a common mode of aggregation.[21] The structure reported for the dication of $\{Me_2(O)S–C[Au(PPh_3)]_4\}^{2+}$ in the bis(perchlorate acetone) solvate shows the same features.[24]

11.4 NMR AND MÖSSBAUER SPECTROSCOPY OF POLYAURATED CARBOCATIONS

All salts with polyaurated carbocations are diamagnetic and give perfectly resolved NMR spectra in di- or trichloromethane solution. The chemical shifts, intensities, and multiplicities of the ligand ^1H, ^{11}B, ^{13}C, ^{19}F, and ^{31}P resonances were found to be in agreement with the proposed composition. It has also been shown that the PR_3 ligands are equivalent in each complex on the NMR timescale in solution, indicating fluxionality where the $AuPR_3$ groups occupy inequivalent vertices of the polyhedron, such as in the trigonal bipyramidal geometry.[27]

However, the ^{13}C NMR signal of the central carbon atom in polyaurated cations is difficult to detect in all cases owing to efficient shielding of this nucleus in an environment of quadrupolar ^{197}Au nuclei ($s = \frac{3}{2}$). It was only with regioselective ^{13}C labeling that these resonances could be detected for several representative compounds.[32]

For no less than four examples of the $[C(AuL)_6]^{2+}$ type [L = PPh_3, $PPh_2(C_6H_4$-NMe_2-4), PEt_3, and P^iPr_3], the signal of the central carbon atom was assigned beyond doubt through its characteristic septet splitting with a coupling constant in the narrow range from 52.0 to 56.1 Hz. For the triarylphosphine complexes the resonance may be masked by the aryl signals, and therefore extreme high-power magnet spectrometers are necessary to achieve complete resolution. For the trialkylphosphine complexes there is no such problem. The septet resonance and the large coupling constant are direct proof for the sixfold connectivity and for the strong covalent C–Au–P bonding of the central carbon atom (Fig. 11.1).[30,32,33]

No similar experiments have been carried out for the homoleptic pentanuclear cations, but data are available for two heteroleptic coordination patterns. The resonance of the methine carbon atom in $[HC(AuL)_4]^+$ with L = PPh_3 has been shown to be a quintet at 88.5 ppm with a $^2J_{CP}$ coupling constant of 65 Hz. The methine hydrogen atom also gives rise to a quintet (at 4.85 ppm, $^2J_{HP}$ 4.0 Hz), suggesting strong covalent bonding for the $HC(AuP)_4$ unit.[29] A similar quintet structure has finally been discovered for the methyl hydrogen atoms in $[MeC(AuL)_4]^+$ with L = PPh_3 ($^4J_{HP}$ 4.5 Hz], which also confirms the pentacoordination at the carbocation in the tetraaurated ethane (Fig. 11.2).[28] It is interesting to note that the ^{13}C resonance of the simple neutral $[C(AuL)_4]$ molecule with L = $P(^cHex)_3$ is also a quintet with $^2J_{(CP)} = 61.0$ Hz.[30] This agreement of the absolute J values may be

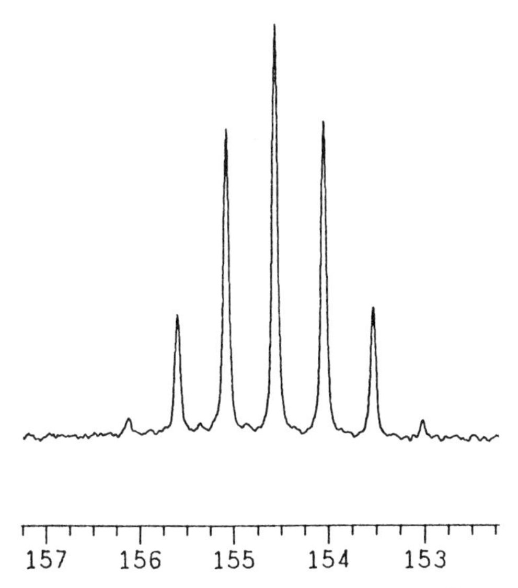

Figure 11.1 Resonance of the central carbon atom in the ^{13}C NMR spectrum of the $\{[(^{i}Pr_3P)Au]_6C\}^{2+}$ dication (BF_4^- salt in CD_2Cl_2 at 25°C; δ scale).

accidental since two different phosphines are present in the cation and in the neutral molecule. The difference of their inductive effects may compensate the expected increase of the coupling constant for the transition from hexa- to penta- and tetra-coordination and the concomitant increase of the *s* character of the carbon orbitals (sp^3d^2, sp^3d, sp^3). It should be pointed out, however, that the general trend of the data follows this suggestion.

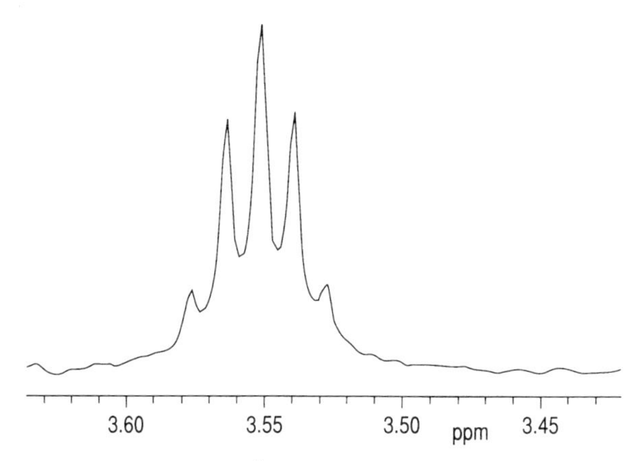

Figure 11.2 Methyl resonance in the ^1H NMR spectrum of the $\{CH_3\text{-}C[Au(PPh_3)]_4\}^+$ cation (BF_4^- salt in CD_2Cl_2 at 25°C; δ scale).

Several of the salts with polyaurated carbocations have been investigated by ^{197}Au Mössbauer spectroscopy.[20] All spectra showed large, well-resolved quadrupole doublets that confirm the +I oxidation state of the metal atoms and their low coordination number. This result is proof that no redox processes are involved in the auration reactions.

11.5 QUANTUM-CHEMICAL CALCULATIONS OF POLYAURATED CARBOCATIONS

Several approaches have been chosen to describe the chemical bonding in polyaurated methanes and the corresponding carbocations, which may also be treated as carbon centered gold clusters. In early work with extended Hückel theory (EHT) by Mingos et al. the existence of a carbon-centered octahedral cluster of gold atoms of the type $[C(AuL)_6]^{2+}$ was actually predicted by simple molecular orbital considerations.[44] After isolation of the first examples of these dications,[4] and of species of the type $[C(AuL)_5]^+$,[27] and $[RC(AuL)_4]^+$, these molecular orbital diagrams[21] proved very useful to rationalize the unexpected findings.[45]

More detailed calculations employing the discrete variational method (relativistic and nonrelativistic), a self-consistent LCAO-MO (linear combination of atomic orbitals–molecular orbital) method within the framework of density functional theory, revealed that the radial C–Au bonding based on both 6*s* and 5*d* orbitals of gold is the dominant contribution to the overall stability. Relativistic effects were demonstrated to have a significant influence, and nonrelativistic calculations lead to erroneous results, suggesting even repulsion of the components.[45,46]

The ligands L were found to have an essential influence on the energy levels of the CAu_6 and CAu_5 core units. For the homoleptic pentacoordinate case, the energy difference between the trigonal bipyramidal and square planar configurations is small, but the former structure is clearly favored (for the isolated carbocation in the gas phase).[47,48]

State-of-the-art theoretical investigations also focused on the heteroleptic pentacoordinate species, in particular on the $[HC(AuL)_4]^+$ type.[49] The experimental square pyramidal structure could be reproduced in MP2 calculations and a proton affinity of the corresponding $[C(AuL)_4]$ molecule was calculated to be no less than 1213 kJ/mol. Notably, tetragoldmethane molecules with small ligands L have been predicted to have a square pyramidal "anti-van't Hoff" ground-state structure (**B**′), at variance with the classical tetrahedral "van't Hoff structure" of methane itself and all its nonstrained derivatives known to date. The experimental observation that tetraauromethanes $[C(AuL)_4]$ with large ligands L have a tetrahedral structure (**B**) is not at variance with the theoretical results, because it is known from related work on the corresponding phosphonium[5] and arsonium cations[47] that these species $[E(AuL)_4]^+$ with E = P, As have square pyramidal "anti-van't Hoff" structures for small ligands L, but tetrahedral "van't Hoff structures" for large ligands L.

All theoretical work of recent years (as of 2004) has confirmed the importance of aurophilic bonding for the stoichiometry, stability, and structure of the polyaurated methane molecules and the carbocations derived therefrom. The phenomenon is based on electron correlation effects strongly affected by relativity.

To summarize, the isolobal relation of $[H]^+$ and $[LAu]^+$ species leads to similar stoichiometries for protonation and auration reactions of alkanes, respectively, but the molecular structures of the resulting carbocations differ owing to the bonding between the closed-shell metal atoms in the polyauromethan(ium) cases. Although the aurated carbocations are formally considered as electron-deficient systems (with eight electrons for five or six covalent bonds), more electrons actually come into play to support the core units. This extra binding energy lends unexpected chemical and thermal stability to the new type of carbocation.

11.6 BINUCLEAR CATIONS IN AROMATIC SYSTEMS

Electrophilic aromatic substitution reactions such as nitration, sulfonation, Friedel–Crafts alkylation/acylation, and proton exchange take place under acidic conditions or in the presence of Lewis acids.[51] The attacking species is an acid or an electrophile, and the mechanism of exchange is via an addition–elimination route. In an intermediate or a transition state the incoming electrophile and atom being substituted are both σ-bound to a common sp^3-hybridized carbon center, affording a positively charged 'benzonium' or "arenonium" ion intermediate (Scheme 1).[51–54]

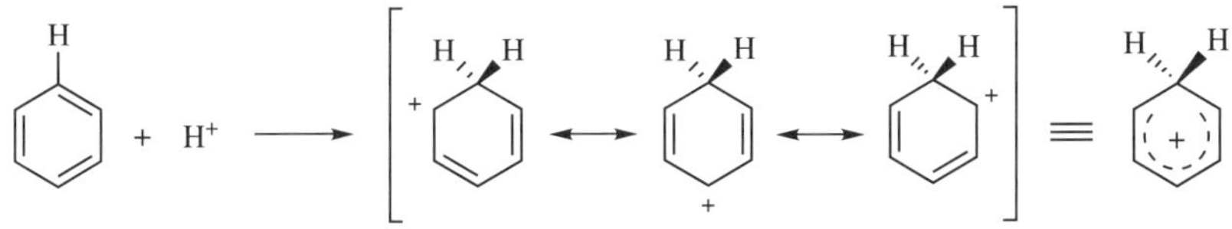

Scheme 1

With attention focused on carbocations in gold(I) chemistry, it is not too difficult to establish a structural relationship between the arenonium ion intermediate predicted in electrophilic aromatic substitution reactions and structures established crystallographically in hypercoordinated σ-organogold(I) compounds of the aromatic series. The $[AuPPh_3]^+$ unit, which is isolobal to H^+, can act as an electrophile toward arenes by employing its 6*s* orbitals for bonding.[5,55] In fact, its addition yields the hypercoordinated σ-bound organogold complexes of type (**F**). Like the aurated alkanes (above), these complexes are stable cationic species that are open to investigation by physical methods.

11.6.1 Metallocene Complexes

Metallocene complexes containing two $[AuPPh_3]^+$ units bound to the same cyclopentadienyl carbon center have been known since the early 1970s.[56,57] Historically, the first known metallocene (and *pseudo*arene) complex of this type was the ferrocenyl derivative.[56,57] Synthesized via treatment of the monoaurated ferrocenyl compound by strong acid, the product is formed together with one equivalent of ferrocene:

AuL Fe $\xrightarrow[L = PPh_3]{HBF_4}$ AuL AuL Fe BF_4^- + Fe (11.10)

The ^{31}P NMR spectrum of the ferrocenyl complex exhibits two signals of the same intensity for the two nonequivalent phosphorus atoms in the $[AuPPh_3]^+$ units, and determination of the crystal structure revealed a complex in which two gold atoms are accommodated at a common carbon center of one cyclopentadienyl ring. Key features of its structure are the short Au···Au bond distance of 2.77 Å and the small Au–C–Au bond angle of 78°, which is significantly less than the standard tetrahedral angle, with the Au-C-Au triangle lying approximately perpendicular to the cyclopentadienyl plane. Ruthenocene was shown to react with $[(Ph_3PAu)_3O][BF_4]$ to afford the analogous 1,1-diaurated complex $[CpRuC_5H_4(AuPPh_3)_2][BF_4]$,[58] and the successful synthesis of the cyclopentadienylcarbonyl complexes $[(CO)_2LMn\ C_5H_4(AuPPh_3)_2][BF_4]$ (L = CO, PPh_3) has provided further evidence that geminal diauration of metal cyclopentadienyls is a general reaction.[59]

Reactions leading to metallocene complexes in which there is an even higher accumulation of positive charge within the same molecule are unusual but not forbidden. The tetranuclear dicationic organogold complex $[\{C_5H_4(AuPPh_3)_2\}Fe\{C_5H_4(AuPPh_3)_2\}][BF_4]_2$ can be synthesized from $[C_5H_4Au(PPh_3)]_2Fe$ as an air stable salt in high yield:[60]

AuL Fe AuL $\xrightarrow[L = PPh_3]{2\,[AuL][BF_4]}$ AuL AuL Fe AuL AuL $2+$ $(BF_4)_2^-$ (11.11)

Free cyclopendiene is readily aurated to give the digold complex $C_4H_4[AuPPh_3]_2$.[61] The tetraphenylated molecule affords the trinuclear cationic complex $[Ph_4C_5\ (Au\ PPh_3)_3][BF_4]$ on reaction of $C_5HPh_4AuPPh_3$ or $Ph_4C_5(AuPPh_3)_2$ with $[AuPPh_3][BF_4]$ in THF.[62] In this cation the third $[AuPPh_3]^+$ unit is clustering side-on to the Au–C–Au triangle (**1**). Note that a second carbon atom becomes tetracoordinate on the third auration, reducing the aromaticity to an allyl carbocation. The loss of energy appears to be compensated in part by stabilization gained from aurophilic contacts.

Ph Ph Ph Ph AuL AuL AuL (*R*) + BF_4^-

L = PPh_3

1

11.6.2 Aryl Complexes

While attention to the chemistry of diaurated carbocations in aromatic systems has initially focussed on metallocene complexes, it is the arene complexes derived from 1,1-diauration of a central benzene unit that most closely resemble the "benzonium" ions observed in electrophilic aromatic substitution reactions.

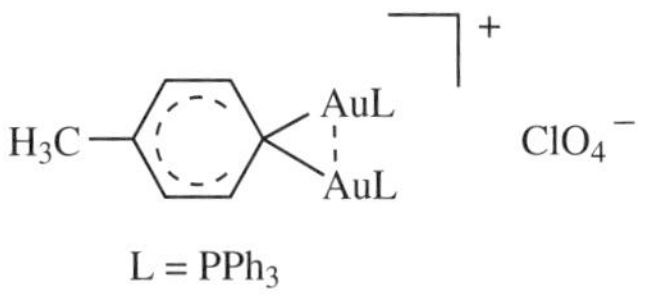

L = PPh_3

2

L = PPh_3

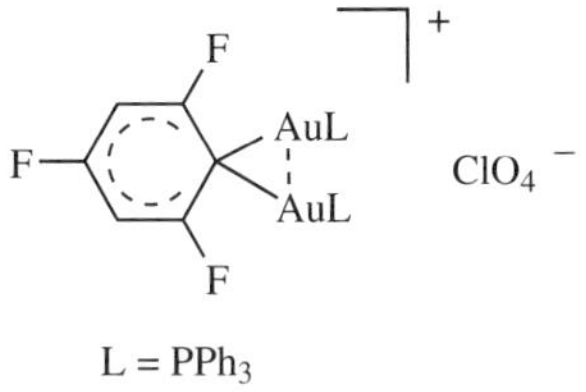

3

The tolyl complex (**2**), in which a single atom acts as a bridging ligand between two gold(I) units, represents the first 1,1-diaurated carbocation derived from benzene.[63] Identified as a white solid with melting point 175–176°C, it exhibits a single phosphorus peak in the NMR spectrum of its solution indicating the equivalence of the $[AuPPh_3]^+$ units. With the use of Mössbauer and NMR spectroscopy a cyclic structure of the cationic tolylgold complex p-$CH_3C_6H_4[AuP(C_6H_5)_2(CH_2)_2]_2+$ was proposed.[64] The tetrafluoroborate can be synthesized by elimination of a molecule of toluene from the binuclear complex $[(p\text{-}CH_3C_6H_4Au)P(C_6H_5)_2]_2[CH_2]_4$ upon addition of HBF_4:

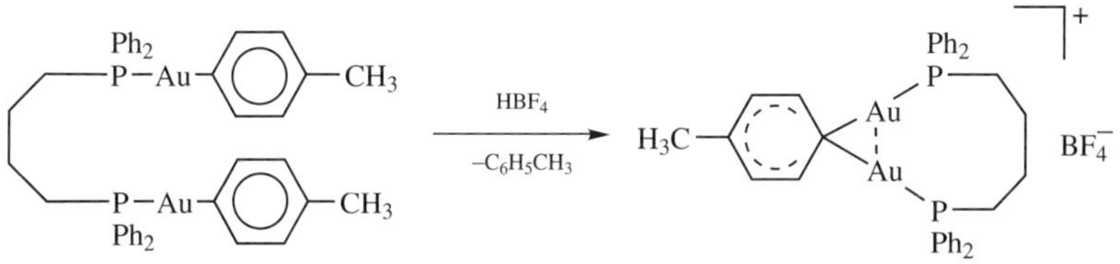

(11.12)

Although an array of aromatic derivatives have been synthesized in which two gold atoms share a common carbon center,[63–69] it was not until the isolation of $[\mu\text{-}2,4,6\text{-}C_6F_3H_2)(AuPPh_3)_2][ClO_4]$ (**3**) that crystallographic evidence was obtained to support 1,1-diauration in cationic complexes derived from benzene.[70] The cation of the perchlorate salt (**3**) exhibits an acute Au–C–Au bond angle of 79.3(3)°, a short Au···Au distance of 2.759 Å, and Au–C bond lengths of 2.16 Å.[71,72] The structural features are again highlighted in the crystal structure of the dication $[(CH_2)_2(C_6H_4\text{-}o)_2(AuPPh_3)_4][BF_4]_2$ (**4**).[73] The CAu_2 triangles are arranged in a perpendicular fashion to the plane of the corresponding aryl rings. Au···H-C agostic interactions with the alkane bridge have also been considered, but the effect may not be significant.

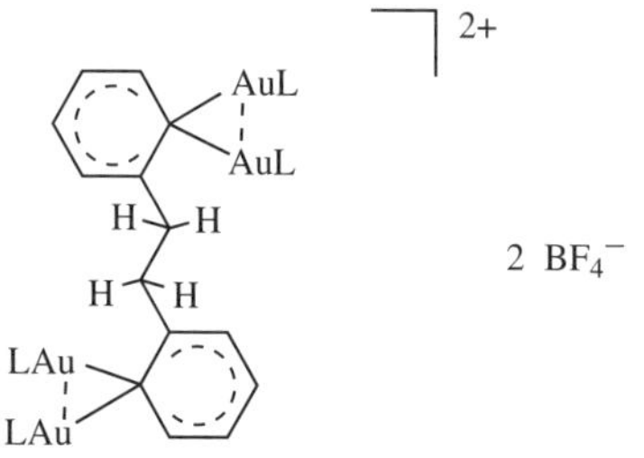

$L = PPh_3$

4

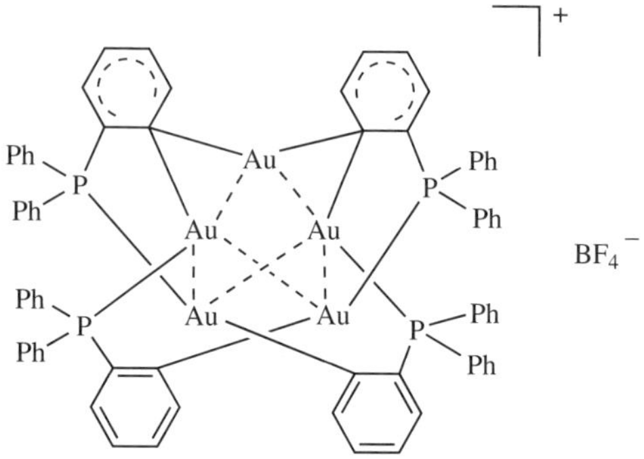

5

An extended X-ray absorption fine structure (EXSAFS) investigation into the binuclear gold(I) derivative $[C_6H_5OC_6H_4(AuPPh_3)_2]^+$ also supports a structure where two $[AuPPh_3]^+$ units are attached to a common bonding carbon center of one of the phenyl rings with a Au···Au bond length of 2.73 Å.[72] The gold complex $[Au_5(C_6H_4PPh_2)_4]^+$ (**5**), a cycloaurated cation containing a pair of *ipso*-carbon digold interactions, is formed by treatment of $Au_2(\mu\text{-}o\text{-}C_6H_4PPh_2)_2$ with acid or other electrophiles. The geometric parameters in the Au–C–Au triangle were reported at 2.709(2) Å for the Au···Au contact and 77(1)° for the Au–C–Au angle.[74] This represents an example of high complexity but with the same basic principles of organization.

11.6.3 Heteroaryl Complexes

The stoichiometry of dinuclear complexes based on a heteroaromatic ring system was first established by Nesmeyanov with the thienyl gold complex $[\alpha\text{-}C_4H_3S(AuPPh_3)_2][BF_4]$:[75]

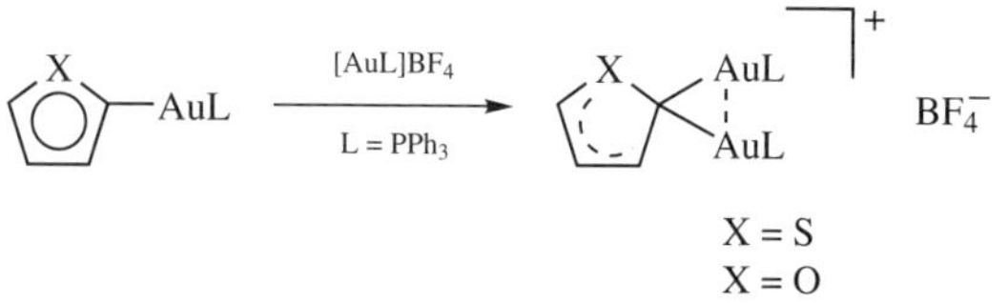

$$\text{C}_4\text{H}_3\text{X–AuL} \xrightarrow[L = PPh_3]{[AuL]BF_4} [\text{C}_4\text{H}_3\text{X}(AuL)_2]^+ \; BF_4^- \qquad (11.13)$$

X = S
X = O

It is only recently, however, that the mode of bonding in these complexes has been ascertained. NMR spectra of a solution of the thienyl complex, synthesized from monoaurated thiophene on addition of $[AuPPh_3][BF_4]$, exhibit a single ^{31}P peak, indicating identical phosphorus environments for the $[AuPPh_3]^+$ units. Structurally, two gold atoms share one of the α-carbon centers of the thienyl ring as a nucleation center. Comparable to the metallocene and arene complexes (described above), a short Au···Au distance of 2.8134 Å and a small Au–C–Au angle of 82.5(2)° are observed. The plane of the Au–C–Au triangle is approximately perpendicular to the plane defined by the thienyl ring, yielding the carbon atom formally

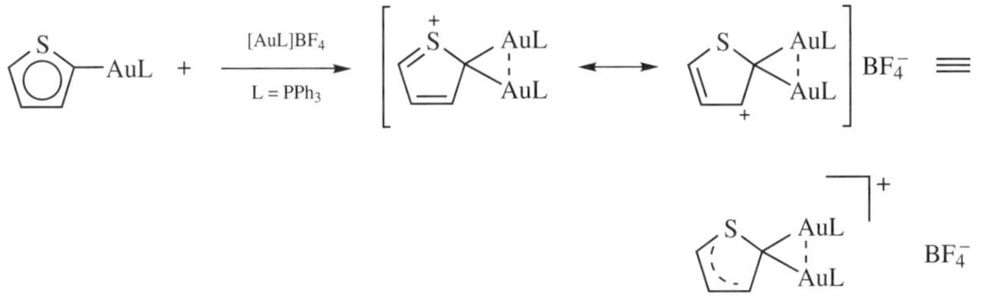

Scheme 2

sp^3-hybridized. The cationic charge can be assigned to the Au–C–Au triangle or delocalized around the thiophenium heterocycle (Scheme 2). The furanyl complex [α-$C_4H_3O(AuPPh_3)_2$][BF_4] [Eq. (11.13)] parallels the structure described for the thienyl complex.[42]

In summary, the isolobal relationship of $[H]^+$ and $[AuL]^+$ species leads to similar structures for the products of auration and protonation reactions of aromatic systems: 1,1-Diauration of a benzene or thiophene unit results in the formation of complexes that contain a tetracoordinated carbon atom formally sp^3-hybridized and with reduced aromaticity of the ring to a pentadienyl or allyl carbocation (Schemes 1 and 2). This parallels the carbocationic "benzonium/thiophenonium" ion intermediate/transition states predicted in electrophilic aromatic substitution reactions.[54]

The present account shows beyond doubt that the polyaurated alkanes, alkenes, and arenes represent true carbocationic species with the central carbon atom engaged in hypercoordinate covalent bonding. The structures arising as the auration progresses gain in stability owing to the growing shell of metal atoms. The aurophilicity phenomenon is a very general feature of gold chemistry, and carbocation chemistry is therefore just another example for the striking influence of gold atoms in molecular and supramolecular arrays.

REFERENCES

1. R. J. Puddephatt, *The Chemistry of Gold*, Elsevier, Amsterdam, 1978.
2. P. G. Jones, *Gold Bull.* **14**, 102, 159 (1981); **16**, 114 (1983); **19**, 46 (1986).
3. H. Schmidbaur, *Gold Bull.* **23**, 11 (1990).
4. F. Scherbaum, A. Grohmann, B. Huber, C. Krüger, and H. Schmidbaur, *Angew. Chem. Int. Ed.* **27**, 1544 (1988).
5. H. Schmidbaur, *Chem. Soc. Rev.* (Lond.) 391 (1995).
6. H. Schmidbaur, *Gold Bull.* **33**, 3 (2000).
7. J. Vicente, M.-T. Chicote, and M. C. Lagunas, *Inorg. Chem.* **32**, 3748 (1993).
8. P. Pyykkö, J. Li, and N. Runeberg, *Chem. Phys. Lett.* **218**, 133 (1994).

9. P. Pyykkö, *Chem. Rev.* **97**, 597 (1997).
10. P. K. Mehrotra and R. Hoffmann, *Inorg. Chem.* **17**, 2187 (1978),
11. Y. Jiang, S. Alvarez, and R. Hoffmann, *Inorg. Chem.* **24**, 749 (1985).
12. P. Pyykkö and Y.-F. Zhao, *Angew. Chem. Int. Ed.* **30**, 604 (1991).
13. P. Schwerdtfeger, *Inorg. Chem.* **30**, 1660 (1991).
14. H. Schmidbaur, W. Graf, and G. Müller, Angew. *Chem. Int. Ed. Engl.* **27**, 417 (1988).
15. J. Zank, A. Schier, and H. Schmidbaur, *J. Chem. Soc. Dalton Trans.* 323 (1998) and references cited therein.
16. C. Hollatz, A. Schier, and H. Schmidbaur, *J. Am. Chem. Soc.* **122**, 8115 (1997), and references cited therein.
17. F. Mendizabal, P. Pyykkö, and N. Runeberg, *Chem. Phys. Lett.* **370**, 733 (2003).
18. D. Braga, F. Grepioni, and G. R. Desiraju, *Chem. Rev.* **98**, 1375 (1998).
19. D. K. Breitinger and W. Kress, *J. Organomet. Chem.* **256**, 217 (1983) and references cited therein.
20. H. Schmidbaur, F. Scherbaum, B. Huber, and G. Müller, *Angew. Chem. Int. Ed.* **27**, 419 (1988).
21. F. Scherbaum, B. Huber, G. Müller, and H. Schmidbaur, *Angew. Chem. Int. Ed.* **27**, 1542 (1988).
22. H. Schmidbaur and O. Gasser, *Angew. Chem. Int. Ed.* **15**, 502 (1976).
23. (a) E. I. Smyslova, E. G. Perevalova, V. P. Dyadchenko, K. I. Grandberg, Yu. I. Slovokhotov, and Yu. T. Struchkov, *J. Organomet. Chem.* **215**, 269 (1981); (b) E. G. Perevalova, Y. T. Struchkov, V. P. Dyadchenko, E. I. Smyslova, Y. L. Slovokhotov, and K. I. Grandberg, *Izv. Akad. Nauk SSSR, Ser. Khim.* 2818 (1983).
24. (a) J. Vicente, M. T. Chicote, R. Guerrero, and P. G. Jones, *J. Am. Chem. Soc.* **118**, 699 (1996); (b) I. J. B. Lin, C. W. Liu, L.-K. Liu, and Y.-Sh. Wen, *Organometallics* **11**, 1447 (1992).
25. J. Vicente, M.-T. Chicote, I. Saura-Llamas, P. G. Jones, K. Meyer-Bäse, and C. F. Erdbrügger, *Organometallics* **7**, 997 (1988).
26. B. Alvarez, E. J. Fernandez, M. C. Gimeno, P. G. Jones, A. Laguna, and J. M. Lopez-de-Luzuriaga, *Polyhedron* **17**, 2029 (1998).
27. F. Scherbaum, A. Grohmann, G. Müller, and H. Schmidbaur, *Angew. Chem. Int. Ed. Engl.* **28**, 463 (1989).
28. O. Steigelmann, P. Bissinger, and H. Schmidbaur, *Z. Naturforsch.* **48b**, 72 (1993).
29. H. Schmidbaur, F. P. Gabbai, A. Schier, and J. Riede, *Organometallics* **14**, 4970 (1995).
30. H. Schmidbaur and O. Steigelmann, *Z. Naturforsch.* **47b**, 1721 (1992).
31. O. Steigelmann, P. Bissinger, and H. Schmidbaur, *Angew. Chem. Int. Ed.* **29**, 1399 (1990).
32. H. Schmidbaur, B. Brachthäuser, and O. Steigelmann, *Angew. Chem. Int. Ed.* **30**, 1488 (1991).
33. H. Schmidbaur, B. Brachthäuser, O. Steigelmann, and H. Beruda, *Chem. Ber.* **125**, 2705 (1992).
34. F. P. Gabbai, A. Schier, J. Riede, and H. Schmidbaur, *Chem. Ber./Recueil* **130**, 111 (1997).
35. A. Grohmann, J. Riede, and H. Schmidbaur, *Nature* **345**, 140 (1990).
36. A. Schier, A. Grohmann, J. M. Lopez-de-Luzuriaga, and H. Schmidbaur, *Inorg. Chem.* **39**, 547 (2000).

37. N. Dufour, A. Schier, and H. Schmidbaur, *Organometallics* **12**, 2408 (1993).
38. H. Schmidbaur, Gold, organic compounds, in *Gmelin Handbook of Inorganic Chemistry*, Springer-Verlag, Berlin, 1980.
39. H. Schmidbaur, A. Grohmann, and M. E. Olmos, Organogold chemistry, in *Gold, Progress in Chemistry, Biochemistry and Technology*, H. Schmidbaur, ed., Wiley, Chichester, UK, 1999, p. 648.
40. C. F. Shaw and R. S. Tobias, *Inorg. Chem.* **12**, 965 (1973).
41. P. D. Gavens, J. J. Guy, M. J. Mays, and G. Sheldrick, *Acta Cryst. B* **33**, 137 (1977).
42. K. A. Porter, A. Schier, and H. Schmidbaur, unpublished results, 2003.
43. B. Djordjevic, K. A. Porter, S. Nogai, A. Schier, and H. Schmidbaur, *Organometallics* **22**, 5336 (2003).
44. (a) D. M. P. Mingos, *J. Chem. Soc. Dalton Trans.* 1163 (1976); (b) D. M. P. Mingos and R. P. F. Kanters, *J. Organomet. Chem.* **384**, 405 (1990); (c) J. K. Burdett, O. Eisenstein, and W. B. Schweizer, *Inorg. Chem.* **33**, 3261 (1994); (d) V. I. Minkin, R. M. Minyaev, and R. Hoffmann, *Russ. Chem. Rev.* **71**, 869 (2003).
45. N. Rösch, A. Görling, D. E. Ellis, and H. Schmidbaur, *Angew. Chem.* **101**, 1410 (1989).
46. A. Görling, N. Rösch, D. E. Ellis, and H. Schmidbaur, *Inorg. Chem.* **30**, 3986 (1991).
47. E. Zeller, H. Beruda, A. Kolb, P. Bissinger, J. Riede, and H. Schmidbaur, *Nature* **352**, 141 (1991).
48. J. Li and P. Pyykkö, *Inorg. Chem.* **32**, 2630 (1993).
49. P. Pyykkö and T. Tamm, *Organometallics* **17**, 4842 (1998).
50. P. Pyykkö, *Angew. Chem. Int. Ed.* (in press) 2004.
51. N. S. Isaacs, *Reactive Intermediates in Organic Chemistry*, Wiley, London, 1975, pp. 127, 129.
52. G. A. Olah and S. J. Khun, *J. Am. Chem. Soc.* **80**, 6535 (1958).
53. T. H. Lowry and K. S. Richardson, *Mechanism and Theory in Organic Chemistry*, R. R. Donnelley, New York, 1987, pp. 629, 632.
54. W. Jones, P. Boissel, B. Chivarino, M. E. Crestoni, S. Fornarini, J. Lemaire, and P. Maitre, *Angew. Chem. Int. Ed. Engl.* **42**, 2057 (2003).
55. R. Hoffmann, *Angew. Chem. Int. Ed. Engl.* **21**, 711 (1982).
56. E. G. Perevalova, D. A. Lememovskii, K. I. Grandberg, and A. N. Nesmeyanov, *Dokl. Akad. Nauk SSSR* 11 (1972).
57. K. I. Grandberg, T. V. Baukova, E. G. Perevalova, and A. N. Nesmeyanov, *Dokl. Akad. Nauk SSSR* 1355 (1972); *Dokl. Chem. Proc. Acad. Sci. USSR* 816 (1972).
58. M. N. Nefedova, I. A. Mamed'yarova, V. I. Sokolov, E. I Smyslova, L. G. Kuz'mina, and K. I Grandberg. *Russ. Chem. Bull.* **43**, 1275 (1994).
59. A. N. Nesmeyanov, E. G. Perevalova, T. V. Baukova, and K. I. Grandberg, *Izv. Akad. Nauk SSSR, Ser Khim.* 866 (1974).
60. E. G. Perevalova, T. V. Baukova, M. M. Sazonenko, and K. I. Grandberg, *Izv. Akad. Nauk SSSR, Ser Khim.* 1877 (1985).
61. E. G. Perevalova, K. I. Grandberg, T. V. Baukova, V. P. Dyadchenko, Y. L. Slovokhotov, and Y. T. Struchkov, *Koord. Khim* (*Russ. Coord. Chem.*) **8**, 1337 (1982).
62. T. V. Baukova, Y. L. Slovokhotov, and Y. T. Struchkov, *J. Organomet. Chem.* **221**, 375 (1981).

63. A. N. Nesmeyanov, E. G. Perevalova, K. I. Grandberg, D. A. Lemenovskii, T. V. Baukova, and O. B. Afanassova, *J. Organomet. Chem.* **65**, 131 (1974).
64. H. Schmidbaur and Y. Inoguchi, *Chem. Ber.* **113**, 1646 (1980).
65. A. N. Nesmeyanov, E. G. Perevalova, T. V. Baukova, and K. I. Grandberg, *Izv. Akad. Nauk SSSR, Ser Khim.* 886 (1974); *Bull. Acad. Sci. USSR, Div. Chem. Sci.* 830 (1974).
66. A. N. Nesmeyanov, E. G. Perevalova, K. I. Grandberg, and D. A. Lemenovskii, *Izv. Akad. Nauk SSSR, Ser Khim.* 1124 (1974); *Bull. Acad. Sci. USSR, Div. Chem. Sci.* 1068 (1974).
67. A. N. Nesmeyanov, E. G. Perevalova, O. B. Afanasova, M. V. Tolstaya, and K. I. Grandberg, *Izv. Akad. Nauk SSSR, Ser Khim.* 1118 (1978); *Bull. Acad. Sci. USSR, Div. Chem. Sci.* 969 (1978).
68. A. N. Nesmeyanov, E. G. Perevalova, O. B. Afanasova, and K. I. Grandberg, *Izv. Akad. Nauk SSSR, Ser Khim.* 1122 (1978); *Bull. Acad. Sci. USSR, Div. Chem. Sci.* 973 (1978).
69. A. N. Nesmeyanov, E. G. Perevalova, O. B. Afanasova, and K. I. Grandberg, *Izv. Akad. Nauk SSSR, Ser Khim.* 1919 (1978); *Bull. Acad. Sci. USSR, Div. Chem. Sci.* 1989 (1978).
70. R. Usón, A. Laguna, E. J. Fernández, A. Mendia, and P. G. Jones, *J. Organomet. Chem.* **350**, 129 (1988).
71. L. G. Kuz'mina, *Metalorg. Khim.* **5**, 744 (1992), *Organomet. Chem. USSR* **5** (1992) (Engl. transl.).
72. T. V. Baukova, N. A. Oleinikova, D. A. Lemenovskii, and L. G. Kuz'mina, *Russ. Chem. Bull.* **43**, 681 (1994)
73. T. V. Baukova, L. G. Kuz'mina, N. A. Oleinikova, D. A. Lemenovskii, and A. L. Blumenfel'd, *J. Organomet. Chem.* **530**, 27 (1997).
74. M. A. Bennett, L. L. Welling, and A. C. Willis, *Inorg. Chem.* **36**, 5670 (1997).
75. A. N. Nesmeyanov, E. G. Perevalova, O. B. Afanasova, M. V. Tolstaya, and K. I. Grandberg, *Izv. Akad. Nauk SSSR, Ser Khim.* 1118 (1978).

12

PROTON EXCHANGE BETWEEN STRONG ACIDS AND ALKANES

Jean Sommer and Alain Goeppert
Laboratoire de Physico-Chimie des Hydrocarbures
Université Louis Pasteur
Strasbourg, France

12.1 INTRODUCTION

Acid-catalyzed alkane conversion is one of the most important reactions related to the transformation of oil into useful chemicals.[1,2,3] Cracking, isomerization, and alkylation are large-scale industrial processes using solid or strong liquid acids,

---

Carbocation Chemistry, Edited by George A. Olah and G. K. Surya Prakash
ISBN 0-471-28490-4

such as H-zeolites, chlorinated aluminas, sulfuric acid, and hydrogen fluoride. High acidity and /or high temperature compensate for the well-known inertness of the starting material.

The earliest suggestion that a proton may react with a saturated hydrocarbon may be ascribed to Bloch et al.[4] in 1946, who observed that *n*-butane isomerized to isobutane under the influence of aluminum chloride, but only in the presence of HCl (Scheme 1).

$$n\text{-}C_4H_{10} + HCl \xrightarrow[AlCl_3]{} sec\text{-}C_4H_9^+, AlCl_4^- + H_2$$

Scheme 1

This interpretation remained controversial for many years as the acids were not strong enough to shift the equilibrium sufficiently to the right.

Definitive evidence was provided in the early 1960s, independently by Olah[5] and the group at Shell[6] generating stable carbenium ions and H_2 under superacidic conditions (Scheme 2).

$$iC_4H_{10} + HF : SbF_5 \longrightarrow t\text{-}C_4H_9^+, SbF_6^- + H_2$$

Scheme 2

Ever since, it was accepted that alkanes do react with acids strong enough to donate a proton to the bonding electron pair, following the concept of σ *basicity* as developed by Olah.[7] The transition state corresponds to a three-center two-electron (3c–2e) bonded structure with pentacoordinated carbon atoms (Scheme 3).

$$R{-}R' + H^+ \rightleftharpoons \left[\begin{matrix} & H & \\ R & & R' \end{matrix} \right]^+ \rightleftharpoons R'{-}H + \text{"}R^{+}\text{"}$$

R : alkyl
R′: alkyl, H

carbonium ion

Scheme 3

The reversibility of both steps of this reaction was demonstrated by using deuterated superacids or alkanes.[8,9] The hypothesis of alkane protonation by solid acids such as zeolites and chlorinated aluminas that are used in petrochemistry or other acids such as heteropolyacids, sulfated zirconias, or supported superacids is still a controversial issue since its proposal by Haag and Dessau in 1984.[10]

Since then, three different approaches, concerning the protonation of sigma-bonds in alkanes can be found in the literature: studies of initial product distribution, isotope exchange measurements, and theoretical calculations. The initial product distribution yielding hydrogen and smaller alkanes can be a good indication for protolysis of C—H and C—C bonds. Isotope exchange between deuterium-labeled alkanes or acids, in the absence of side reaction and kinetic isotope effects, provide a good support for studying the mechanism and the structure of reaction intermediates. Theoretical approaches have been favored by the strong development in recent years of easily accessible high-level ab initio and DFT methods and studies concerning the interaction of liquid as well as solid acids with alkanes have been published.

In this chapter we separately review H/D exchange in methane and ethane, because of the absence of side reactions under usual experimental conditions.

12.2 METHANE AND ETHANE

Both compounds are main components of natural gas and are used mainly as a cheap energy source. However, because of the enormous reserves of natural gas in the world, there is a strong economic incentive to develop processes that would convert these alkanes into more valuable chemicals. One of the tracks is acid catalysis, and despite the fact that methane and ethane are the weakest σ bases available and also the least reactive alkanes, both undergo H/D exchange with a variety of liquid and solid acids.

12.2.1 Hydrogen/Deuterium (H/D) Exchange with Liquid Acids

The first quantitative studies of proton exchange with liquid acids have been made by Hogeveen[11] using monodeuteromethane and -ethane in HF/SbF_5 (1 : 1 molar ratio). The reactions, which proceeded in a very clean manner between -20 and $+25°C$, are the simplest example of electrophilic aliphatic substitution. No hydrogen, carbenium ions, or higher alkanes were detected, and the acidity of the acid system remained unchanged. The activation energy was estimated to be about 70–75 kJ/mol with a surprisingly small difference between CDH_3 and C_2DH_5, suggesting that only a minor fraction of positive charge was located on the carbon atom in the transition state.

Recently the behavior of methane in a large variety of liquid and solid acids has been reinvestigated by various authors. Liquid acids, able to exchange protons with methane, ranged from concentrated sulfuric acid to $HF{:}SbF_5$.

Sulfuric Acid When methane was allowed to react in a recirculation system with a large excess of D_2SO_4, H/D exchange was already noticeable at 250°C, in the absence of any side reaction.[12] The rates increased with temperature and extensive H/D exchange was observed at the boiling temperature of the acid (338°C) (Fig.12.1).

As can be seen from Figure 12.1, mono-, bis-, ter-, and tetradeuterated methane isotopologs were formed in a consecutive manner. From the slope of the Arrhenius plot in the temperature range 270–315°C the apparent energy of activation was estimated to be 176 kJ/mol. However, at temperatures above 300°C a simultaneous

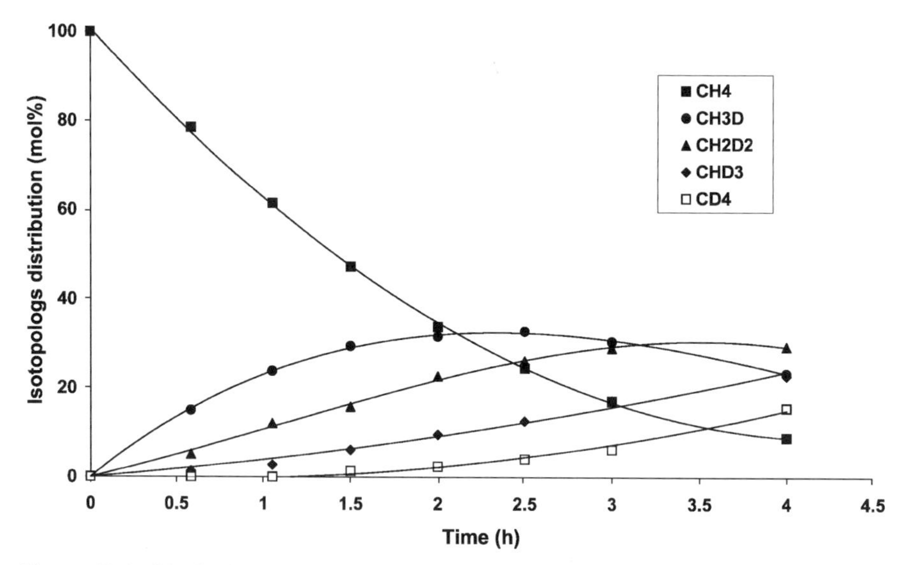

Figure 12.1 Distribution of the methane isotopologs (%) as a function of time in D_2SO_4 at 338°C. (From Ref. 12).

formation of CO_2 was observed attesting an oxidation reaction generating SO_2 and water following the reaction shown in Scheme 4.

$$4\,H_2SO_4 + CH_4 \longrightarrow CO_2 + 4\,SO_2 + 6\,H_2O$$

Scheme 4

DFT studies of H/H exchange between methane and a dimer of H_2SO_4 show that the transition state involved is bifunctional. While one proton is transferred from a hydroxy group in sulfuric acid to methane, another proton is abstracted from methane by a nonhydroxy oxygen atom in sulfuric acid (Fig. 12.2).

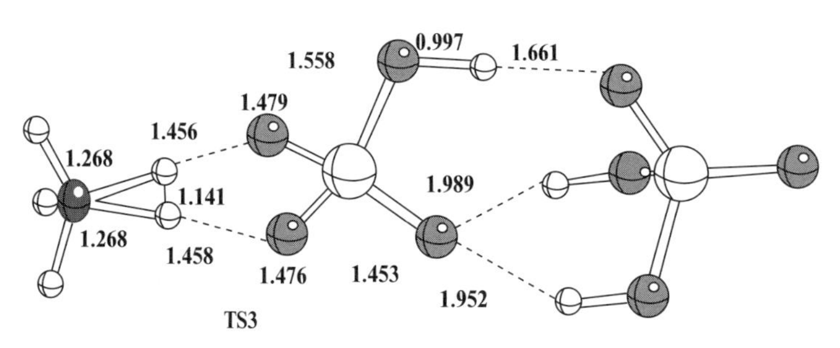

Figure 12.2 Geometry the hydron exchange transition state between methane and a dimer of sulfuric acid (From Ref. 12).

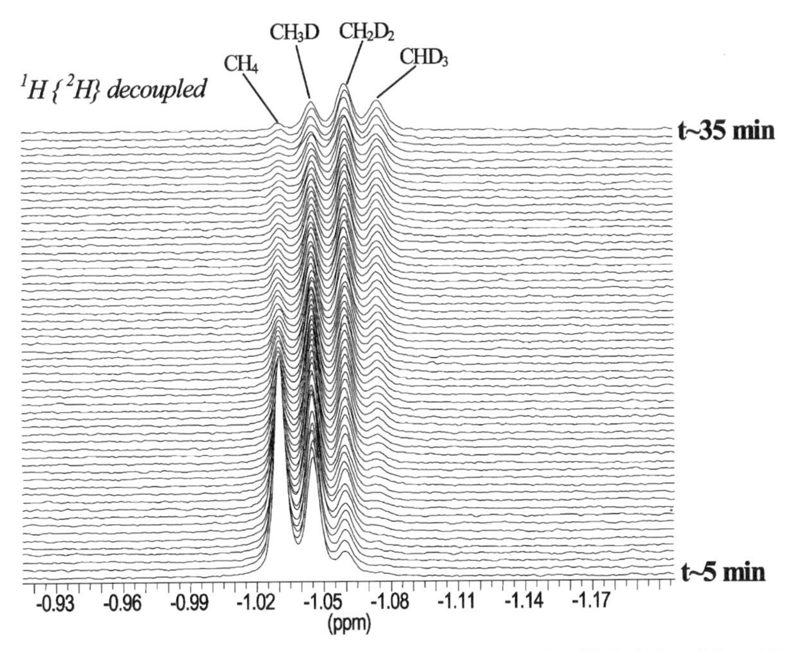

Figure 12.3 ^{2}H-decoupled ^{1}H NMR spectra of methane in DSO_3F/SbF_5 (36 mol%) at 25°C as a function of time.

FSO₃H-Based Superacids Liquid superacids are not only extremely strong acids but also under many circumstances, strong oxidants. The relative importance of these two factors on reactivity depends on both the system and on the substrate.

Up to the boiling point (163°C) of HSO_3F, methane does not react and no H/D exchange was detected even after long contact times. Increasing the temperature under pressure was not tested. However, when the acidity of the system was increased by addition of SbF_5, isotope exchange could be monitored either by GCMS analysis or directly by ^{1}H NMR spectroscopy of the acid solution of dissolved methane. Figure 12.3 shows the time dependence of the NMR spectrum of methane dissolved in DSO_3F/SbF_5,(36 mol%).

The influence of acidity on the rate of exchange is apparent on Figure 12.4.

In the so-called magic acid (1 : 1 molar HSO_3F/SbF_5) an activation energy of 65 kJ/mol^{-1} was reported.[26b]

HF-Based Superacids Methane is also slightly soluble in HF/SbF_5 (5×10^{-3} M) which is enough for direct studies by NMR. The rates are higher in this system, and the exchange studies were performed at −20°C by Ahlberg and coworkers.[13] The transition states for methane activation in this medium have been studied by experimental determination of secondary kinetic isotope effects (SKIEs) and by computational chemistry. In order to achieve high accuracy in comparing the relative H/D exchange rates of methane isotopologs, the rate constants for pairs of isotopologs were measured simultaneously in the same superacid solution by

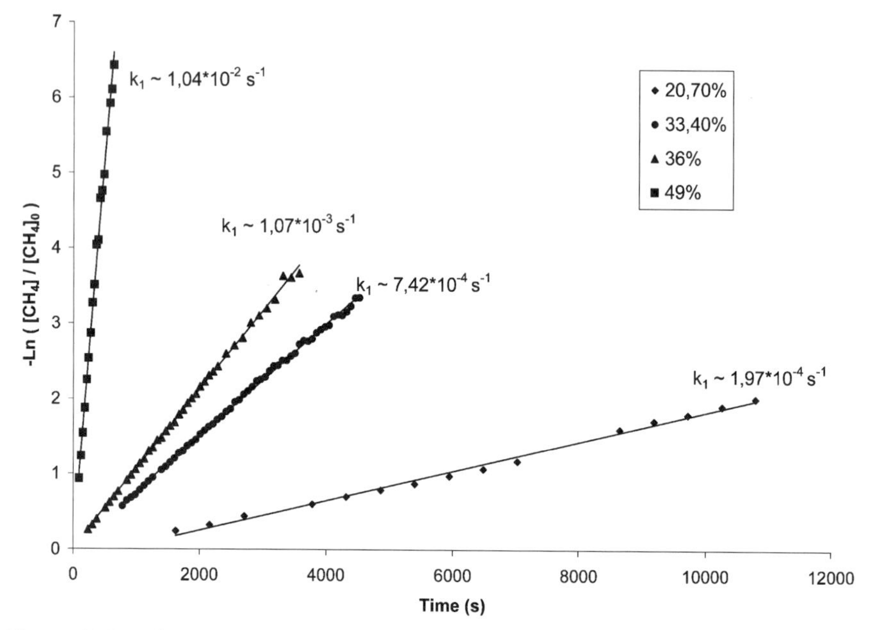

Figure 12.4 Influence of acidity (SbF_5%) on H/D exchange between DSO_3F/SbF_5 and methane.

^{2}H-decoupled ^{1}H NMR spectroscopy. The two components of the isotopolog mixture were distinguishable as one was ^{13}C-labeled.

Figure 12.5 shows the time-dependent 600-MHz ^{1}H NMR spectra obtained from a 1 : 1 mixture of $^{13}CH_4$ and CH_3D.

Both isotopologs ($^{13}CH_4$ and CH_3D) disappear by first order reactions whereas the more deuterated new products, appear upfield from the starting materials.

Due to ^{13}C–H coupling, the peaks of ^{13}C-labeled methanes appear as doublets ($J_{C\text{–}H} = 180$ Hz) and do not interfere with the ^{12}C isotopologs. With this technique the rates of following reactions have been determined in DF:SbF_5 (15 mol% SbF_5) (Scheme 5).

The rate constants $4k_a$ and $3k_b$ were defined as rate constants per proton available for exchange in each isotopolog.

$$^{13}CH_4 + DF/SbF_5 \xrightarrow{4k_a} {}^{13}CH_3D + HF/SbF_5$$

$$CH_3D + DF/SbF_5 \xrightarrow{3k_b} CH_2D_2 + HF/SbF_5$$

$$CH_2D_2 + DF/SbF_5 \xrightarrow{2k_c} CHD_3 + HF/SbF_5$$

$$CHD_3 + DF/SbF_5 \xrightarrow{k_d} CH_4 + HF/SbF_5$$

Scheme 5

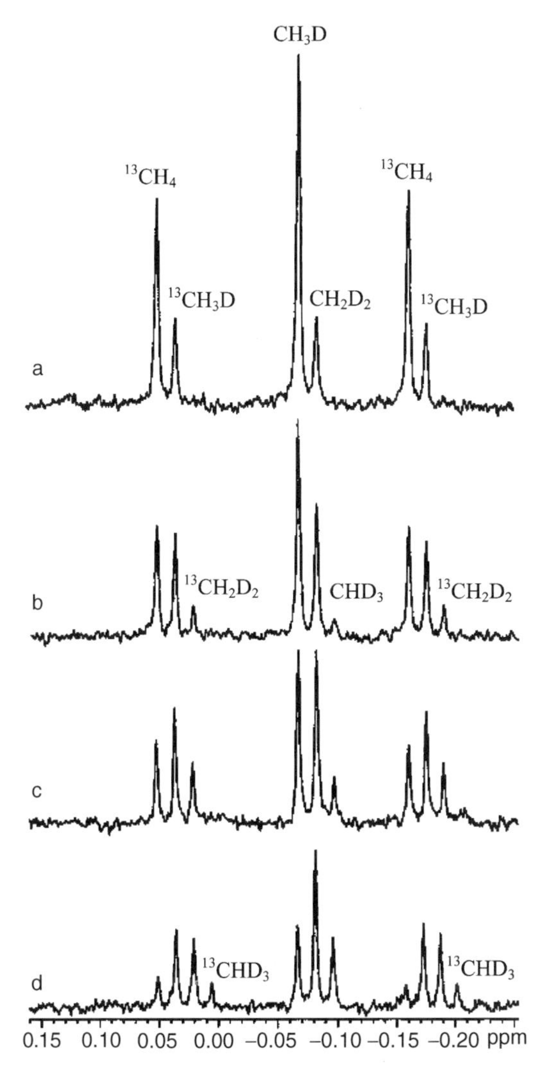

Figure 12.5 ^1H NMR spectra of the isotopolog mixture at different time intervals (From Ref. 13).

Similar experiments were conducted, using the reagent pairs $^{13}CH_4/CH_2D_2$ and $^{13}CH_4/CHD_3$ as starting materials. The estimated SKIEs, 1,00±0,02, were surprisingly small in view of the expected dramatic structural change going from methane to carbonium-ion-like activated complex but in line with the early suggestion by Hogeveen[14] that little charge was present in the transition state. This statement was confirmed by the ab initio and DFT studies.[13] The assumption was that the superacidic species involved in the proton exchange was the H_2F^+ cation more or less solvated by HF. In fact, only unsolvated H_2F^+ was found to be strong enough

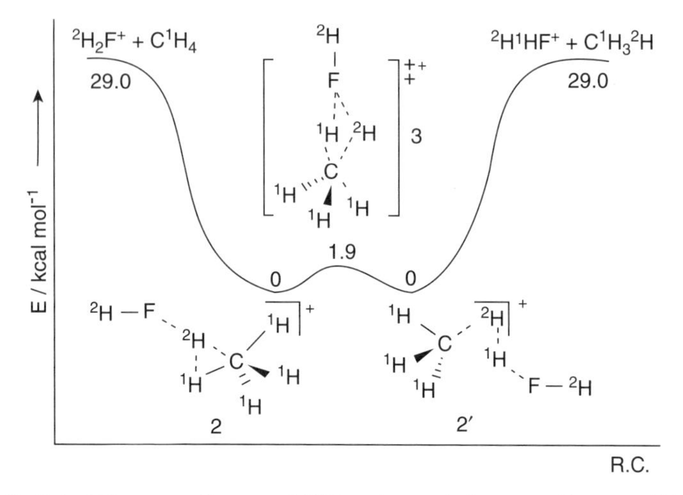

Figure 12.6 Potential energy diagram of H/H exchange between methane and H_2F^+ (From Ref. 13).

to protonate methane, yielding the HF-solvated methonium ion as a potential energy minimum (Fig. 12.6)

This suggests that even in the strongest HF–SbF_5 superacid system the methonium ion is not an intermediate in the H/D exchange process but acts as a part of the activated complex strongly solvated by $(HF)_x$. Other high-level ab initio studies have been performed by Esteves and coworkers[15] by including the fluorinated anion. In this case the carbonium ion was interacting with the $Sb_2F_{11}^-$ moiety as shown on Figure 12.7.

The free-energy profiles for the protonation of methane in HF/SbF_5 employing the same system as Esteves et al. have also been studied by molecular dynamics

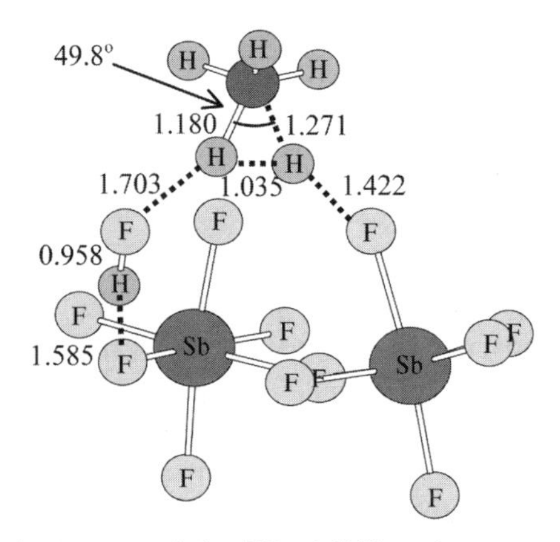

Figure 12.7 Optimized structure of the TS of H/H exchange reaction of methane with HF:Sb_2F_{10} (From Ref. 15c).

simulation.[16] The results were similar to those found using the B3LYP/6-31^{++}G** method. In all transition states there were two protons forming an H_2 moiety and thus a 3c–2e bond with the carbon atom. The free energy of activation was increasing with an increase in solvation by HF.

12.2.2 H/D Exchange with Solid Acids

The proton/deuterium exchange reaction between methane-d_4 and the hydroxyl groups of γ-alumina was reported by Larson and Hall in 1965.[17] The reaction that occurred already at room temperature was catalyzed by a small number of active sites (only 1% of the catalysts hydroxyl groups). The activation energy was estimated at 17 ± 4 kJ/mol. A primary kinetic isotope effect was measured, suggesting that the breaking of a C—H bond was rate-determining. As methane was found to have the highest rate of exchange in comparison with larger alkanes, the reaction mechanism was suggested to involve intermediate carbanions, where CH_3^- was the most stable.

Later, Kemball and coworkers[18] studied the exchange behaviour of methane over γ-aluminas in the presence of deuterium and interpreted their results also in terms of carbanionic intermediates. Similar interpretation was given more recently by Engelhardt and Valyon,[19] who studied H/D exchange between CD_4 and the OH groups of various H-zeolites, as well as γ-alumina. The activation energy was found to be independent of the Brønsted acidity of the solid and the authors suggested that methane was activated by dissociative adsorption over Lewis acid:Lewis base pair sites similar to the mechanism of H/D exchange between methane and potassium amide[20] under very basic conditions. Another study of deuterium exchange between methane and a range of metal oxides by Hargreaves and coworkers[21] concluded that basicity arguments could not be used to rationalize all their data.

However, these interpretations are in strong contrast with those of other groups studying H/D exchange between methane and H-zeolites or other strong acidic oxides. The first theoretical and experimental work in this field was published in *Nature* by Kramer and coworkers[22] comparing experimental results obtained with H-zeolites of faujasite and ZSM-5 structure, with those obtained using ab initio calculations with a small SiH_3–OH–$Al(OH)_2$–SiH_3 clusters representing the acid zeolite. The study showed the importance of Brønsted acid sites in the exchange process. Despite the covalent bondsharing character of the transition state (Fig. 12.8), the barrier was proportional to the difference in proton affinity between donor and acceptor sites.

In 2002, a more complete theoretical approach was published by Vollmer and Truong,[23] who employed an accurate ab initio embedded-cluster model to study the mechanism of proton exchange between CH_4 and the H–Y zeolite. They found that the transition state had carbonium-ion-like characteristics from both geometric and electron density considerations, ascribing the increased carbonium ion character to the inclusion of the Madelung potential, neglected in previous cluster calculations (Fig. 12.9).

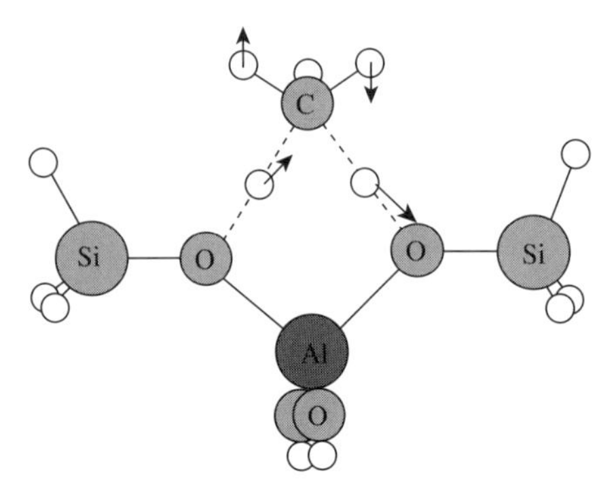

Figure 12.8 Transition state of H/H exchange between methane and a zeolite cluster (From Ref. 22).

Experimental data obtained by Gates and Jentoft[24] on the H/D exchange reaction between CD_4 and each of a family of strong solid acids (HZSM-5, sulfated zirconias, and $AlCl_3$/sulfonic acid resin) showed that the rates were faster with the strongest acids, in line with a carbonium-type intermediate. Schoofs and coworkers[25] came to the same conclusion when measuring the rates of H/D exchange reactions between methane and deuterated acid FAU and MFI zeolites in the 450–550°C range. As the rate constant of MFI was 3 times larger than that of FAU, and considering the higher acid strength of the bridging hydroxyls in MFI, the authors suggested using the activity per site in the H/D exchange as a measure of acidity of the bridging hydroxyl groups. Nevertheless, on the basis of the measured kinetic isotope effect, the authors rejected the earlier proposed concerted reaction mechanism.

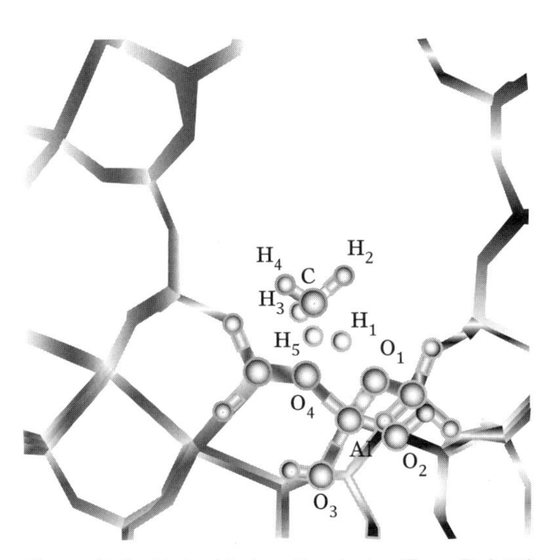

Figure 12.9 Embedded zeolite cluster (From Ref. 23).

The experimental work by Hua and Goeppert[26] on H/D exchange between methane and various solid acids such as SZ, SZA (aluminium-doped SZ), and HZSM-5 confirmed the trend of increased rate constants with increased acidity of the solid: SZA > SZ > HZSM-5. It thus appears that the H/D exchange reaction between solid acids and methane deserves further investigation related to the difficulty in properly assessing the acidity of solids.

12.3 SMALL ALKANES WITH MORE THAN TWO CARBON ATOMS

With alkanes having three or more carbon atoms, competitive secondary reactions may be expected, such as cleavage of C—H and C—C bonds yielding secondary carbenium ions. These ions may deprotonate, yielding alkenes at the origin of oligomerization, cracking, and disproportionation reactions, toward which methane and ethane are protected under the usual reaction conditions as the formation of primary carbenium ions is energetically very unfavorable.

Between alkanes with more than two carbon atoms and solid or liquid acids, the exchange reaction may proceed by three competitive pathways: (1) direct reversible protonation of C—H bonds, (2) deprotonation of the intermediate carbenium ion and reprotonation, and (3) hydride (deuteride) transfer from higher-molecular-weight material such as polyenes generated by oligomerization of alkene intermediates. The relative importance of pathways 1 and 2 depends greatly on the acidity and on the acid system and can be measured by trapping the carbenium ions (pathway 2) by carbon monoxide. Pathway 3, generally absent in the initial stages, appears after longer reaction times in acid systems favoring the buildup of oligomeric material.

12.3.1 H/D Exchange with Liquid Acids

The most convenient model for mechanistic studies of alkane activation is isobutane as it generally leads to a simple product distribution due to the large difference in reactivity of the tertiary and primary C—H bonds. For this reason it has been used in almost all H/D exchange studies.

Sulfuric Acid In a landmark paper published in 1951, Otvos[27] and coworkers described the H/D exchange reaction, observed when isoalkanes were contacted with deuterosulfuric acid at room temperature. They noticed that the hydrogens, exchanging with the acid, were always adjacent to the tertiary carbon and suggested that the exchange of those more distant resulted from isomerization of the intermediate ions. The only hydrogen that was not exchanged was that attached to the tertiary carbon. The authors considered these results as proof that carbenium ion intermediates had a finite lifetime. A reaction mechanism was proposed involving deprotonation and redeuteration of carbenium ion intermediates as predicted by Markovnikov's rule[28] and followed by hydride transfer from unreacted isoalkane in a catalytic cycle (Scheme 6).

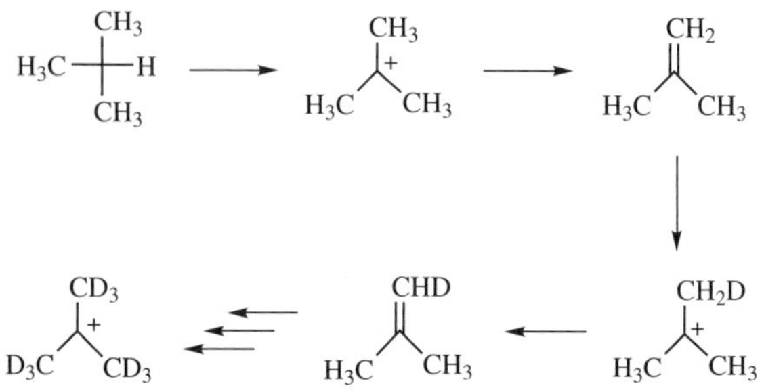

Last step: hydride transfer

$$(CD_3)_3C^+ + (CH_3)_3CH \longrightarrow (CH_3)_3C^+ + (CD_3)_3CH$$

Scheme 6

In relation to our interest in H/D exchange on solid acids, we have reinvestigated by NMR, UV, and GCMS this reaction to study similarities and differences between liquid and solid acid alkane activation.[29]

When isobutane was bubbled at a rate of 4 mL/min through 5 mL of 98% D_2SO_4 (99.9% D), at room temperature during the first 30 min only SO_2 was detected besides undeuterated starting material. During the induction period, *tert*-butyl cations are formed by oxidation of the isoalkane by sulfuric acid:

$$i\text{-}C_4H_{10} + 2D_2SO_4 \rightarrow t\text{-}C_4H_9^+DSO_4^- + HDO + D_2O + SO_2$$

After the first 10 h, during which 50% of the primary protons were exchanged, deuterium started to appear also in the tertiary position where up to 10 atom% of D were found after an additional 10 h. At the same time the acid turned slightly yellow. Investigation of the solution by UV spectroscopy indicated the presence of delocalized aromatic and polyenylic ions described earlier by Deno and co-workers.[30] The explanation can be found in Scheme 7.

With longer times on stream, due to water formation (Scheme 4), the acidity of the medium decreases and deprotonation of the *t*-butyl cations to isobutylene is favored. Under these conditions cationic oligomerization is occurring generating deuterated dienyl and polyenyl cations. These heavy and delocalized ions lead to polyenes that may ionize preferentially by deuteride transfer to the *t*-butyl cations. At the same time the acid slowly looses its activity in a manner very similar to the deactivation process of solid acids.[31]

Triflic Acid Triflic acid is known to be one of the strongest simple protic superacids.[32] As it lacks the sulfonating properties of HSO_3F, it has gained a wide range of applications as a catalyst in Friedel–Crafts reactions[33] and rare-earth triflate salts have also found a wide application in organic synthesis.[34] However, H/D exchange occurring between CF_3SO_3D and isobutane is as regiospecific as observed in the

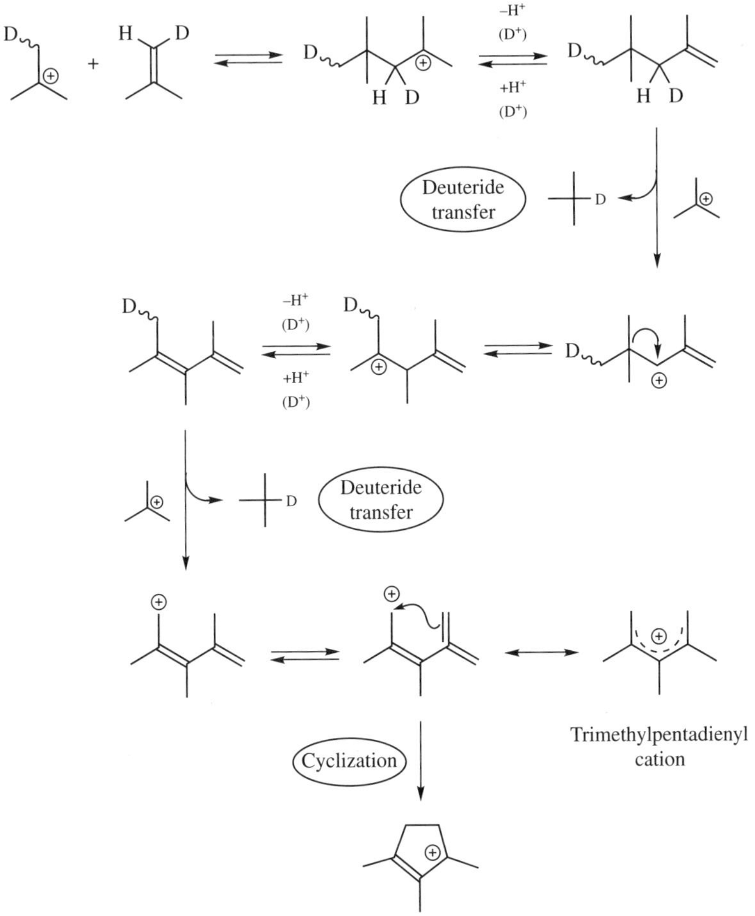

Deuteride transfer

Deuteride transfer

Trimethylpentadienyl cation

Cyclization

Scheme 7

weaker D_2SO_4 acid and hindered in the presence of carbon monoxide.[35] But in contrast to the observation in sulfuric acid, the reaction was accompanied, even at room temperature, by oligomerization of isobutane and cracking of higher-molecular-weight ions or alkanes yielding propane and isopentane as initial side products.[35]

Flurosulfonic Acid Fluorosulfonic acid is also a superacid by itself ($H_0 = -15.1$[32]) slightly more acidic than triflic acid. It is employed as catalyst and chemical reagent in various processes, including alkylation, acylation, and sulfonation, its acidity can be substantially enhanced by the addition of fluorinated Lewis acids such as SbF_5. Also, when isobutane was circulated through DSO_3F, at 5°C under nitrogen, extensive deuterium incorporation was observed in the primary position (53 atom% in a one-pass experiment) and to a minor extent (6%) in the tertiary position. The same reaction conducted under CO at the same temperature showed

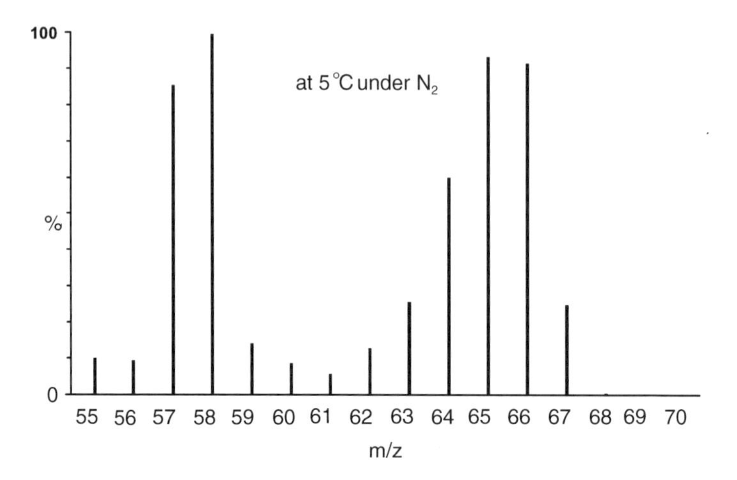

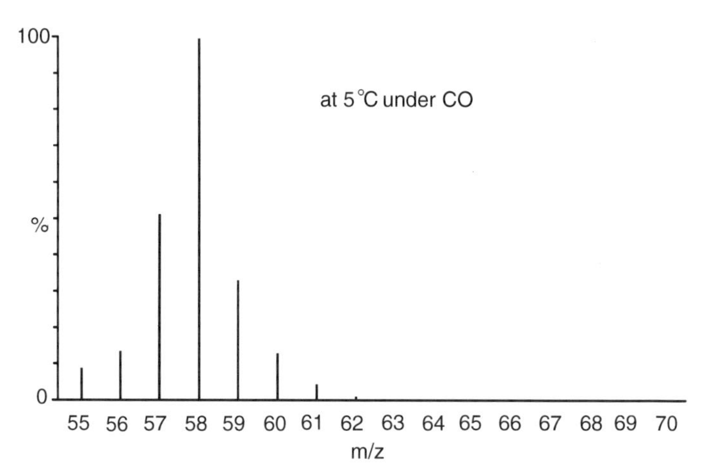

Figure 12.10 GC-MS spectra of isobutane after H/D exchange with DSO_3F at 5°C in the presence and absence of CO (From Ref. 36).

a much lower deuteration and even at 25°C, despite an extensive H/D exchange in the primary position no exchange was observed in the tertiary position (Fig. 12.10).

This is in accord with the classical (described above) Otvos-type exchange process, involving carbenium intermediates that are in equilibrium with oxocarbenium ions in the presence of CO.

To obtain more information on the reaction mechanism, especially on the rate-limiting step, KIE (kinetic isotope effect) measurements were performed by comparing the rates of isobutane with isobutane-2*d*. A KIE of 2.06 ± 0.2 was found even in the presence of 2% isobutylene and was thus assigned to the hydride transfer step in the catalytical cycle.[36]

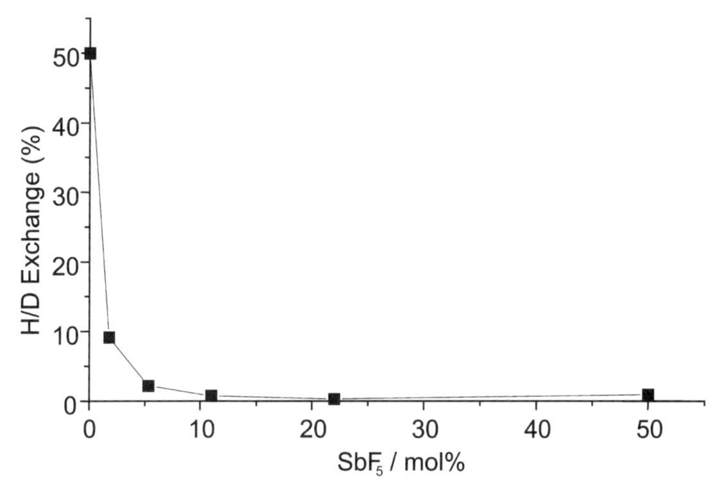

Figure 12.11 Isobutane H/D exchange in DSO_3F/SbF_5 at 25°C as a function of the SbF_5 concentration (From Ref. 36).

DSO_3F-SbF_5 (at Various Concentrations in SbF_5) To avoid deuteration via hydride transfer to the tertiary position, all reactions were conducted in the presence of carbon monoxide.

In contrast to the increasing deuteration of the alkane expected with increasing acidity of the deuterated acid system, the opposite was observed: deuteration was exponentially decreasing with increasing SbF_5 concentration (Fig. 12.11). In fact, ionization was the major reaction as demonstrated by the disappearance of isobutane when the reaction was conducted in the presence of carbon monoxide. In magic acid (1 : 1 molar ratio), the reaction produced a small amount of hydrogen that seems to indicate that a minor protolytic pathway competes with oxidative ionization.

DF-SbF_5 In this acid system, which is the strongest superacid available,[37] up to 18 mol% SbF_5, the predominant reaction for all small alkanes, beside minor protolytic cleavage, is rapid isotope exchange via reversible protonation of all C—H bonds. The relative rate of exchange follows Olah's *σ-basicity* concept (tertiary > secondary > primary) independently of the further reactivity of these bonds. At higher concentrations of SbF_5, the exchange process gives way to increasing ionization of the alkane, first via protolysis and later via an oxidative process with concomitant reduction of SbF_5. We will limit our discussion to the reactions of propane, isobutane, and isopentane.

Isobutane was first found to exchange initially its tertiary proton[38] on the basis of MS analysis of the recovered material after reaction with DF:SbF_5 at −78°C. Later on, the exchange reaction was reinvestigated in the presence of carbon monoxide at 0°C.[39] The deuterium distribution observed in isobutane recovered after short contact times with the DF:SbF_5 superacid was in accord with a very fast reversible protonation of all C—H bonds before ionization of the alkane following the Olah *σ-basicity* concept.

In order to distinguish between ionization and the H/D exchange process, all reactions between small alkanes and DF:SbF_5 were conducted in the presence of carbon monoxide, which converts the reactive, initially produced, carbenium ions into stable oxocarbenium ions. This procedure also inhibited inter- and intramolecular hydride transfers, which would have made the interpretation more difficult. Scheme 8 shows the reaction scheme of propane, and Figure 12.12 shows a comparison between exchange rates and ionization rates measured by quenching the oxocarbenium ions with excess ethanol followed by neutralization.

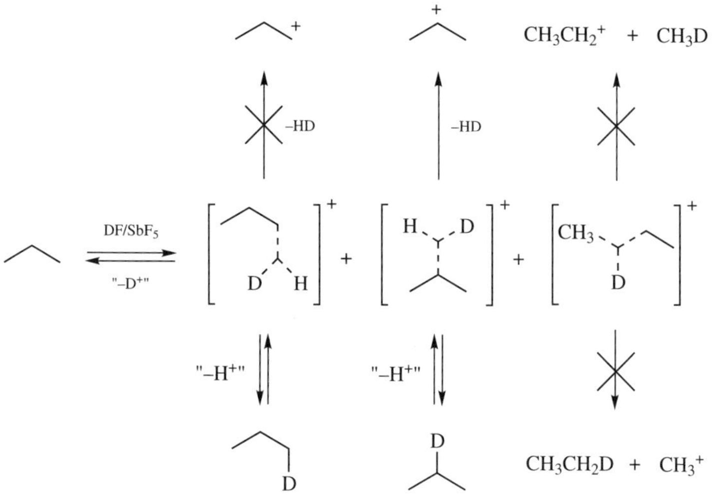

$CH_3CH_2^+$ + CH_3D

–HD

–HD

DF/SbF_5

"–D^+"

"–H^+"

"–H^+"

CH_3CH_2D + CH_3^+

Scheme 8

The product distribution depends heavily on the acid composition: When the concentration of SbF_5 is increased from 10 to 18 mol%, the exchange rate increased rapidly, but above 18 mol% it decreased because of the predominance of the ionization process involving reduction of SbF_5. This pattern was also observed for *n*- and isobutane[39,40] as well as isopentane.[41]

In isopentane, the exchange rates, normalized to 1 for primary protons, were in the sequence: 1.0 : 1,34 : 1,60 for primary : secondary : tertiary protons, respectively. This is in full agreement with relative σ-basicities, whereas the relative reactivities of primary versus secondary or tertiary C—H bonds show a difference of several orders of magnitude.

Esteves and coworkers have calculated the potential energy surface of propane protonation leading to the $C_3H_9^+$ cation[15b] at the MP4(SQTQ)/6-311^{++}6**/MP2 full/6-316** level. They found that the C-proponium cation was the lowest in energy but the van der Waals complex of the isopropyl cation plus hydrogen was only 0.3 kcal/mol above.

In order to determine whether any intramolecular scrambling was occurring during the H/D exchange, 1,1,1,3,3,3-propane-d_6, 2,2-propane-d_2, 2-^{13}C propane, and 2-(methyl-d_3) 1,1,1,3,3,3-propane-d_6 were used as starting material. Under the

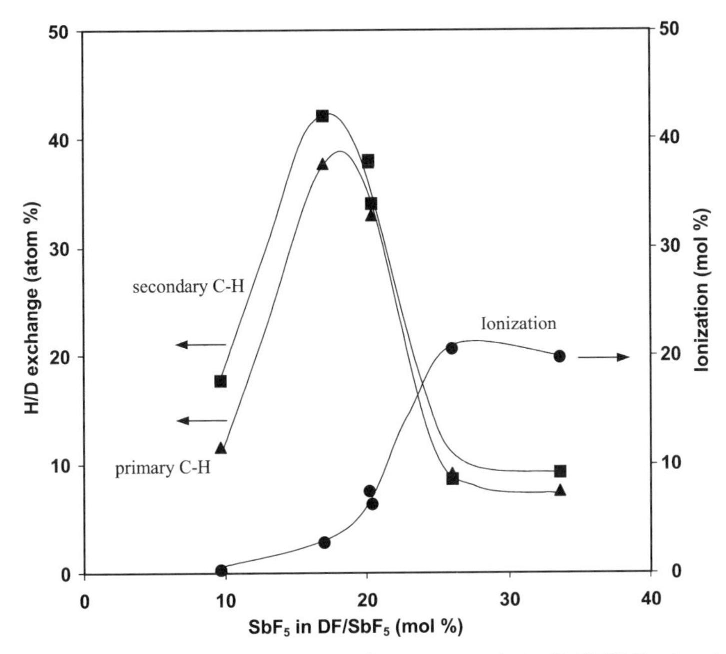

Figure 12.12 H/D exchange and conversion of propane reacted with DF/SbF_5 at various SbF_5 concentrations (From Ref. 41).

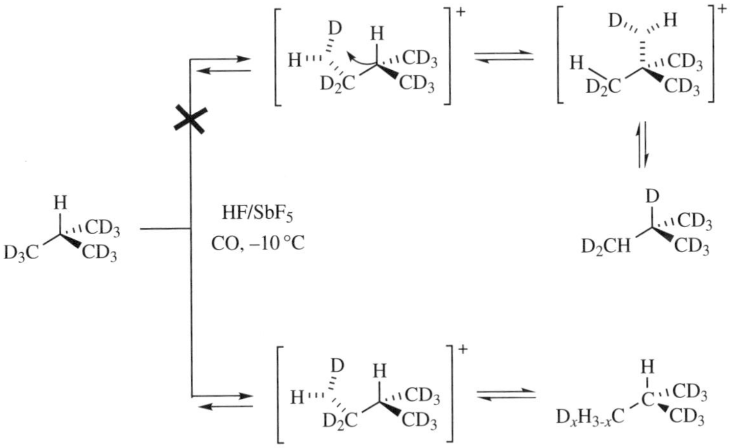

HF/SbF_5

CO, −10 °C

Scheme 9

conditions of H/D exchange, no proton or ^{13}C scrambling was detected[42] even after longer contact times and extensive H/D exchange (Scheme 9).

These results show that in contrast with well-documented carbenium ion rearrangements,[32,43–45] even in small cations such as *t*-butyl or *s*-propyl, intramolecular proton scrambling or skeletal rearrangement does not occur in protonated alkanes.

12.3.2 H/D Exchange between Small Alkanes and Solid Acids

There is a general agreement that carbenium ion intermediates play a major role in industrial acid–catalyzed alkane transformations. However, the nature of the initial step is still a matter of controversy, and a variety of suggestions can be found in the literature, including protolysis of C—H and C—C bonds via carbonium-ion-like transition states.[10,46,47] Solid acids such as zeolites, acidic aluminas, heteropoly-acids, sulfated zirconias, or acid-supported solids are generally oxides containing various OH groups that may function as proton donors. The deuteronation of the OH groups or studying their reactivity in the presence of deuterated alkanes allows mechanistic isotope tracer studies. Moreover, H/D exchange measurements between solid acids and D_2O allows a quantitative determination of Brønsted acid sites of the solid.[48] H/D exchange between cracking catalysts and butane has already been reported in the early 1950s by Hindin et al.,[49] who emphasized that the rate was increased by the addition of small amounts of alkene.

Since the mid-1990s, this reaction was reinvestigated by several groups[50–53] showing that the exchange occurs following the Otvos scheme via deprotonation

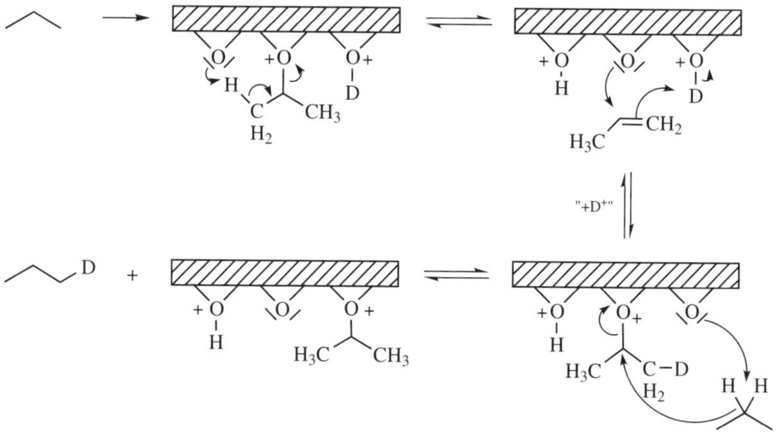

Scheme 10

of the adsorbed carbenium and very fast reprotonation of alkene intermediates. (Scheme 10).

When isoalkanes were contacted with D_2O-exchanged solid acids such as zeolites, heteropolyacids, or sulfated zirconias, proton–deuterium exchange took place even at room temperature but only at the position vicinal to branching (Scheme 11).

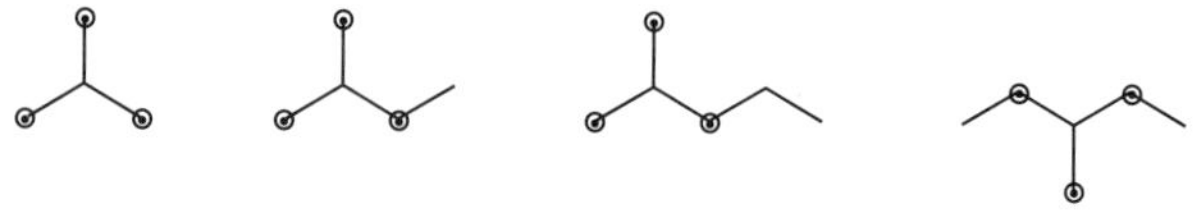

Acids: HUSY, HBEA, HEMT, HZSM-5, HMAZ, SZ, FMZS, HPA (etc.)

Scheme 11

Linear alkanes, however, needed higher temperatures to undergo this exchange. Isomerization of *n*-butane on deuterated sulfated zirconia at 150°C showed H/D exchange both in the starting material and in the product isobutane in accord with the intermediacy of olefins tied up as "carbenium ions" or surface alkoxy groups.[54] The occurence of carbenium ions as reaction intermediates is strongly supported by the observation that the exchange is inhibited in the presence of carbon monoxide.[55] Moreover, trapping of the intermediates has been monitored directly by in-situ solid-state NMR when isobutane, CO, and water reacted on HZSM-5 zeolite to form pivalic acid.[56]

For propane, however, on the basis of ^{13}C scrambling and extensive deuteration observed at higher temperatures, carbonium ion intermediates were suggested.[57,58] A careful reinvestigation of the exchange process with selectively labeled 1-^{13}C propane, 2,2-propane d_2 and 1,1,1,3,3,3-propane-d_6[59] showed that the initial H/D exchange was regioselective as predicted on the basis deprotonation of carbenium ions. At higher temperature, ^{13}C scrambling was due to the formation of protonated cyclopropanes in equilibrium with the adsorbed carbenium ions, following the

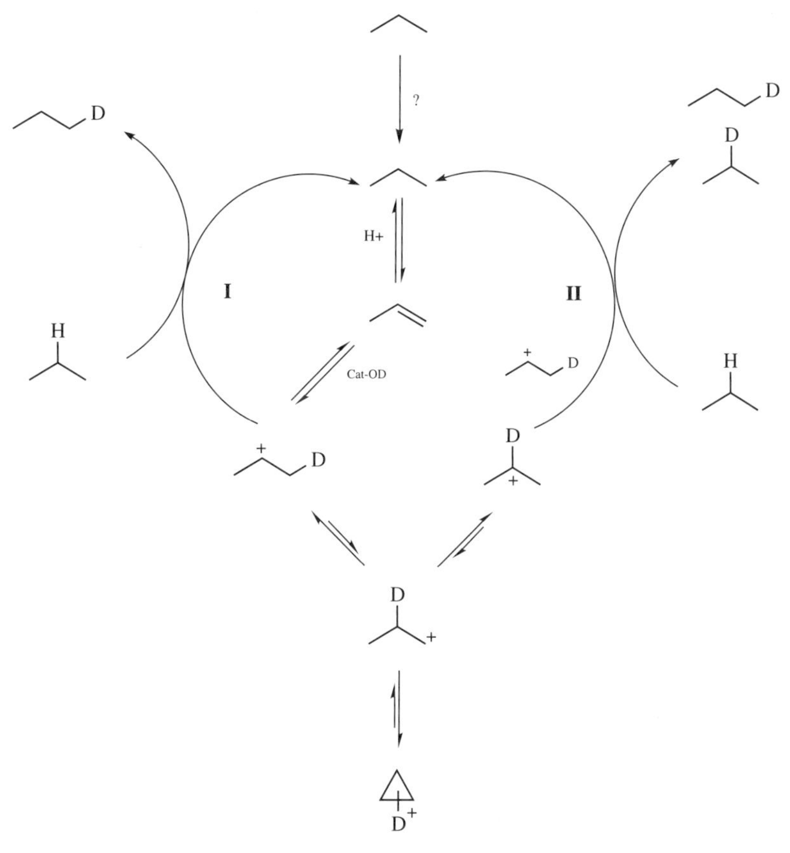

Scheme 12

model described by Saunders and coworkers.[43] Based on these results, a general scheme was proposed in which propene was a common reaction intermediate for both the regioselective H/D exchange (route I in Scheme 12) and the skeletal rearrangement reaction (route II in Scheme 12).

These findings are in accord with earlier results[60] and confirm that, at temperatures up to 200°C, small alkanes with more than two carbon atoms exchange their protons with solid Brønsted acids via deprotonation of adsorbed carbenium ion intermediates and fast reprotonation of the resulting alkenes.

However, this still leaves open the nature of the initial step producing the ions from alkanes, and it does not exclude the possibility that at higher temperatures (450°C and above) fast reversible protonation may occur in relation to the initial product distribution in accord with C—H and C—C bond protolysis.[10]

12.4 CONCLUSION

With the exception of HF–SbF_5 in which reversible protonation of small alkanes can be very fast in comparison with other reactions, despite the variety of experimental techniques and procedures, a high degree of ressemblance can be noticed between liquid acids or superacids and solid acids with a wide range of structural differences.

Methane and ethane undergo H/D exchange via direct reversible protonation in liquid as well as on solid acids. Alkanes with more than two carbon atoms generate secondary or tertiary carbenium ions that are bona fide reaction intermediates in the proton exchange process. These ions (adsorbed on solids or solvated in liquids) are in equilibrium with alkenes, the oligomerization of which is at the origin of the deactivation of the acid systems.

REFERENCES

1. H. Pines, *The Chemistry of Catalytic Hydrocarbon Conversion*, Academic Press, New York, 1981.
2. G. A. Olah and A. Molnar, *Hydrocarbon Chemistry*, Wiley-Interscience, New-York, 1995.
3. A. Corma, *Chem. News* **95**, 559 (1995).
4. H. S. Bloch, H. Pines, and L. Schmerling, *J. Am. Chem. Soc.* **68**, 153 (1946).
5. G.A. Olah and R. H. Schlosberg, *J. Am. Chem. Soc.* **90**, 2726 (1968).
6. H. Hogeveen and A. F. Bickel, *J. Chem. Soc., Chem. Commun.* 635 (1967).
7. G. A. Olah, *Angew. Chem., Int. Ed. Engl.* **12**, 173 (1973).
8. G. A. Olah, G. Klopman, and R. H. Schlossberg, *J. Am. Chem. Soc.* **91**, 3261(1969)
9. (a) D. M. Brouwer and J. M. Oelderick, *Rec. Trav. Chim. Pays-Bas* **87**, 721 (1968); (b) H. Hogeveen and A. F. Nickel, *Rec. Trav. Chim. Pays-Bas* **86**, 1313 (1967).
10. W. O. Haag and R. M. Dessau, in *Proc. 8th Int. Congress on Catalysis*, Berlin, Vol. **2**, Dechema, Frankfurt am Main, 1989, p. 305.

11. H. Hogeveen and A. F. Bickel, *Rec. Trav. Chim. Pays-Bas* **88**, 371 (1969).
12. A. Goeppert, P. Dinér, P. Ahlberg, and J. Sommer, *Chem. Eur. J.* **8**, 3277–3283 (2002).
13. P. Ahlberg, A. Karlsson, A. Goeppert, S.O. Nilsson Lill, P. Dinér, and J. Sommer, *Chem. Eur. J.* **7**, 1936–1943 (2001).
14. H. Hogeveen and J. Gaasbeck, *Rec. Trav. Chim. Pays-Bas* **87**, 319 (1968).
15. (a) P. M. Esteves, G. G. P. Alberto, A. Ramírez-Solís, and C. J. A. Mota, *J. Am. Chem. Soc.* **121**, 7345 (1999); (b) P. M. Esteves, C. J. A. Mota, A. Ramírez-Solís, and R. Hernández-Lamoneda, *J. Am. Chem. Soc.* **120**, 3213 (1998); (c) C. J. A. Mota, P. M. Esteves, and A. Ramírez-Solís, and R. Hernández-Lamoneda, *J. Am. Chem. Soc.* **119**, 5193 (1997); (d) N. Okulik, N. M. Peruchena, P. M. Esteves, C. J. A. Mota, and A. Jubert, *J. Phys. Chem. A* **103**, 8491 (1999).
16. S. Raugei and M. Klein, *J. Phys. Chem. B* **106**, 11596–11605 (2002).
17. J. G. Larson and W. H. Hall, *J. Phys. Chem.* **69**, 3080 (1965).
18. P. J. Robertson, M. S. Scurrell, and C. Kemball, *Trans. Faraday Soc.* **71**, 903 (1975).
19. J. Engelhardt and J. Valyon, *React. Kinet. Catal. Lett.* **74**, 217–224 (2001).
20. H. Handa, T. Baba, and Y. Ono, *J. Chem. Soc., Faraday Trans.* **94**, 451–454 (1998).
21. J. S. J. Hargreaves, G. J. Hutchings, R. W. Joyner, and S. H. Taylor, *Appl. Catal. A General* **227**, 191–200 (2002).
22. G. M. Kramer, R. A. van Santen, C. A. Eneis, and A. K. Nowak, *Nature* **363**, 529 (1993).
23. J. M. Vollmer and T. N. Truong, *J. Phys. Chem. B* **104**, 6308 (2000).
24. R. E. Jentoft and B. C. Gates, *Catal. Lett.* **72**, 129 (2001).
25. B. Schoofs, J. A. Martens, P. A. Jacobs, and R. Schoonheydt, *J. Catal.* **183**, 355 (1999).
26. (a) W. Hua, A. Goeppert, and J. Sommer, *J. Catal.* **197**, 406 (2001); (b) A. Goeppert, Ph.D. thesis, Univ. Louis Pasteur, Strasbourg, France, April 2002.
27. (a) J. W. Otvos, D. P. Stevenson, C. D. Wagner, and O. Beeck, *J. Am Chem. Soc.* **73**, 5741 (1951); (b) D. P. Stevenson, C. D. Wagner, O. Beeck, and J. W. Otvos, *J. Am. Chem. Soc.* **74**, 3269 (1952).
28. V. Markovnikov (1838–1904), Moscow Univ.
29. J. Sommer, A. Sassi, M. Hachoumy, R. Jost, A. Karlson, and P. Ahlberg, *J. Catal.* **171**, 391–397 (1997).
30. (a) N. C. Deno, in *Carbonium Ions*, G. A. Olah and P. v. R. Schleyer, eds., Wiley, New York, 1970, Vol. II, p. 783; (b) T. S. Soerensen, in *Carbonium Ions*, G.A. Olah and P. v. R. Schleyer, eds., Wiley, New York, 1970, Vol. II, p. 807.
31. D. Spielbauer, G. A. H. Mekhemer, M. I. Zaki, and H. Knözinger, *Catal. Lett.* **40**, 71 (1996).
32. G. A. Olah, G. K. S. Prakash, and J. Sommer, *Superacids*, Wiley-Interscience, New York, 1985.
33. R. D. Howells and J. D. McCown, *Chem. Rev.* **77**, 69 (1977).
34. S. Kobayashi, M. Sugihara, H. Kitagawa, and W. L. Lam, *Chem. Rev.* **102**, 2227–2302 (2002).
35. A. Goeppert, B. Louis, and J. Sommer, *Catal. Lett.* **56**, 43–48 (1998).
36. A. Goeppert, D. L. Behring, J. Sommer, and C. J. A. Mota, *J. Phys. Org. Chem.* **15**, 869–873 (2002).
37. R. Jost and J. Sommer, *Rev. Chem. Intermed.* **9**, 171 (1988).

38. G. A. Olah, Y. Halpern, J. Shen, and Y. K. Mo, *J. Am. Chem. Soc.* **93**, 1251 (1971).
39. J. Sommer, J. Bukala, M. Hachoumy, and R. Jost, *J. Am. Chem. Soc.* **119**, 3274–3279 (1997).
40. J. Sommer, J. Bukala, and M. Hachoumy, *Res. Chem. Intermed.* **22**(8), 753–766 (1996).
41. A. Goeppert and J. Sommer, *New J. Chem.* **26**, 1335–1339 (2002).
42. A. Goeppert, A. Sassi, J. Sommer, P. M. Esteves, C. J. Mota, and A. Karlson, *J. Am. Chem. Soc.* **121**, 10628–10629 (1999).
43. M. Saunders and E. L. Hagen, *J. Am. Chem. Soc.* **90**, 6881 (1968).
44. M. Saunders, P. Vogel, E. L. Hagen, and J. Rosenfeld, *Acc. Chem. Res.* **6**, 53 (1973).
45. M. Saunders, A. P. Hewett, and O. Kronja, *Croat. Chem. Acta* **65**, 673 (1992).
46. S. Kotrel, H. Knözinger, and B. C. Gates, *Microporous Mesoporous Mater.*, **35–36**, 11–20 (2000).
47. J. Lercher, R.A. van Santen, and H.Vinek, *Catal. Lett.* **27**, 91–96 (1996).
48. R. Olindo, A. Goeppert, D. Habermacher, J. Sommer, and F. Pinna, *J. Catal.* **197**, 344–349 (2001).
49. S. G. Hindin, G. A. Mills, and A. G. Oblad, *J. Am. Chem. Soc.* **73**, 278 (1951).
50. J. Engelhardt and W. K. Hall, *J. Catal.* **151**, 1 (1995).
51. B. Schoofs, J. Schuermans, and R. A. Schoonheidt, *Microporous Mesoporous Mater.* **35–36**, 99–111 (2000).
52. T. F. Narbeshuber, M. Stockenhuber, A. Brait, K. Seshan, and J. A. Lercher, *J. Catal.* **160**, 183–189 (1996).
53. J. Sommer and R. Jost, *Pure Appl. Chem.* **72**, 2309–2318 (2000).
54. B. Q. Xu and M. H. Sachtler, *J. Catal.* **165**, 231–240 (1997).
55. J. Sommer, D. Habermacher, R. Jost, A. Sassi, A. G. Stepanov, M. V. Luzgin, D. Freude, H. Ernst, and J. Martens, *J. Catal.* **181**, 265–270 (1999).
56. M. V. Luzgin, A. G. Stepanov, A. Sassi, and J. Sommer, *Chem. Eur. J.* **6**, 2368–2376 (2000).
57. I. I. Ivanova, E. B. Pomakina, A. I. Rebrov, and E. G. Derouane, *Top. Catal.* **6**, 49 (1998).
58. A. G. Stepanov, H. Ernst, and D. Freude, *Catal. Lett.* **59**, 51 (1999)
59. (a) M. Haouas, S. Walspurger, and J. Sommer, *J. Catal.* **215**, 122–128 (2004); (b) M. Haouas, S. Walspurger, F. Taulelle, and J. Sommer, *J. Amer. Chem. Soc.* **126**, 599–606 (2004).
60. J. Sommer, D. Habermacher, M. Hachoumy, R. Jost, and A. Reynaud, *Proc. DMGK Conf.*, Berlin, March 14–16, 1996, pp. 33–40.

13

ELECTROPHILICITY SCALES FOR CARBOCATIONS

Herbert Mayr and Armin R. Ofial
Department of Chemistry
Ludwig Maximilian University Munich
Munich, Germany

13.1 CARBOCATION STABILITY

In 1902, Baeyer recognized the carbocationic character of the triphenylmethyl species,[1] which had previously been observed by Norris[2,3] and Kehrmann[4] when treating chlorotriphenylmethane with aluminum trichloride or triphenylmethanol with sulfuric acid.[5] Less than three decades later, Ingold's investigations of the mechanisms of nucleophilic substitutions revealed the first quantitative relationships

Carbocation Chemistry, Edited by George A. Olah and G. K. Surya Prakash
ISBN 0-471-28490-4

between structures and stabilities of carbocations.[6] The fact that *tert*-butyl chloride solvolyzes approximately 1500 times faster than 2-butyl chloride (ethanol, 25°C) has been explained by the higher stability of the *tert*-butyl cation **2** compared to the *sec*-butyl cation **1**.[7] In accord with this conclusion, the rearrangement **1** → **2** has been observed under superacidic conditions,[8–11] and the difference in stability between the two ions can in principle be calculated from the equilibrium constant for the isomerization of Eq. (13.1) according to $\Delta G^\circ = -RT \ln K$:[10]

$$\underset{\mathbf{1}}{H_3CCH_2CH^+CH_3} \underset{}{\overset{K}{\rightleftharpoons}} \underset{\mathbf{2}}{(CH_3)_3C^+} \tag{13.1}$$

The fact that the *sec*-butyl cation **1** is not observable in equilibrium along with the *tert*-butyl cation **2** unequivocally demonstrates that the *tert*-butyl cation **2** is much more stable than its isomer **1**, and one can define Hess cycles to determine the precise difference in stability between the isomeric carbocations **1** and **2** in the gas phase or in solution.[11]

Analogously, the higher S_N1 reactivity of *tert*-butyl chloride compared to isopropyl chloride has been considered as evidence for the greater stability of the *tert*-butyl cation **2** relative to the isopropyl cation **3**. Since now two nonisomeric cations are considered, a direct comparison of their stabilities is not possible, and one may employ any of the equations (13.2),[12,13] (13.3),[14–16] or (13.4)[17] (or similar ones) to quantify the difference in stability between the secondary and the tertiary carbenium ion (ΔH° referring to the gas phase):

$$\underset{\mathbf{3}}{(CH_3)_2C^+H} + H{-}C(CH_3)_3 \xrightarrow[\text{hydride transfer}]{\Delta H^\circ = -62.8\ \text{kJ/mol}} (CH_3)_2CH_2 + \underset{\mathbf{2}}{(CH_3)_3C^+} \tag{13.2}$$

$$\underset{\mathbf{3}}{(CH_3)_2C^+H} + Cl{-}C(CH_3)_3 \xrightarrow[\text{chloride transfer}]{\Delta H^\circ = -55.2\ \text{kJ/mol}} (CH_3)_2CHCl + \underset{\mathbf{2}}{(CH_3)_3C^+} \tag{13.3}$$

$$\underset{\mathbf{3}}{(CH_3)_2C^+H} + HO{-}C(CH_3)_3 \xrightarrow[\text{hydroxide transfer}]{\Delta H^\circ = -53.1\ \text{kJ/mol}} (CH_3)_2CHOH + \underset{\mathbf{2}}{(CH_3)_3C^+} \tag{13.4}$$

Equations (13.2)–(13.4) lead to the conclusion that the *tert*-butyl cation is 53–63 kJ/mol more stable than the isopropyl cation. While the relative stability of cations **1** and **2** is unequivocally defined by the equilibrium constant K of Eq. (13.1), the exact difference in stability between the isopropyl and the *tert*-butyl cation depends on the arbitrary choice of the reference reaction. The inappropriateness of the term "stability" for comparing nonisomeric carbocations is perhaps best demonstrated by comparing the *tert*-butyl cation **2** with the benzenium ion **4**. Both Eqs. (13.5) and (13.6) appear to be reasonable choices for defining the relative stability of **2** and **4**. The first comparison [Eq. (13.5)] indicates the benzenium ion **4** to be 83.7 kJ/mol "more stable,"[18,19] the second comparison [Eq. (13.6)], 55.6 kJ/mol "less stable"[20,21] than the *tert*-butyl cation **2**; thus, the term "stability" is meaningless without referring to the reference reaction:

$$\mathbf{4} + \text{H}-\text{C}(\text{CH}_3)_3 \xrightarrow[\text{hydride transfer}]{\Delta H^\circ = +83.7\ \text{kJ/mol}} \text{C}_6\text{H}_8\ (\text{1,4-cyclohexadiene}) + (\text{CH}_3)_3\text{C}^+\ \mathbf{2} \quad (13.5)$$

$$\mathbf{4} + \text{H}_2\text{C}=\text{C}(\text{CH}_3)_2 \xrightarrow[\text{proton transfer}]{\Delta H^\circ = -55.6\ \text{kJ/mol}} \text{C}_6\text{H}_6 + (\text{CH}_3)_3\text{C}^+\ \mathbf{2} \quad (13.6)$$

What is the main difference between Eqs. (13.5) and (13.6)? In the forward and the backward reactions of Eq. (13.5), the carbocations accept a base (H^-); that is, they act as Lewis acids, while they donate a proton in Eq. (13.6); thus, they now act as Brønsted acids. Consequently, the benzenium ion is a weaker Lewis, but a stronger Brønsted acid, than the *tert*-butyl cation.

As explicitly pointed out by Hine,[22] "stability scales" of carbocations are often Lewis acidity scales [Eq. (13.7)] or Brønsted acidity scales [Eq. (13.8)]:

$$\text{H}_3\text{C}-\text{C}^+\text{R}_2 + \text{B}^- \rightleftharpoons \text{H}_3\text{C}-\text{CR}_2-\text{B} \quad (13.7)$$

Lewis acid *electrophile* — Lewis base *nucleophile*

$$\text{H}_3\text{C}-\text{C}^+\text{R}_2 + \text{B}^- \rightleftharpoons \text{H}_2\text{C}=\text{CR}_2 + \text{BH} \quad (13.8)$$

Brønsted acid — Brønsted base

Lewis acidity scales of carbocations are derived by determining the equilibrium constants K of Eq. (13.7) for a given reference base B^-. Well-known examples are the hydride affinity scales,[13,23] chloride affinity scale,[11,24–27] and the pK_{R^+} scale,[26,28–32] which represents a Lewis acidity scale with respect to OH^- as the reference base. Many common trends in these Lewis acidity scales are due to the fact that variation of the substituents R usually affects the carbocations of Eq. (13.7) to a greater extent than the corresponding Lewis acid–Lewis base adducts.

"Stability scales" based on the rates of S_N1 reactions[27,33,34] refer to the breakup of Lewis acid–Lewis base adducts [the reverse of Eq. (13.7)] and, therefore, are also related to the scales of Lewis acidity mentioned above.

Whereas Lewis acidity and Lewis basicity refer to the equilibrium constants of Eq. (13.7), *electrophilicity* and *nucleophilicity* refer to the corresponding rate constants. The Bell–Evans–Polanyi principle[35] postulates that an increase of Lewis basicity of B^- usually results in an increase in nucleophilicity. For the same reason, the kinetic term "electrophilicity" is related to its thermodynamic counterpart, "Lewis acidity".

13.2 EXPERIMENTAL APPROACHES

The systematic development of electrophilicity scales started in the early 1970s, when Ritchie reported about the kinetics of the reactions of highly stabilized carbocations, such as tritylium and tropylium ions, with anions and amines:[32,36]

$$Ar_3C^+ + Nu^- \xrightarrow{k} Ar_3C-Nu \qquad (13.9)$$

This work has been extended by Jencks and Richard, who developed the azide clock method.[37] Since many solvolytically generated carbocations react with azide ions with the constant diffusion-controlled rate constant of $5 \times 10^9\ M^{-1}\ s^{-1}$, the rate constants for the reactions of solvolytically generated carbocations with solvents can be calculated from the product ratio $[RN_3]/[ROS]$ produced by solvolysis of RX in solvents containing N_3^- salts (Scheme 1).

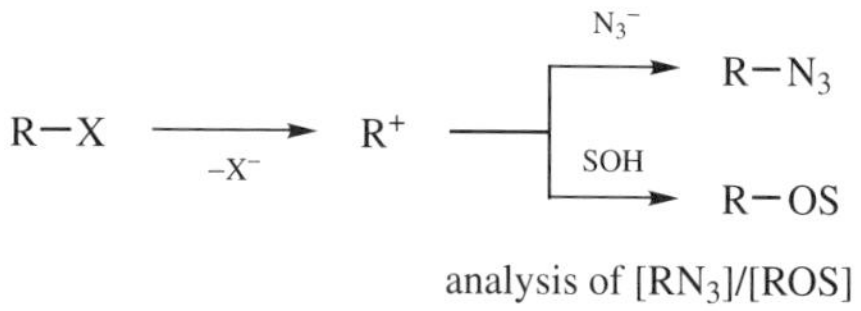

$$R-X \xrightarrow[-X^-]{} R^+ \begin{cases} \xrightarrow{N_3^-} R-N_3 \\ \xrightarrow{SOH} R-OS \end{cases}$$

analysis of $[RN_3]/[ROS]$

Scheme 1 Competition experiments employing the azide clock method.[37]

A corroboration of the rate constants thus derived came through kinetic investigations of laser flash photolytically generated carbocations.[38,39] Laser flash photolysis of benzyl halides and related derivatives results in the homolytic and heterolytic cleavage of the C–X bonds, yielding radicals and carbocations concomitantly.[39i] When these photolyses are performed in the presence of variable concentrations of nucleophiles, the rate constants for the combinations of carbocations with nucleophiles can be derived from the rates of decay of the carbocation absorbances [Eq. (13.10)]. In this way, rate constants from $10^7\ M^{-1}\ s^{-1}$ up to the diffusion limit can be determined efficiently.[39]

$$R-X \xrightarrow[-X^-]{h\nu} R^+ \xrightarrow[Nu]{k} R-Nu^+ \qquad (13.10)$$

It has been found that solvolysis reactions may proceed via rapid, almost quantitative ionization of the substrates, followed by slow reactions of the carbocations with the solvents (Eq. (13.11)].[40] In such cases, which can be expected to be quite common in solvolyses in fluorinated alcohols, relative electrophilicities of the intermediate carbocations can be derived from the rates of decay of the carbocations in the nucleophilic solvent:

$$Ar_2CHCl \xrightarrow[\text{fast}]{CF_3CH_2OH} Ar_2CH^+ \xrightarrow[\text{slow}]{CF_3CH_2OH} Ar_2CH{-}OCH_2CF_3 \quad (13.11)$$

Electrophilicities of a large variety of positively charged metal π complexes have been determined by Kane-Maguire and Sweigart,[41,42] who studied the kinetics of the reactions of metal π complexes with phosphanes [Eq. (13.12)], amines, and some aromatic π systems, predominantly by stopped-flow methods. In analogy to the investigations by Ritchie,[32] the relative electrophilicities of these metal π complexes have been reported to be almost independent of the nature of the nucleophilic reaction partners:[42]

$$(\pi\text{-ligand})^+{-}ML_n + PR_3 \xrightarrow{k} (^+PR_3\text{-ligand}){-}ML_n \quad (13.12)$$

In the 1980s, we succeeded in controlling Lewis acid–induced reactions of alkyl halides with alkenes such that 1 : 1 addition products were formed selectively.[43] By proper choice of counterions and reaction conditions, it became possible to determine the rates of the reactions of carbocations with alkenes and other π systems:[44–46]

$$R^+ \; MX^-_{n+1} + \text{alkene} \xrightarrow{k} R{-}C{-}C^+ \; MX^-_{n+1} \xrightarrow[-MX_n]{\text{fast}} R{-}C{-}C{-}X \quad (13.13)$$

More recently, we showed that the rates of the reactions of benzhydrylium ions with various nucleophiles can be used for a direct comparison of π-,[47–49] σ-,[47,50] and *n*-nucleophiles,[51] including carbanions[52,53] (Scheme 2).

In this report we will show how these data can be employed for deriving electrophilicity scales of carbocations.[54] We will further compare these data with the electrophilicities of electron-deficient neutral π systems,[53,55,56] which will allow us to view carbocations as a subgroup of the large family of carbon electrophiles.

13.3 DETERMINATION OF THE ELECTROPHILICITY PARAMETERS *E*

As discussed in the first paragraph, Lewis acidity scales, which are based on equilibrium constants, depend on the nature of the reference Lewis base. Likewise,

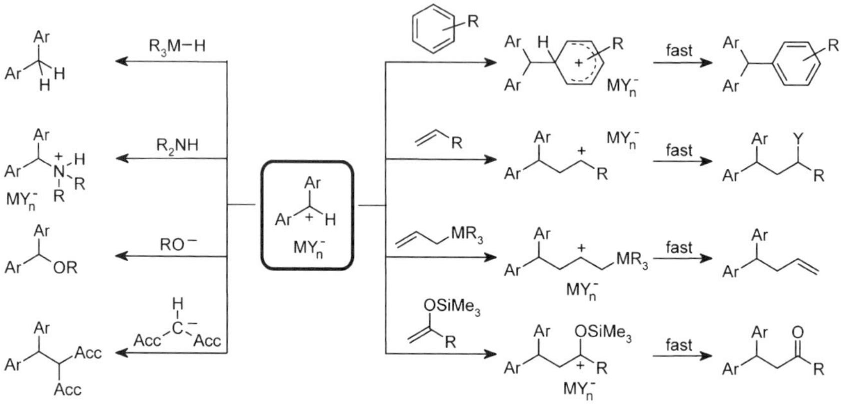

Scheme 2 Reactions of benzhydrylium ions Ar_2CH^+ with various types of neutral and negatively charged nucleophiles.

electrophilicity scales, which are based on rate constants, depend on the nature of the reference nucleophile. One, therefore, might select a certain nucleophile and specific reaction conditions when trying to develop an electrophilicity scale. For a simple reason, this approach cannot be used for constructing a comprehensive electrophilicity scale; on one side, second-order rate constants smaller than $10^{-5}\,M^{-1}\,s^{-1}$ (half-life 12 days for $c_0 = 0.1$ M) are hard to determine; on the other side, diffusion control prevents bimolecular reactions faster than $k = 10^9$–$10^{10}\,M^{-1}\,s^{-1}$. As shown for 1-(trimethylsiloxy)cyclohexene as a reference nucleophile, only electrophiles that differ by less than 15 orders of magnitude in reactivity can be compared with respect to a single nucleophile (Scheme 3). Although an extension of this range appears feasible by performing kinetic experiments at variable temperature, in practice this range is even smaller.

Statistical treatment of the rate constants of 209 reactions of 23 benzhydrylium ions with 39 carbon nucleophiles showed that the whole set of rate constants can be described with a standard deviation of factor 1.19 by employing a single parameter, *E*, for electrophiles and two parameters, the nucleophilicity parameters *Nu* or *N* and the slope parameter *s* for nucleophiles.[47] As shown in Fig. 13.1, the parameters $E((p\text{-MeOC}_6\text{H}_4)_2\text{CH}^+) = 0$ and s(2-methyl-1-pentene) $= 1.0$ were kept constant in the optimization procedure.

Equations (13.14) and (13.15) are two simple analytical expressions for these correlations. The conventional linear free-energy equation (13.14) defines nucleophilicity *Nu* as the intercept with the ordinate ($E = 0$); that is, *Nu* corresponds to $\log k$, the real or extrapolated rate constant for the reaction of the corresponding nucleophile with $(p\text{-MeOC}_6\text{H}_4)_2\text{CH}^+$. Equation (13.15) defines nucleophilicity *N* as the intercepts of the correlation lines on the abscissa ($\log k = 0$); thus, $N = Nu/s$ equals

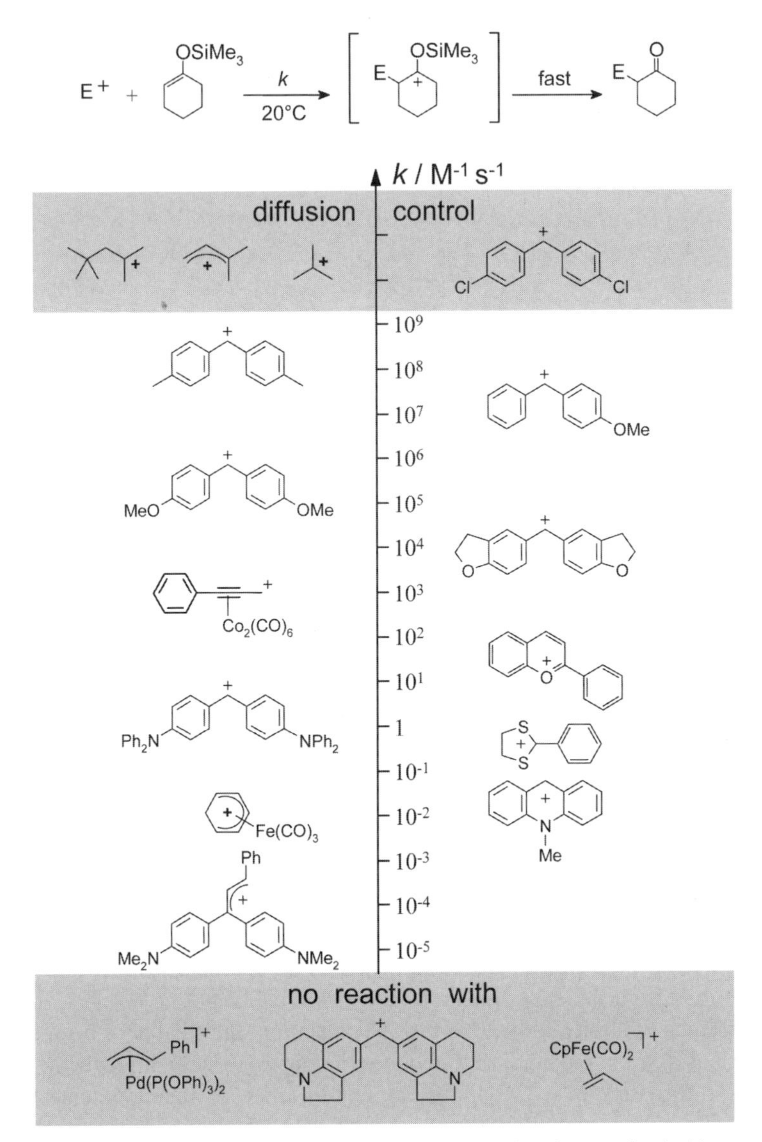

Scheme 3 Only a limited range of electrophilic reactivity can be characterized with respect to a certain nucleophile.

the negative value of the electrophilicity parameter E for the electrophile, which reacts with the corresponding nucleophile with a second-order rate constant of $1\ \mathrm{M^{-1}\ s^{-1}}$ at 20°C:

$$\log k_{20^\circ\mathrm{C}} = Nu + sE \tag{13.14}$$

$$\log k_{20^\circ\mathrm{C}} = s(N + E) \tag{13.15}$$

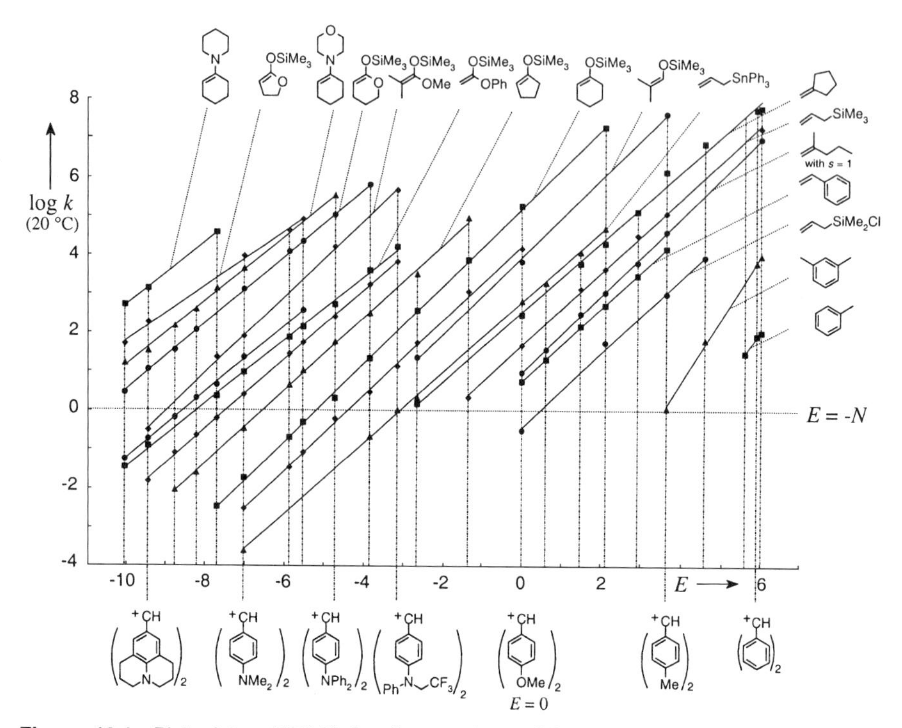

Figure 13.1 Plot of log k(20°C) for the reactions of benzhydryl cations with several π-nucleophiles versus the E parameters of the benzhydryl cations.[47] Seven of the 23 reference electrophiles are depicted below the abscissa; the whole list of reference electrophiles is shown in Scheme 4.

While both expressions are mathematically equivalent, definition of nucleophilicity by Eq. (13.15) is preferred, since N can generally be obtained without long-ranging extrapolations from the experimentally accessible data.[47,54c]

For the sake of clarity, only 20 of the 39 correlation lines that have been obtained with the basis set compounds are depicted in Figure 13.1. Although the s parameters for the various nucleophiles do not differ very much, intersections of the correlation lines would occur if all 209 data points had been presented in Figure 13.1.

All data points, that have been employed for the determination of E, N, and s of the basis set compounds can be shown in a plot of $(\log k)/s$ versus N. Figure 13.2 presents this plot for the previously mentioned basis set reactions (uncharged π-nucleophiles + benzhydrylium ions)[47] supplemented by carbanions (in DMSO)[52,53,56] as nucleophiles and six quinone methides as noncharged structural analogues of benzhydrylium ions. While the rate constants on the left side of Figure 13.2 (carbocations + neutral π systems), which have been shown to be almost independent of solvent polarity, refer mostly to dichloromethane solution, the reactions with carbanions refer to DMSO at concentrations where ion pairing

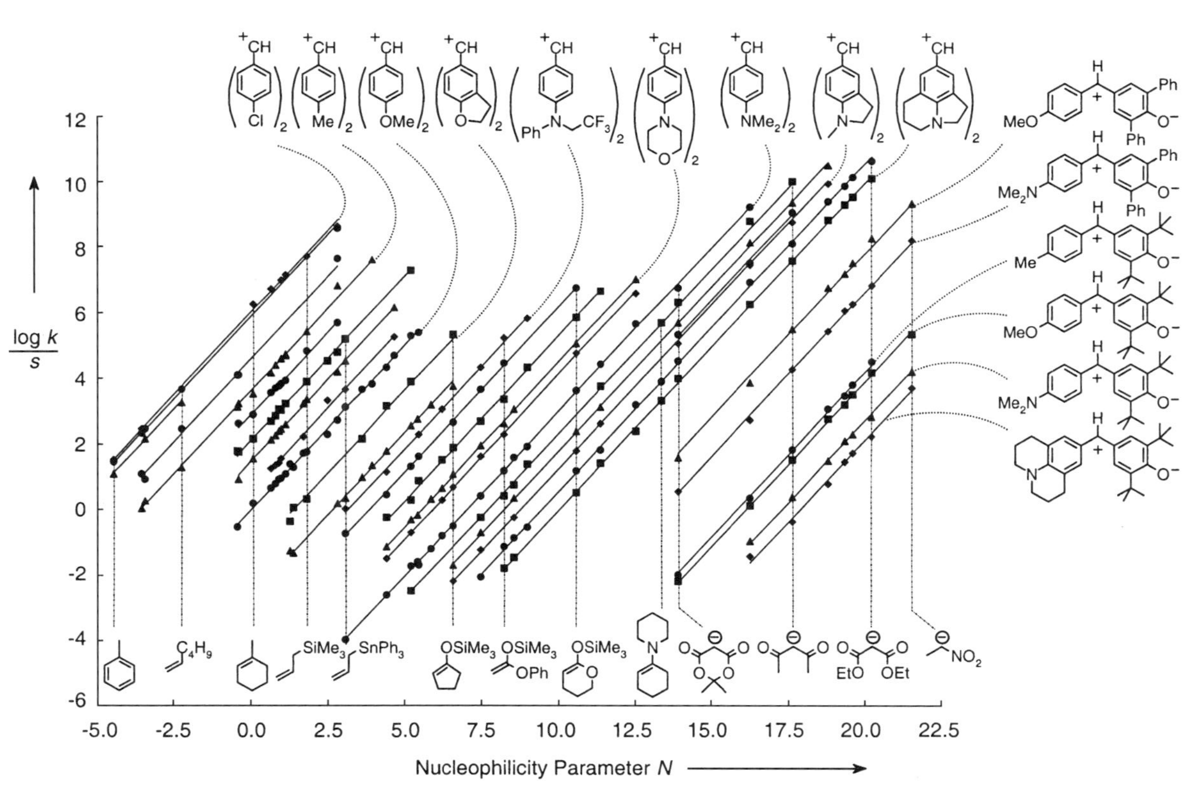

Figure 13.2 Plot of (log k)/s versus the nucleophilicity parameter N for the reactions of benzhydryl cations and quinone methides with π-nucleophiles and carbanions.

is negligible. Since the same E parameters for benzhydrylium ions were employed to describe their reactivities toward neutral nucleophiles in CH_2Cl_2 and toward carbanions in DMSO, E parameters have been defined as solvent-independent parameters. This approximation holds well for the highly stabilized carbocations investigated in DMSO, but certainly cannot be generalized because highly electrophilic carbenium ions will interact with DMSO. Since the rates of carbocation–carbanion combinations are known to depend on solvent polarity,[57] the definition of solvent-independent E parameters requires N and s parameters of anions to be defined with respect to a certain solvent.

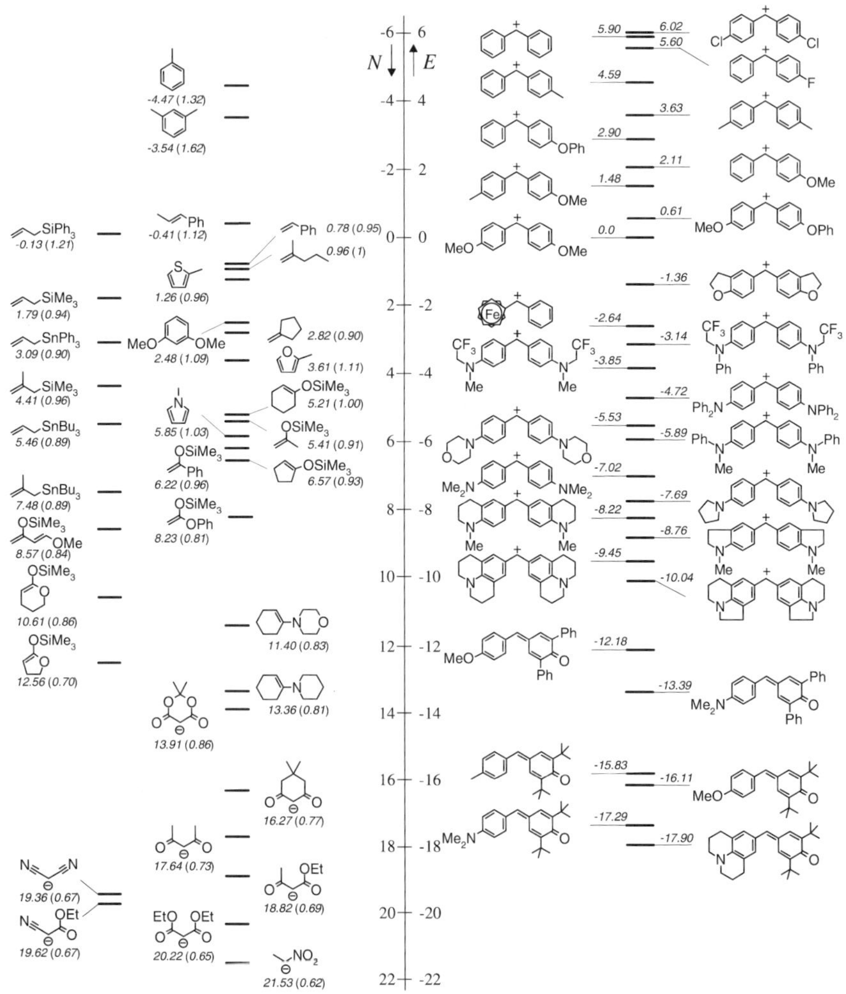

Scheme 4 Reactivity parameters N (s in parentheses) and E of the recommended reference nucleophiles and electrophiles.[47,53,56] The N and s parameters of the carbanions refer to DMSO as the solvent. See text for solvent effects in general.

Scheme 4 presents a summary of the reactivity parameters of reference electrophiles and reference nucleophiles presently used.[47,53,56] It should be noted that only 26 of the 39 neutral nucleophiles that have been used for the initial parameterization are recommended as reference nucleophiles for the characterization of further electrophiles.[47,48] While the reactivity parameters for the electrophiles with $E > -11$ and for the nucleophiles with $N < 13.5$ in Scheme 4 are considered as "final" because they have been derived from a broad data basis, the E parameters for the quinone methides and the N and s parameters for the carbanions[53,56] are not yet equally well supported and may undergo slight changes in the future. Nevertheless, the reference compounds in Scheme 4 now provide an efficient tool for determining reactivity parameters for new electrophiles and nucleophiles.

Although the electrophilicity scale depicted in Scheme 4 has been derived from reactions with carbon nucleophiles, Figure 13.3 demonstrates that it is equally suitable for describing reactions with hydrides[47,50,58–60] as well as with n-nucleophiles.[51] One can, therefore, conclude that the electrophilicity scale of Scheme 4 allows one to describe reactivities toward any types of nucleophiles.

In analogy to the frequently discussed relationships between nucleophilicity and basicity, electrophilicity is related to Lewis acidity. Accordingly, calculated hydride, methyl anion, and hydroxide affinities of benzhydrylium ions correlate linearly with the empirical electrophilicity parameters E. Figure 13.4 shows that this correlation holds in a range of 195 kJ/mol in reaction enthalpy corresponding to 16 logarithmic units in the kinetic parameter E.[61] The fact that the three lines in Figure 13.4 have exactly the same slopes implies that the *relative* Lewis acidities of benzhydrylium ions are independent of the nature of the Lewis base.

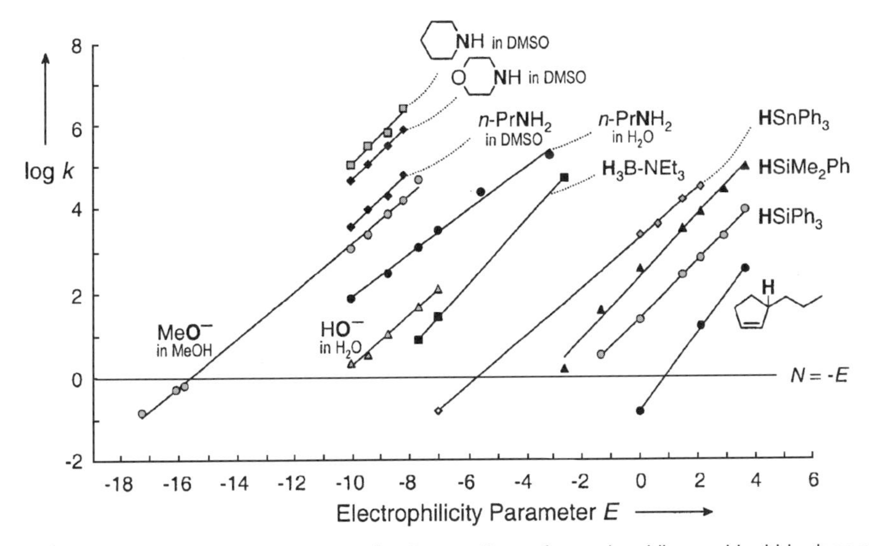

Figure 13.3 Plot of the rate constants for the reactions of n-nucleophiles and hydride donors with benzhydrylium ions and quinone methides versus their electrophilicity parameters E (bold face print indicates: center of nucleophilicity).

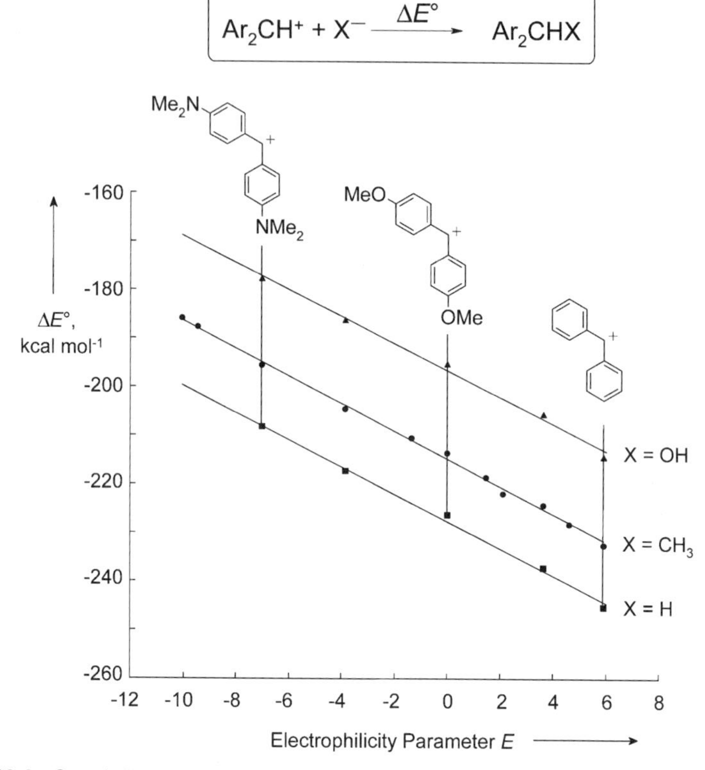

Figure 13.4 Correlation of the calculated H^-, CH_3^-, and OH^- affinities [B3LYP/6-31G(d,p)] of benzhydrylium ions with the empirical electrophilicity parameters E.[61]

In order to determine electrophilicity parameters E of other cationic or neutral electrophiles, one can employ the correlation lines shown in Figure 13.1 and search for a least-squares fit of the rate constants (log k) of the reactions of these electrophiles with some reference nucleophiles as shown for some examples in Figure 13.5.[56,62–64]

Since electrophiles are characterized by only one parameter, single-point calibrations are less problematic than in characterizations of further nucleophiles.[48] If reactions of structurally related electrophiles with the corresponding nucleophiles have been demonstrated to follow Eq. (13.15), a single rate constant for a reaction of an electrophile with a reference nucleophile is sufficient to determine its E parameter, as shown for the arylidenemalonates in Figure 13.5.[56,64]

It is obvious that the fit of the new data in Figure 13.5 (open symbols) is of lower quality than for the reactions with benzhydrylium ions (solid symbols), because now the surroundings of the reaction center in both electrophile *and* nucleophile are varied. However, the reported standard deviation between k_{exp} and k_{calc} of a factor of 2.05 for reactions of that type[47] is tolerable, particularly since ranges of more

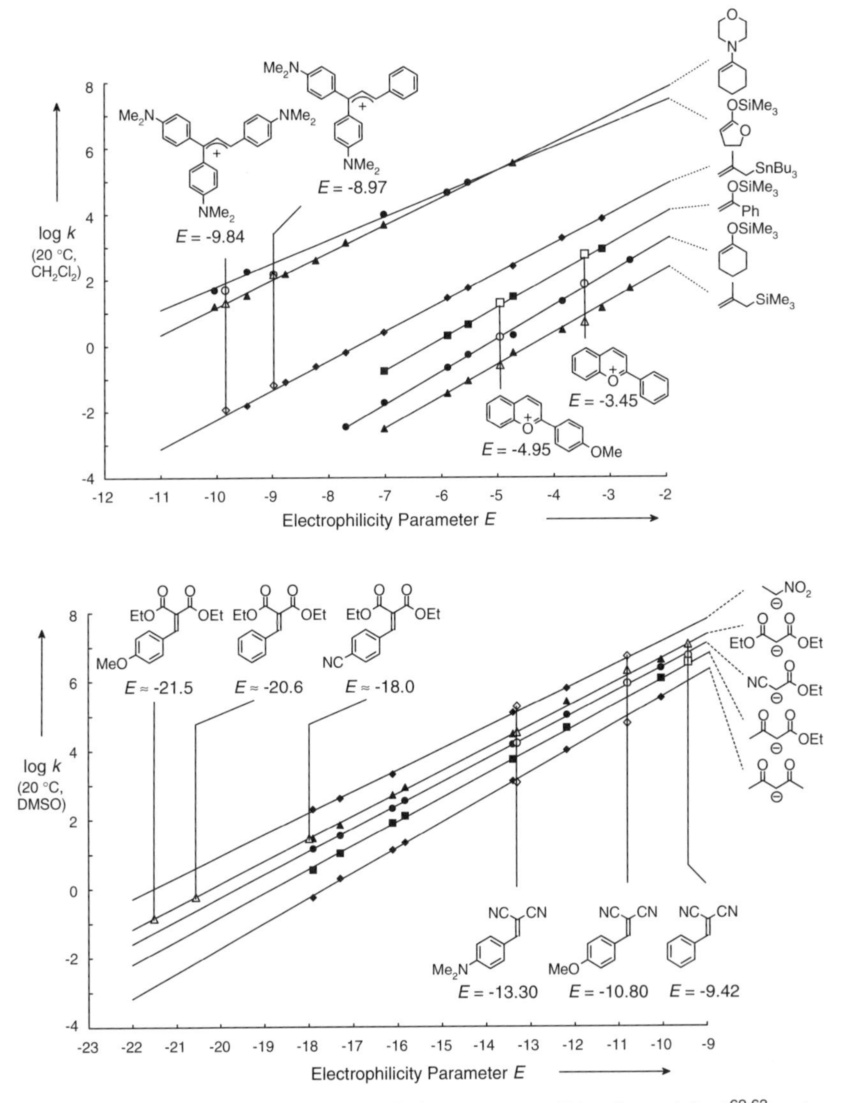

Figure 13.5 Determination of the electrophilicity parameters *E* for charged (top)[62,63] and non-charged electrophiles (bottom)[56,64] by fitting log *k* for the reactions of the electrophiles to be characterized (open symbols) to the correlation lines obtained with reference electrophiles (solid symbols).

than 25 orders of magnitude are covered by the reactivity scales for electrophiles and nucleophiles.

Preferably, the rate constants used in Figure 13.5 are obtained by direct rate measurements. If this were not possible because of the low intrinsic stability of the carbocation under consideration, one could determine approximate rate constants by

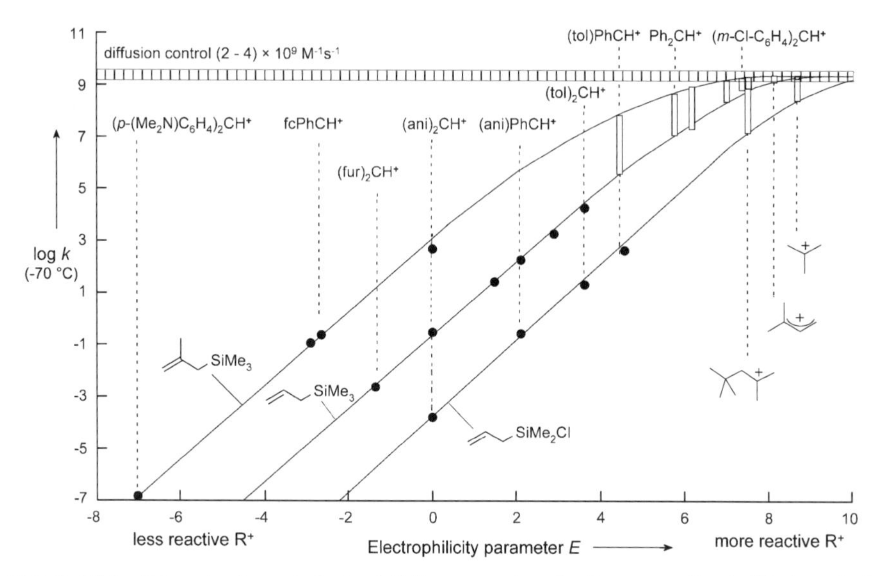

Figure 13.6 Determination of electrophilicity parameters by the generalized diffusion clock method[65] (data points refer to absolute rate constants and shaded bars to reactivity ratios determined by competition experiments; fc—ferrocenyl, fur—2,3-dihydrobenzofuran-5-yl, ani—4-methoxyphenyl, tol—4-methylphenyl).

competition experiments. As illustrated for the *tert*-butyl cation in Scheme 5, the carbocation under consideration may be generated in situ in the presence of two nucleophiles, and the ratio k_a/k_b can be calculated from the observed product ratio. If one of these rate constants is known to correspond to diffusion control ($\sim 3 \times 10^9\,M^{-1}\,s^{-1}$ for carbocations + neutral nucleophiles in acetonitrile at 20°C), the other can be derived from the product ratio.[65]

From Figure 13.6 one can deduce that (2-methylallyl)trimethylsilane reacts approximately 10^5 times faster with carbocations of low electrophilicity than allyl-chlorodimethylsilane.[47] The reactivity ratio $k_a/k_b = 9$, derived from the competition experiment of Scheme 5,[65] therefore, indicates that the more reactive of the two nucleophiles reacts with the *tert*-butyl cation with diffusion control ($k_a = 3 \times 10^9\,M^{-1}\,s^{-1}$), which allows one to calculate k_b as $3 \times 10^8\,M^{-1}\,s^{-1}$. This method is an extension of the azide clock method (see above) and has been employed for determining E parameters for a variety of in situ–generated carbocations.[65,66] Since the E parameters determined in this way are less reliable, these data are not included in the following compilation.

Scheme 6 lists the electrophilicity parameters derived from absolute rate constants,[48,51,55,56] and indicates whether they are derived from a series of experiments or from a single rate constant. Equation (13.15) does not specifically treat steric effects. For that reason it cannot be employed for predicting rate constants for reactions where steric factors become dominant, such as for reactions of tritylium ions with π-nucleophiles. Since tritylium ions are an important class of carbocations, some representatives are also included in Scheme 6. It has to be borne in mind,

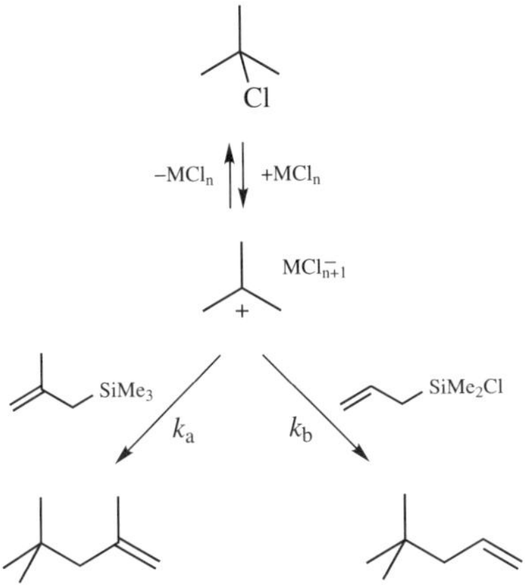

Cl

$-MCl_n$ $+MCl_n$

MCl_{n+1}^-

SiMe$_3$ k_a k_b SiMe$_2$Cl

Scheme 5

however, that the E parameters of tritylium ions may be used only for calculating the rates of their reactions with hydride donors and n-nucleophiles, which have been employed for deriving $E(Ar_3C^+)$.[51] Reactions of tritylium ions with π-nucleophiles are several orders of magnitude slower[67] than predicted by Eq. (13.15) using the E, N, and s parameters listed in Schemes 4 and 6.

13.4 APPLICATIONS OF THE ELECTROPHILICITY PARAMETERS *E*

Unlike the other "stability parameters" for carbocations mentioned in the first paragraph, the electrophilicity parameters E as defined by Eq. (13.15), can be directly employed for predicting the rates of polar organic reactions.[53,68] In Figure 13.7, electrophiles are ordered with increasing E parameters from top to bottom, and nucleophiles are ordered with increasing N parameters from left to right. The diagonal from bottom left to top right corresponds to electrophile nucleophile combinations with $E + N = 0$, corresponding to $k = 1\ M^{-1}\ s^{-1}$ at 20°C according to Eq. (13.15). Since the half-life of bimolecular reactions with equal initial concentrations of the reactants (c_0) is given by $\tau_{1/2} = 1/(kc_0)$, electrophile nucleophile combinations on the diagonal will be complete within a couple of minutes; for $c_0 = 0.1$ M, the first half-life will be 10 s.

Three domains are differentiated by color, and the diffuse borders separating these domains are partially a result of the variable magnitude of s. The blue sector (cold) identifies combinations of electrophiles with nucleophiles that are too slow to occur at room temperature ($k < 10^{-6}\ M^{-1}\ s^{-1}$). The red (hot) region denotes diffusion-controlled reactions that proceed with rate constants of 10^9–$10^{10}\ M^{-1}\ s^{-1}$. In between, those reactions that have rate constants in the range from 10^{-6} to $10^{10}\ M^{-1}\ s^{-1}$ are found in the green and yellow zones. When moving from left

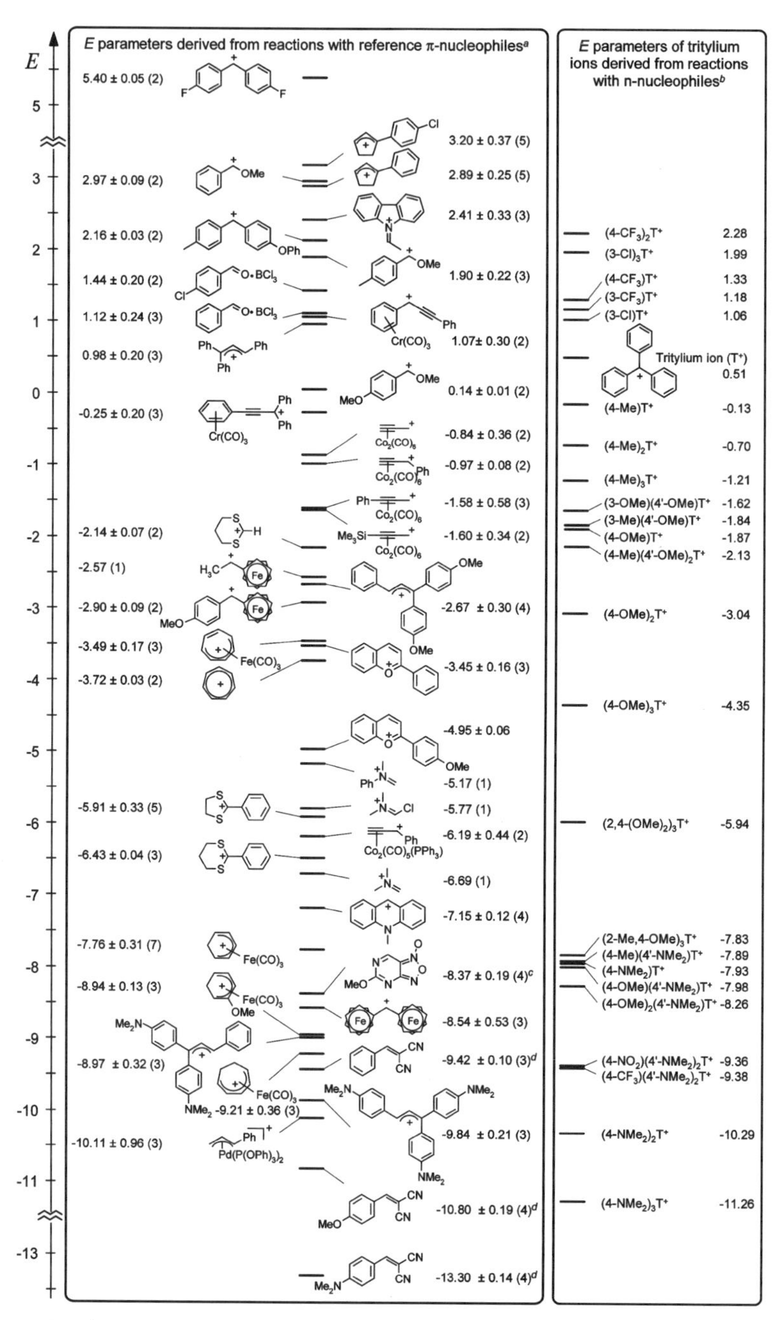

Scheme 6 Electrophilicity parameters E for carbocationic and neutral electrophiles with standard deviations and numbers of experiments in parentheses. Superscript letters denote the following: (a) from Ref. 48 if not stated otherwise; (b) the E parameters for tritylium ions given in this scheme (from Ref. 51) must not be used for predicting reactivities toward π-nucleophiles (see text); (c) from Ref. 55; (d) from Ref. 56.

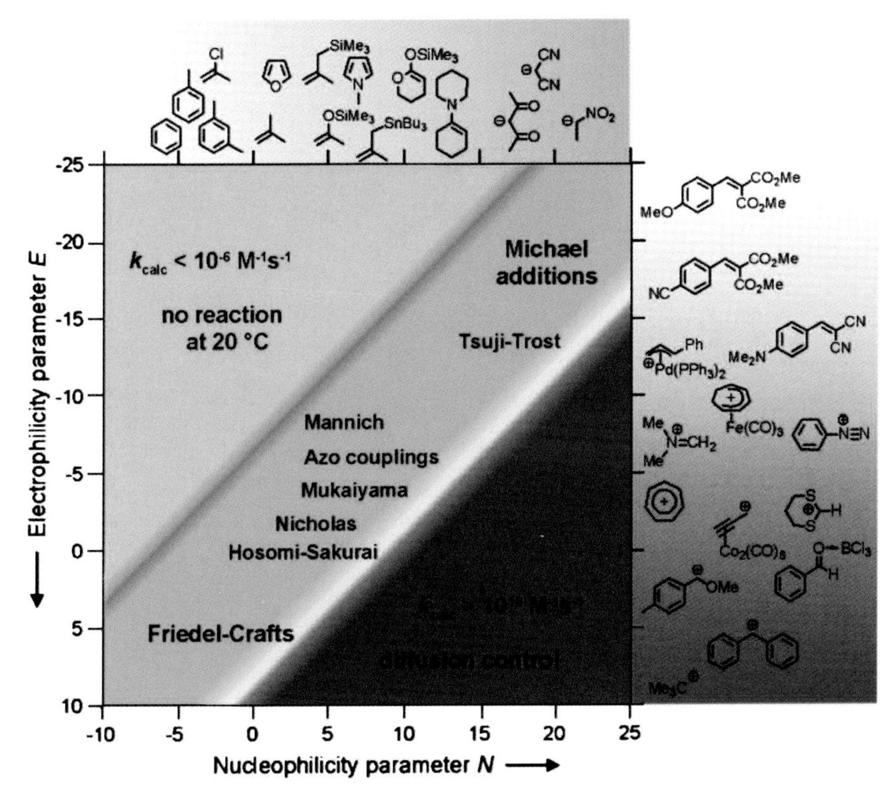

Figure 13.7 Most known organic reactions are in the green corridor: electrophile-nucleophile combinations above (blue) do not take place, combinations below (red) are diffusion controlled and often unselective.

to right or from top to bottom in the green zone, the increase in reactivity is predominantly a result of a decrease in $\Delta H^{\ddagger}$,[69b] which becomes zero in the yellow strip. The yellow domain encompasses reactions with $\Delta H^{\ddagger} = 0$ that are not diffusion-controlled (entropy-controlled reactions).[69]

As indicated by some name reactions in Figure 13.7, most synthetically employed reactions are electrophile–nucleophile combinations in the green and yellow zone as well as in the borderline region of the red zone. Electrophile–nucleophile combinations in the bottom right corner of the red zone are seldom used synthetically. Apart from the fact that it is difficult to find common solvents for the carbocations and carbanions in this range, control of these reactions would be rather difficult because of their extremely high driving force. Furthermore, the expected products can more easily be obtained with synthetic equivalents, for example, with alkyl halides instead of carbocations and with enamines or silyl enol ethers instead of carbanions.

Previously, we have suggested the rule of thumb that electrophile nucleophile combinations can be expected to take place if $E + N > -5$.[54] This rule defines

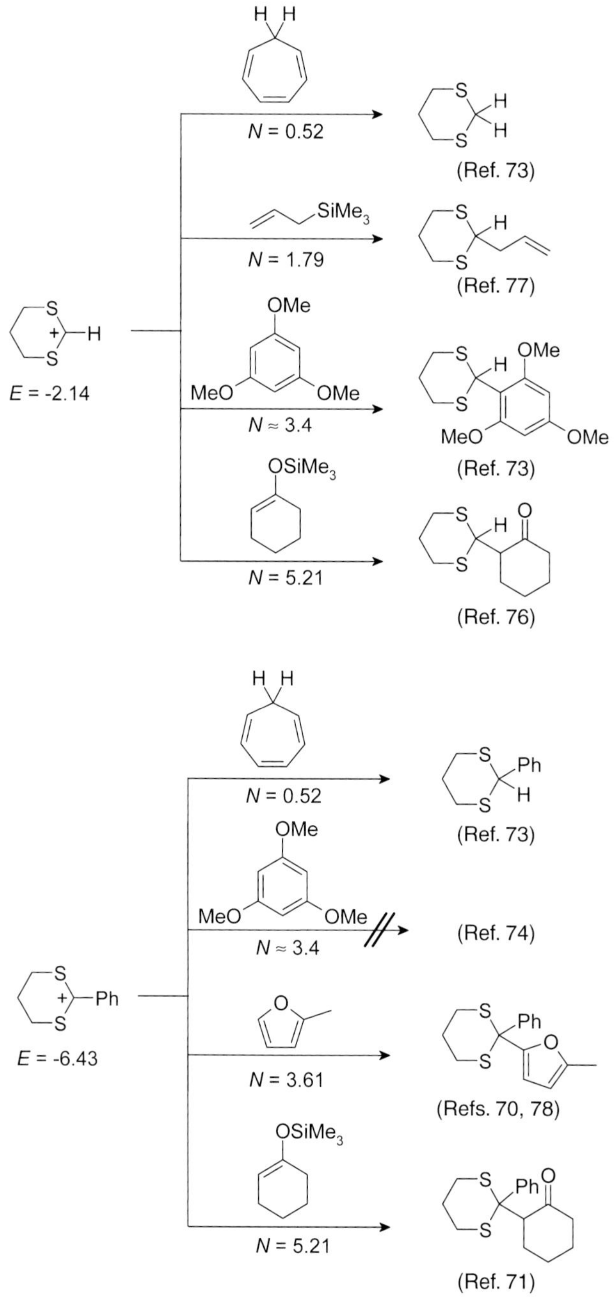

H H
N = 0.52
S
H
H
S
(Ref. 73)
$SiMe_3$
N = 1.79
S
H
S
(Ref. 77)
S
+ —H
S
E = -2.14
OMe
MeO
OMe
$N \approx$ 3.4
S
H
OMe
S
MeO
OMe
(Ref. 73)
$OSiMe_3$
N = 5.21
S
H
O
S
(Ref. 76)

H H
N = 0.52
S
Ph
H
S
(Ref. 73)
OMe
MeO
OMe
$N \approx$ 3.4
(Ref. 74)
S
+ —Ph
S
E = -6.43
O
N = 3.61
S
Ph
O
S
(Refs. 70, 78)
$OSiMe_3$
N = 5.21
S
Ph
O
S
(Ref. 71)

Scheme 7 Reactions of dithiocarbenium ions with nucleophiles. Reactivity parameters E and N were taken from Refs. 48 and 50.

the borderline between the blue and the green zones of Figure 13.7. In this way, we have been able to rationalize the reported reactivities of dithiosubstituted carbocations. In order to define the synthetic potential of dithiosubstituted carbenium ions, we studied the kinetics of their reactions with suitable nucleophiles, namely, allylsilanes, allylstannanes, and silylated enol ethers, and derived the corresponding E parameters for these cations.[47,48,70] Published reactions of these dithiosubstituted carbenium ions with π-nucleophiles[70–77] and the hydride donor 1,3,5-cycloheptatriene[73] are collected in Scheme 7, and readers may convince themselves that the nucleophiles reported to react with these cations are in the green range, fulfilling the condition $E + N > -5$. However, Stahl reported the failure of the 2-phenyl-1,3-dithian-2-ylium ion to react with 1,3,5-trimethoxybenzene,[74] although Eq. (13.15) predicts a slow reaction at room temperature. We repeated Stahl's experiment, and indeed did not find any reaction even on warming.[70,78] We therefore agree with Stahl's suggestion that in this case steric effects cannot be neglected. In accord with this interpretation, 2-methylfuran, which has a nucleophilicity parameter N similar to that of 1,3,5-trimethoxybenzene, readily reacted with the 2-phenyl-1,3-dithian-2-ylium ion (Scheme 7).[70,78]

Analogously, one can explain why the bis-ferrocenylmethylium ion ($E = -8.54$)[48] reacts with silyl enol ethers and ketene acetals, but not with ordinary alkenes and allylsilanes.[79]

The E parameters of hexacarbonyldicobalt-coordinated propargyl cations (Nicholas cations)[47,48,80] have been used to summarize their synthetic potential.[81–84] While reactions of these electrophiles with alkylated benzenes have not been reported [N(m-xylene) = -3.54[47]], many reactions with nucleophiles of $N > -2$ are known (Scheme 8).[85–89]

Nicholas cations have also been applied for the synthesis of [7]metacyclophane–diyne complexes from hexacarbonyldicobalt-coordinated alkynyl ethers and electron-rich arenes in the presence of a Lewis acid (Scheme 9). After decomplexation of the initial product, the smallest [n]metacyclophane containing a triple bond reported so far could be obtained.[90] Whereas furan ($N = 1.36$),[48] 1,3-dimethoxybenzene ($N = 2.48$),[47] and pyrrole ($N = 4.63$)[49] underwent analogous reactions as described in Scheme 9, thiophene ($N = -1.01$)[48] did not react,[90] possibly because of incomplete ionization of the cation precursor by the boron trifluoride-diethylether complex.

The electrophilicity parameters E of tricarbonyliron-coordinated cycloalkadienyl cations[48,91] and tropylium ions[48,92] have analogously been used to rationalize their use in organic synthesis.[54b]

Figure 13.7 shows that the combination of the N,N-dimethyliminium ion ($E = -6.7$)[48,93] with benzene ($N \approx -6.3$)[48] is located in the blue sector, in agreement with the fact that Mannich reactions take place only with electron-rich arenes and not with benzene.[94] The N,N-dimethyliminium ion of cinnamaldehyde can be assumed to be less electrophilic than the N,N-dimethyliminium ion. In accord with the conclusions deduced from Figure 13.7, Shudo and coworkers reported that the N,N-dimethylcinnamaldiminium ion does not react with benzene in trifluoroacetic acid, whereas in superacidic media this reaction was observable due to an

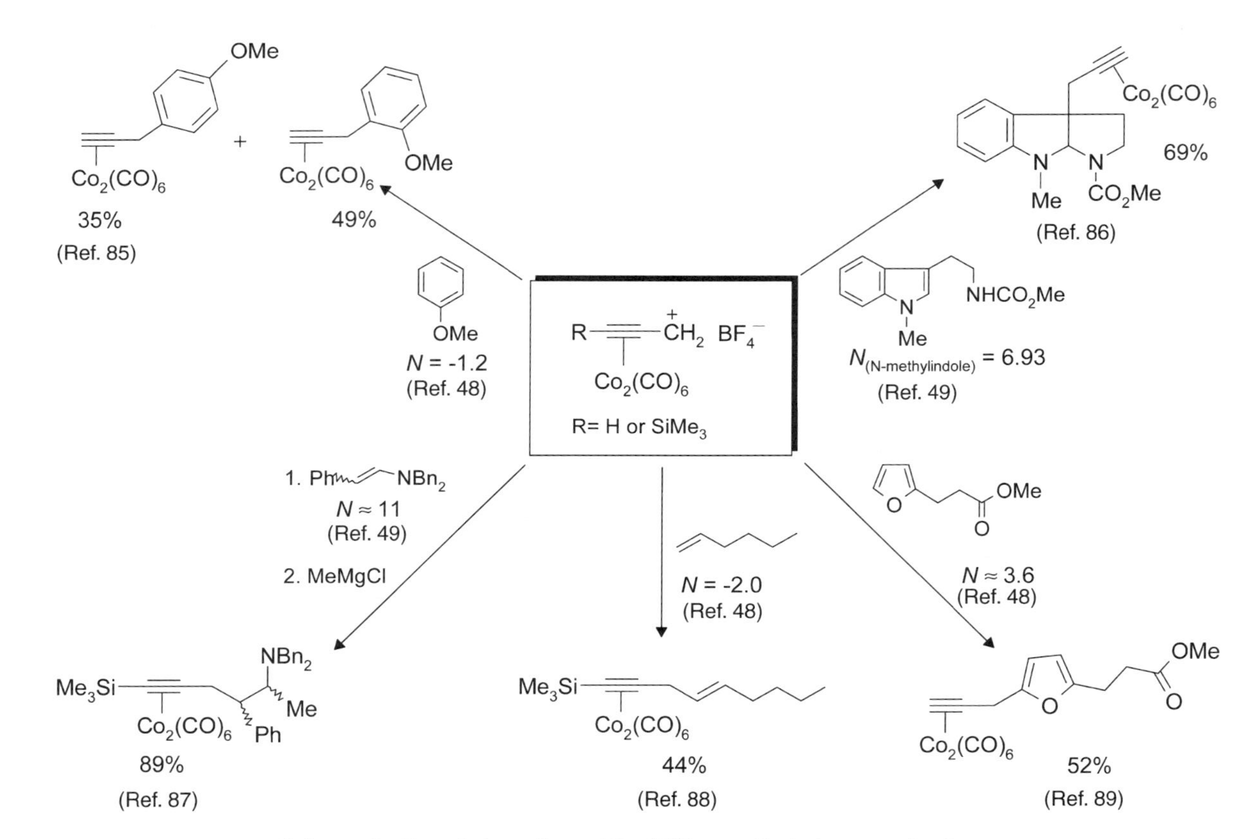

Scheme 8 Reported reactions of $Co_2(CO)_6$ coordinated propargyl cations.

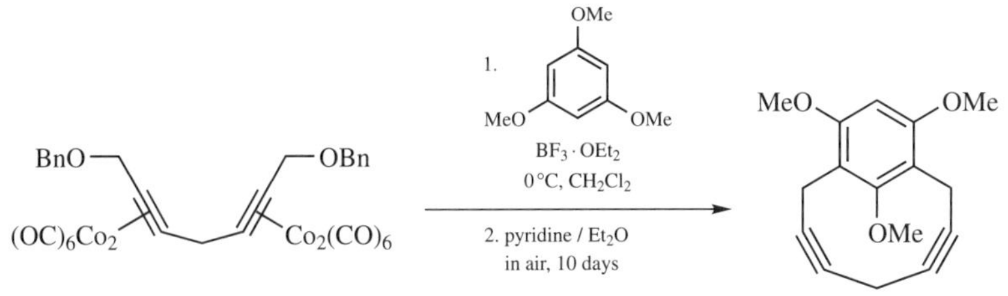

Scheme 9 Formation of [7]metacyclophane.[90]

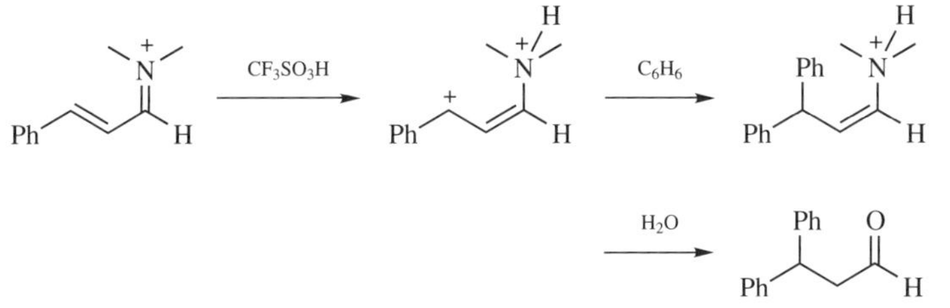

Scheme 10

additional protonation at the nitrogen atom and thus the formation of a dicationic species with enhanced electrophilicity (Scheme 10).[95–98]

A three-parameter equation that does not treat steric effects specifically can certainly not be expected to provide accurate rate constants. However, it has been shown for reactions of carbocations with π-nucleophiles that the deviation between calculated and experimental rate constants seldom exceeds a factor of 30 when bulky reagents are excluded.[47,99] A remarkable agreement in view of the many orders of magnitude covered by this model! Since this precision is sufficient for planning organic syntheses, a common set of reactivity parameters E, N, and s can be used to describe reactions as different as the name reactions mentioned in Figure 13.7.

Although it is unlikely that there will be no reaction between electrophiles and nucleophiles if Eq. (13.15) predicts a combination rate constant of $k > 10^{-3}\ M^{-1}\ s^{-1}$, one cannot be sure that the desired product can be isolated. Consecutive reactions[43a] or faster alternative reaction pathways may occur.

On the other hand, sometimes reactions also occur if the electrophile nucleophile combination is in the blue range of Figure 13.7. In such cases, namely, when the observed reaction is considerably faster than predicted by Eq. (13.15), a change of mechanism is indicated. Thus, we have proved that the unexpectedly fast reactions of allylpalladium complexes with allyl stannanes are due to an attack of the allyl stannanes at the palladium and not at the allyl ligand of the palladium complex.[100]

It is obvious that Eq. (13.15) cannot be used to predict the rate constants of concerted cycloadditions between electrophilic and nucleophilic reaction partners. We have discussed, however, that in such cases the deviation between calculated and observed rate constants can be used to calculate the energy of concert.[101–104]

While most applications of Eq. (13.15) have so far been directed to elucidate the feasibility of electrophile nucleophile combinations, that is, to answer the question as to whether a certain reaction refers to the blue or green zone, the differentiation between the green/yellow and the red domains is of equal importance, because this is the borderline where selectivity can be expected to break down. Furthermore, Figure 13.7 gives direct information about the feasibility of the formation of stable encounter complexes. The allocation of the benzene/*tert*-butyl cation combination in the green area implies a barrier for the reaction of benzene with the *tert*-butyl cation. Accordingly, MP2/6-31+G** calculations have characterized the π complexes between the *tert*-butyl cation and benzene as a local energy minimum.[105] The same calculations were unable to locate an energy minimum for a π-complex between the isopropyl cation and benzene.[105] Since the isopropyl cation can be estimated to be more electrophilic than the *tert*-butyl cation by three to four units of E,[106] one expects the combination of $(CH_3)_2CH^+$ with benzene in the yellow or red zone, which corresponds to the absence of a barrier for the formation of the Wheland intermediate.[107]

13.5 CONCLUSION

Electrophilicity E, as defined by Eq. (13.15), is a useful parameter for ordering carbocations according to their reactivities toward nucleophiles. Since it is possible to study the reactivities of carbocations and noncharged electrophiles toward the same nucleophiles, it becomes possible to directly link the chemistry of carbocations to that of electron-deficient noncharged aromatic and nonaromatic π systems. The main advantage of the E scale is the possibility to combine it with the nucleophilicity scale N for predicting reactivities and selectivities of a large variety of polar organic reactions.

ACKNOWLEDGMENT

Financial support by the Deutsche Forschungsgemeinschaft and the Fonds der Chemischen Industrie is gratefully acknowledged.

REFERENCES AND NOTES

1. (a) A. Baeyer and V. Villiger, *Ber. Dtsch. Chem. Ges.* **35**, 1189–1201 (1902); (b) A. Baeyer and V. Villiger, *Ber. Dtsch. Chem. Ges.* **35**, 3013–3033 (1902).
2. J. F. Norris and W. W. Sanders, *Am. Chem. J.* **25**, 54–62 (1901).
3. J. F. Norris, *Am. Chem. J.* **25**, 117–122 (1901).

4. F. Kehrmann and F. Wentzel, *Ber. Dtsch. Chem. Ges.* **34**, 3815–3819 (1901).

5. For a review on the history of carbocations, see G. A. Olah, *J. Org. Chem.* **66**, 5943–5957 (2001).

6. (a) C. K. Ingold and E. Rothstein, *J. Chem. Soc.* 1217–1221 (1928); (b) E. D. Hughes, C. K. Ingold, and C. S. Patel, *J. Chem. Soc.* 526–530 (1933); (c) C. K. Ingold, *J. Chem. Soc.* 1120–1127 (1933); (d) E. D. Hughes and C. K. Ingold, *J. Chem. Soc.* 1571–1576 (1933); (e) J. L. Gleave, E. D. Hughes, and C. K. Ingold, *J. Chem. Soc.* 236–244 (1935); (f) E. D. Hughes and C. K. Ingold, *J. Chem. Soc.* 244–255 (1935); (g) E. D. Hughes, *J. Chem. Soc.* 255–258 (1935); (h) L. C. Bateman, K. A. Cooper, E. D. Hughes, and C. K. Ingold, *J. Chem. Soc.* 925–935 (1940); (i) C. K. Ingold, *Structure and Mechanism in Organic Chemistry*, 2nd ed., Cornell Univ. Ithaca, NY, 1969.

7. E. M. Arnett, C. Petro, and P. v. R. Schleyer, *J. Am. Chem. Soc.* **101**, 522–526 (1979).

8. M. Saunders, E. L. Hagen, and J. Rosenfeld, *J. Am. Chem. Soc.* **90**, 6882–6884 (1968).

9. G. A. Olah and J. A. Olah, in *Carbonium Ions*, G. A. Olah and P. v. R. Schleyer, eds., Wiley-Interscience, New York, 1968–, Vol. II, Chapter 17, pp. 715–782.

10. E. W. Bittner, E. M. Arnett, and M. Saunders, *J. Am. Chem. Soc.* **98**, 3734–3735 (1976).

11. E. M. Arnett and C. Petro, *J. Am. Chem. Soc.* **100**, 5408–5416 (1978).

12. Calculated from the hydride ion affinities for 2-$C_3H_7^+$ (251.8 kcal/mol) and *t*-$C_4H_9^+$ (236.8 kcal/mol) given in Ref. 13.

13. D. H. Aue, in *Dicoordinated Carbocations*, Z. Rappoport and P. J. Stang, eds., Wiley, New York, 1997, Chapter 3, pp 105–156.

14. Calculated from the heats of formation $\Delta_f H^\circ{}_{298}$ for **3** (192.1 kcal/mol, from Ref. 13), *t*-C_4H_9Cl (−43.0 kcal/mol, from Ref. 15), **2** (169.9 kcal/mol, from Ref. 16), and 2-C_3H_7Cl (−34.0 kcal/mol, from Ref. 15).

15. *NIST Chemistry WebBook, NIST Standard Reference Database Number 69*, P. J. Linstrom and W. G. Mallard, eds., March 2003, National Institute of Standards and Technology, Gaithersburg MD, 20899 (http://webbook.nist.gov).

16. J. W. Keister, J. S. Riley, and T. Baer, *J. Am. Chem. Soc.* **115**, 12613–12614 (1993).

17. Calculated from the heats of formation $\Delta_f H^\circ{}_{298}$ for **3** (192.1 kcal/mol, from Ref. 13), *t*-C_4H_9OH (−74.7 kcal/mol, from Ref. 15), **2** (169.9 kcal/mol, from Ref. 16), and 2-C_3H_7OH (−65.2 kcal/mol, from Ref. 15).

18. Calculated from the hydride ion affinities for **4** (216.8 kcal/mol, from Ref. 19) and **2** (236.8 kcal/mol, from Ref. 13).

19. E.-U. Würthwein, G. Lang, L. H. Schappele, and H. Mayr, *J. Am. Chem. Soc.* **124**, 4084–4092 (2002).

20. Calculated from the proton affinities for benzene (178.4 kcal/mol, from Ref. 21) and isobutene (191.7 kcal/mol, from Ref. 16).

21. D. H. Aue, M. Guidoni, and L. D. Betowski, *Int. J. Mass Spectrom.* **201**, 283–295 (2000).

22. J. Hine, *Structural Effects on Equilibria in Organic Chemistry*; Robert E. Krieger, Huntingon, NY, 1981, Chapter 7.

23. (a) D. H. Aue and M. T. Bowers, in *Gas Phase Ion Chemistry*, M. T. Bowers, ed., Academic Press, New York, 1979, Vol. 2, Chapter 9; (b) J.-L. Abboud, I. Alkorta, J. Z. Dávalos, P. Müller, and E. Quintanilla, *Adv. Phys. Org. Chem.* **37**, 57–135 (2002).

24. H. H. Freedman, S. V. McKinley, J. W. Rakshys, Jr., and A. E. Young, *J. Am. Chem. Soc.* **93**, 4715–4724 (1971).

25. E. M. Arnett and C. Petro, *J. Am. Chem. Soc.* **100**, 2563–2564 (1978).
26. E. M. Arnett and T. C. Hofelich, *J. Am. Chem. Soc.* **105**, 2889–2895 (1983).
27. C. Schade, H. Mayr, and E. M. Arnett, *J. Am. Chem. Soc.* **110**, 567–571 (1988).
28. N. C. Deno, J. J. Jaruzelski, and A. Schriesheim, *J. Am. Chem. Soc.* **77**, 3044–3051 (1955).
29. N. C. Deno and A. Schriesheim, *J. Am. Chem. Soc.* **77**, 3051–3054 (1955).
30. H. H. Freedman, in *Carbonium Ions*, G. A. Olah and P. v. R. Schleyer, eds., Wiley, New York, 1973, Vol. 4, pp 1501–1578.
31. K. Okamoto, K. Takeuchi, K. Komatsu, Y. Kubota, R. Ohara, M. Arima, K. Takahashi, Y. Waki, and S. Shirai, *Tetrahedron* **39**, 4011–4024 (1983).
32. C. D. Ritchie, *Can. J. Chem.* **64**, 2239–2250 (1986).
33. A. Streitwieser, Jr., *Solvolytic Displacement Reactions*, McGraw-Hill, New York, 1962.
34. H. Mayr, in *Cationic Polymerization: Mechanisms, Synthesis and Applications*, K. Matyjaszewski, ed., Marcel Dekker, New York, 1996, Chapter 2, pp. 51–136.
35. M. J. S. Dewar and R. C. Dougherty, *The PMO Theory of Organic Chemistry*, Plenum Press, New York, 1975.
36. (a) C. D. Ritchie, *Acc. Chem. Res.* **5**, 348–354 (1972); (b) C. D. Ritchie, *J. Am. Chem. Soc.* **97**, 1170–1179 (1975); (c) C. D. Ritchie and M. Sawada, *J. Am. Chem. Soc.* **99**, 3754–3761 (1977); (d) C. D. Ritchie, J. E. Van Verth, and P. O. I. Virtanen *J. Am. Chem. Soc.* **104**, 3491–3497 (1982); (e) C. D. Ritchie, *J. Am. Chem. Soc.* **106**, 7187–7194 (1984).
37. (a) J. P. Richard and W. P. Jencks, *J. Am. Chem. Soc.* **104**, 4689–4691 (1982); (b) J. P. Richard and W. P. Jencks, *J. Am. Chem. Soc.* **104**, 4691–4692 (1982); (c) J. P. Richard, M. E. Rothenberg, and W. P. Jencks, *J. Am. Chem. Soc.* **106**, 1361–1372 (1984).
38. R. A. McClelland, V. M. Kanagasabapathy, N. S. Banait, and S. Steenken, *J. Am. Chem. Soc.* **113**, 1009–1014 (1991).
39. For recent (as of 2004) laser flash photolysis studies of highly reactive carbocations, see (a) T. V. Pham, and R. A. McClelland, *Can. J. Chem.* **79**, 1887–1897 (2001); (b) F. L. Cozens, V. M. Kanagasabapathy, R. A. McClelland, and S. Steenken, *Can. J. Chem.* **77**, 2069–2082 (1999); (c) J. P. Pezacki, D. Shukla, J. Lusztyk, and J. Warkentin, *J. Am. Chem. Soc.* **121**, 6589–6598 (1999); (d) S. Steenken, M. Ashokkumar, P. Maruthamuthu, and R. A. McClelland, *J. Am. Chem. Soc.* **120**, 11925–11931 (1998); (e) S. Kobayashi, Y. Hori, T. Hasako, K.-i. Koga, and H. Yamataka, *J. Org. Chem.* **61**, 5274–5279 (1996); (f) E. MacKnight and R. A. McClelland, *Can. J. Chem.* **74**, 2518–2527 (1996); (g) F. L. Cozens, N. Mathivanan, R. A. McClelland, and S. Steenken, *J. Chem. Soc. Perkin Trans.* 2, 2083–2090 (1992); (h) R. A. McClelland, C. Chan, F. Cozens, A. Modro, and S. Steenken, *Angew. Chem.* **103**, 1389–1391 (1991); *Angew. Chem. Int. Ed. Engl.* **30**, 1337–1339 (1991); (i) J. Bartl, S. Steenken, H. Mayr, and R. A. McClelland, *J. Am. Chem. Soc.* **112**, 6918–6928 (1990); (j) J. Bartl, S. Steenken, and H. Mayr, *J. Am. Chem. Soc.* **113**, 7710–7716 (1991); (k) R. A. McClelland, N. Mathivanan, and S. Steenken *J. Am. Chem. Soc.* **112**, 4857–4861 (1990). For reviews, see (l) R. A. McClelland, *Tetrahedron* **52**, 6823–6858 (1996); (m) R. A. McClelland, in *Organic Reactivity: Physical and Biological Aspects*, B. T. Golding, R. J. Griffin, and H. Maskill, eds., The Royal Society of Chemistry, Cambridge, UK, 1995, pp. 301–319; (n) P. K. Das, *Chem. Rev.* **93**, 119–144 (1993).

40. (a) H. Mayr and S. Minegishi, *Angew. Chem.* **114**, 4674–4676 (2002); *Angew. Chem. Int. Ed.* **41**, 4493–4495 (2002); (b) B. Denegri, S. Minegishi, O. Kronja, and H. Mayr, *Angew. Chem.* **116**, 2353–2356 (2004); *Angew. Chem. Int. Ed.* **43**, 2302–2305 (2004).

41. (a) L. A. P. Kane-Maguire and C. A. Mansfield, *J. Chem. Soc. Chem. Commun.* 540–541 (1973); (b) L. A. P. Kane-Maguire and C. A. Mansfield, *J. Chem. Soc. Dalton Trans.* 2192–2196 (1976); (c) G. R. John, C. A. Mansfield, and L. A. P. Kane-Maguire, *J. Chem. Soc. Dalton Trans.* 574–578 (1977); (d) G. R. John and L. A. P. Kane-Maguire, *J. Chem. Soc. Dalton Trans.* 1196–1199 (1979); (e) T. I. Odiaka and L. A. P. Kane-Maguire, *Inorg. Chim. Acta* **37**, 85–87 (1979); (f) G. R. John and L. A. P. Kane-Maguire, *Inorg. Chim. Acta* **48**, 179–183 (1981); (g) G. R. John, L. A. P. Kane-Maguire, T. I. Odiaka, and C. Eaborn, *J. Chem. Soc. Dalton Trans.* 1721–1727 (1983).

42. For reviews, see (a) L. A. P. Kane-Maguire, E. D. Honig, and D. A. Sweigart, *Chem. Rev.* **84**, 525–543 (1984); (b) R. D. Pike and D. A. Sweigart, *Coord. Chem. Rev.* **187**, 183–222 (1999).

43. (a) H. Mayr, in *Selectivities in Lewis Acid Promoted Reactions*, D. Schinzer, ed., NATO ASI Series C, Vol. 289, 1989, pp. 21–36; (b) H. Mayr, C. Schade, M. Rubow, and R. Schneider, *Angew. Chem.* **99**, 1059–1060 (1987); *Angew. Chem. Int. Ed. Engl.* **26**, 1029–1030 (1987); (c) H. Mayr, *Angew. Chem.* **102**, 1415–1428 (1990); *Angew. Chem. Int. Ed. Engl.* **29**, 1371–1384 (1990).

44. H. Mayr, R. Schneider, C. Schade, J. Bartl, and R. Bederke, *J. Am. Chem. Soc.* **112**, 4446–4454 (1990).

45. H. Mayr, R. Schneider, B. Irrgang, and C. Schade, *J. Am. Chem. Soc.* **112**, 4454–4459 (1990).

46. H. Mayr, R. Schneider, and U. Grabis, *J. Am. Chem. Soc.* **112**, 4460–4467 (1990).

47. H. Mayr, T. Bug, M. F. Gotta, N. Hering, B. Irrgang, B. Janker, B. Kempf, R. Loos, A. R. Ofial, G. Remennikov, and H. Schimmel, *J. Am. Chem. Soc.* **123**, 9500–9512 (2001).

48. H. Mayr, B. Kempf, and A. R. Ofial, *Acc. Chem. Res.* **36**, 66–77 (2003).

49. B. Kempf, N. Hampel, A. R. Ofial, and H. Mayr, *Chem. Eur. J.* **9**, 2209–2218 (2003).

50. H. Mayr, G. Lang, and A. R. Ofial, *J. Am. Chem. Soc.* **124**, 4076–4083 (2002).

51. S. Minegishi and H. Mayr, *J. Am. Chem. Soc.* **125**, 286–295 (2003).

52. R. Lucius and H. Mayr, *Angew. Chem.* **112**, 2086–2089 (2000); *Angew. Chem. Int. Ed.* **39**, 1995–1997 (2000).

53. (a) R. Lucius, R. Loos, and H. Mayr, *Angew. Chem.* **114**, 97–102 (2002); *Angew. Chem. Int. Ed.* **41**, 91–95 (2002); (b) T. Bug and H. Mayr, *J. Am. Chem. Soc.* **125**, 12980–12986 (2003).

54. (a) H. Mayr and M. Patz, *Angew. Chem.* **106**, 990–1010 (1994); *Angew. Chem. Int. Ed. Engl.* **33**, 938–957 (1994); (b) H. Mayr, O. Kuhn, M. F. Gotta, and M. Patz, *J. Phys. Org. Chem.* **11**, 642–654 (1998); (c) H. Mayr, M. Patz, M. F. Gotta, and A. R. Ofial, *Pure Appl. Chem.* **70**, 1993–2000 (1998).

55. G. Ya. Remennikov, B. Kempf, A. R. Ofial, K. Polborn, and H. Mayr, *J. Phys. Org. Chem.* **16**, 431–437 (2003).

56. T. Lemek and H. Mayr, *J. Org. Chem.* **68**, 6880–6886 (2003).

57. E. M. Arnett and K. E. Molter, *Acc. Chem. Res.* **18**, 339–346 (1985).

58. M.-A. Funke and H. Mayr, *Chem. Eur. J.* **3**, 1214–1222 (1997).

59. H. Mayr and N. Basso, *Angew. Chem.* **104**, 1103–1105 (1992); *Angew. Chem. Int. Ed. Engl.* **31**, 1046–1048 (1992).
60. H. Mayr, N. Basso, and G. Hagen, *J. Am. Chem. Soc.* **114**, 3060–3066 (1992).
61. C. Schindele, K. N. Houk, and H. Mayr, *J. Am. Chem. Soc.* **124**, 11208–11214 (2002).
62. C. Fichtner, G. Remennikov, and H. Mayr, *Eur. J. Org. Chem.* 4451–4456 (2001).
63. H. Mayr, C. Fichtner, and A. R. Ofial, *J. Chem. Soc. Perkin Trans. 2* 1435–1440 (2002).
64. R. Lucius, *Kinetische Untersuchungen zur Nucleophilie stabilisierter Carbanionen*, Dissertation, Ludwig-Maximilian Univ. Munich, 2001.
65. M. Roth and H. Mayr, *Angew. Chem.* **107**, 2428–2430 (1995); *Angew. Chem. Int. Ed. Engl.* **34**, 2250–2252 (1995).
66. H. Mayr, M. Roth, and G. Lang, in *Cationic Polymerization, Fundamentals and Applications*, R. Faust and T. D. Shaffer, eds., ACS Symposium Series, Vol. 665, 1997, pp. 25–40.
67. S. Fukuzumi, K. Ohkubo, and J. Otera, *J. Org. Chem.* **66**, 1450–1454 (2001). For a discussion, see Ref. 51.
68. For more E, N, and S parameters, see www.cup.uni-muenchen.de/oc/mayr/.
69. (a) K. N. Houk, N. G. Rondan, and J. Mareda, *Tetrahedron* **41**, 1555–1563 (1985); (b) M. Patz, H. Mayr, J. Bartl, and S. Steenken, *Angew. Chem.* **107**, 519–521 (1995); *Angew. Chem. Int. Ed. Engl.* **34**, 490–492 (1995); (c) H. Mayr, in *Ionic Polymerizations and Related Processes*, J. E. Puskas, A. Michel, S. Barghi, and C. Paulo, eds., NATO Science Series E, Applied Sciences, Vol. 359, Kluwer Academic Publishers, Dordrecht, 1999, pp. 99–115.
70. H. Mayr, J. Henninger, and T. Siegmund, *Res. Chem. Intermed.* **22**, 821–838 (1996).
71. I. Stahl, *Chem. Ber.* **118**, 1798–1808 (1985).
72. I. Stahl, *Chem. Ber.* **118**, 3159–3165 (1985).
73. I. Stahl, *Chem. Ber.* **118**, 4857–4868 (1985).
74. I. Stahl, *Chem. Ber.* **120**, 135–139 (1987).
75. For further reactions of dithiosubstituted carbenium ions with highly nucleophilic carbanions ($N > 13$), see M. Linker, G. Reuter, G. Frenzen, M. Maurer, J. Gosselck, and I. Stahl, *J. Prakt. Chem.* **340**, 63–72 (1998).
76. I. Paterson and L. Price, *Tetrahedron Lett.* **22**, 2833–2836 (1981).
77. C. Westerlund, *Tetrahedron Lett.* **23**, 4835–4838 (1982).
78. J. Henninger, *Reaktionen von Nitrilium-Ionen und Schwefel-stabilisierten Carbenium-Ionen mit Nucleophilen—Kinetik und Mechanismen*, Dissertation, Ludwig-Maximilian Univ. Munich, 1999.
79. H. Mayr and D. Rau, *Chem. Ber.* **127**, 2493–2498 (1994).
80. O. Kuhn, D. Rau, and H. Mayr, *J. Am. Chem. Soc.* **120**, 900–907 (1998).
81. K. M. Nicholas, *Acc. Chem. Res.* **20**, 207–214 (1987).
82. A. J. M. Caffyn and K. M. Nicholas, in *Comprehensive Organometallic Chemistry II*, E. W. Abel, F. G. A. Stone, and G. Wilkinson, eds., Pergamon Press, New York, 1995, Vol. 12, Chapter 7.1.
83. J. R. Green, *Curr. Org. Chem.* **5**, 809–826 (2001).
84. B. J. Teobald, *Tetrahedron* **58**, 4133–4170 (2002).
85. R. F. Lockwood and K. M. Nicholas, *Tetrahedron Lett.* 4163–4166 (1977).

86. M. Nakagawa, J. Ma, and T. Hino, *Heterocycles* **30**, 451–462 (1990).
87. K. D. Roth, *Synlett* 435–438 (1992).
88. M. E. Krafft, Y. Y. Cheung, C. Wright, and R. Cali, *J. Org. Chem.* **61**, 3912–3915 (1996).
89. H. J. Jaffer and P. L. Pauson, *J. Chem. Res.* (*S*) 244 (1983); *J. Chem. Res.* (*M*) 2201–2218 (1983).
90. R. Guo and J. R. Green, *J. Chem. Soc., Chem. Commun.* 2503–2504 (1999).
91. H. Mayr, K.-H. Müller, and D. Rau, *Angew. Chem.* **105**, 1732–1734 (1993); *Angew. Chem. Int. Ed. Engl.* **32**, 1630–1632 (1993).
92. H. Mayr, K.-H. Müller, A. R. Ofial, and M. Bühl, *J. Am. Chem. Soc.* **121**, 2418–2424 (1999).
93. H. Mayr and A. R. Ofial, *Tetrahedron Lett.* **38**, 3503–3506 (1997).
94. For reviews on Mannich aminoalkylations, see (a) H. Böhme and M. Haake, in *Iminium Salts in Organic Chemistry, Part 1*, H. Böhme and H. G. Viehe, eds., Interscience, New York, 1976, pp. 107–223; (b) M. Tramontini, *Synthesis* 703–775 (1973); (c) M. Tramontini and L. Angiolini, *Tetrahedron* **46**, 1791–1837 (1990); (d) I. Fleming, J. Dunoguès, and R. Smithers, *Org. React.* **37**, 57–575 (1989); (e) E. F. Kleinman, in *Comprehensive Organic Synthesis*, B. M. Trost, I. Fleming, and C. H. Heathcock, eds., Pergamon Press, Oxford, 1991, Vol. 2, pp. 893–951; (f) H. Heaney, in *Comprehensive Organic Synthesis*, B. M. Trost, I. Fleming, and C. H. Heathcock, eds., Pergamon Press, Oxford, 1991, Vol. 2, pp. 953–973; (g) E. F. Kleinman and R. A. Volkmann, in *Comprehensive Organic Synthesis*, B. M. Trost, I. Fleming, and C. H. Heathcock, eds., Pergamon Press, Oxford, 1991, Vol. 2, pp. 975–1006; (h) L. E. Overman and D. J. Ricca, in *Comprehensive Organic Synthesis*, B. M. Trost, I. Fleming, and C. H. Heathcock, eds., Pergamon Press, Oxford, 1991, Vol. 2, pp. 1007–1046; (i) G. Pandey, *Top. Curr. Chem.* **168**, 175–221 (1993); (j) M. Arend, B. Westermann, and N. Risch, *Angew. Chem.* **110**, 1096–1122 (1998); *Angew. Chem. Int. Ed.* **37**, 1044–1070 (1998).
95. T. Ohwada, N. Yamagata, and K. Shudo, *J. Am. Chem. Soc.* **113**, 1364–1373 (1991).
96. K. Shudo and T. Ohwada, in *Stable Carbocation Chemistry*, G. K. S. Prakash and P. v. R. Schleyer, eds., Wiley, New York, 1997, Chapter 16, pp 525–548.
97. For a review on superelectrophiles, see G. A. Olah, *Angew. Chem.* **105**, 805–827 (1993); *Angew. Chem. Int. Ed. Engl.* **32**, 767–788 (1993).
98. For a recent ab initio study on the structure of superelectrophilic protonated dimethylmethyleneiminium cations, see G. A. Olah, G. K. S. Prakash, and G. Rasul, *J. Org. Chem.* **67**, 8547–8551 (2002).
99. See Ref. 48. Steric factors cannot be neglected in reactions with bulky reagents, for example, tritylium (see Refs. 69c and 51) or alkyldiaryl carbenium ions: M. Roth, H. Mayr, and R. Faust, *Macromolecules* **29**, 6110–6113 (1996).
100. O. Kuhn and H. Mayr, *Angew. Chem.* **111**, 356–358 (1999); *Angew. Chem. Int. Ed.* **38**, 343–346 (1999).
101. C. Fichtner and H. Mayr, *J. Chem. Soc. Perkin Trans. 2*, 1441–1444 (2002).
102. H. Mayr, A. R. Ofial, J. Sauer, and B. Schmied, *Eur. J. Org. Chem.* 2013–2020 (2000).
103. H. Mayr and J. Henninger, *Eur. J. Org. Chem.* 1919–1922 (1998).

104. For analogous investigations on the cycloaddition reactions of arenediazonium ions with 1,3-dienes, see M. Hartnagel, K. Grimm, and H. Mayr, *Liebigs Ann./Recueil* 71–80 (1997).

105. D. Heidrich, *Angew. Chem.* **114**, 3343–3346 (2002); *Angew. Chem. Int. Ed.* **41**, 3208–3210 (2002).

106. From the correlation of *E* with ethanolysis rate constants of the corresponding alkyl chlorides, see Refs. 54c and 34.

107. D. Lenoir, *Angew. Chem.* **115**, 880–883 (2003); *Angew. Chem. Int. Ed.* **42**, 854–857 (2003).

14

ORGANIC SYNTHESIS IN SUPERACIDS

Jean-Claude Jacquesy
Faculty of Sciences
University of Poitiers
Poitiers Cedex, France

In superacids such as fluoroantimonic acid HF/SbF_5 functionalized organic substrates are mono- (or poly)protonated. Under these highly acidic conditions novel reactions (isomerization of saturated or unsaturated compounds, functionalization

Carbocation Chemistry, Edited by George A. Olah and G. K. Surya Prakash
ISBN 0-471-28490-4

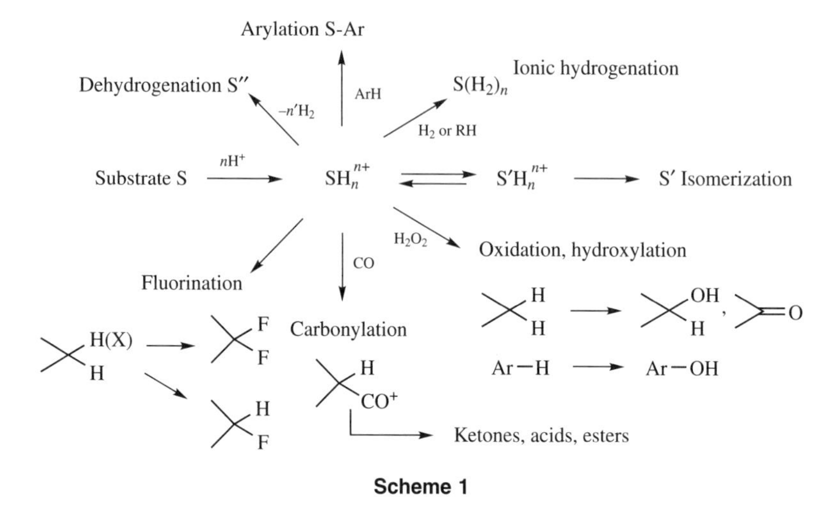

Scheme 1

of nonactivated C—H bonds, ionic hydrogenation, dehydrogenation, carbonylation, arylation, hydroxylation, etc.) can be carried out efficiently (Scheme 1).

14.1 FUNCTIONALIZATION OF NONACTIVATED C—H BONDS

14.1.1 Acyclic Ketones, Imines, and Amides

Olah and coworkers discovered the electrophilic oxygenation of alkanes in superacids using hydrogen peroxide or ozone. The resulting protonated hydrogen peroxide $H_3O_2^+$ or ozone O_3H^+ react on C—H or C—C bonds through pentacoordinated carbonium ions to yield alcohols and ketones on hydrolysis.[1]

Alcohols, ethers, and ketones are more selectively oxidized, with insertion of protonated hydrogen peroxide (or protonated ozone) occurring into the farthest secondary C—H bond from the protonated oxygen group.[2]

This reaction has been applied to the oxidation of 3-keto steroids with various functional groups at C17 to yield the corresponding oxo derivatives at either C6, C7, C11, or C12, with functionalization occurring as for as possible from the two protonated groups at C3 and C17 (see Fig. 14.1).[3]

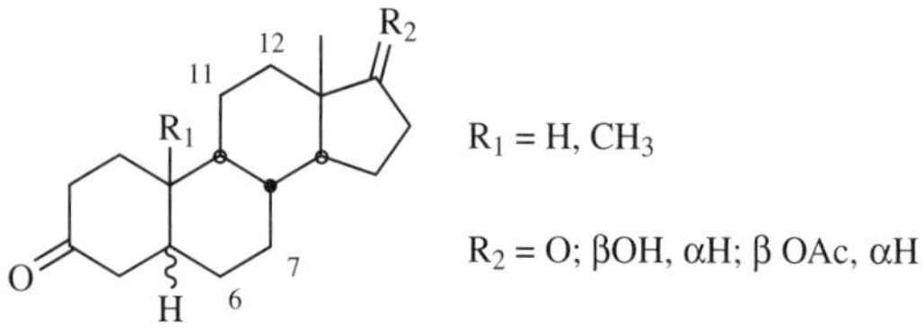

R_1 = H, CH_3

R_2 = O; βOH, αH; β OAc, αH

Figure 14.1 Formation of oxo derivatives.

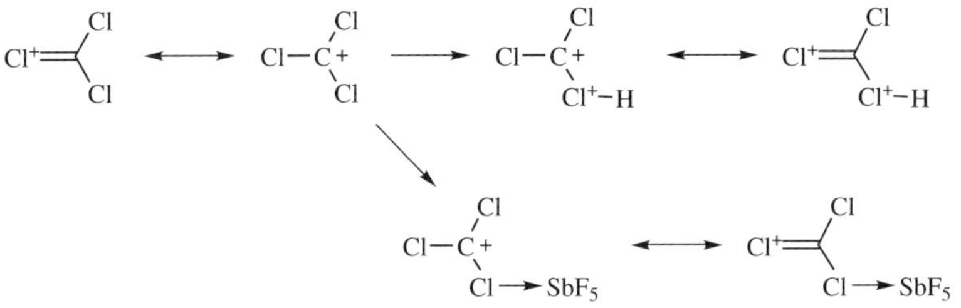

Figure 14.2 Activation of trichloromethyl cation.

Efficient hydride abstraction at a nonactivated C–H bond can also be carried out using CCl_4 source of the electrophilic trichloromethyl cation CCl_3^+. This ion previously observed by Olah using NMR[4] has been shown by Sommer[5] to rapidly ionize propane; the reaction in superacid is extremely slow in absence of CCl_4. It has been suggested as a "superelectrophilic" activation of trichloromethyl cation by protosolvatation or complexation by the Lewis acid SbF_5 in superacids (see Fig. 14.2).[6]

Similar hydride abstraction, possibly followed by rearrangement, is observed with ketones and imines that are *Ó* and *N*-protonated, respectively, in superacids.[7–9]

Compounds **1a** and **1b** (listed in Table 14.1) are too deactivated to react, both yielding **1a** after hydrolysis. For all the other substrates hydride abstraction occurs at the more reactive C–H bonds, located far from the protonated functional group.

Table 14.1 Reaction of ketones and imines with $HF/SbF_5/CCl_4$[a]

Substrate	Temperature (0°C)	Time (min)	Quenching Conditions[b]	Products (%)
1a	−30	30	A or B	**1a** (qqtive)
1b	−30	30	A or B	**1a** (qqtive)
2a	−30	5	A or B	**7** + **8**(67) (1 : 3 ratio)
2a	0	30	A or B	**9**(44)
5	−30	10	A or B	**9**(82)
2b	0	30	A	**9**(21)+**10**(46)+**2a**(9)
2b	0	30	B	**10**(68)
3a	−30	5	A or B	**11**(80)
6	−30	5	A or B	**11**(92)
3b	−30	30	A	**11**(14)+**12**(24)
3b	−30	30	B	**12**(67)
4a	−30	3	A	**13**(50)+**14**(12)
4a	−30	3	B	**13**(32)+**14**(20)
4b	−30	30	A	**13**(20)+**14**(20)
4b	−30	30	B	**13**(19)+**14**(36)

[a] HF/SbF_5/substrate molar ratio 20/1/0.05.

[b] A—Na_2CO_3/ice/H_2O; B (1) excess PPHF (HF/pyridine molar ratio 70/30); (2) Na_2CO_3/ice/H_2O.

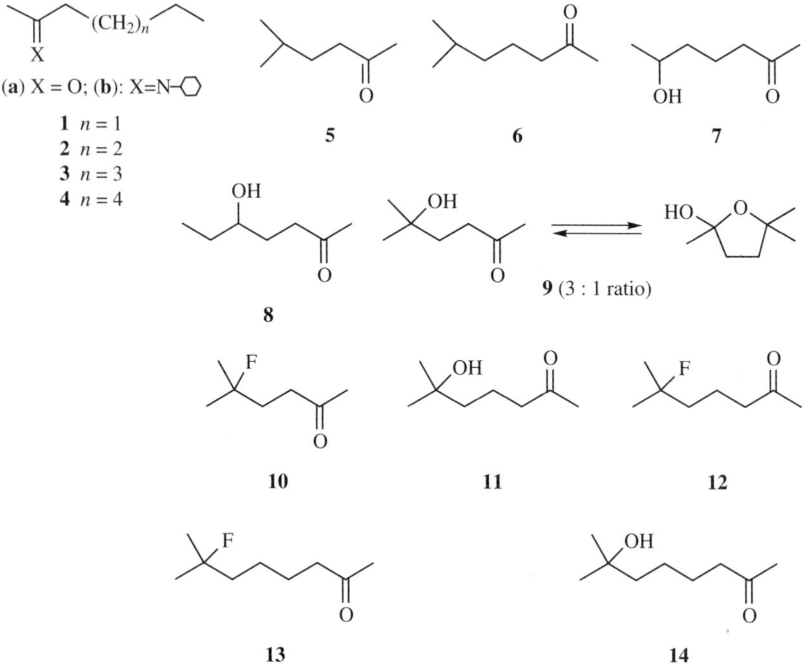

Figure 14.3 Formation of hydroxy or fluoro derivatives.

Tertiary C–H in compounds **5** and **6** ionize directly to give the tertiary carbenium ions, whereas for compounds **2a**-**4a** and **2b**-**4b** hydride abstraction at the (ω-1) carbon atom yields a secondary ion which may isomerize into a tertiary one.

Ketones **2a**, **3a**, **5** and **6** yield hydroxy derivatives regardless of the quenching conditions (see Fig. 14.3). This result can be accounted for by intermediacy of a five or six-membered ring carboxonium ion which leads to hydroxy derivatives after hydrolysis (Scheme 2).

It should be pointed out that ketone **4a** yields a mixture of hydroxy or fluoroderivatives, formation of a seven-membered ring carboxonium ring is more difficult and favors equilibrium with the diprotonated acyclic species that is trapped by a fluoride ion.

In the reaction conditions, fluorination is a reversible step; fluoroketone **13**, placed in $HF/SbF_5/CCl_4$, gives ketones **13** and **14**, with yields similar to those from ketone **4a**.

In HF/SbF_5 imines are *N*-protonated; the resulting iminium ion is less prone to deprotonation than the corresponding protonated ketone. After hydride abstraction by ion CCl_3^+, the open carbenium ion is either fluorinated in the medium or hydroxylated after quenching. Hydrolysis of the iminium yields the corresponding ketone (Scheme 3).

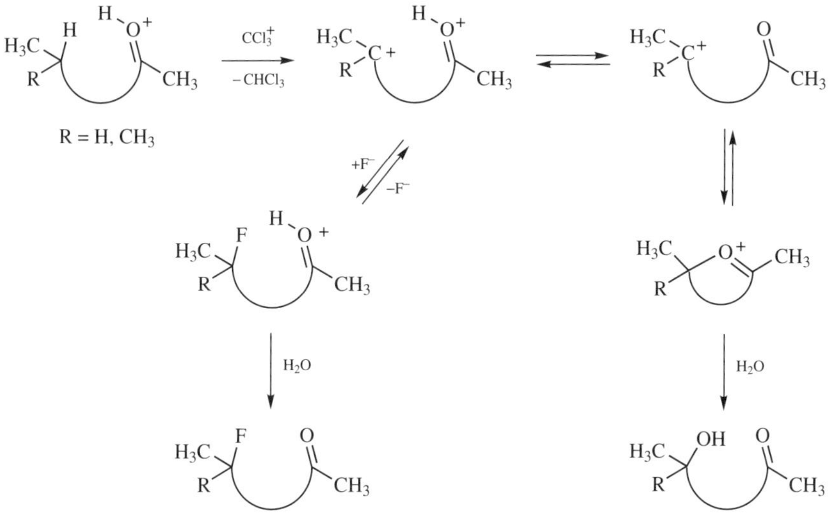

$R = H, CH_3$

Scheme 2

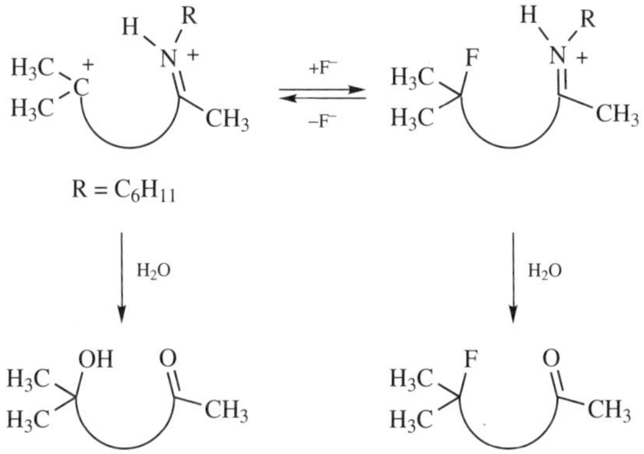

$R = C_6H_{11}$

Scheme 3

Table 14.1 shows that the yields of fluoroketones **10**, **12**, and **14** are improved when the reaction mixture is treated with PPHF.

Similar hydroxylation or fluorination of amides is observed in HF/SbF_5 in the presence of CCl_4 as no protolytic activation is operative in HF/SbF_5 alone (see Table 14.2).[10]

14.1.2 Carbonylation (Carboxylation) of Cyclic Ketones

Synthesis of carboxylic acids from alkenes can be carried out using the Koch–Haaf reaction in aqueous sulfuric acid.[12]

Table 14.2 Reaction of amides with $HF/SbF_5/CCl_4$ at −30°C

Substrate	Time (min)	Quenching Conditions[a]	Products (Yield %)
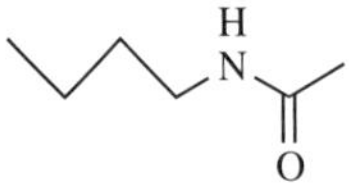	30	A	No reaction
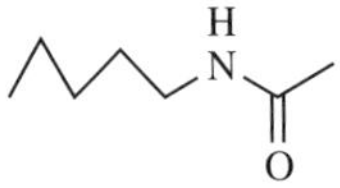	20	A	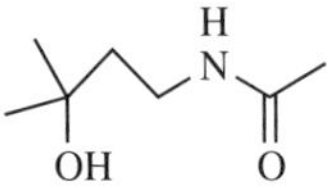 (55) + 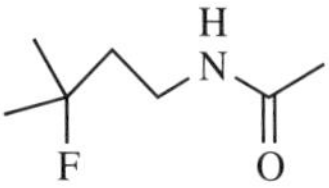 (10)
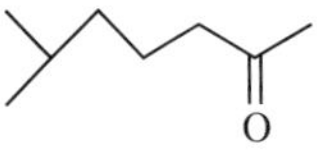	20	A	
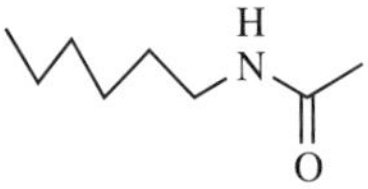	5	A	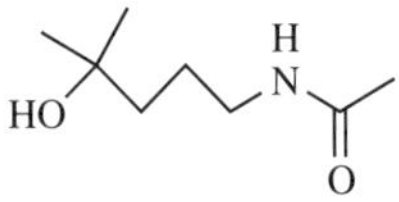 (31) + 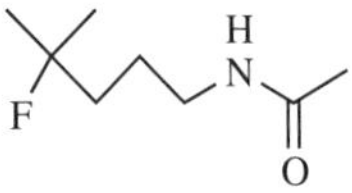 (26)
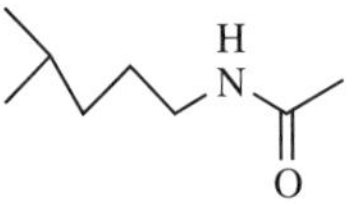	5	B	(76)
	5	B	(80)

[a]A—Na_2CO_3/ice/H_2O; B (1) excess PPHF (HF/pyridine molar ratio 70/30) at −30°C for 2 minutes; (2) Na_2CO_3/ice/H_2O.

Carboxylation can be achieved in superacid from alkanes, alcohols, and alkyl methyl ketones at ambient temperature and atmospheric pressure.[13] Selective carbonylation of bicyclic enones **15** and **16** in HF–SbF_5 with CO and quenching the resulting acylium ion with methanol yields esters **17**, **18**, and **19**, respectively.[14]

These results imply diprotonation of the enone moiety and isomerization to more stable species that are trapped by carbon monoxide (Scheme 4).

Ketones **18** and **19** can also been obtained in similar yields, starting from *trans*-ketone **20** in HF/SbF_5 in the presence of CCl_4 and CO followed by quenching with methanol. Hydride abstraction, as far as possible from the protonated carbonyl group, generates the same carbenium ions **21** and **22**.

Selective formation of ketone **17**, **18**, and **19** is regio- and stereoselective; in the indane series, carbonylation is observed on the less hindered face of the carbenium ion, whereas in the decaline series, carbonylation of the carbenium ion is axial.

14.1.3 Dehydrogenation of Polycyclic Ketones to Dienones

Trichloromethyl cation CCl_3^+ is an extremely reactive hydride abstracting reagent with hydrocarbons, ketones, imides, and amides to yield carbocations that can be quenched by fluoride, carbon monoxide, and water.

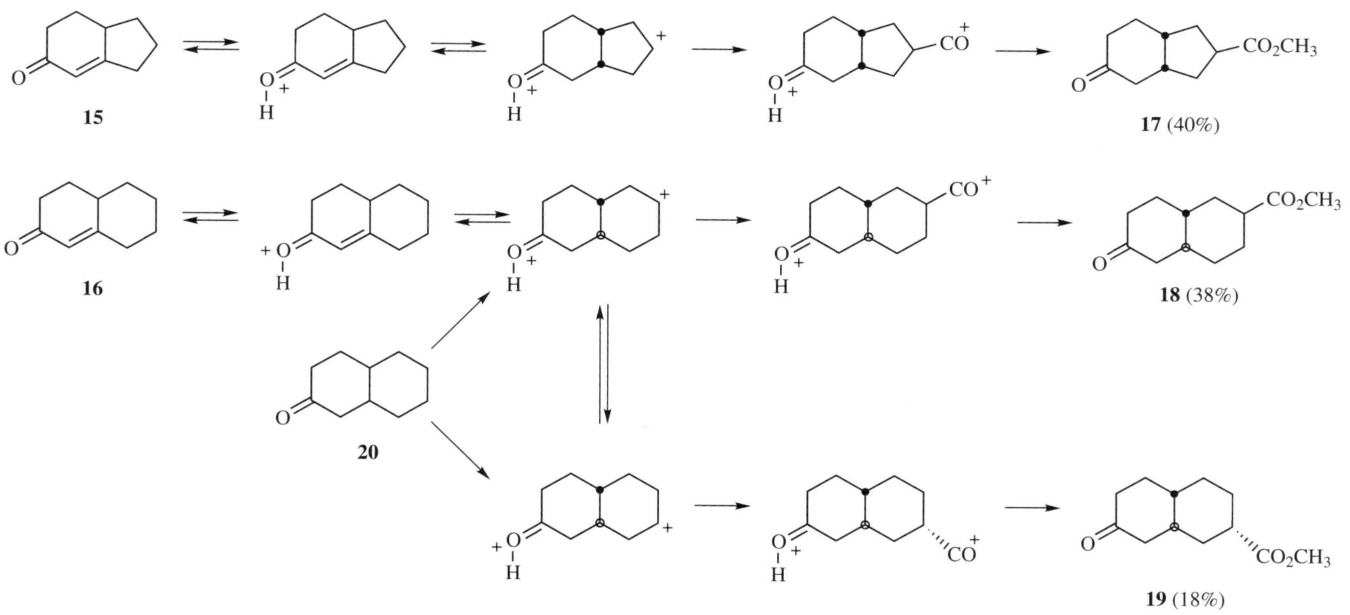

15

16

20

17 (40%)

18 (38%)

19 (18%)

Scheme 4

It has been shown that ketone **20** in $HF/SbF_5/CCl_4/CO$ yields acylium ions after trapping of carbenium ions with carbon monoxide. Table 14.3 shows that the reaction can go further in the absence of carbon monoxide to yield dienones after usual workup.[15]

This implies, after hydride abstraction at the more reactive C—H bonds, far from the protonated carbonyl group, the formed cyclohexyl cation(s) rearrange to methylcyclopentylcation(s).[10] Formation of dienones **23** and **24** (Scheme 5) implies a second hydride abstraction leading ultimately to allylic ions **33** and **34**, precursors of the dienones. Ions **33** and **34** have been previously characterized by NMR.[16]

Steroidal ketones **28–30** (see Table 14.3) are less reactive and dehydrogenated in $HF/SbF_5/CCl_4$ into dienones **31** (8αH + 8βH) and **32** (8αH + 8βH). This result

Table 14.3 Dehydrogenation of polycyclic ketones[a,b]

Substrate	CCl_4 (equiv)	Temperature (°C)	Time (min)	Products (Yield %)
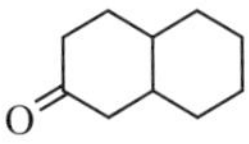 **20**	3.6	0	30	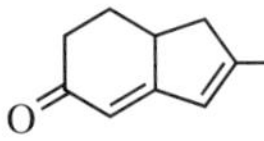 **23** (25) + 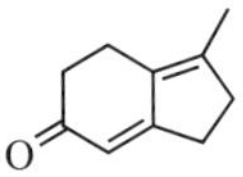 **24** (12)
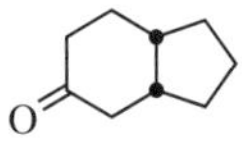 **25**	1.2	−30	45	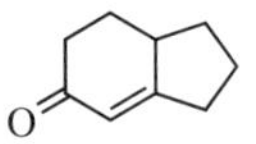 **26** (31) + 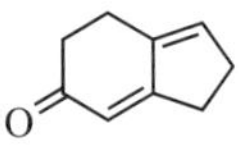 **27** (46)
	3.6	−30	45	**27** (63)
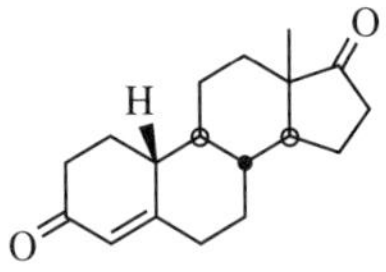 **28**	2.4	0	10	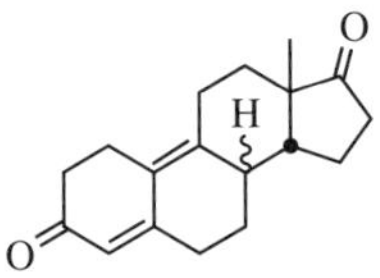 **31** (8αH, 8βH) (35) (1 : 1 ratio) +
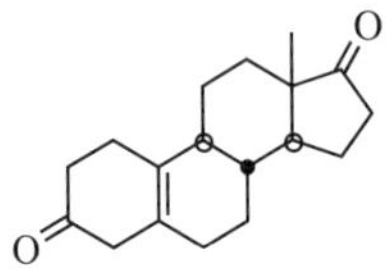 **29**	2.4	0	10	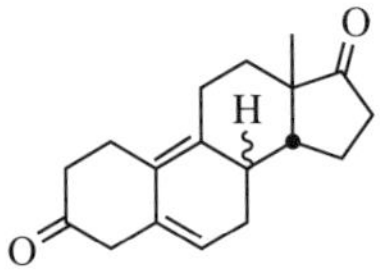
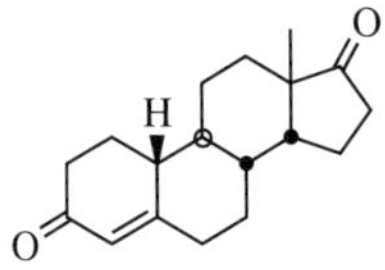 **30**	2.4	0	10	**32** (8αH, 8βH) (34) (1 : 1 ratio)

[a] HF/SbF_5/substrate molar ratio 20/1/0.05.

[b] Ketone **20**: mixture of *cis/trans* isomers molar ratio 5/95.

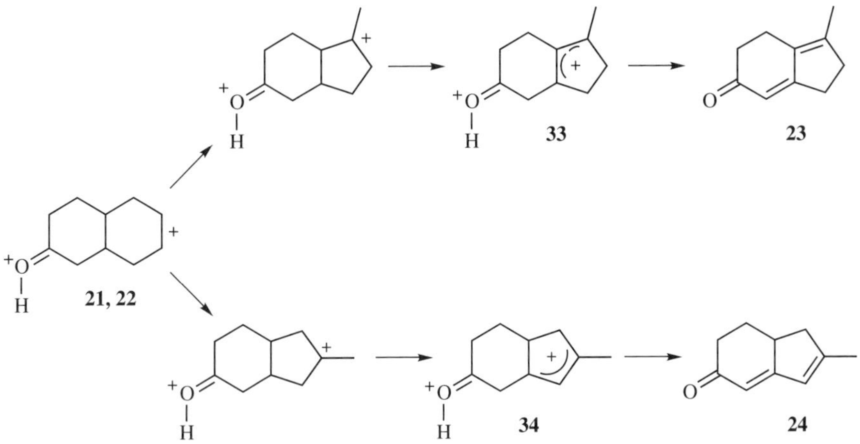

Scheme 5

implies hydride abstraction, probably occurring in the B ring, as far as possible from the two protonated carbonyl groups.

14.2 ELECTROPHILIC HYDROXYLATION OF AROMATIC AMINES, INDOLES, AND ALKALOIDS

14.2.1 Aromatic Amines

Olah reported the electrophilic hydroxylation of aromatic hydrocarbons in superacids with hydrogen peroxide, which is protonated to give the hydroperoxonium ion, a source of electrophilic hydroxyle OH^+.[17] The reaction has been extended to substituted phenols to give resorcinols in fair to excellent yields.[18]

Hydroxylation is not limited to oxygenated aromatics, as illustrated by the reaction of hydrogen peroxide with anilines, indolines, and tetrahydroquinoline in superacids (Scheme 6).[19]

Ring hydroxylation yields phenols in good yields, with the *meta* always as the major product. Whereas hydroxylation of aniline yields three hydroxy derivatives, with all other aromatic amines the electronic effect of both substituents (ammonium group and alkyl group) favors regioselective substitution *meta* to the nitrogen group. In these reactions, proton acts as a protecting group, preventing any oxidation at the nitrogen substituent.

14.2.2 Indoles and Alkaloids

The indole ring system is common to many natural products, and many of them exhibit physiological properties. Olah has shown that indoles **35** (Scheme 7) are

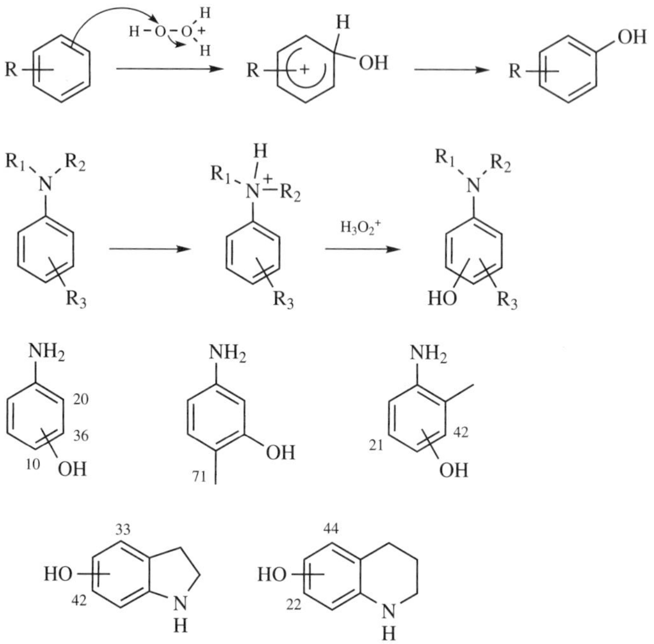

[Numeric data are the yield (%) of hydroxy derivative at the corresponding site.]

Scheme 6

protonated in superacids at the carbon β of the pyrrole ring to give iminium ions **36**.[20] Delocalization of the positive charge in the resulting ions accounts for the high reactivity of indoles (compared to indolines) and for the poor regioselectivity observed in the electrophilic hydroxylation.[21]

With tryptamine and tryptophane derivatives **37** and **38**, additional protonation can be expected on the amide and ester groups. Hydroxylation yields monohydroxylated derivatives; serotonine and pretonine derivatives are the major products, in 38% and 42%, respectively. This reaction constitutes a novel access to serotonine and pretonine, as these two compounds have powerful physiological and psychological properties.

Aspidoderma alkaloid vincadifformine **41** in superacid is expected to be polyprotonated to ion **42**, the indolenine moiety leading to an iminium ion, equivalent to a protonated indole.[22] As a result, hydroxylation in HF/SbF_5 using H_2O_2 yields a mixture of 15- and 16-hydroxy derivatives (66% overall yield). On the other hand, starting from 2,3-dihydrovincadifformine, hydroxylation of the indoline system leads selectively to the sole 16-hydroxy derivative (40% yield).

A completely different reaction is observed on tabersonine **43**, the 6,7-didehydro analog of vincadifformine, yielding a mixture of fluorhydrins **44**, by *anti* addition on the 6,7 double bond.[23]

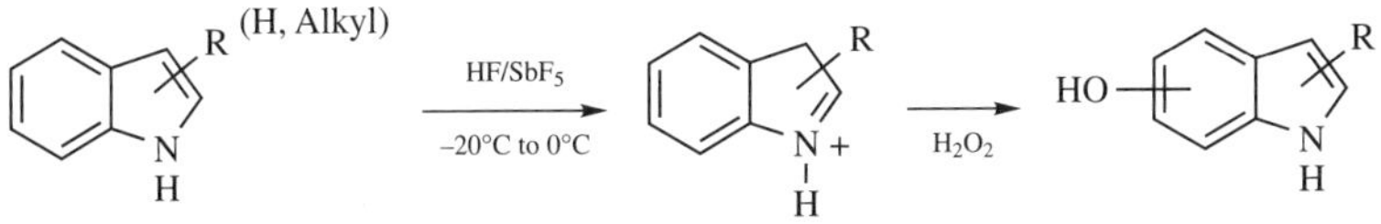

R (H, Alkyl)

35 — HF/SbF_5, –20°C to 0°C → **36** — H_2O_2 → HO-substituted indole

yield 35–96%

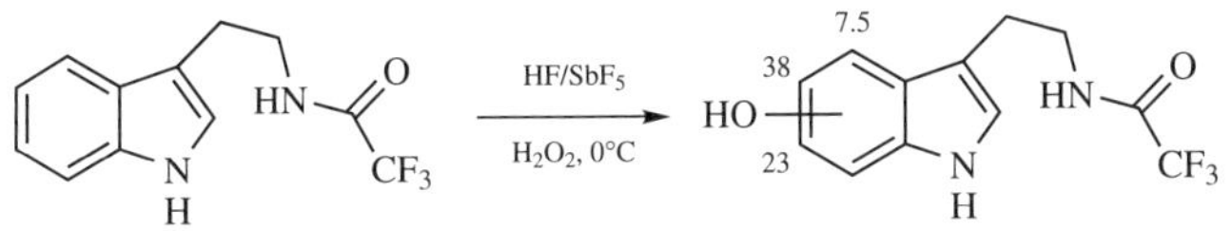

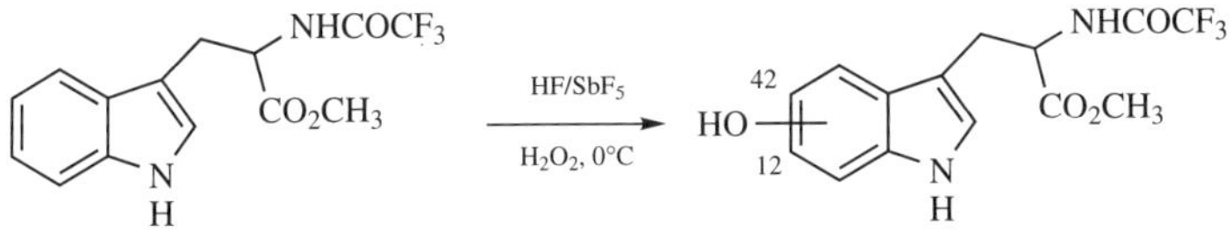

37 — HF/SbF_5, H_2O_2, 0°C → **39** (7.5, 38, 23)

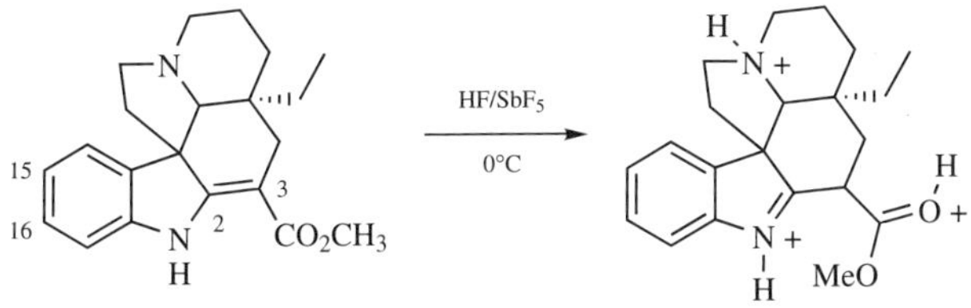

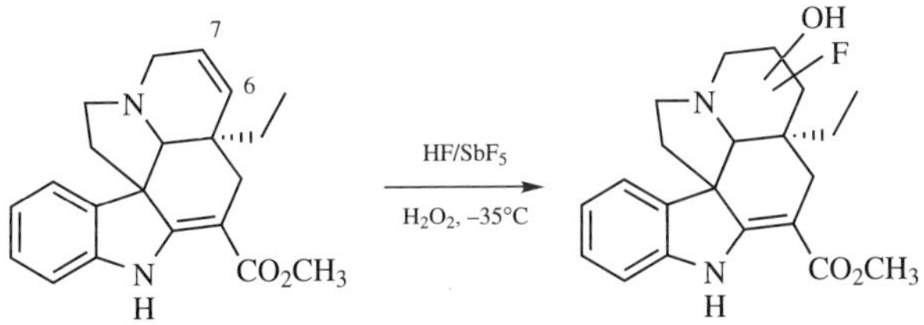

38 — HF/SbF_5, H_2O_2, 0°C → **40** (42, 12)

41 — HF/SbF_5, 0°C → **42**

43 — HF/SbF_5, H_2O_2, –35°C → **44**

45 — HF/SbF_5, $Na_2S_2O_8$ or H_2O_2, –35°C → **46**

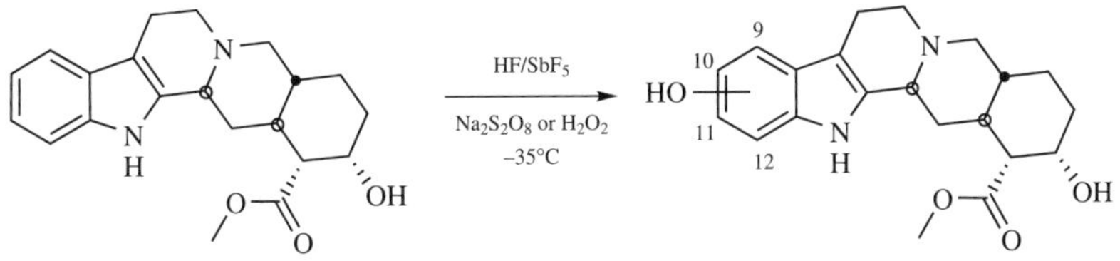

9-OH 11%
10-OH 32%
11-OH 23%
12-OH 3%

Scheme 7

Indole alkaloid yohimbine **45** is known for its activity against male impotence and female sexual disorders. Two human metabolites, 10- and 11-hydroxyyohimbine, were previously identified[24] and prepared in poor yield by demethylation of methoxy compounds[25] isolated from natural sources.[26]

In superacid in the presence of sodium persulfate or of hydrogen peroxide, the four positions of the benzene ring are hydroxylated; the human metabolites 10- and 11-hydroxy derivatives are by far the major products (32% and 23% yields, respectively). In this reaction, protonation of all the functional groups present in the molecule appears to protect the substrate from any degradation, especially from dehydration of the fragile β-hydroxy ester moiety.[27]

14.3 FLUORINATION AND IONIC HYDROGENATION OF ALKALOIDS

Vindoline **47** (Scheme 8) is one of the major alkaloids present in the Madagascan periwinkle *Catharanthus roseus* (L.) G. Don. and constitutes the lower half of antimitotic agents vinblastine **48** and semisynthetic vinorelbine (Navelbine) **49** used in cancer chemotherapy.

Numerous derivatives of vinblastine have been evaluated with the aim of identifying more active and less toxic drugs exhibiting a wider spectrum of anticancer potency.[28] An original approach giving possible access to new derivatives was to study the reactivity of these alkaloids in superacids.

First, the reactivity of vindoline **47** in HF/SbF_5 was examined in the presence of NBS, NCS, or H_2O_2.[29] Unexpected **7**β-substituted-20-fluoro-6,7-dihydrovindolines **51a–c** was obtained, and the (20*S*) stereoisomer was by far the major product. This surprising reaction implies addition of the electrophile (Br^+, Cl^+, and OH^+ equivalent) on the β face of the C6–C7 double bond followed by a 1,3-hydride shift from C20 to C6 and fluorination at C20. Unfortunately, condensation with catharanthine according to Potier's procedure gave a modified anhydrovinblastine displaying pharmacological activity lower than that exhibited by the parent compound.[30]

Vinorelbine **49** and anhydrovinblastine **50** possess two unsaturated piperidine nuclei, and the reactivity of these comppounds in superacids could be expected at both cleavamine (upper part) and vindoline moieties (lower part).

Vinorelbine **49** was reacted at −40°C in HF/SbF_5 with NBS to yield a very complex mixture containing 20′,20′-difluoro-3′,4′-dihydrovinorelbine **52**. Starting from either vinblastine **48** or anhydrovinblastine **50** gave the difluoro analog **53**.

This surprising result, corresponding to an oxidative difluorination, was unexpected. Since chloromethanes are precursors in superacids of chloromethyl ions such as CCl_3^+ and $CHCl_2^+$, which possess a strong hydride abstraction power and therefore act as oxidizing agents, anhydrovinblastine **50** was reacted in the HF/SbF_5 in the presence of $CHCl_3$ at −50°C. After usual workup, chlorofluorinated derivative **54** (6%), *gem*-difluoro compound **53** (>50%), and epimeric analog **55** (15%) were isolated. Careful kinetic study permitted the isolation of intermediates

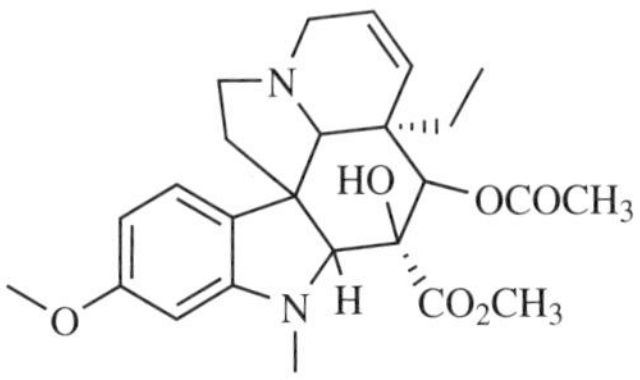

47

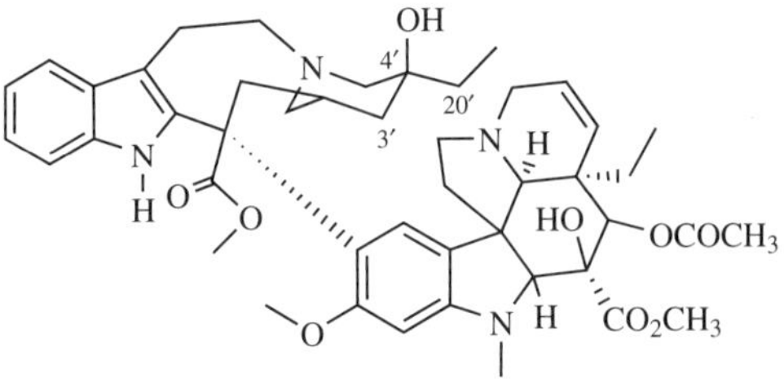

48

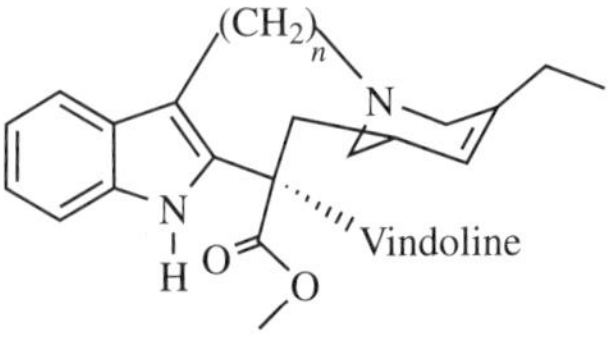

49 ($n = 1$)
50 ($n = 2$)

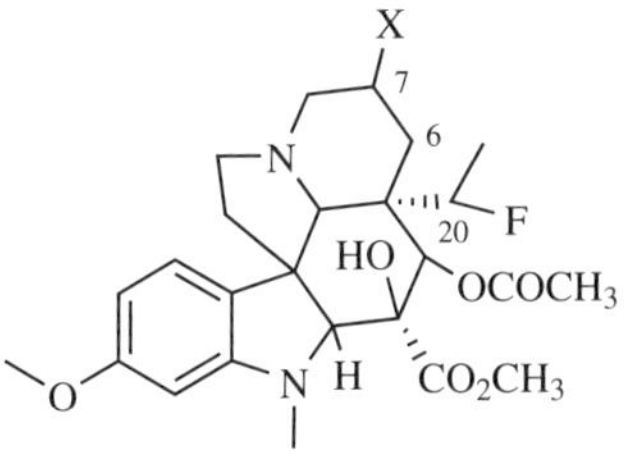

51 **a** X = Br
b X = Cl
c X = OH

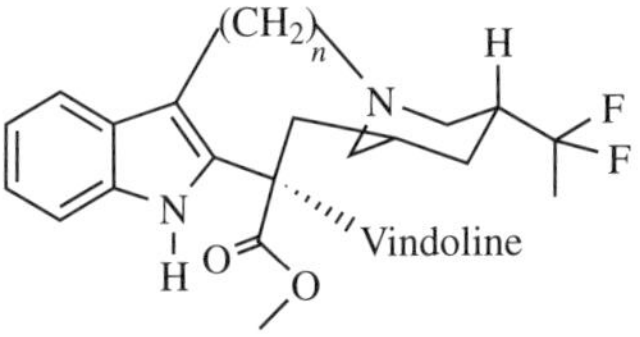

52 ($n = 1$)
53 ($n = 2$

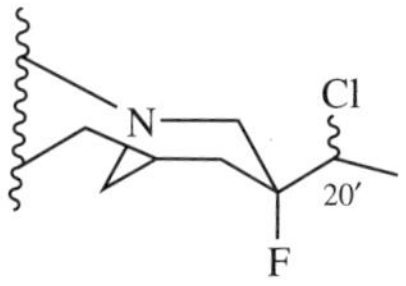

54

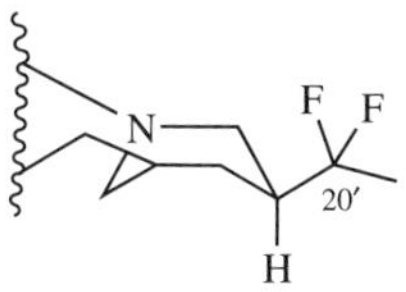

55

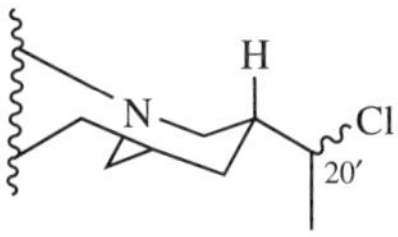

56

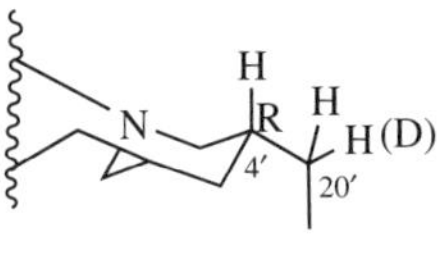

57 ($n = 1$)
58 ($n = 2$)

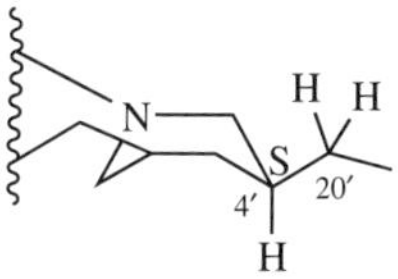

59 ($n = 1$)
60 ($n = 2$)

Scheme 8

C20′ diastereoisomers of 20′-chloro-4′deoxyvinblastine **56**, disappearing simultaneously with the formation of the products **53–55**.[31] The same products were obtained with vinblastine **48** under the same conditions; dehydration of the tertiary alcohol function probably occurred in the first step. Starting from vinorelbine **49**, the main isolated product was the *gem*-difluoro derivative **52** (35%).

Taking into account all these findings, the postulated mechanism is depicted in Scheme 9. The main reaction pathway implies formation of cation **58** resulting most likely from protonation of the 3′,4′ double bond at C′3 followed by a hydride shift from C20′. Trapping of ion **58** by a complex anion SbF_5Cl^- leads to the intermediates **56**.

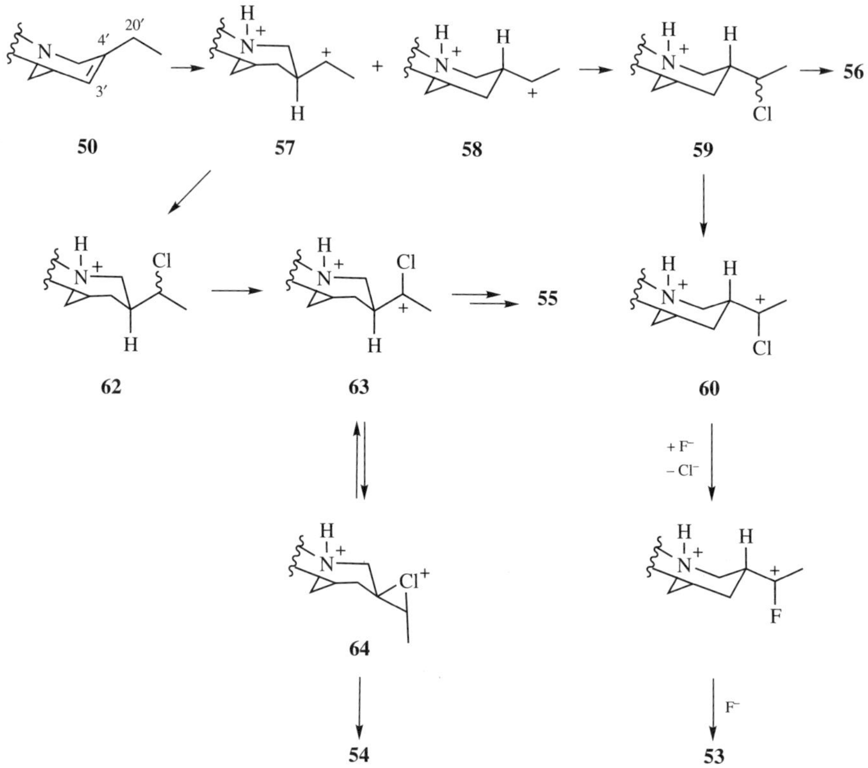

50 → **57** + **58** → **59** → **56**

50 → **62** → **63** ⇀ **55**; **63** ⇌ **64** → **54**

59 → **60** ($+F^-$, $-Cl^-$) → → (F^-) **53**

Scheme 9

Hydride abstraction at C20′ yields ion **60** which by halogen exchange finally gives compound **53**. A similar process starting from ion **57** through ions **62** and **63** (Scheme 9) yields epimer **55**. Isomerization of ion **63** to the ethylene chloronium ion **64**, explains the formation of chlorofluoro derivative **54**.[32] Structure of compound **53** was established by NMR experiments and confirmed by X-ray crystallography.

Furthermore applying Polonovski reaction formerly used by Potier to prepare vinorelbine **49** from anhydrovinblastine **50**, compound **53** yielded the corresponding C′-ring-contracted analog vinflunine **52** (90% yield), thus confirming the structure of the latter.[33]

Vinflunine has been selected for its promising antitumor activity, superior to vinorelbine in several experimental tumor model systems, and in now synthesized on a kilogram scale.[34]

The nonfluorinated analogs **57** and **58** of compounds **52** and **53,** respectively, could be prepared in HF/SbF_5 by ionic hydrogenation, using cyclohexane or methylcyclopentane as hydride donors.[35] Vinorelbine **49** yielded the reduced analog **57** (70% yield), whereas either vinblastine **48** or anhydrovinblastine **50** gave the dihydro analog **58** (65%) (Scheme 10). The site of reduction could be determined to

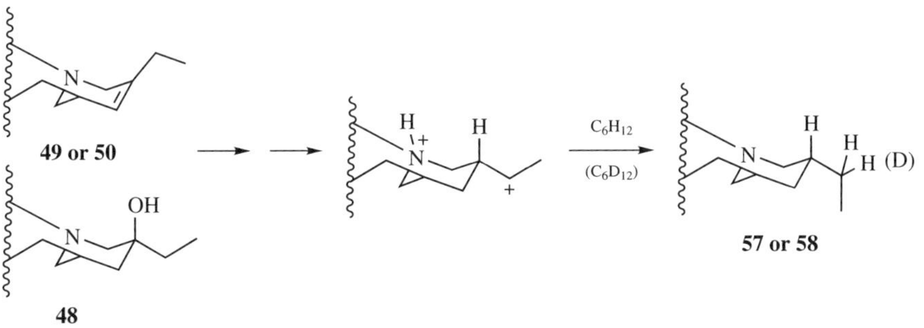

49 or 50 → → ; **48**; C_6H_{12} (C_6D_{12}); H (D); **57 or 58**

Scheme 10

be at C20′, using perdeuterocyclohexane C_6D_{12}. It should be pointed out that catalytic hydrogenation of these substrates gave only the 4′S reduced compounds **59** and **60**.

Experiments conducted on the P388 (IV/IP) in vivo model showed that compound **57** is significantly less active than vinflunine **57**.

14.4 SUPERACID-CATALYZED REARRANGEMENT OF QUININE

Molecular rearrangements that take place in nature under the influence of enzymes, can be performed in the laboratory under the action of various acids.

Cinchona alkaloid quinine **65** and quinidine cleave in acetic acid to yield quinicine **66** (Scheme 11).[36] The reaction can be accounted for by protonation of the nitrogen atom of the quinuclidyl group, followed by the cleavage of the N1–C8 bond and deprotonation to give the enol of quinicine. Surprisingly in HF–SbF_5 at −30°C, quinine **65** yielded, after usual workup, compound **67** in 89% yield.[37] In the reaction conditions quinine **65** is polyprotonated at the nitrogen and at the oxygen atom of the quinoline moiety and at the nitrogen atom of the quinuclidyl group. As a result, formation of the benzylic ion by dehydration of the protonated hydroxyl (in equilibrium with the neutral form) is highly disfavored. Furthermore,

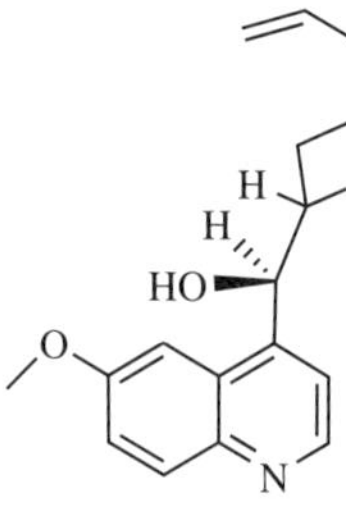

65 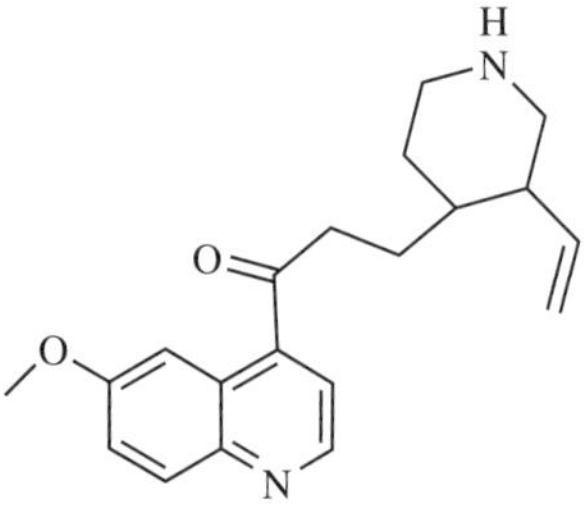 66 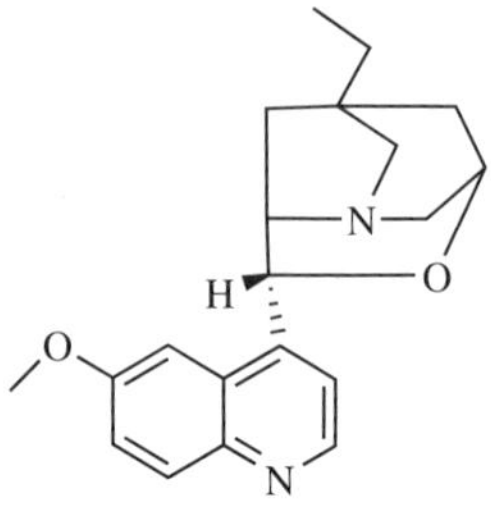 67

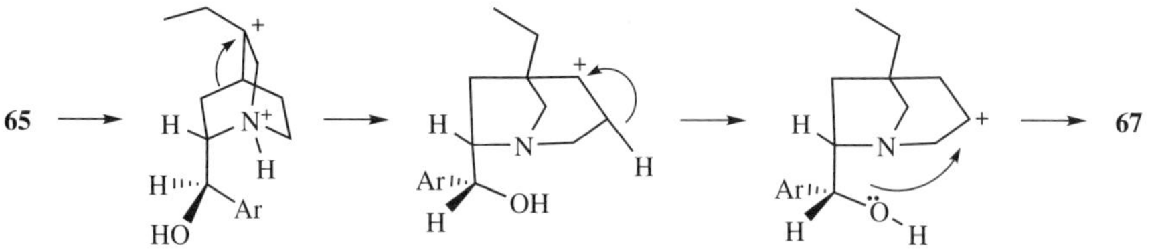

Scheme 11

formation of the enol of quinicine implying a deprotonation process is not observed under the reaction conditions. A concerted mechanism, outlined in Scheme 11, may be operative in the rearrangement of quinine **65** to compound **67**. *Cinchona* alkaloids are presently used as catalysts in a variety of enantioselective reactions. The novel chiral system of compound **67** might be a potent transmitter for such reactions.

14.5 CONCLUSION

The new and sometimes unexpected reactions presented in this chapter even with acid-sensitive substrates confirm the importance of superacids in synthetic organic chemistry. With polyfunctional compounds, such as alkaloids, proton acts as an efficient protecting group, favoring reaction at nonactivated sites.

ACKNOWLEDGMENTS

Thanks are due to my colleagues, referred to in the references, for their contributions in this stimulating field.

I thank the CNRS, La Société Pierre Fabre, and La Ligue Contre le Cancer for their financial support.

REFERENCES

1. (a) G. A. Olah, N. Yoneda, and D. G. Parker, *J. Am. Chem. Soc.* **99**, 483 (1977); (b) G. A. Olah, D. G. Parker, and N. Yoneda, *Angew. Chem. Int. Ed. Engl.* **17**, 909 (1978).

2. (a) G. A. Olah, N. Yoneda, and R. Ohnishi, *J. Am. Chem. Soc.* **98**, 7341 (1976); (b) N. Yoneda, H. Sato, T. Fukuhara, Y. Takahashi, and A. Susuki, *Chem. Lett.* 19 (1983).
3. J. C. Jacquesy, R. Jacquesy, L. Lamandé, C. Narbonne, J. F. Patoiseau, and Y. Vidal, *Nouv. J. Chim.* **6**, 589 (1982).
4. G. A. Olah, L. Heiliger, and G. K. S. Prakash, *J. Am. Chem. Soc.* **111**, 8020 (1989).
5. J. C. Culmann, M. Simon, and J. Sommer, *J. Chem. Soc., Chem. Commun.* 1098 (1990).
6. G. A. Olah, *Angew. Chem. Int. Ed. Engl.* **32**, 767 (1993).
7. G. A. Olah and P. Kreinbühe, *J. Am. Chem. Soc.* **89**, 4756 (1967).
8. G. A. Olah, M. Cahin, and D. H. O'Brien, *J. Am. Chem. Soc.* **89**, 3568 (1967).
9. S. Thibaudeau, A. Martin-Mingot, M.-P. Jouannetaud, and J.-C. Jacquesy, *Tetrahedron Lett.* **58**, 6643 (2002).
10. A. Martin, M.-P. Jouannetaud, and J.-C. Jacquesy, *Tetrahedron Lett.* **37**, 2967 (1996).
11. (a) A. B. Moodie, P. N. Thomas, and K. Schofield, *J. Chem. Soc., Perkin Trans. II* 1693 (1977); (b) R. J. Gillespie and T. Birchall, *Can. J. Chem.* **41**, 148, 2642 (1963).
12. H. Koch and N. Haaf, *Org. Synth.* **44**, 1 (1964).
13. (a) N. Yoneda, Y. Takahashi, and A. Susuki, *Chem. Lett.* 1151 (1978); (b) R. Paatz and G. Weisberger, *Chem. Ber.* **100**, 984 (1967); (c) N. Yoneda, H. Sato, T. Fukuhara, Y. Takahashi, and A. Susuki, *Chem. Lett.* 19 (1983).
14. J.-M. Coustard and J.-C. Jacquesy, *J. Chem. Res.* (*S*) 280 (1977).
15. A. Martin, M.-P. Jouannetaud, and J.-C. Jacquesy, *Tetrahedron Lett.* **37**, 7731 (1996).
16. J.-M. Coustard and J.-C. Jacquesy, *Bull. Soc. Chim. Fr.* 2098 (1973).
17. (a) G. A. Olah and R. Ohniski, *J. Org. Chem.* **43**, 865 (1978); (b) G. A. Olah, T. Keumi, J.-C. Lecoq, A. P. Fung, and J. A. Olah, *J. Org. Chem.* **56**, 6148 (1991).
18. J.-P. Gesson, L. Di Giusto, and J.-C. Jacquesy, *Tetrahedron* **34**, 1715 (1978).
19. (a) J.-C. Jacquesy, M.-P. Jouannetaud, G. Morellet, and Y. Vidal, *Bull. Soc. Chim. Fr.* 625 (1986); (b) C. Berrier, J.-C. Jacquesy, M.-P. Jouannetaud, and A. Renoux, *New J. Chem.* **11**, 611 (11987).
20. G. A. Olah, D. L. Brydon, and R. D. Porter, *J. Org. Chem.* **37**, 317 (1970).
21. C. Berrier, J.-C. Jacquesy, M.-P. Jouannetaud, and A. Renoux, *New J. Chem.* **11**, 611 (1987).
22. C. Berrier, J.-C. Jacquesy, M.-P. Jouannetaud, and Y. Vidal, *Tetrahedron* **46**, 827 (1990).
23. C. Berrier, J.-C. Jacquesy, M.-P. Jouannetaud, and Y. Vidal, *Tetrahedron*, **46**, 815 (1990).
24. R. Le Verge, P. Le Corre, F. Chevanne, M. Döe de Maindreville, R. Royer, and J. Levy, *J. Chromatogr. Biomed. Appl.* **574**, 283 (1992).
25. J. Levy, personal communication.
26. C. Miet, G. Croquelois, and J. Poisson, *J. Phytochem.* **16**, 803 (1977).
27. A. Duflos, F. Redoules, J. Fahy, J.-C. Jacquesy, and M.-P. Jouannetaud, *J. Nat. Prod.* **64**, 193 (2001).
28. For reviews, see A. Brossi and M. Suffness, eds., Antitumor Bisindole alkaloids from *Catharanthus roseus* (L.), in *The Alkaloids*, Academic Press, San Diego, CA, 1990, Vol. 37.
29. C. Berrier, J.-C. Jacquesy, M.-P. Jouannetaud, C. Lafitte, Y. Vidal, F. Zunino, J. Fahy, and A. Duflos, *Tetrahedron* **54**, 13761 (1998).

30. P. Mangeney, R. Z. Andriamialisoa, N. Langlois, Y. Langlois, and P. Potier, *J. Am. Chem. Soc.* **101**, 2243 (1979).
31. J. Fahy, A. Duflos, J.-P. Ribet, J.-C. Jacquesy, C. Berrier, M.-P. Jouannetaud, and F. Zunino, *J. Am. Chem. Soc.* **119**, 8576 (1997).
32. G. A. Olah, J. M. Bollinger, and J.-M. Brinich, *J. Am. Chem. Soc.* **90**, 2587 (1968).
33. P. Mangeney, R. Z. Andriamialisoa, J.-Y. Lallemand, N. Langlois, Y. Langlois, and P. Potier, *Tetrahedron* **35**, 2175 (1979).
34. A. Kruczynski, F. Colpaert, J.-P. Tarayre, P. Mouillard, J. Fahy, and B. T. Hill, *Cancer Chemother. Pharmacol.* **41**, 437, (1998).
35. C. Lafitte, M.-P. Jouannetaud, J.-C. Jacquesy, J. Fahy, and A. Duflos, *Tetrahedron Lett.* **39**, 8281 (1998).
36. R. Verpoorte, J. Schripsema, and T. V. Der Leer, *Cinchona* alkaloids, in *The Alkaloids, Chemistry and Pharmacology*, A. Brossi, ed., Academic Press, San Diego, CA, 1988, Vol. 34, pp. 331–398.
37. S. Thibaudeau, B. Violeau, A. Martin-Mingot, M.-P. Jouannetaud, and J.-C. Jacquesy, *Tetrahedron Lett.* **43**, 8773 (2002).

INDEX

Carbocation Chemistry, Edited by George A. Olah and G. K. Surya Prakash
ISBN 0-471-28490-4